ACTION TO LANGUAGE VIA THE MIRROR NEURON SYSTEM

Mirror neurons may hold the brain's key to social interaction: each coding not only a particular action or emotion, but also the recognition of that action or emotion in others. The Mirror System Hypothesis adds an evolutionary arrow to the story: from the mirror system for hand actions, shared with monkeys and chimpanzees, to the uniquely human mirror system for language. In this volume, written to be accessible to a wide audience, experts from child development, computer science, linguistics, neuroscience, primatology, and robotics present and analyze the mirror system and show how studies of action and language can illuminate each other. Topics discussed in the 15 chapters include the following. What do chimpanzees and humans have in common? Does the human capability for language rest on brain mechanisms shared with other animals? How do human infants acquire language? What can be learned from imaging the human brain? How are sign- and spoken-language related? Will robots learn to act and speak like humans?

MICHAEL A. ARBIB is the Fletcher Jones Professor of Computer Science, as well as a Professor of Biological Sciences, Biomedical Engineering, Electrical Engineering, Neuroscience and Psychology at the University of Southern California (USC), which he joined in 1986. He has been named as one of a small group of University Professors at USC in recognition of his contributions across many disciplines.

ACTION TO LANGUAGE VIA THE MIRROR NEURON SYSTEM

edited by

MICHAEL A. ARBIB

University of Southern California

CAMBRIDGE
UNIVERSITY PRESS

CAMBRIDGE UNIVERSITY PRESS
Cambridge, New York, Melbourne, Madrid, Cape Town, Singapore,
São Paulo, Delhi, Dubai, Tokyo, Mexico City

Cambridge University Press
The Edinburgh Building, Cambridge CB2 8RU, UK

Published in the United States of America by Cambridge University Press, New York

www.cambridge.org
Information on this title: www.cambridge.org/9780521182683

First published 2006
First paperback edition 2010

A catalogue record for this publication is available from the British Library

ISBN 978-0-521-84755-1 Hardback
ISBN 978-0-521-18268-3 Paperback

Contents

Contributors

Michael A. Arbib
Computer Science Department
Neuroscience Program and USC
Brain Project
University of Southern California
Los Angeles, CA 90089, USA

Aude Billard
Autonomous Systems Laboratory
Swiss Institute of Technology
Lausanne (EPFL)
1015 Lausanne, Switzerland

Mihail Bota
Neuroscience Program and
USC Brain Project
University of Southern California
Los Angeles, CA 90089, USA

Nina Bradley
Department of Biokinesiology and
Physical Therapy
University of Southern California
Los Angeles, CA 90033, USA

Dani Byrd
Department of Linguistics
University of Southern California
Los Angeles, CA 90089, USA

Karen Emmorey
School of Speech, Language, and
Hearing Sciences
San Diego State University
San Diego, CA 92120, USA

Ioana D. Goga
Center for Cognitive and
Neural Studies
400504 Clvj-Napoca, Romania

Louis Goldstein
Department of Linguistics
Yale University
New Haven, CT 06511, USA

Andrew S. Gordon
Institute for Creative Technologies
University of Southern California
Marina del Rey, CA 90292, USA

Patricia Greenfield
Department of Psychology
FPR-UCLA Center for
Culture, Brain, and
Development
University of Southern California
Los Angeles
CA 90095, USA

Jerry R. Hobbs
USC Information Sciences Institute
University of Southern California
Marina del Rey, CA 90292, USA

Laurent Itti
Department of Computer Science,
University of Southern California
Los Angeles
CA 90089, USA

David Kemmerer
Department of Speech, Language,
and Hearing Sciences
Purdue University
West Lafayette
IN 47907, USA

Howard C. Nusbaum
Department of Psychology and
the Brain Research Imaging Center
The University of Chicago
Chicago
IL 60637, USA

Erhan Oztop
JST-ICORP Computational Brain Project
ATR
Computational Neuroscience
Laboratories
Kyoto 619-0288, Japan

Elliot Saltzman
Department of Physical Therapy
Boston University
Boston, MA 02215, USA

Stefan Schaal
Department of Computer Science
University of Southern California
Los Angeles, CA 90089, USA

Jeremy I. Skipper
Department of Neurology
The University of Chicago
Chicago, IL 60637, USA

Steven L. Small
Departments of Psychology
and Neurology and the Brain
Research Imaging Center
The University of Chicago
Chicago, IL 60637, USA

Craig B. Stanford
Department of Anthropology
University of Southern California
Los Angeles, CA 90089, USA

Patricia Zukow-Goldring
Linguistics Department
University of Southern California
Los Angeles, CA 90089, USA

Preface

There are many ways to approach human language – as a rich human social activity, as a formal system structured by rules of grammar, and as a pattern of perception and production of utterances, to name just a few. The present volume uses this last concern – with the perception and production of utterances – as its core. The aim is not to ignore the other dimensions of language but rather to enrich them by seeking to understand how the use of language may be situated with respect to other systems for action and perception.

The work is centered on, but in no way restricted to, the Mirror System Hypothesis (introduced by Arbib and Rizzolatti in 1997). This is the hypothesis that the mirror neuron system for the recognition of movements of the hands in praxic action – which is present both in monkey and in human in a number of areas including Broca's area (generally considered to be the frontal speech area) – provides the evolutionary basis for the brain mechanisms which support language. The Mirror System Hypothesis sees the ancestral action recognition system being elaborated through the evolution of ever more capable neural mechanisms supporting imitation of hand movements, then pantomime emerging on the basis of displacement of hand movements to imitate other degrees of freedom. A system of "protosign" emerges as conventionalized codes extend the range of manual communication, and serves as scaffolding for "protospeech." An expanding spiral of protosign and protospeech yields a brain able to support both action and language.

The arguments pro and con the Mirror System Hypothesis and the more general issue of how the studies of action and language can illuminate each other are developed in 15 chapters by experts in child development, computer science, linguistics, neuroscience, primatology, and robotics.

Part I of the book provides *Two perspectives* on the evolution of language. I discuss "The Mirror System Hypothesis on the linkage of action and languages," presenting the essential data and modeling of the mirror system for grasping in the macaque brain as well as related human brain imaging data to set the stage for the Mirror System Hypothesis on the evolution of the language-ready brain. Two controversial hypotheses are discussed: the view that the path to protospeech was indirect, depending on the scaffolding of protosign rather than evolving directly from primate vocalizations within the vocal domain; and the view that protolanguage was "holophrastic" and that the emergence of

ix

language rested on the simultaneous fractionation of "unitary utterances" into words and the development of varied syntactic strategies to put the pieces back together again. Jerry Hobbs then offers a view of "The origin and evolution of language: a plausible, strong-AI account" which may be set against the Mirror System Hypothesis. He presents a computational approach in which abductive logic is realized in the structured connectionist networks of the SHRUTI model of language processing, and on this he bases an account of the evolution of language mechanisms. He uses the development of folk psychology to explain the evolution of Gricean non-natural meaning, i.e., the notion that what is conveyed is not merely the content of the utterance, but also the intention of the speaker to convey that meaning by means of that specific utterance. He also presents an account of syntax arising out of discourse, and uses that to argue against the holophrasis hypothesis.

Part II of the book presents three chapters addressing the theme of *Brain, evolution, and comparative analysis*. Craig Stanford provides, in "Cognition, imitation, and culture in the great apes," a comparative view of humans and the great apes. The "culture of apes" is discussed, especially variations in behavior in chimpanzees in different areas of Africa. The capacity (or lack of it) for imitation in apes and the relation between "communication in the wild" and simple forms of "language" taught to apes in captivity leads to a discussion of which of the cognitive abilities that make human language possible are possessed by non-human primates; while an emphasis on social behavior is the basis for an evaluation of claims for Theory of Mind (Machiavellian intelligence) in non-human primates. Karen Emmorey provides a comparative analysis of language modalities within humans, focusing on "The signer as an embodied mirror neuron system: neural mechanisms underlying sign language and action." Like mirror neurons, signers must associate the visually perceived manual actions of another signer with self-generated actions of the same form. However, unlike grasping and reaching movements, sign articulations are structured within a phonological system of contrasts. She relates this to language evolution, showing how languages escape their pantomimic and iconic roots and develop duality of patterning, and probes the similarities and differences between the neural systems for production and perception that support speech, sign, and action. Finally for Part II, Mihail Bota joins me in the chapter "Neural homologies and the grounding of neurolinguistics" to use comparative neurobiology of the monkey and human to establish homologies between brain regions of the two species related to the mirror system as well as communication and (precursors of) language to ground claims as to the brain of the common ancestor of monkeys and humans of perhaps 20 million years ago, and thus evaluate the Mirror System Hypothesis on how such brains changed to become language-ready. Of particular importance to charting the similarities and differences between human speech and the vocalizations of non-human primates is a set of data on the macaque auditory system and on the role of anterior cingulate cortex in both motivation and vocalization to better distinguish mechanisms that support language from those that support quite different forms of communication.

Part III analyzes *Dynamic systems in action and language*. In "Dynamic systems: brain, body, and imitation," Stefan Schaal focuses primarily on technological approaches to learning "motor primitives" by building up dynamical systems by using combinations of basis functions in the system description. Learning methods can set the weights of these combinations both to yield imitation of behaviors expressed in a form that uses all the relevant dynamic variables and to yield recognition of actions as well. The discussion is extended to superposition of movements and, briefly, to sequential and hierarchical behavior. A short discussion relates these concepts to "what the brain really does." With this we turn to speech as a dynamical system. Louis Goldstein, Dani Byrd, and Elliot Saltzman discuss "The role of vocal tract gestural action units in understanding the evolution of phonology," thus helping us think through the issue of to what extent speech production shares mechanisms with motor control more generally. Like Emmorey, they stress the central role of duality of patterning – the use of a set of non-meaningful arbitrary discrete units that allows word creation to be productive – in phonology. They analyze vocal tract action gestures in terms of dynamical systems which are discrete and combinable. They see the iconic aspects of manual gestures as critical to evolution of a system of symbolic units whereas phonological evolution crucially requires the emergence of effectively non-meaningful combinatorial units, with the two systems ultimately converging in a symbiotic relationship. Continuing in this vein, Jeremy Skipper, Howard Nusbaum, and Steven Small discuss "Lending a helping hand to hearing: another motor theory of speech perception." They describe an active model of speech perception that involves mirror neurons as the basis for inverse and forward models used in the recognition of speech, with special emphasis on audiovisual speech in which facial movements (e.g., of the lips) provide extra cues for speech recognition. According to this model, the mirror neuron system maps observed speech production and manual gestures to abstract representations of speaking actions that would have been activated had the observer been the one producing the action (inverse models). These representations are then mapped in a somatotopically appropriate manner to pre- and primary motor cortices. The resulting motor commands have sensory consequences (forward models) that are compared to processing in the various sensory modalities. This aids in the recognition of a particular acoustic segment of an utterance by constraining alternative linguistic interpretations. Brain imaging experiments are adduced in support of these ideas.

Part IV takes us *From mirror system to syntax and Theory of Mind*. Laurent Itti and I discuss "Attention and the minimal subscene," showing how perception of a "minimal subscene" linking an agent and an action to one or more objects may underlie processes of scene description and question-answering, linking the schematic structure of visual scenes to language structure. The approach integrates studies of the role of salience in visual attention with "top–down" attention, cooperative computation models of vision and speech processing, and theories and data linking sentence processing to eye movements. David Kemmerer then offers an integrative view of "Action verbs, argument structure constructions, and the mirror neuron system." Here the route from mirror neurons to syntax is via the constructionist framework which regards syntax as an extension of the lexicon. Construction Grammar provides the setting for discussing the major semantic properties of action verbs and argument structure constructions. This in turn sets the stage for analysis of the neuroanatomical substrates of action verbs and argument structure constructions in support of the proposal that the linguistic representation of action is grounded in the mirror neuron system. The discussion is then broadened to consider the emergence of language during ontogeny, history, and phylogeny. Andrew Gordon looks at "Language evidence for changes in a Theory of Mind", with Theory of Mind viewed as the set of abilities that enable people to reflect on their own reasoning, to empathize with other people by imagining what it would be like to be in their position, and to generate reasonable expectations and inferences about mental states and processes. He analyzes a corpus of English and American novels in search of a "Freudian shift" and finds an increasing use of language about the unconscious in the decades *preceding* Freud. He thus concludes that a major shift has occurred in Theory of Mind, strengthening the case that its roots are cultural rather than biological.

The final part of the book, Part V, examines *Development of action and language*. Erhan Oztop, Nina Bradley, and I study "The development of grasping and the mirror system" from a modeling perspective comprising the Infant Learning to Grasp Model (ILGM), the Grasp Affordance Emergence Model (GAEM), and the mirror neuron system (MNS) model of the development of the mirror neuron system. The account has strong links to the literature on infant motor development and makes clear that the range of mirror neuron responses is highly adaptive rather than being "hard-wired." This paves the way for an understanding of how imitation builds on the mirror system. Iona D. Goga and Aude Billard provide a broad conceptual framework for the study of "Development of goal-directed imitation, object manipulation, and language in humans and robots." Social abilities, such as imitation, turn-taking, joint attention, and intended body communication, are fundamental for the development of language and human cognition. Inspired by this perspective, they offer a composite model of the mechanisms underlying the development of action, imitation, and language in human infants. A recent trend of robotics research seeks to equip artifacts with social capabilities. Thus the model also sets the stage for reproducing imitation and language in robots and simulated agents. They validate the model through a dynamic simulation of a child–caregiver pair of humanoid robots. Patricia Zukow-Goldring discusses "Assisted Imitation: affordances, effectivities, and the mirror system in child development." She takes a mirror-system oriented view of cognitive development in the child, showing how caregivers help the child learn the effectivities of her own body and the affordances of the world around her. The basic idea is that a shared understanding of action grounds what individuals know in common. In particular, this perspective roots the ontogeny of language in the progression from action and gesture to speech and supports the view that the evolutionary path to language also arises from perceiving and acting, leading to gesture, and eventually to speech. Finally, Patricia Greenfield explores "Implications of mirror neurons for the ontogeny and phylogeny of cultural processes: the examples of tools and language." Mirror neurons underlie the ability of the monkey and human to respond both to their own acts and to the same act performed by another, selectively responding to intentional or goal-directed action rather than to movement per se. Greenfield argues that this neural substrate underwrites phylogenetic and ontogenetic development of two key aspects of human culture, tool use and language. She stresses that ontogeny does not recapitulate phylogeny, but that infant behavior is more likely to be conserved than adult behavior across phylogeny. The analysis of both ontogeny and phylogeny draws on comparison of chimpanzees, bonobos, and humans to derive clues as to what foundations of human language may have been present in the common ancestor 5 to 7 million years ago.

This multi-author volume gains unusual coherence through its emergence from a year-long seminar led by the Editor in which eleven of the authors met every 2 to 3 weeks to debate and build upon the Mirror System Hypothesis. We thank the Center for Interdisciplinary Research of the University of Southern California which made this seminar, and thus this book, possible.

Part I
Two perspectives

1

The Mirror System Hypothesis on the linkage of action and languages

Michael A. Arbib

1.1 Introduction

Our progress towards an understanding of how the human brain evolved to be ready for language starts with the mirror neurons for grasping in the brain of the macaque monkey. Area F5 of the macaque brain is part of premotor cortex, i.e., F5 is part of the area of cerebral cortex just in front of the primary motor cortex shown as F1 in Fig. 1.1 (left). Different parts of F5 contain neurons active during manual and orofacial actions. Crucially for us, an anatomically segregated subset of these neurons are *mirror neurons*. Each such mirror neuron is active not only when the monkey performs actions of a certain kind (e.g., a precision pinch or a power grasp) but also when the monkey observes a human or another monkey perform a more or less similar action. In humans, we cannot measure the activity of single neurons (save when needed for testing during neurosurgery) but we can gather comparatively crude data on the relative blood flow through (and thus, presumably, the neural activity of) a brain region when the human performs one task or another. We may then ask whether the human brain also contains a "mirror system for grasping" in the sense of a region active for both execution and observation of manual actions as compared to some baseline task like simply observing an object. Remarkably, such sites were found in frontal, parietal, and temporal cortex of the human brain. Most significantly for this book, the frontal activation was found in or near Broca's area (Fig. 1.1 right), a region which in most humans lies in the left hemisphere and is traditionally associated with speech production. Moreover, macaque F5 and human Broca's area are considered to be (at least in part) homologous brain regions, in the sense that they are considered to have evolved from the same brain region of the common ancestor of monkeys and humans (see Section 1.2 as well as Arbib and Bota, this volume).

But why should a neural system for language be intimately related with a mirror system for grasping? Pondering this question led Giacomo Rizzolatti and myself to formulate the Mirror System Hypothesis (MSH) (Arbib and Rizzolatti, 1997; Rizzolatti and Arbib, 1998) which will be presented more fully in Section 1.3.1. We view the mirror system for

Action to Language via the Mirror Neuron System, ed. Michael A. Arbib. Published by Cambridge University Press.

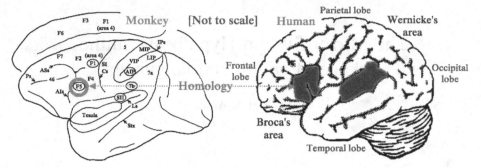

Figure 1.1 A comparative side view of the monkey brain (left) and human brain (right), not to scale. The view of the monkey brain emphasizes area F5 of the frontal lobe of the monkey; the view of the human brain emphasizes two of the regions of cerebral cortex, Broca's area and Wernicke's area, considered crucial for language processing. The homology between monkey F5 and human Broca's area is a key component of the present account.

grasping as a key neural "missing link" between the abilities of our non-human ancestors of 20 million years ago and the modern human capability for language, with manual gestures rather than a system for vocal communication providing the initial seed for this evolutionary process. The present chapter, however, goes "beyond the mirror" to offer hypotheses on evolutionary changes within and outside the mirror system which may have occurred to equip *Homo sapiens* with a language-ready brain, a brain which supports language as a behavior in which people communicate "symbolically" through integrated patterns of vocal, manual and facial movements. Language is essentially *multi-modal*, not just a set of sequences of words which can be completely captured by marks on the printed page.

1.1.1 Protolanguage defined

In historical linguistics (e.g., Dixon, 1997), a *protolanguage* for a family of extant human languages is the posited ancestral language – such as proto-Indo-European for the family of Indo-European languages – from which all languages of the family are hypothesized to descend historically with various modifications. In this chapter, however, we reserve the term *protolanguage* for a system of utterances used by a particular hominid species (possibly including *Homo sapiens*) which we would recognize as a precursor to human language (if only we had the data!), but which is not itself a human language in the modern sense.

Bickerton (1995) hypothesized that the protolanguage of *Homo erectus* was a system whose utterances took the form of a string of a few words much like those of today's language, but which conveyed meaning without the aid of syntax. On this view, language just "added syntax" through the evolution of Universal Grammar. Contrary to Bickerton's hypothesis, I argue (Section 1.5) that the protolanguage of *Homo erectus* and early *Homo sapiens* was "holophrastic", i.e., composed mainly of "holophrases" or "unitary utteran-aces" (phrases in the form of semantic wholes with no division into meaningful subparts) which symbolized frequently occurring situations and that words as we know them then co-evolved culturally with syntax through fractionation. My view is shared, e.g., by Wray (2002) and opposed by Hobbs (this volume). Bickerton (2005) and Wray (2005) advance the debate within the context of the Mirror System Hypothesis.

1.1.2 Relating language to the vocalizations of non-human primates

Humans, chimpanzees, and monkeys share a general physical form and a degree of manual dexterity, but their brains, bodies, and behaviors differ. Humans have abilities for bipedal locomotion and learnable, flexible vocalization that are not shared by other primates. Monkeys exhibit a primate call system (a limited set of species-specific calls) and an orofacial (mouth and face) gesture system (a limited set of gestures expressive of emotion and related social indicators). Note the linkage between the two systems: communication is inherently multi-modal. This communication system is *closed* in the sense that it is restricted to a specific repertoire. This is to be contrasted with the open nature of human languages which can form endlessly many novel sentences from the current word stock and add new words to that stock. Admittedly, chimpanzees and bonobos (great apes, not monkeys) can be trained to acquire a form of communication – based either on the use of hand signs or objects that each have a symbolic meaning – that approximates the complexity of the utterances of a 2-year-old human child, in which a "message" generally comprises one or two "lexemes." However, there is no evidence that an ape can reach the linguistic ability of a 3-year-old human. Moreover, such "ape languages"[1] are based on hand–eye coordination rather than vocalization whereas a crucial aspect of human biological evolution has been the emergence of a vocal apparatus and control system that can support speech.

It is tempting to hypothesize that certain species-specific vocalizations of monkeys (such as the "snake call" and "leopard call" of vervet monkeys) provided the basis for the evolution of human speech, since both are in the vocal domain (see Seyfarth (2005) for a summary of arguments supporting this view[2]). However, combinatorial properties for the openness of communication are virtually absent in basic primate calls and orofacial communication, though individual calls may be graded. Moreover, Jürgens (1997, 2002) found that voluntary control over the initiation and suppression of such vocalizations relies on the mediofrontal cortex including anterior cingulate gyrus (see Arbib and Bota, this volume). MSH explains why it is F5, rather than the cingulate area involved in macaque vocalization, that is homologous to the human's frontal substrate for language by asserting that a *specific* mirror system – the primate mirror system for grasping – evolved into a

---

[1] I use the quotes because these are not languages in the sense in which I distinguish languages from protolanguages.

[2] For some sense of the debate between those who argue that protosign was essential for the evolution of the language-ready brain and those who take a "speech only" approach, see Fogassi and Ferrari (2004), Arbib (2005b), and MacNeilage and Davis (2005).

key component of the mechanisms that render the human brain language-ready. It is this specificity that will allow us to explain below why language is multi-modal, its evolution being rooted in the execution and observation of hand movements and extended into speech.

Note that the claim is not that Broca's area is genetically preprogrammed for language, but rather that the development of a human child in a language community normally adapts this brain region to play a crucial role in language performance. Zukow-Goldring (this volume) takes a mirror-system-oriented view of cognitive development in the child, showing how caregivers help the child learn the effectivities of her own body and the affordances of the world around her. As we shall see later in this chapter, the notion of *affordance* offered by Gibson (1979) – that of information in the sensory stream concerning opportunities for action in the environment – has played a critical role in the development of MSH, with special reference to neural mechanisms which extract affordances for grasping actions. Not so much has been made of *effectivities*, the range of possible deployments of the organism's degrees of freedom (Turvey *et al.*, 1981), but Oztop *et al.* (this volume) offer an explicit model (the Infant Learning to Grasp Model, ILGM) of how effectivities for grasping may develop as the child engages in new activities. Clearly, the development of novel effectivities creates opportunities for the recognition of new affordances, and vice versa.

1.1.3 Language in an action-oriented framework

Neurolinguistics should emphasize performance, explicitly analyzing both perception and production. The framework sketched in Fig. 1.2 leads us to ask: to what extent do language mechanisms exploit existing brain mechanisms and to what extent do they involve biological specializations specific to humans? And, in the latter case, did the emergence of language drive the evolution of such mechanisms or exploit them?

Arbib (2002, 2005a) has extended the original formulation of MSH to hypothesize a number of (possibly overlapping) stages in the evolution of the language-ready brain. As we shall see in more detail in Section 1.3.2, crucial early stages extend the mirror system for grasping to support imitation, first so-called "simple" imitation such as that found in chimpanzees (which allows imitation of "object-oriented" sequences only as the result of extensive practice) and then so-called "complex" imitation found in humans which combines the perceptual ability to recognize that a novel action was in fact composed of (approximations to) known actions with the ability to use this analysis to guide the more or less successful reproduction of an observed action of moderate complexity.[3] Complex imitation was a crucial evolutionary innovation in its own right, increasing the ability to learn and transmit novel skills, but also provides a basis for the later emergence of the language-ready brain – it is crucial not only to the child's ability to acquire

---

[3] See the discussion of research by Byrne (2003) on "program-level imitation" in gorillas (cf. Stanford, this volume). A key issue concerns the divergent time course of acquisition of new skills in human versus ape.

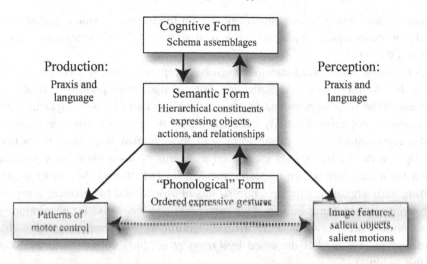

Figure 1.2 The figure places production and perception of language within a framework of action and perception considered more generically. Language production and perception are viewed as the linkage of Cognitive Form, Semantic Form, and Phonological Form, where the "phonology" may involve vocal or manual gestures, or just one of these, with or without the accompaniment of facial gestures.

language and social skills, but the perceptual ability within it is essential to the adult use of language.

But it requires a further evolutionary breakthrough for the complex imitation of action to yield pantomime, as the purpose comes to be to communicate rather than to manipulate objects. This is hypothesized to provide the substrate for the development of protosign, a combinatorially open repertoire of manual gestures, which then provides the scaffolding for the emergence of protospeech (which thus owes little to non-human vocalizations), with protosign and protospeech then developing in an expanding spiral. Here, I am using "protosign" and "protospeech" for the manual and vocal components of a protolanguage, in the sense defined earlier. I argue that these stages involve biological evolution but that the progression from protosign and protospeech to languages with full-blown syntax and compositional semantics was a historical phenomenon in the development of *Homo sapiens*, involving few if any further biological changes.

For *production*, the notion is that at any time we have much that we could possibly talk about which might be represented as cognitive structures (Cognitive Form; schema assemblages) from which some aspects are selected for possible expression. Further selection and transformation yields semantic structures (hierarchical constituents express-ing objects, actions and relationships) which constitute a Semantic Form enriched by linkage to schemas for perceiving and acting upon the world. Finally, the ideas in the Semantic Form must be expressed in words whose markings and ordering reflect the relationships within Semantic Form. These words must be conveyed as "phonological" structures – where here I extend phonological form to embrace a wide range of ordered expressive gestures which may include speech, sign, and orofacial expressions (and even writing and typing).

For *perception*, the received utterance must be interpreted semantically with the result updating the "hearer's" cognitive structures. For example, perception of a visual scene may reveal "Who is doing what and to whom/which" as part of a non-linguistic *action–object frame* in cognitive form. By contrast, the *verb–argument structure* is an overt linguistic representation in semantic form – in most human languages, the action is named by a verb and the objects are named by nouns or noun phrases. A production grammar for a language is then a specific mechanism (whether explicit or implicit) for converting verb–argument structures into strings of words (and hierarchical compounds of verb–argument structures into complex sentences) and vice versa for perception.

These notions of "forward" and "inverse" grammars may be compared with the forward and inverse models discussed by Oztop *et al.* (this volume) and by Skipper *et al.* (this volume).

Emmorey (this volume) discusses the neural basis for parallels between praxis (including spatial behavior), pantomime, sign production, and speech production, establishing what is modality-general for brain mechanisms of language and what is modality-specific. Here I would suggest a crucial distinction between signed language and speech. In both speech and signing, I claim, we recognize a novel utterance as in fact composed of (approximations to) known actions, namely uttering words, and – just as crucially – the stock of words is open-ended. However, signed language achieves this by a very different approach to speech. Signing exploits the fact that the signer has a very rich repertoire of arm, hand, and face movements, and thus builds up vocabulary by variations on this multidimensional theme (move a hand shape (or two) along a trajectory to a particular position while making appropriate facial gestures). By contrast, speech employs a system of articulators that are specialized for speech – there is no rich behavioral repertoire of non-speech movements to build upon.[4] Instead evolution "went particulate" (Studdert-Kennedy, 2000; Goldstein *et al.*, this volume), so that the spoken word is built (to a first approximation) from a language-specific stock of phonemes (actions defined by the coordinated movement of several articulators, but with only the goal of "sounding right" rather than conveying meaning in themselves).

But if single gestures are the equivalent of phonemes in speech or words in sign, what levels of motor organization correspond to derived words, compound words, phrases, sentences, and discourse? Getting to derived words seems simple enough. In speech, we play variations on a word by various morphological changes which may modify internal phonemes or add new ones. In sign, "words" can be modified by changing the source and

---

[4] However, it is interesting to speculate on the possible relevance of the oral dexterity of an omnivore – the rapid adaptation of the coordination of chewing with lip, tongue, and swallowing movements to the contents of the mouth – to the evolution of these articulators. Such oral dexterity goes well beyond the "mandibular cyclicities" posited by MacNeilage (1998) to form the basis for the evolution of the ability to produce syllables as consonant–vowel pairs.

origin, and by various modifications to the path between. For everything else, it seems enough – for both action and language – that we can create hierarchical structures subject to a set of transformations from those already in the repertoire. For this, the brain must provide a computational medium in which already available elements can be composed to form new ones, irrespective of the "level" at which these elements were themselves defined. When we start with words as the elements, we may end up with compound words or phrases, other operations build from both words and phrases to yield new phrases or sentences, etc., and so on recursively. Similarly, we may learn arbitrarily many new motor skills based on those with which we are already familiar.

With this, let me give a couple of examples (Arbib, 2006) which suggest how to view language in a way which better defines its relation to goal-directed action. Consider a conditional, hierarchical motor plan for opening a child-proof aspirin bottle:

While holding the bottle with the non-dominant hand, grasp the cap, push down and turn the cap with the dominant hand; then repeat (release cap, pull up cap, and turn) until the cap comes loose; then remove the cap.

This hierarchical structure unpacks to different sequences of action on different occasions, with subsequences conditioned on the achievement of goals and subgoals. To ground the search for similarities between action and language, I suggest we view an action such as this one as a "sentence" made up of "words" which are basic actions. A "paragraph" or a "discourse" might then correspond to, e.g., an assembly task which involves a number of such "sentences."

Now consider a sentence like "Serve the handsome old man on the left.", spoken by a restaurant manager to a waiter. From a "conventional" linguistic viewpoint, we would apply syntactic rules to parse this specific string of words. But let us look at the sentence not as a structure to be parsed but rather as the result of the manager's attempt to achieve a *communicative goal*: to get the waiter to serve the intended customer. His *sentence planning strategy* repeats the "loop"

<add adjective or prepositional phrase>

until (he thinks) ambiguity is resolved:

(1) Serve the old man.

Still ambiguous?

(2) Serve the old man on the left.

Still ambiguous?

(3) Serve the handsome old man on the left.

Still ambiguous?

Apparently not. So the manager "executes the plan" and says to the waiter

"Serve the handsome old man on the left."

Here, a noun phrase NP may be expanded by adding a prepositional phrase PP after it (as in expanding (1) to (2) above) or an adjective Adj before it (as in expanding (2) to (3)). The suggestion is that syntactic rules of English which I approximate by NP → NP PP and

NP → Adj NP are abstracted from procedures which serve to reduce ambiguity in reaching a communicative goal. This example concentrates on a noun phrase – and thus exemplifies ways in which reaching a communicative goal (identifying the right person, or more generally, object) may yield an unfolding of word structures in a way that may clarify the history of syntactic structures.

While syntactic constructions can be usefully analyzed and categorized from an abstract viewpoint, the pragmatics of what one is trying to say and to whom one is trying to say it will drive the goal-directed process of producing a sentence. Conversely, the hearer has the inferential task of unfolding multiple meanings from the word stream (with selective attention) and deciding (perhaps unconsciously) which ones to meld into his or her cognitive state and narrative memory.

1.2 Grasping and the mirror system

1.2.1 Brain mechanisms for grasping

Figure 1.3 shows the brain of the macaque (rhesus monkey) with its four lobes: frontal, parietal, occipital, and temporal. The parietal area AIP is near the front (anterior) of a groove in the parietal lobe, shown opened up in the figure, called the intraparietal sulcus – thus AIP (for anterior region of the intraparietal sulcus).[5] Area F5 (the fifth region in a numbering for areas of the macaque frontal lobe) is in what is called ventral premotor cortex. The neuroanatomical coordinates refer to the orientation of the brain and spinal cord in a four-legged vertebrate: "dorsal" is on the upper side (think of the dorsal fin of a shark) while "ventral" refers to structures closer to the belly side of the animal. Together, AIP and F5 anchor the cortical circuit in macaque which transforms visual information on intrinsic properties of an object into hand movements for grasping it. AIP processes visual information to implement perceptual schemas for extracting grasp parameters (affordances) relevant to the control of hand movements and is reciprocally connected with the so-called *canonical neurons* of F5. Discharge in most grasp-related F5 neurons correlates with an action rather than with the individual movements that form it so that one may relate F5 neurons to various *motor schemas* corresponding to the action associated with their discharge. By contrast, primary motor cortex (F1) formulates the neural instructions for lower motor areas and motor neurons.

The FARS (Fagg–Arbib–Rizzolatti–Sakata) model (Fagg and Arbib, 1998) provides a computational account of the system (it has been implemented, and various examples of grasping simulated) centered on this pathway: the dorsal stream via AIP does not know "what" the object is, it can only see the object as a set of possible affordances, whereas the ventral stream from primary visual cortex to inferotemporal cortex (IT), by contrast, is able to recognize what the object is. This information is passed to prefrontal cortex (PFC)

---

[5] The reader uncomfortable with neuroanatomical terminology can simply use AIP and similar abbreviations as labels in what follows, without worrying about what they mean. On the other hand, the reader who wants to know more about the neuroanatomy should turn to Arbib and Bota (this volume) in due course.

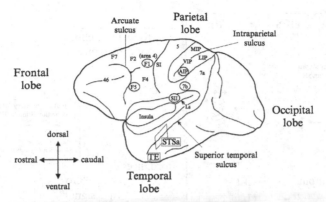

Figure 1.3 A side view of the left hemisphere of the macaque brain. Area 7b is also known as area PF. (Adapted from Jeannerod *et al.*, 1995.)

which can then, on the basis of the current goals of the organism and the recognition of the nature of the object, bias the affordance appropriate to the task at hand. The original FARS model suggested that the bias was applied by PFC to F5; subsequent neuroanatomical data (as analyzed by Rizzolatti and Luppino, 2003) suggest that PFC and IT may modulate action selection at the level of parietal cortex rather than premotor cortex.

Figure 1.4 gives a partial view of "FARS Modificato," the FARS model updated to show this modified pathway. AIP may represent several affordances initially, but only one of these is selected to influence F5. This affordance then activates the F5 neurons to command the appropriate grip once it receives a "go signal" from another region, F6, of PFC.

We now turn to those parts of the FARS model which provide mechanisms for sequencing actions: the circuitry encoding a sequence is postulated to lie within the supplementary motor area called pre-SMA (Rizzolatti *et al.*, 1998), with administration of the sequence (inhibiting extraneous actions, while priming imminent actions) carried out by the basal ganglia. In Section 1.3, I will argue that the transition to early *Homo* coincided with the transition from a mirror system used only for action recognition and "simple" imitation to more elaborate forms of "complex" imitation: I argue that what sets hominids apart from their common ancestors with the great apes is the ability to rapidly exploit novel sequences as the basis for immediate imitation or for the immediate construction of an appropriate response.

Here, I hypothesize that the macaque brain can generate sequential behavior on the basis of overlearned coordinated control programs (cf. Itti and Arbib, this volume), but that only the human brain can perform the inverse operation of observing a sequential behavior and inferring the structure of a coordinated control program that might have generated it. In the FARS model, Fagg and Arbib (1998) modeled the interaction of AIP and F5 in the sequential task employed by Sakata in his studies of AIP. In this task, the monkey sits with his hand on a key, while in front of him is a manipulandum. A change in an LED signals him to prepare to grasp; a further change then tells him to execute the

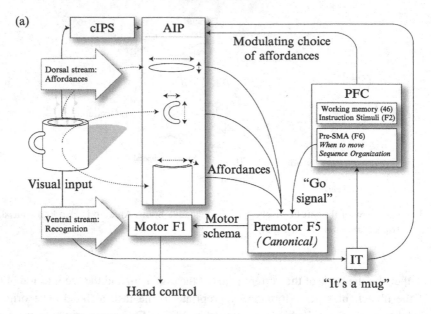

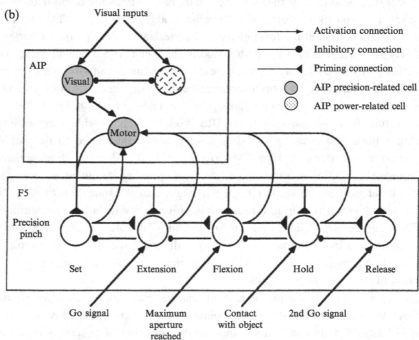

Figure 1.4   (a) "FARS Modificato": the original FARS diagram (Fagg and Arbib, 1998) is here modified to show PFC acting on AIP rather than F5. The idea is that AIP does not "know" the identity of the object, but can only extract affordances (opportunities for grasping for the object consider as an unidentified solid); prefrontal cortex uses the IT identification of the object, in concert with task analysis and working memory, to help AIP select the appropriate action from its

grasp. He must then hold the manipulandum until the onset of another LED signals him to release the manipulandum. Successful completion of the task yields a juice reward.

Figure 1.4b simplifies how the FARS model presents the interaction of AIP and F5 in the Sakata Task. Three AIP cells are shown: a visual-related and a motor-related cell that recognize the affordance for a precision pinch, and a visual-related cell for power grasp affordances. The five F5 units participate in a common program (in this case, a precision grasp), but each cell fires during a different phase of the program. However, the problem with the diagram of Fig. 1.4b is that it suggests that each action coded by F5 neurons is hard-wired to determine a unique successor action – hardly appropriate for adaptive behavior. Indeed, the Sakata Task is invariant save that the nature of the object, and thus the appropriate grasp, may change between trials. One would thus like to understand how "grasp switching" can occur within the one overall pattern of behavior.

Karl Lashley (1951) raised the problem of serial order in behavior in a critique of the behaviorist view that each action must be triggered by some external stimulus. If we tried to learn a sequence like $A \to B \to A \to C$ by reflex chaining, what is to stop A triggering B every time, to yield the performance $A \to B \to A \to B \to A \to ...$ (or we might get $A \to B + C \to A \to B + C \to A \to ...$)? One solution is to store the "action codes" (motor schemas) A, B, C, ... in one part of the brain and have another area hold "abstract sequences" and learn to pair the right action with each element. Fagg and Arbib posited that F5 holds the "action codes" (motor schemas) for grasping, while the pre-SMA holds the code for the overlearned sequence. We further posited that the basal ganglia (BG) manage priming and inhibition (cf. Arbib *et al.* (1998), for modeling, and Lieberman (2000) for the role of human BG in language).

This analysis emphasizes that F5 alone is not the "full" mirror system. Behavior, and its imitation, involves not only the "unit actions" encoded in the F5 mirror system but also sequences and more general patterns. The FARS model sketches how pre-SMA and BG may cooperate in generating a sequence. But this is a model of the performance of *overlearned* sequences. Thus, if we are to extend the FARS model to include complex imitation, we must not only model a monkey-like mirror system which relates observed grasping actions to ones which the observer can himself perform, but must also go "beyond the mirror" to show how the units of a sequence *and* their order/interweaving can be recognized and how novel performances (imitation) can be based on this

Caption for Figure 1.4 (cont.)
"menu." Area cIPS is a part of the intraparietal sulcus that appears to be involved in recognizing the orientation of surfaces, information useful for both the dorsal stream (extracting affordances for grasping) and the ventral stream (using shape cues to recognize objects). (b) Interaction of AIP and F5 in the Sakata Task. (In the Sakata Task, the animal places its hand on a key once an LED signals it to do so; a second cue instructs the animal to reach for, grasp, and hold a manipulandum; while a third cue instructs the animal to release its hold and return to the resting position.) Three AIP cells are shown: a visual-related and a motor-related cell that recognize the affordance for a precision pinch, and a visual-related cell for power grasp affordances. The five F5 units participate in a common program (in this case, a precision grasp), but each cell fires during a different phase of the program.

recognition. This *future* model will require recognition of a complex behavior on multiple occasions with increasing success in recognizing component actions and in linking them together. Arbib and Bota (this volume; see also Itti and Arbib, this volume) offer related discussion, viewing sequences as the expression of hierarchical structures.

1.2.2 A mirror system for grasping in macaques

Gallese *et al.* (1996) discovered that, among the F5 neurons related to grasping there are neurons, which they called *mirror neurons*, that are active not only when the monkey executes a specific hand action but also when it observes a human or other monkey carrying out a similar action. Mirror neurons do not discharge in response to simple presentation of objects even when held by hand by the experimenter. They require a specific action – whether observed or self-executed – to be triggered. Moreover, mirror neurons do not fire when the monkey sees the hand movement unless it can also see the object or, more subtly, if the object is not visible but is appropriately "located" in working memory because it has recently been placed on a surface and has then been obscured behind a screen behind which the experimenter is seen to be reaching (Umiltà *et al.*, 2001). In either case, the trajectory and handshape must match the affordance of the object. All mirror neurons show visual generalization. They fire when the instrument of the observed action (usually a hand) is large or small, far from or close to the monkey. They fire whether the action is made by a human or monkey hand. A few neurons respond even when the object is grasped by the mouth.

The majority of mirror neurons respond selectively to one type of action. For some mirror neurons, the congruence between the observed and executed actions which are accompanied by strong firing of the neuron is extremely strict, with the effective motor action (e.g., precision grip) coinciding with the action that, when seen, triggers the neuron (again precision grip, in this example). For other neurons the congruence is broader. For them the motor requirement (e.g., precision grip) is usually stricter than the visual (e.g., any type of hand grasping, but not other actions). These neurons constitute the "mirror system for grasping" in the monkey and we say that these neurons provide the neural code for matching execution and observation of hand movements. By contrast, the *canonical neurons* are those grasp-related neurons that are not mirror neurons; i.e., they are active for execution but not observation. They constitute the F5 neurons simulated in the FARS model. The populations of canonical and mirror neurons appear to be spatially segregated in F5 (see Rizzolatti and Luppino (2001) for an overview and further references).

Perrett *et al.* (1990) and Carey *et al.* (1997) found that STSa, in the rostral part of the superior temporal sulcus (STS, itself part of the temporal lobe), has neurons which discharge when the monkey observes such biological actions as walking, turning the head, bending the torso, and moving the arms. Of most relevance to us is that a few of these neurons discharged when the monkey observed goal-directed hand movements, such as grasping objects (Perrett *et al.*, 1990) – though STSa neurons do not seem to discharge during movement execution as distinct from observation.

STSa and F5 may be indirectly connected via inferior parietal area PF (Brodmann area (BA) 7b) (Matelli *et al.*, 1986; Cavada and Goldman-Rakic, 1989; Seltzer and Pandya, 1994). About 40% of the visually responsive neurons in PF are active for observation of actions such as holding, placing, reaching, grasping, and bimanual interaction. Moreover, most of these action observation neurons were also active during the execution of actions similar to those for which they were "observers," and were thus called PF mirror neurons (Fogassi *et al.*, 1998).

In summary, area F5 and area PF include an observation/execution matching system: when the monkey observes an action that resembles one in its movement repertoire, a subset of the F5 and PF mirror neurons is activated which also discharges when a similar action is executed by the monkey itself.

We next develop the conceptual framework for thinking about the relation between F5, AIP, and PF. For now we focus on the role of mirror neurons in the visual recognition of hand movements. However, Section 1.4 expands the mirror neuron database, reviewing the findings by Kohler *et al.* (2002) that 15% of mirror neurons in the hand area of F5 can respond to the distinctive sound of an action (breaking peanuts, ripping paper, etc.) as well as viewing the action, and by Ferrari *et al.* (2003) that the orofacial area of F5 (adjacent to the hand area) contains a small number of neurons tuned to communicative gestures (lip-smacking, etc.). Clearly, these data must be assessed carefully in weighing the relative roles of protosign and protospeech in the evolution of the language-ready brain.

The Mirror Neuron System (MNS) model of Oztop and Arbib (2002; described further by Oztop *et al.*, this volume) provides some insight into the anatomy while focusing on the learning capacities of mirror neurons. Here, the task is to determine whether the shape of the hand and its trajectory are "on track" to grasp an observed affordance of an object using a known action. The model is organized around the idea that the AIP → F5$_{canonical}$ pathway emphasized in the FARS model (Fig. 1.4) is complemented by PF → F5$_{mirror}$. As shown in Fig. 1.5, the MNS model can be divided into three parts.

- **Top diagonal (upper)** Recognizing the location of the object provides parameters to the motor programming area F4 which computes the reach. The information about the reach and the grasp is taken by the motor cortex M1 to control the hand and the arm.
- **Top diagonal (lower)** Object features are processed by AIP to extract grasp affordances; these are sent on to the canonical neurons of F5 that choose a particular grasp.
- **Essential elements for the mirror system** The third part of the figure provides components that can learn and apply key criteria for activating a mirror neuron, recognizing that (a) the preshape that the monkey is seeing corresponds to the grasp that the mirror neuron encodes; (b) the preshape that the observed hand is executing is appropriate to the object that the monkey can see (or remember); and (c) that the hand is moving on a trajectory that will bring it to grasp the object. The two schemas at bottom left recognize the shape of the hand of the actor being observed by the monkey whose brain we are interested in, and how that hand is moving. Just to the right of these is the schema for hand–object spatial relation analysis. It takes information about object features, the motion of the hand, and the location of the object to infer the relation between hand and object. Just above this is the schema for associating object affordances and hand state. Together with F5 canonical neurons, this last schema (in PF = 7b) provides the input to the F5 mirror neurons.

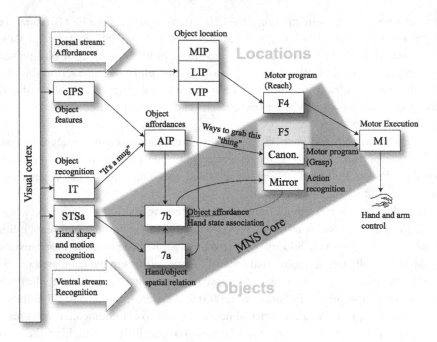

Figure 1.5   The Mirror Neuron System (MNS) model (Oztop and Arbib, 2002).

In the MNS model, the *hand state* was defined as a vector whose components represented the movement of the wrist relative to the location of the object and of the handshape relative to the affordances of the object. Oztop and Arbib (2002) showed that an artificial neural network corresponding to PF and $F5_{mirror}$ could be trained to recognize the grasp type from the *hand state trajectory*, with correct classification often being achieved well before the hand reached the object. The modeling assumed that the neural equivalent of a grasp being in the monkey's repertoire is that there is a pattern of activity in the F5 canonical neurons that commands that grasp. During training, the output of the F5 canonical neurons, acting as a code for the grasp being executed by the monkey at that time, was used as the training signal for the F5 mirror neurons to enable them to learn which hand–object trajectories corresponded to the canonically encoded grasps. As a result of this training, the appropriate mirror neurons come to fire in response to viewing the appropriate trajectories even when the trajectory is not accompanied by F5 canonical firing.

Because the input to the F5 mirror neurons encodes the trajectory of the relation of parts of the hand to the object rather than the visual appearance of the hand in the visual field, this training prepares the F5 mirror neurons to respond to hand–object relational trajectories even when the hand is of the "other" rather than the "self." What makes the modeling worthwhile is that the trained network responded not only to hand state trajectories from the training set, but also exhibited interesting responses to novel hand–object relationships. Such learning models, and the data they address, make clear that *mirror neurons are not restricted to recognition of an innate set of actions but can be recruited to recognize and encode an expanding repertoire of novel actions.*

1.2.3 From mirror neurons to understanding

Studies of monkey mirror neurons (e.g., Rizzolatti *et al.*, 1996) show the monkey attending to an action of the experimenter and responding not with an overt action, but rather with neural activity in mirror neurons that are active when the monkey itself performs a similar action. The monkey does not imitate. What then is the adaptive value of mirror neuron activity? Most writers have noted the adaptive advantage that such a system could have for social interaction, allowing one monkey to "understand" the actions of another, and thus position itself to compete or cooperate with the other monkey more effectively. However, I hypothesize that the adaptive pressure for the initial evolution of mirror neurons was not so much social as to serve as part of a feedback system for increasingly dexterous manual control – thus yielding extraction of the hand–object relations posited in the MNS model to provide the key input to mirror neurons. But once a neural representation that could extract such relations existed, it could provide the opportunity for "social circuitry" to evolve so as to best exploit this information to more subtly delineate the behavior of conspecifics. However, I suggest that further evolution was required to go from mirror neurons that are part of the feedback system for motor control to having motor neurons serve as part of the circuitry that mediates understanding.

Arbib and Rizzolatti (1997) asserted that what makes a movement into an action is that (a) it is associated with a goal, and (b) initiation of the movement is accompanied by the creation of an expectation that the goal will be met. To the extent that the unfolding of the movement departs from that expectation, to that extent will an error be detected and the movement modified. In other words, the brain of an individual performing an action is able to predict its consequences and, therefore, the action representation and its consequences are associated. Thus a "grasp" involves not only a specific cortical activation pattern for the preshape and enclose movements of the hand but also expectations concerning making appropriate contact with a specific object (cf. Fig. 12.9 of Oztop *et al.*, this volume). However, we must distinguish "an individual making an action has neural activity that predicts the consequences of his pattern of movement" from "the individual has conscious awareness of what he is doing."

The data from the Rizzolatti laboratory show that a macaque (the data are silent about other creatures) can "recognize" certain manual and orofacial actions made by others in the very special sense that the neural pattern elicited by those actions in the F5 mirror neurons is similar to that generated when he performs a similar action himself. However this form of "recognition" *in itself* may lack the rich subjective dimensions that often accompany a human's recognition of an act or situation. Note that I am *not* denying that the monkey's recognition of action may be quite rich (though I would argue that human language and other cognitive abilities make human awareness very different from the monkey's; Arbib, 2001). What I *do* deny is that the mere activity of F5 mirror neurons alone suffices to provide such richness, or to constitute "understanding" the action. Consider a pattern recognition device that can be trained to classify pixel patterns from its camera into those which resemble a line drawing of a circle and those which do not (with the degree of resemblance cut off at some arbitrary threshold). I would argue that this device does not *understand* circles. However, to the extent that this recognition could be linked to circuitry for drawing a circle, or for forming associations like "the outline of the sun or an orthogonal cut through a cone yields an appropriate stimulus," one might say that the *system* of which the pattern recognizer is part does exhibit understanding. Understanding is thus not a binary concept but rather a matter of degree; some things may be encoded appropriately yet not understood at all, others may be understood in great richness because their neural encoding is linked to many other behaviors and perceptions.

In any case, the hypothesis is that the mirror system in macaques is able both to provide feedback for manual control and to make useful information available for interacting with others. Alas, there are no data on recording mirror neurons while monkeys engage in natural social interactions, so more needs to be done to verify that the macaque brain does indeed support "action understanding." To crudely simplify the necessary processes, we have a system that can recognize an action A, recall (perhaps implicitly) that action A is most likely to have consequences B or C, determine from the context that B is more likely to occur, and then use the expectation of B's occurrence to speed the choice of a course of action appropriate to B. However, such circuitry in and of itself cannot support the converse, namely imitation: observe that a novel action A achieves goal B in context W, master the skill of performing A, and then use it next time B is one's goal in a context not unrelated to W. The ability to learn from the actions of others (which may thus yield, but not be restricted to, imitation) takes us beyond the capabilities we share with monkeys and (as we shall see in Section 1.3) on to capabilities we share with chimpanzees. (Stanford (this volume) may be read as suggesting that imitation is more widespread among non-human primates than I would concede, but this may depend on my using "imitation" for a narrower range of behaviors than would Stanford. In any case, the literature on imitation in non-human primates is not without controversy.)

I thus argue that the mirror neurons of PF and F5 must be embedded in a larger neural context if "understanding" is to occur, extending "knowing" from the individual action to the social context. Indeed, we saw earlier that STSa neurons are active during movement observation though perhaps not during movement execution, and that STSa and F5 may be indirectly connected via area PF which, like F5, contains mirror neurons. As Rizzolatti *et al.* (2001) observe, STSa is also part of a circuit that includes the amygdala and the orbitofrontal cortex and so may be involved in the elaboration of affective aspects of social behavior (Adolphs, 1999; Allison *et al.*, 2000). My current position is that the "mirror systems for action" and "systems for understanding action" overlap but that a number of different subsystems had to develop before each gained the capability seen in humans. Further evolution was also required for such a system to mediate imitation. In other words, the claim is that both imitation and understanding build, at least in part, upon the mirror system, whatever the relationship of one to the other. In the next section, I will discuss the importance of imitation not only in and of itself, but also as a crucial step toward the skills needed to mediate language (evolving a "language-ready brain").

Indeed, the FARS model contains mechanisms which involve cortical regions outside F5 for creating and monitoring expectations (Fagg and Arbib, 1998; not shown in Fig. 1.4), such as "the feel of the object when grasped," even though for F5 itself it only models canonical neurons, not mirror neurons. But these expectations are "private" to the animal. The mirror system is based on "public" symptoms of the ongoing movement. The key to the mirror system is that it brings those symptoms for self-movement together with those for other-movement in generating a code for "action" (movement + goal) and not just for movement alone. Recall the MNS assumption that relations between hand and object are the key to mirror neuron function in the macaque (and recall Umiltà *et al.* (2001) for supporting data). However, there is a subsidiary problem here, namely recognizing which "symptoms" of self-movement correspond to which symptoms of other-movement, since the retinal display for, say, the hand movement of one's self or another is radically different. The MNS model addresses this problem by linking F5 canonical activity (the internal code for the action) with data on the relation of hand and object in training the F5 mirror neurons.

1.2.4 Linking a mirror system for grasping in humans to imitation and language

Before presenting the Mirror System Hypothesis in detail in Section 1.3.1, we first look, briefly, at evidence from human brain imaging that there is a mirror system for grasping in humans, and that it is related to mechanisms for imitation and language.

The notion that a mirror system might exist in humans was first tested by positron emission tomography (PET) experiments which showed that both the action and the observation of grasping significantly activated the superior temporal sulcus (STS), the inferior parietal lobule, and the inferior frontal gyrus (containing Broca's area). As already noted (and see Arbib and Bota, this volume), F5 in macaques is generally considered to be the homologue of Broca's area in humans. Thus, the cortical areas active during action observation in humans and macaques correspond very well, indicating that there is a fundamental primate mechanism for recognition of manual actions. There is now a rapidly growing literature which relates human brain mechanisms for action recognition, imitation, and language. A full review of such literature is beyond the scope of this volume, but the chapters by, e.g., Emmorey, Skipper *et al.*, and Kemmerer provide a sample.

Several studies provide behavioral evidence supporting the hypothesis that the system involved in observation and preparation of grasp movements has strong links with cortical areas involved in speech production. Gentilucci (2003) had subjects pronounce either the syllable "ba" or "ga" while observing motor acts of hand grasp directed to objects of two sizes, and found that both lip aperture and voice peak amplitude were greater when the observed hand grasp was directed to the large object. Conversely, Glover and Dixon (2002; see Glover *et al.*, 2004 for related results) presented subjects with objects on which were printed either the word "LARGE" or "SMALL." More recently, Gentilucci, Roy and colleagues, have further investigated the tight link between manual actions and speech production (Gentilucci *et al.*, 2004a, 2004b), showing that manual gestures relevant to communication can have natural vocal concomitants that may have helped the further development of intentional vocal communication. Their results suggest that the emergence of voice modulation and thus of an articulatory movement repertoire could have been associated with, or even prompted by, the pre-existing manual action repertoire.

A number of insights have been gleaned from the study of signed language. Corina *et al.* (2003) used PET to examine deaf users of American Sign Language (ASL) as they generated verb signs independently with their right dominant and left non-dominant hands (compared to the repetition of noun signs). Nearly identical patterns of left inferior frontal and right cerebellum activity were observed, and these were consistent with patterns that have been reported for spoken languages. Thus lexical–semantic processing in production relies upon left hemisphere regions regardless of the modality in which a language is realized, and, in signing, no matter which hand is used. Horwitz *et al.* (2003) studied the activation of Broca's area during the production of spoken and signed language. They showed that BA 45, not BA 44, was activated by both speech and signing during the production of language narratives in bilingual subjects (fluent from early childhood in both ASL and English) with the generation of complex movements and sounds as control. Conversely, BA 44, not BA 45, was activated by the generation of complex articulatory movements of oral–laryngeal or limb musculature. They thus conclude that BA 45 is the part of Broca's area that is fundamental to the modality-independent aspects of language generation.

Much more must be done to take us up the hierarchy from elementary actions to the recognition and generation of novel compounds of such actions. Nonetheless, the above preliminary account strengthens the hypothesis that no powerful syntactic mechanisms need have been encoded in the brain of the first *Homo sapiens*. Rather it was the extension of the imitation-enriched mirror system to support intended communication that enabled human societies, across tens of millennia of invention and cultural evolution, to achieve human languages in the modern sense.

Arbib *et al.* (2000) relate relevant human brain imaging data, e.g., from PET and functional magnetic resonance imaging (fMRI) to the underlying neural networks. Models tied to human brain imaging data often focus on a few "boxes" based on brain regions associated with significantly (though rather little, in percentage terms) enhanced blood flow, rather than analyzing the cooperative computation of multiple brain regions. As Arbib *et al.* (2000) show, one can link brain imaging to neurophysiological data by using synthetic PET imaging (Arbib *et al.*, 1994; see also Tagamets and Horwitz, 1998). This method uses computational models of biological neural circuitry based on animal data to predict and analyze the results of human PET studies. This technique makes use of the hypothesis that regional cerebral blood flow (rCBF) is correlated with the integrated synaptic activity in a localized brain region. Arbib *et al.* (2000) exemplify this general research program with two case studies, one on visuomotor processing for control of grasping (applying synthetic PET to the FARS model) and the other to imitation of motor skills.

1.3 The mirror system hypothesis: complex imitation

1.3.1 The hypothesis defined

What turns a movement into an action is that it is associated with a goal, so that initiation of the movement is accompanied by the creation of an expectation that the goal will be met. We distinguish *praxic action*, in which the hands are used to interact physically with objects or other creatures, from *communicative action* (both manual and vocal). Our assumption is that macaques use hand movements primarily for praxic actions. The mirror system allows other macaques to understand these actions and act on the basis of this understanding. Similarly, the macaque's orofacial gestures register emotional state, and primate vocalizations can also communicate something of the current situation of the macaque. However, building on the idea that the mirror system in macaques is the homologue of Broca's area in humans, Arbib and Rizzolatti (1997) and Rizzolatti and Arbib (1998) developed the following hypothesis.

The Mirror System Hypothesis (MSH): The mechanisms which support language in the human brain evolved atop a basic mechanism *not* originally related to communication. Instead, the *mirror system for grasping* with its capacity to generate *and* recognize a set of actions, provides the evolutionary basis for *language parity* – i.e., an utterance means roughly the same for both speaker and hearer. In particular, human Broca's area contains a mirror system for grasping which is homologous to the F5 mirror system of macaque.

Arbib (2002, 2005a) has amplified the original account of Rizzolatti and Arbib to hypothesize seven stages in the evolution of language, with imitation grounding two of the stages.[6] The first three stages are pre-hominid:

S1 Grasping;
S2 A mirror system for grasping shared with the common ancestor of human and monkey; and
S3 A simple imitation system for grasping shared with the common ancestor of human and chimpanzee.

---

[6] When I speak of a "stage" in phylogeny, I do not have in mind an all-or-none switch in the genotype that yields a discontinuous change in the phenotype, but rather the coalescence of a multitude of changes that can be characterized as forming a global pattern that may emerge over the course of tens or even hundreds of millennia.

The next three stages then distinguish the hominid line from that of the great apes:

S4 A complex imitation system for grasping;

S5 *Protosign*, a manual-based communication system, breaking through the fixed repertoire of primate vocalizations to yield an open repertoire;

S6 *Protospeech*, resulting from the ability of control mechanisms evolved for protosign coming to control the vocal apparatus with increasing flexibility.

The final stage takes us to language:

S7 *Language*: the change from action–object frames to verb–argument structures to syntax and semantics; the co-evolution of cognitive and linguistic complexity.

I claimed that the transition to language involved little if any biological evolution, but instead resulted from cultural evolution (historical change) in *Homo sapiens*.

MSH is simply the assertion that the mechanisms which get us to the role of Broca's area in language depend in a crucial way on the mechanisms established in stage S2. The above seven stages provide just one set of hypotheses on how this dependence may have arisen. However, MSH is silent on *why* the members of a species would want to communicate. Its study is thus usefully complemented by the discussion of social intelligence by Stanford (this volume).

1.3.2 Imitation

Let me first review some evidence for the claim implicit in stages S2–S4 that monkeys have a mirror system for grasping shared with the common ancestor of human and monkey and that chimpanzees have only a "simple" imitation system, whereas humans have a "complex" imitation system for grasping.[7] The idea is that *complex imitation* presupposes the capacity (a) for *complex action analysis*, the ability to analyze another's performance as a combination of actions (approximated by variants of) actions already in the repertoire, and (b) to add new, complex actions to one's repertoire on this basis. I view complex imitation as an evolutionary step of great advantage independent of its implications for communication. However, in modern humans it undergirds the child's ability to acquire language, while complex action analysis is essential for the adult's ability to comprehend the novel compounds of "articulatory gestures" that constitute language.

Visalberghi and Fragaszy (2002) review data on attempts to observe imitation in monkeys, including their own studies of capuchin monkeys. They conclude that there is a huge difference between the major role that imitation plays in learning by human children, and the very limited role, if any, that imitation plays in social learning in monkeys (and, to a lesser extent, apes). There is little evidence for vocal imitation in

---

[7] Meanwhile, reports abound of imitation in many species, including dolphins and orangutans, and even tool use in crows (Hunt and Gray, 2002). Thus, I accept that the demarcation between the capability for imitation of humans and non-humans is problematic. Nonetheless, I still think it is clear that humans can master feats of imitation beyond those possible for other primates.

monkeys or apes (Hauser, 1996), but it is generally accepted that chimpanzees are capable of some forms of imitation (Tomasello and Call, 1997).

Myowa-Yamakoshi and Matsuzawa (1999) observed in a laboratory setting that chimpanzees typically took 12 trials to learn to "imitate" a behavior, and in doing so paid more attention to where the manipulated object was being directed than to the actual movements of the demonstrator. The chimpanzees focused on using one or both hands to bring two objects into relationship, or to bring an object into relationship with the body, rather than the actual movements involved in doing this. Chimpanzees do use and make tools in the wild, with different tool traditions found in geographically separated groups of chimpanzees: Boesch and Boesch (1983) have observed chimpanzees in Taï National Park, Ivory Coast, using stone tools to crack nuts open, although Goodall has never seen chimpanzees do this in the Gombe in Tanzania. The Taï chimpanzees crack harder-shelled nuts with stone hammers and stone anvils. They live in a dense forest where suitable stones are hard to find. The stone anvils are stored in particular locations to which the chimpanzees continually return. To open soft-shelled nuts, chimpanzees use thick sticks as hand hammers, with wood anvils. The nut-cracking technique is not mastered until adulthood. Young chimpanzees first attempt to crack nuts at age 3 years, and require at least four years of practice before any benefits are obtained. Mothers may or may not correct and instruct their young. Tomasello (1999) comments that, over many years of observation, Boesch observed only two possible instances in which the mother *appeared* to be actively attempting to instruct her child, and that even in these cases it is unclear whether the mother had the goal of helping the young chimp learn to use the tool. We may contrast the long and laborious process of acquiring the nut-cracking technique with the rapidity with which human adults can acquire novel sequences, and the crucial role of caregivers in assisting complex imitation (Zukow-Goldring, this volume).

Many critics have dismissed MSH on the grounds that monkeys do not have language and so the mere possession of a mirror system for grasping cannot suffice for language. But the key phrase in MSH is "evolved atop" – the mirror system expanded its roles in concert with other brain regions as the human brain evolved. The MSH provides a neurological basis for the oft-repeated claim that hominids had a (proto)language based primarily on manual gestures before they had a (proto)language based primarily on vocal gestures (e.g., Hewes, 1973; Kimura, 1993; Armstrong *et al.*, 1995; Stokoe, 2001) – but see Section 1.4 for a somewhat nuanced expression of this hypothesis. In this regard, I stress that the "openness" or "generativity" which some see as the hallmark of language (i.e., its openness to new constructions, as distinct from having a fixed repertoire like that of monkey vocalizations) is present in manual behavior which can thus supply part of the evolutionary substrate for its appearance in language.

The issue is how evolution took us from *praxis*, "practical actions" in which the hands are used to interact physically with objects or the bodies of other creatures, to *gestures* (both manual and vocal) whose purpose is communication. MSH stresses the notion of parity and provide a neurological basis for the parity of meaning for language by rooting it in the parity of meaning for actions, such as when both actor and observer recognize that reaching for a raisin with a precision pinch is probably the prelude to eating it, while reaching for a broken branch with a power grasp may be the prelude to using it as a club. The mirror system is presumed to allow other monkeys to use this understanding of praxic actions as a basis for cooperation, teamwork, averting a threat, etc. One might say that this is *implicitly* communicative, or that communication here is a side effect of conducting an action for non-communicative goals. Similarly, the monkey's orofacial gestures register emotional state, and primate vocalizations can also communicate something of the current situation of the monkey – these "devices" evolved to signal certain aspects of the monkey's current internal state or situation.

Having a mirror system for grasping is *not* in itself sufficient for copying grasping actions. It is one thing to recognize an action using the mirror system; it is another thing to use that representation as a basis for repeating the action. Moreover, language evolution must be studied in concert with the analysis of cognitive evolution more generally. In using language, we make use of, for example, negation, counterfactuals, and verb tenses. But each of these linguistic structures is of no value unless we can understand that the facts contradict an utterance, or can recall past events and imagine future possibilities.

The next few sections will examine protosign, protospeech, and the emergence of language in more detail. The remainder of the book will provide many different contributions which enrich this perspective.

For example, Itti and Arbib (this volume) show how visual perception of a "minimal subscene" linking an agent and an action to one or more objects may ground processes of scene description and question-answering, linking salience and top–down attention to language structure.

A key question is what changes as we go from "simple" imitation to "complex" imitation. The answer has links to neural mechanisms underlying sequential behavior as well as to data on primate behavior (as reviewed by Stanford, this volume), suggesting ways in which hierarchical behaviors may relate to the hierarchical structure of language. Such a treatment calls into question the assertion of Hauser *et al.* (2002) that the faculty of language in the narrow sense (FLN) includes only recursion and is the one uniquely human component of the faculty of language. Recursion is in no way restricted to language. Recall the "waiter example" in Section 1.2 which linked the nested structure of a sentence to the nested behaviors revealed in goal-directed behavior more generally. The study of animal behavior is replete with examples of how an animal can analyze a complex sensory scene and, in relation to its internal state, determine a course of action. When a frog faced with prey and predators and a barrier ends up, say, taking a path around the barrier to escape the predator (see Cobas and Arbib (1992) for a model) it exemplifies the ability to analyze the spatial relations between objects in planning its action. However, there is little evidence of recursion here – once the frog rounds the barrier, it seems to need to see the prey anew to trigger its prey-catching behavior. By contrast, the flow diagram given by Byrne (2003) shows that the processing (from getting a nettle plant to putting a folded handful of leaves into the mouth) used by a mountain gorilla when preparing bundles of nettle leaves to eat is indeed recursive. Gorillas (like many other species, and not only mammals) have the working memory to refer their next action not only to sensory data but also to the state of execution of some current plan. Thus when we refer to the monkey's grasping and ability to recognize similar grasps in others (stages S1 and S2) it is a mistake to treat the individual grasps in isolation – the F5 system is part of a larger system that can direct those grasps as part of a recursively structured plan.

1.4 Protolanguage as protosign + protospeech

1.4.1 The doctrine of the expanding spiral

For most humans, language is heavily intertwined with speech. MSH offers a compelling explanation of why F5, rather than the medial cortex already involved in monkey vocalization, is homologous to the Broca's area substrate for language.[8] I argue that the human brain evolved to support protosign as well as protospeech, that it is a historical fact that spoken languages have predominated over sign languages, but that the brain mechanisms that support human language are not specialized for speech but rather support communication in an integrated manual–facial–vocal multi-modal fashion. Here, I focus on three stages of the Arbib (2002, 2005a) version of MSH:

S4 A complex imitation system for grasping. This involves both a perceptual component – involving the ability to recognize that a novel action was in fact composed of (approximations to) known actions – and the ability to master new actions on the basis of such an analysis.

S5 *Protosign*, a manual-based communication system, breaking through the fixed repertoire of primate vocalizations to yield an open repertoire.

S6 *Protospeech*, resulting from the ability of control mechanisms evolved for protosign coming to control the vocal apparatus with increasing flexibility.[9]

My hypothesis is that complex imitation for hand movements evolved because of its adaptive value in supporting the increased transfer of manual skills and thus preceded the emergence of protolanguage in whatever modality. Donald (1994, 1999), however, argues that *mimesis* (which, I think, is similar to what I term complex imitation) was a general-purpose adaptation, so that a capacity for *vocal–auditory* mimesis would have emerged simultaneously with a capacity for *manual* mimesis. Indeed, it is clear that stage S6 does require vocal–auditory mimesis. However, my core argument does not rest on the debate over whether or not manual mimesis preceded vocal–auditory mimesis. The doctrine of the expanding spiral is that (a) protosign exploits the ability for complex imitation of hand

---

[8] Although we stress the parallels between F5 and Broca's area in the present chapter, Arbib and Bota (this volume) will develop the view of Broca's area as just one part of the "language network" of the human brain, and relate parietal and temporal areas in mediating the monkey mirror system (recall Fig. 1.5) with the role of temporal and parietal areas in language.

[9] The present section is based in part on Arbib (2005b), which includes a critique of MacNeilage's (1998) argument that the speech system evolved without the support of protosign. MacNeilage and Davis (2005) offer a spirited argument against the critique. The crux of the argument is that I say that they provide too little insight into how sounds came to convey complex meanings, and they note that they offer a rich developmentally based account that offers insights into the evolution of the vocal system that I cannot match. In summary, I believe that some elements of their account might enrich the analysis of stage S6, but they would see stage S5 as being completely superfluous.

movements to support an open system of communication; (b) the resulting protosign provides scaffolding for protospeech; but that (c) protosign and protospeech develop together thereafter – stages S5 and S6 are intertwined. The strong hypothesis here is that protosign is essential to this process, so that the full development of protospeech was made possible by the protosign scaffolding.

To develop the doctrine, we must distinguish two roles for imitation in the transition from stage S4 to stage S5:

1. The transition from *praxic action* directed towards a goal object to *pantomime* in which similar actions are produced away from the goal object.
2. The emergence of conventionalized gestures to ritualize or disambiguate pantomimes.

As Stokoe (2001) and others emphasize, the power of pantomime is that it provides open-ended communication that works without prior instruction or convention. However, I would emphasize that pantomime per se is not a form of protolanguage; rather it provides a rich scaffolding for the emergence of protosign. In the present theory, one crucial ingredient for the emergence of symbolization is the extension of imitation from the imitation of hand movements to the ability to project the degrees of freedom of quite different movements into hand movements which evoke something of the original in the mind of the observer. This involves not merely changes internal to the mirror system but its integration with a wide range of brain regions involved in perception and action.

When pantomime is of praxic hand actions, that pantomime directly taps into the mirror system for these actions. However, as the pantomime begins to use hand movements to mime different degrees of freedom (as in miming the flying of a bird), or to evoke an object either by miming a related action or by movements which suggest tracing out the characteristic shape of the object, a dissociation begins to emerge. The mirror system for the pantomime (based on movements of face, hand, etc.) is now different from the recognition system for the action that is pantomimed, and – as in the case of flying – the action may not even be in the human action repertoire. However, the system is still able to exploit the praxic recognition system because an animal or hominid must observe much about the environment that is relevant to its actions but is not in its own action repertoire. Nonetheless, this dissociation now underwrites the emergence of actions which are defined only by their communicative impact, not by their praxic goals. Here it is important to distinguish the mirror system for the sign (phonological form) from the linkage of the sign to the neural schema for the signified (Section 1.4.3).

Imitation is the generic attempt to reproduce movements performed by another, whether to master a skill or simply as part of a social interaction. By contrast, pantomime is performed with the intention of getting the observer to think of a specific action, object, or event. It is essentially communicative in its nature. The imitator observes; the panto-mimic intends to be observed.

The transition to pantomime does seem to involve a genuine neurological change. Mirror neurons for grasping in the monkey will fire only if the monkey sees a hand movement directed towards an observed (or just observed) object appropriate as target for the grasp (Umiltà *et al.*, 2001). By contrast, in pantomime, the observer sees the movement in isolation and *infers* (a) what non-hand movement is being mimicked by the hand movement, and (b) the goal or object of the action. This is an evolutionary change of key relevance to language readiness. The very structure of these sequences can serve as the basis for immediate imitation or for the immediate construction of an appropriate response, as well as contributing to the longer-term enrichment of experience.

A further critical change en route to language emerges from the fact that in pantomime it might be hard, for example, to distinguish a movement signifying "bird" from one meaning "flying." This inability to adequately convey shades of meaning using "natural" pantomime would favor the invention of gestures which could in some way combine with (e.g., sequentially, or by modulation of some kind) the original pantomime to disambiguate which of its associated meanings was intended. Note that whereas a pantomime can freely use any movement that might evoke the intended observation in the mind of the observer, a disambiguating gesture must be conventionalized.[10]

Because a conventionalized gesture is not iconic, it can only be used within a community that has negotiated or learned how it is to be interpreted. This use of non-pantomimic gestures requires extending the use of the mirror system to attend to a whole new class of hand movements, those with conventional meanings agreed upon by the protosign community to reduce ambiguity and extend semantic range. With this, we have moved from pantomime to protosign. *Pantomime is not itself part of protosign but rather a scaffolding for creating it.* Pantomime involves the production of a motoric representation through the transformation of a recalled exemplar of some activity. As such, it can vary from actor to actor, and from occasion to occasion. By contrast, the meaning of conventional gestures must be agreed upon by a community.

I have separated stage S6, the evolution of protospeech, from stage S5, the evolution of protosign, since the role of F5 in grounding the evolution of a protolanguage system would work just as well if we and all our ancestors had been deaf. However, primates do have a rich auditory system which contributes to species survival in many ways of which communication is just one (Ghazanfar, 2003; Ghazanfar and Santos, 2004). The hypothesis here, then, is not that the protolanguage system had to create the appropriate auditory and vocal–motor system "from scratch" but rather that it could build upon the existing mechanisms to derive protospeech. My hypothesis is that protosign grounded the crucial innovation of using arbitrary symbolic gestures to convey novel meanings, and that this in turn provided the scaffolding for protospeech. I suggest that the interplay between protospeech and protolanguage was an expanding spiral which yielded a brain that was ready for language in the multiple modalities of gesture, vocalization,[11] and facial expression and a vocal–motor system able to accommodate an ever-expanding proto-speech vocabulary. As we shall see below, monkeys already have orofacial communicative gestures and these may certainly support a limited communicative role. The

---

[10] Note that dogs, for example, can learn to recognize conventionalized signs – whether spoken, or hand signals – uttered by their owners. However, they lack the ability to incorporate these utterances into their own action repertoire.

[11] Goldstein *et al.* (this volume) offer an extensive discussions of the actions that constitute speech.

essential point here, though, is that they lack the ability of pantomime to convey a rich, varied, and *open* repertoire of meanings without prior conventionalization.

It is not claimed that stage S5 (protosign) was "completed" before stage S6 (proto-speech) was initiated or that protosign attained the status of a full language prior to the emergence of early forms of protospeech. Rather, once hominids had come to employ pantomime and discovered how to use conventional gestures to increasingly augment, ritualize and in some part replace the use of pantomime, then stage S6 followed naturally as vocal gestures entered the mix.

1.4.2 Mirror neurons are not just for grasping

Non-human primates do have a set of species-specific vocalizations, but such a monkey call system is *closed* in the sense that it is (more or less) restricted to a specific repertoire. This is to be contrasted with human languages which are *open* both as to the creation of new vocabulary (a facility I would see as already present in protosign and protospeech) and the ability to combine words and grammatical markers in diverse ways to yield an essentially unbounded stock of sentences (a facility I would see as peculiar to language as distinct from protolanguage).

It has been found that macaque F5 mirror neurons are not limited to visual recognition of hand movements. Kohler *et al.* (2002) studied mirror neurons for actions which are accompanied by characteristic sounds (e.g., breaking a peanut in half), and found that 15% of mirror neurons in the hand area of F5 can respond to the distinctive sound of an action as well as viewing the action. However, monkeys do not use their vocal apparatus to mimic the sounds they have heard, thus weakening any case that these neurons might serve vocal communication but demonstrating that mirror neurons do receive auditory input that could be relevant to the protosign–protospeech transition.

Complementing earlier studies on hand neurons in macaque F5, Ferrari *et al.* (2003; see also Fogassi and Ferrari, 2004) found that about one-third of mouth motor neurons in F5 also discharge when the monkey observes another individual performing mouth actions. The majority of these "mouth mirror neurons" become active during the execution and observation of mouth actions related to ingestive functions such as grasping with the mouth, sucking, or breaking food. Another population of mouth mirror neurons also discharges during the execution of ingestive actions, but the most effective visual stimuli in triggering them are communicative mouth gestures (e.g., lip-smacking). This fits with the hypothesis that neurons learn to associate patterns of neural firing rather than being committed to learn specifically pigeonholed categories of data. Thus a potential mirror neuron is in no way committed to become a mirror neuron in the strict sense, even though it may be more likely to do so than otherwise. The observed communicative actions (with the effective executed action for different "mirror neurons" in parentheses) include lip-smacking (sucking and lip-smacking); lips protrusion (grasping with lips, lips protrusion, lip-smacking, grasping with mouth and chewing); tongue protrusion (reaching with tongue); teeth-chatter (grasping with mouth); and lips/tongue protrusion (grasping with lips and reaching with tongue; grasping). Ferrari *et al.* (2003, p.1713). state that "the knowledge common to the communicator and the recipient of communication *about food and ingestive action* became the common ground for social communication. Ingestive actions are the basis on which communication is built" (my italics). Thus their strong claim "Ingestive actions are the basis on which communication is built" might better be reduced to "Ingestive actions are the basis on which communication about feeding is built" which complements but does not replace communication about manual skills.

Does this suggest that protospeech mediated by the F5 homologue in the hominid brain could have evolved without the scaffolding provided by protosign? My answer is negative. I have argued that imitation is crucial to grounding pantomime in which a movement is performed in the absence of the object for which such a movement would constitute part of a praxic action. However, the sounds studied by Kohler *et al.* (2002) cannot be created in the absence of the object and there is no evidence that monkeys can use their vocal apparatus to mimic the sounds they have heard. Moreover, manual skills support pantomime in a way which the facial actions described by Ferrari *et al.* (2003) cannot. In any case, the observed communicative actions for their neurons – lip-smacking, lip protrusion, tongue protrusion, teeth-chatter, and lips/tongue protrusion – are a long way from the sort of vocalizations that occur in speech.

Evidence of a linkage between manual activity and vocalization has been reported in macaques by Hihara *et al.* (2003). They trained two Japanese monkeys to use a rake-shaped tool to retrieve distant food. After training, the monkeys spontaneously began vocalizing coo-calls in the tool-using context. Hihara *et al.* then trained one of the monkeys to vocalize to request food or the tool:

Condition 1 When the monkey produced a coo-call (call A), the experimenter put a food reward on the table, but out of his reach. When the monkey again vocalized a coo-call (call B), the experimenter presented the tool within his reach. The monkey was then able to retrieve the food using the tool.

Condition 2 Here the tool was initially presented within the monkey's reach on the table. When the monkey vocalized a coo-call (call C), the experimenter set a food reward within reach of the tool.

The monkey spontaneously differentiated its coo-calls to ask for either food or tool during the course of this training, i.e., calls A and C were similar to each other but different from call B. Hihara *et al.* (2003) speculate that this process might involve a change from emotional vocalizations into intentionally controlled ones by associating them with consciously planned tool use. However, I view this as another example of the unconscious linkage between limb movement and vocal articulation demonstrated in humans by Gentilucci *et al.* (2004a, 2004b, reviewed above in Section 1.2.4).

My preferred hypothesis, then, is that:

- manual gesture is primary in the early stages of the evolution of language-readiness, but
- the brain creates some unconscious synergies between manual and orofacial movements and vocalizations,

- audiomotor neurons lay the basis for later extension of protosign to protospeech, and
- the protospeech neurons in the F5 precursor of Broca's area may be rooted in ingestive behaviors.

The usual evolutionary caveat: macaques are not ancestral to humans. What is being said here is shorthand for the following: (a) there are ingestion-related mirror neurons observed in macaque; (b) I hypothesize that such neurons also existed in the common ancestor of human and macaque of 20 million years ago; (c) noting with Fogassi and Ferrari (2004) that there is little evidence of voluntary control of vocal communication in non-human primates, I further hypothesize that evolution along the hominid line (after the split 5 million years ago between the ancestors of the great apes and those of humans) expanded upon this circuitry to create the circuitry for protospeech.

The notion, then, is that the manual domain supports the expression of meaning by sequences and interweavings of gestures, with a progression from "natural" to increasingly conventionalized gesture to speed and extend the range of communication within a community. I then argue that stage S5 (protosign) provides the scaffolding for stage S6 (protospeech). We have already seen that some mirror neurons in the monkey are responsive to auditory input. We now note that there are orofacial neurons in F5 that control movements that could well affect sounds emitted by the monkey. The speculation here is that the evolution of a system for voluntary control of intended communication based on F5/Broca's area could then lay the basis for the evolution of creatures with more and more prominent connections from F5/Broca's area to the vocal apparatus. This in turn could provide conditions that lead to a period of co-evolution of the vocal apparatus and the neural circuitry to control it. My essential claim is that complex imitation of hand movements was crucial to the development of an open system of communication. Related data concerning the trade-off in roles of the supplementary motor area and the anterior cingulate gyrus between monkey vocalizations and human speech are taken up by Arbib and Bota (this volume).

In conclusion, I have argued that it was the discovery in protosign that arbitrary gestures could be combined to convey novel meanings that provided the essential scaffolding for the transition from the closed set of primate vocalizations to the limited openness of protospeech. Association of vocalization with manual gestures allowed them to assume a more open referential character, and to exploit the capacity for imitation of the underlying brachiomanual system. I do not claim that meaning cannot evolve within the orofacial domain, but only that the range of "obvious" meanings is impoverished compared with those expressible by pantomime. I claim that it became easy to share a wide range of meanings once the mirror system for grasping evolved in such a way as to support pantomime, and that the need for disambiguation then created within a community a shared awareness of the use of conventional gestures as well as iconic gestures – whereas onomatopoeia seems to be far more limited in what can be conveyed. I thus hypothesize that *Homo habilis* and even more so *Homo erectus* had a "proto-Broca's area" based on an F5-like precursor mediating communication by manual and orofacial gesture. This made possible a process whereby this "proto-Broca's area" gained primitive control of the vocal machinery, thus yielding increased skill and openness in vocalization, moving from the fixed repertoire of primate vocalizations to the unlimited (open) range of vocalizations exploited in speech. Speech apparatus and brain regions could then co-evolve to yield the configuration seen in modern *Homo sapiens*.

This approach is motivated in great part by the multimodal features of facial and manual gestures which accompany real human spoken language communication, as opposed to the idealized view which equates language with speech alone. McNeill (1992) has used videotape analysis to show the crucial use that people make of gestures synchronized with speech. Pizzuto *et al.* (2004) stress the interaction of vocalization and gesture in early language development (see also Iverson and Goldin-Meadow, 1998). Modern sign languages are fully expressive human languages (Emmorey, this volume), and so must not be confused with protosign – we use signing here to denote manually based linguistic communication, as distinct from the reduced use of manual gesture that is the frequent accompaniment of speech.

A "speech only" evolutionary hypothesis leaves mysterious the availability of this vocal–manual–facial complex which not only supports a limited gestural accompaniment to speech but also the ease of acquisition of signed languages for those enveloped within it in infancy. However, the "protosign scaffolding" hypothesis has the problem of explaining why speech became favored over gestural communication. Standard arguments (see, e.g., Corballis, 2002) include the suggestions that, unlike speech, signing is not omnidirectional, does not work in the dark, and does not leave the hands free. But does the advantage of communication in the dark really outweigh the choking hazard of speech? Moreover Falk (2004; but see also MacNeilage and Davis, 2005) argues that vocal "motherese" in chimpanzees serves to bond parent and child and other kin and thus could provide a direct precursor to speech.

This still leaves open the question "If protosign was so successful why did spoken languages come to predominate over signed languages?" As in much of evolutionary discussion, the answer must be *post hoc*. One can certainly imagine a mutation which led to a race of deaf humans who nonetheless prospered mightily as they built cultures and societies on the rich adaptive basis of signed languages. So the argument is not that speech *must* triumph, any more than that having a mirror system *must* lead to (proto)language. Let me just note that the doctrine of the expanding spiral is far more hospitable to the eventual answer to this question than is the view that protosign yielded full signed languages preceded the emergence of speech. For, indeed, if hominid protolanguage combined protosign and protospeech there is no issue of how a fully successful system of signed language could become displaced by speech. Signers are perfectly good at signing while showing someone how to use a tool, and sign might actually be better than speech when the tool is not present. Karen Emmorey (this volume) thus argues against the expanding spiral on the grounds that "If communicative pantomime and protosign preceded protospeech, it is not clear why protosign simply did not evolve into sign language" and pre-empt the evolution of spoken language.

Having said this, I think the demand for an all-or-none quasi-biological answer ("this is why speech must predominate over signing in human evolution") may be inappropriate. The answer may actually lie in the vagaries of human history. It is a historical fact that humans have evolved writing systems that may be ideographic (Chinese), or based on a syllabary (Japanese kana), or alphabetic (as in English). It is also a historical fact that alphabetic systems are in the ascendancy and that the Chinese move towards an alphabetic system (pinyin) in the 1970s was probably reversed only because word processing removed the problems posed by trying to make efficient use of typewriters for a stock of thousands of characters. The advantage of writing (of course!) is that it leaves a written record that can be used in teasing apart the historical ebb and flow in the development and propagation of different writing systems (see, e.g., Coulmas, 2003). No such record is available for the ebb and flow of speech and signing. We do know that some aboriginal Australian tribes (Kendon, 1988), and some native populations in North America (Farnell, 1995) use both signing and speech; and note Kendon's republication of the classic text on Neapolitan signing (de Jorio, 2000). In other words, I am not convinced that I need to make the case for driving the spiral *toward* speech – it is enough to have shown how protosign created a link to the use of arbitrary gestures to convey meaning as a scaffolding to creating such a link to protospeech, and that thereafter advances in either system could create a space for advances in the other.

1.4.3 The Saussurean sign

Protosign may lose the ability of the original pantomime to elicit a response from someone who has not seen it before. However, the price is worth paying in that the simplified form, once agreed upon by the community, allows more rapid communication with less neural effort. In the same way, I suggest that pantomime is a valuable crutch for acquiring a modern sign language, but that even signs which resemble pantomimes are conventionalized and are thus distinct from pantomimes. Interestingly, signers using ASL show a dissociation between the neural systems involved in sign language and those involved in conventionalized gesture and pantomime. For example, Corina *et al.* (1992) described patient WL with damage to left hemisphere perisylvian regions. WL exhibited poor sign language comprehension, and his sign production had phonological and semantic errors as well as reduced grammatical structure. Nonetheless, WL was able to produce stretches of pantomime and tended to substitute pantomimes for signs, even when the pantomime required more complex movement.

Emmorey (2002) sees such data as providing neurological evidence that signed languages consist of linguistic gestures and not simply elaborate pantomimes (see also Emmorey, this volume). Consistent with this, I would argue that the evolution of neural systems involving Broca's area is adequate bilaterally to support conventionalized gestures when these are to be used in isolation. However, the predisposition for the skilled weaving of such gestures into complex wholes which represent complex meanings which are *novel* (in the sense of being created on-line rather than being well rehearsed) has

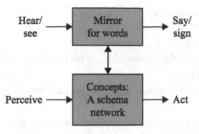

Figure 1.6 The bidirectional sign relation links (neural codes for) words and concepts (Arbib, 2004, inspired by Hurford, 2004).

become lateralized in humans. In other words, conventionalized gestures that can stand alone can still be used when the adult left hemisphere is damaged, but those which form part of the expression of a larger meaning depend essentially on left hemisphere mechanisms. Yet if the left hemisphere is removed early enough, then the child can acquire far greater language capacity with the right hemisphere than is possible for an adult who has a left hemispherectomy. This suggests that what evolved is not a human brain whose left hemisphere is specialized for language but rather one for which both hemispheres have the capacity to provide the mechanisms for language in a child raised in a language-rich environment. Nonetheless, despite this shared potential of both hemispheres, the asymmetry between the two hemispheres would seem to favor the left hemisphere over the right for syntactic processing under normal developmental circumstances, with the right hemisphere sharing the load by taking on the role of prosodic and emotional modulation of utterances.

Elsewhere, I have stated (Arbib, 2005a) that the transition from pantomime to protosign does not seem to require a biological change beyond that limned in the previous paragraph. However, the transition from slow and sporadic symbol use to fast and frequent habitual use might have been both the drive for and fruit of the development of the ability to segregate (whether structurally or just functionally) the neural code for the symbols from the neural code for the actions or objects they represent (and recall the discussion of Fig. 1.4b). Then, in pantomime mode, one's brain seeks to relate the observed performance directly to one's stock of representations of actions and objects, whereas in protosign mode, the representation of the protosign's symbol stock provide a necessary bridge between observation and interpretation.

Thus, irrespective of whether we are talking about the words of a language or the "protowords" of a protolanguage, we must (in the spirit of Saussure) distinguish the "sign" from the "signified." In Fig. 1.6, inspired by the comments of Hurford (2004), we distinguish the "neural representation of the signifier" (top row) from the "neural representation of the signified" (bottom row). The top row of the figure makes explicit the result of the progression (Arbib, 2002, 2005a) of mirror systems for:[12]

---

[12] Barrett *et al.* (2005) usefully review evidence for differential localization of neuronal systems controlling limb praxis, speech and language, and emotional communication, and assert that such data justify rejection of an evolutionary relationship

1. grasping and manual praxic actions,
2. imitation of grasping and manual praxic actions,
3. pantomime of grasping and manual praxic actions,
4. pantomime of actions outside the panto-mimic's own behavioral repertoire (e.g., flapping the arms to mime a flying bird),
5. conventional gestures used to formalize and disambiguate pantomime (e.g., to distinguish "bird" from "flying"),
6. conventionalized manual, facial, and vocal communicative gestures ("protowords") separate from pantomime.

Arbib (1981) distinguished *perceptual schemas* which determine whether a given "domain of interaction" is present in the environment and provide parameters concerning the current relationship of the organism with that domain, and *motor schemas* which provide the control systems which can be coordinated to effect a wide variety of actions (cf. Itti and Arbib, this volume). Recognizing an object (an apple, say) may be linked to many different courses of action (to place the apple in one's shopping basket; to eat the apple; to discard a rotten apple, etc.). In this list, some items are apple-specific whereas others invoke generic schemas for reaching and grasping. Only for some actions (run, walk, hit, ...) or expressions of emotion will perceptual and motor schemas be integrated into a "mirror schema." Viewing Fig. 1.6 as an anatomical diagram, patterns of activity encoding "words" in the top row mirror system become associatively linked with various activity patterns in the various parts of the brain serving action and perception. I do not see a "concept" as corresponding to one word, but rather to a graded set of activations of the schema network.

1.5 The inventions of languages

The divergence of the Romance languages from Latin took about 1000 years. The divergence of the Indo-European languages to form the immense diversity of Hindi, German, Italian, English, etc., took about 6000 years (Dixon, 1997). How can we imagine what has changed since the emergence of *Homo sapiens* some 200 000 years ago? Or in 5 000 000 years of prior hominid evolution? I claim that the first *Homo sapiens* were language-ready but did not have language in the modern sense.

I have already mentioned the hypothesis (e.g., Arbib, 2002, 2005a; Wray, 1998, 2000) that the "protolanguage" of *Homo erectus* and early *Homo sapiens* had neither words in the sense of modern human languages nor the syntax to combine them freely. Rather, the protolanguage of *Homo erectus* and early *Homo sapiens* were composed mainly of "holophrases" or "unitary utterances" which symbolized frequently occurring situations

---

between the underlying mechanisms. However, studies of mammalian brain evolution suggest that increasing complexity of behavior is paralleled by increases in the number of functional subdivisions of neocortex and that reduplication of circuitry may form the basis for differential evolution of copies of a given system, with differing connectivities, etc., to serve a variety of functions.

(in a general sense) without being decomposable into distinct words denoting components of the situation or their relationships, and that words as we know them then co-evolved culturally with syntax through fractionation. On this view, many ways of expressing relationships that we now take for granted as part of language were the discovery of *Homo sapiens*, and were "post-biological" in origin.

My hypothesis is that stage S7, the transition from protolanguage to language, is the culmination of manifold discoveries in the history of mankind: I suggest that language emerged from protolanguage through the repeated *discovery* that one could gain expressive power by *fractionating* such utterances into shorter utterances conveying components of the scene or command. As fractionation yielded a new symbol as a sequence of gestures that could convey the same meaning when embedded in a variety of hitherto unitary utterances, it required further invention to regularize the placement of the gestures in both utterances – and in the process words crystallized at the same time as the protosyntax which combined them. Clearly, such fractionation could apply to protosign as well as to protospeech.

A "just-so story" will flesh out the latter assertion. Unitary utterances such as "grooflook" or "koomzash" might have encoded quite complex descriptions such as "The alpha male has killed a meat animal and now the tribe has a chance to feast together. Yum, yum!" or commands such as "Take your spear and go around the other side of that animal and we will have a better chance together of being able to kill it." On this view, "protolanguage" grew by adding arbitrary novel unitary utterances to convey *complex but frequently important* situations, and it was a major later discovery en route to language as we now understand it that one could gain expressive power by *fractionating* such utterances into shorter utterances conveying components of the scene or command. Put differently, the utterances of prelanguage were more akin to the "calls" of modern primates – such as the "leopard call" of the vervet monkey which is emitted by a monkey who has seen a leopard and which triggers the appropriate escape behavior in other monkeys – than to sentences as defined in a language like English, but they differed *crucially* from the primate calls in that new utterances could be invented and acquired through learning within a community, rather than emerging only through biological evolution. Thus the set of such unitary utterances was open, whereas the set of calls was closed.

The following, hypothetical but instructive, example is similar to examples offered at greater length by Wray (1998, 2000) to suggest how the fractionation of unitary utterances might occur (and see Kirby (2000) for a computer simulation based on this idea). Imagine that a tribe has two unitary utterances concerning fire which, by chance, contain similar substrings which become regularized so that for the first time there is a sign for "fire." Now the two original utterances are modified by replacing the similar substrings by the new regularized substring. Eventually, some tribe members regularize the complementary gestures in the first string to get a sign for "burns"; later, others regularize the complementary gestures in the second string to get a sign for "cooks meat." However, because of the arbitrary origin of the sign for "fire," the placement of the gestures that have come to denote "burns" relative to "fire" differs greatly from those for "cooks meat" relative to "fire." It thus requires a further invention to regularize the placement of the gestures in both utterances – and in the process words are crystallized at the same time as the protosyntax which combines them. Clearly, such fractionation could apply to protosign as well as to protospeech.

Other mechanisms could also produce composite structures. For example, a tribe might, over the generations, develop different signs for "sour apple," "ripe apple," "sour plum," "ripe plum," etc., but not have signs for "sour" and "ripe" even though the distinction is behaviorally important. Thus $2n$ signs are needed to name n kinds of fruit. Occasionally, someone will eat a piece of sour fruit by mistake and make a characteristic face and intake of breath when doing so. Eventually, some genius pioneers the innovation of getting a conventionalized variant of this gesture accepted as the sign for "sour" by the community, thus extending the protolanguage.[13] A step towards language is taken when another genius gets people to use the sign for "sour" + the sign for "ripe X" to replace the sign for "sour X" for each kind X of fruit. This innovation allows new users of the proto-language to simplify learning fruit names, since now only $n + 1$ names are required for the basic vocabulary, rather than $2n$ as before. More to the point, if a new fruit is discovered, only one name need be invented rather than two. I stress the invention of "sour" is a great discovery in and of itself. It might take hundreds of such discoveries distributed across centuries or more before someone could recognize the commonality across all these constructions and thus invent the precursor of what we would now call adjectives.[14] In this spirit, I posit that over the millennia, features added as "patches" to a protolanguage became more or less regularized, with general "rules" emerging both consciously and unconsciously only as generalizations could be imposed on, or discerned in, a population of *ad hoc* mechanisms.

This story also exemplifies possibilities for the emergence of protospeech on a proto-sign scaffolding. The posited symbol for "sour" is a vocal–facial gesture, not a manual gesture, and thus could contribute to a process whereby protolanguage began to use protospeech symbols to enrich protosign utterances, whereafter an increasing number of symbols would have become vocalized, freeing the hands to engage in both praxis and communication as desired by the "speaker." Similarly, the ability to create novel sounds to match degrees of freedom of manual gestures (for example, rising pitch might represent an upward movement of the hand), coupled with a co-evolved ability to imitate novel

---

[13] I use the word "genius" advisedly. I believe that much work on language evolution has been crippled by the inability to imagine that things we take for granted were in no way a priori obvious, or to see that current generalities were by no means easy to discern in the particularities that they embrace. Consider, for example, that Archimedes (*c.*287–212 BCE) had the essential idea of the integral calculus, but that it took almost 2000 years before Newton (1642–1727) and Leibniz (1646–1716) found notations that could express the generality implicit in his specific examples, and thus unleash an explosion of mathematical innovation. I contend that language, like mathematics, has evolved culturally by such fits and starts.

[14] Indeed, adjectives are not the "natural category" they may appear to be. As Dixon (1997, pp.142 *et seq.*) observes, there are two kinds of adjective classes across human languages: (a) an open class with hundreds of members (as in English), (b) a small closed class. Languages with small adjective classes are found in every continent except Europe. Igbo, from west Africa, has just eight adjectives: *large* and *small*; *black/dark* and *white/light*; *new* and *old*; and *good* and *bad*. Concepts that refer to physical properties tend to be placed in the verb class (e.g., "the stone heavies") and words referring to human propensities tend to be nouns (e.g., "she has cleverness").

sound patterns with onomatopoeia yielding vocal gestures not linked to manual gestures, could have helped create the early vocal repertoire. However, articulatory gestures alone do not have the rich *ad hoc* communicative potential that pantomime provides with the manual gesture system.

The point about both these mechanisms is that there is no single step from protolanguage to language. In particular, the latter example is meant to indicate how a sign for "sour" could be added to the protolanguage vocabulary with no appeal to an underlying "adjective mechanism." Instead, one would posit that the features of language emerged by bricolage (tinkering) which added many features as "patches" to a protolanguage, with general "rules" emerging both consciously and unconsciously only as generalizations could be imposed on, or discerned in, a population of *ad hoc* mechanisms. There was thus no point at which one could say of a tribe "Until now they used protolanguage but because they have added this new mechanism they will henceforth use language."

1.5.1 From phonology to grammar

Research in motor control tends to be at the level of *phonology* – how do effectors produce a basic action, what "co-articulation" may modify one action on the basis of another? – than at the level of *syntax and semantics* which analyzes the structure of sentences, and related issues. MSH claims that mechanisms supporting language evolved atop the mirror system for grasping, not that circuits for praxis and communication are identical. At least two different changes in the brain were required to support the emergence of language: (a) the availability of complex action analysis and complex imitation at "both ends of the arrow" of Fig. 1.6, abilities in which the evolution of the mirror system plays a role; and (b) the complementary ability for situational analysis which links sentence production and perception to assemblages of conceptual schemas.

The hearer's processes for understanding (more or less) what the speaker intends, and the speaker's processes for conveying the intended message with (more or less) reduced ambiguity must, to succeed, be *approximately* inverse to each other. I distinguished *production grammar* – getting from a communicative goal to words that express it – from *perception grammar* – getting from a sequence of words to the goal behind it. Syntax in the normal sense is a compact answer to the question: "In this community, what regularities seem to be shared by the sentences that are produced and understood?" In this way, the linguist may define a *single* grammar to represent regularities common to both perception and production of utterances – but I deny that there is a single grammar represented in the brain that is consulted by separate processes of perception and production.

The neuroscience of action must go far beyond the neurophysiology of controlling muscles. Successful planning or interpretation in general requires the interweaving of many motor activities to achieve an overall goal through a hierarchy of subgoals. "Action" at this relevant level integrates perceptual, conceptual and motor activities. Miller *et al.* (1960) related motor sequences to plans defined as hierarchical structures of elementary "TOTE (Test–Operate–Test–Exit) units" to make explicit the complex contingencies within a plan. Arbib (1981) offered the assemblage of available schemas to form new schemas as *coordinated control programs* which could control the interweaving and overlapping of perceptual and motor acts.

Linguists view the sentence as a basic locus of well-formedness and focus syntax on the characterization of sentence structure and the constituents that yield it. However, sentences only approximate the way words are gathered in real speech (Iacoboni, 2004) as different thoughts vie for expression, or as partial expressions do the job of a whole sentence. One may employ run-on phrases which cumulatively convey some message, yet do not cohere into a well-formed sentence. Thus in some sense a sentence is a theoretical construct sanctioned by a specific choice of grammar for a language. Written language reflects the lessons of grammar more than does spoken language. I say this not to disparage the utility of the notions of sentence and grammar but rather to note that they describe approximations to oral language use rather than its full complexity.

It is thus somewhat arbitrary how one inserts the punctuation to turn a speech stream into a sequence of sentences. It is similarly arbitrary to break the "action stream" into a sequence of actions. In our bottle cap example, one must recognize that the correct analysis would not include, say, "turn the cap three times" but "turn the cap repeatedly till it comes loose." The ability to perform and recognize skills rests on the ability to extract "constituents" by linking action patterns to subgoals.

I have argued that the full syntax of any modern language is the result of bricolage – a historical accumulation of piecemeal strategies for achieving a wide range of communicative goals complemented by a process of generalization whereby some strategies become unified into a single general strategy. In correspondence related to Arbib (2005a), David Kemmerer notes that the formal criteria used by linguists to identify grammatical categories in one language may be absent or employed very differently in another: verbs are often marked for tense, aspect, mood, and transitivity but some languages, like Vietnamese, lack all such inflection, while Makah applies aspect and mood markers not only to words that are translated into English as verbs, but also to words that are translated into English as nouns or adjectives. Croft (2001) uses "construction grammar," seeking to identify the grammatical categories of individual languages according to the constructions employed in those languages. Of course, one may still relate these to semantic and pragmatic prototypes, but in a language-specific way. Both Kemmerer (this volume) and Itti and Arbib (this volume) take up this theme of construction grammar within the general project of linking action, perception, and language.

Just as we see the compositionality of language at work in the elaboration of the sentence "He picked a berry" to "He pulled the branch to pick the berry" so can we see the ability to take the "action constructions" <If a berry is within reach, pick it to eat it> and <If something not within reach is attached to something within reach, then grasp the latter and pull it to bring the former within reach> and compose them to plan the action which combines pulling a branch with one hand to permit picking a berry with the other.

Informed by this example, we can suggest how a "grammar for action" may relate to a "grammar for language," without blurring the distinction between the two.

a. A "grammar for action" would start from multiple "action constructions" and seek to replace *ad hoc* verbal descriptions by a more formal typology whereby groups of individual constructions could be revealed as instances of a single more abstract construction, and then attempt a principled account of how those structures may be combined.
b. Praxic action requires temporal and spatial coordination of simultaneously occurring actions whereas language encodes "conceptual overlap" in a linear ordering of words and morphemes. Words are required to make explicit both more and less than what is simply "there" in vision and action – we have to say "put the apple on the plate," but our act simply puts the apple in one place in a whole continuum of resting places (and this language can only handle by "number plug-ins" such as "place the apple 3.15 cm to the left of the daisy in the plate pattern").
c. Schema assemblage is not only "somewhat syntactic" as in (a) but also involves "tuning." In learning to drive, one may rapidly acquire the "syntax" that relates turning direction to motion of the steering wheel, etc., but this novel coordinated control program will be of little use in heavy traffic unless tuned by extensive practice.
d. The genius of language is that it supports metaphorical extension whereby constructions emerge that can be linked step by step back to the concrete but in the end gain their independence as a domain of expression. Just consider how much experience is required to go from a 5-year-old mind to one which can see the previous sentence as meaningful.

1.5.2 Even abstract language has roots in embodiment

I have suggested that syntactic structures are somehow scaffolded on pre-existing understanding of object–action schemas. But most sentences (like this one) do not describe action–object events. I would see the explanation of the range of sentence structures as more the task of historical and comparative linguistics and cognitive grammar than of an action-oriented linguistics, evolutionary or otherwise. However, a critique of an argument of Bickerton (1995) may indicate why I think that the transition from object–action frames to verb–argument structures may be seen as grounding the development of sentences of increasing abstraction.

Bickerton (1995, p.22) notes that a sentence like "The cat sat on the mat" is far more abstract than the image of a particular cat sitting on a particular mat. An image does not bring in the sense of time distinguishing "The cat sat on the mat" from "The cat is sitting on the mat" or "The cat will sit on the mat." An image does not distinguish "The cat is sitting on the mat" from "The mat is underneath the cat." All this is true, and we must reflect these distinctions in characterizing language. For example, we might relate the focus of a sentence (where prosody plays a crucial role not obvious in the written words) to the focus of attention in vision. However, Bickerton creates a false dichotomy when he asserts that "it is not true that we build a picture of the world and dress it out in language. Rather, language builds us the picture of the world that we use in thinking and communicating." The idea that language builds our picture of the world – rather than contributing to its richness – is misguided for it ignores the role of visual experience and then of *episodic memory* (linking episodes in temporal and other relationships) and *expectations* in building the rich perceptions and cognitions (Cognitive Form) of which sentences (Phonological Form) are just a précis. There is no claim that the relationship is one-to-one. Bickerton's approach leaves little room for understanding how the ability to *mean* that a cat is on the mat could be acquired in the first place. The language of the brain or schema network is vastly richer than a linear sequence of words. This does not deny that language can express what pictures cannot – or vice versa. Perception is *not* invertible – even if I see an actual cat on an actual mat, I am unlikely to recall more than a few details. And what one sees is knowledge-based: e.g., a familiar cat vs. a generic cat, or recognizing a specific subspecies. There is an intimate relation between naming and categorization.

Bickerton (1995, pp.22–24) argues that one cannot picture "My trust in you has been shattered forever by your unfaithfulness." because no picture could convey the uniquely hurtful sense of betrayal the act of infidelity provokes if you did not know what trust was, or what unfaithfulness was, or what it meant for trust to be shattered. "In the case of trust or unfaithfulness, there can be nothing beneath the linguistic concept except other linguistic representations, because abstract nouns have no perceptual attributes to be attached to them and therefore no possible representation outside those areas of the brain devoted to language." This is wrong on three levels:

1. The words themselves (i.e., the sequences of letters on the page or spoken phonemes) do not convey "the uniquely hurtful sense of betrayal." It is only if they "hook into" an appropriate body of experience and association, which not all people will share – each word is the tip of the schema iceberg. Words must link into the network which itself links to non-verbal experience, both perceptual and behavioral (cf. the discussion of a person's knowledge as a "schema encyclopedia" in Arbib, 1985, p.43).
2. Given this, an image (whether static like a picture, or extended in time like a video clip) may tap a similar network of experience, such as seeing one person turning away with an expression of disillusionment and despair from the sight of another engaged in lovemaking. The words and the images have complementary strengths – the words make explicit the key relationships, the image provides a host of details that could be only supplied (if indeed they were deemed relevant) by the piling on of more and more sentences. If one recalls a beautiful sunset, then it may be that "The sunset where we saw the green flash at Del Mar" will *index* the scene in my own thoughts or for communication with others, but the words alone do not recapture the beauty of the scene by forming an image of the setting and the colors of the sky.
3. Many would argue that one does not fully understand "hurtful sense of betrayal" unless one to some extent feels something of the emotion concerned, a feeling which involves a multiplicity of brain regions unlinked to language (see Fellous and Arbib (2005) and, in particular, Jeannerod (2005) for discussion of the possible role of a mirror system in human empathy).

As an exercise, let me try to link the sentence "My trust in you has been shattered forever by your unfaithfulness" back to the schema network anchored in my action and perception. I look at the definitions of the words and see how they are – eventually – rooted in behavior, noting the necessary role of metaphor in the use of "shattered," and in the use of "your" to indicate both possession of an object and possession of a disposition:

My trust in you expresses the objectification of the behavioral schema *Trust (I, You)*, where *Trust(A, B)* means "For all C, B tells A that C is the case $\Rightarrow$ A acts on the assumption that C is true." I do *not* argue that my mental states need exploit representations expressing such a formalism. Rather, the above formula is a shorthand for a whole range of behaviors and expectations that constitute the mental state of "trusting."

B is faithful to A is defined socially by a set of behaviors *prescribed* and *proscribed* for B by nature of his/her relationship to A. Infidelity is then detected by, perhaps, repeated failure in a prescribed behavior, or possibly even one example of proscribed behavior.

That an object is *broken* is, in the grounding case, testable either perceptually – the recognizable structure has been disrupted – or behaviorally – the object does not behave in the expected way. *Repairing* is acting upon an object in such a way as to make it look or perform as it is expected to. An object is *shattered* if it is broken into many pieces – implying that repairing the damage (making the object functional again) will be difficult or impossible.

Shattered forever then asserts that repair is impossible – there is no set of operations such that at any future time the object will function again, introducing the element of time and a hypothetical, involving the semantic extension of schemas from the here and now of action and perception. But note too that planning and expectations are implicit in behavior, and relate to the notion of an internal model of the world. Moreover, our notions of future time rest on extrapolation from our experience of past times in relation to the expectations we held at even earlier times.

Having said all this, note the many "inventions" required to go, historically, from simple wants and actions to a language + thought system rich enough to express the above sentence; and note, too, the long path a child must go through in coming to understand what these words mean. Of course, the formal aspects sketched above do not begin to exhaust the meaning of the above sentence, and this can only be done by consideration of the embodied self. To say my "trust is shattered" also implies a state of emotional devastation that needs empathy of another human to understand.

This account is little more than a caricature, but serves to reinforce the view that the use of language is rooted in our experience of action within the world, enriched by our ability to recall past events or imagine future ones and expanded by the cultural history of our society as reflected in our own personal experience as embodied and social beings. The ability to understand "My trust in you has been shattered forever by your unfaithfulness." is not the expression of some standalone linguistic faculty, but expresses the fruits of our individual cognitive and linguistic development within a community that has built a rich set of linguistic devices through an expanding spiral increasing the range of language and cognition through their mutual interaction. (Consider the debate about the relation between language and Theory of Mind in human evolution as set forth by Gordon, this volume.)

1.6 Discussion

I argue that human language and action both build on the evolutionary breakthrough that gives us a brain capable of complex action analysis and complex imitation. However, the challenges of "linearizing thought" posed by language are sufficiently different from those of spatial interaction with the world that I accept that there is some specialization of neural circuitry to support language. But specialization in what sense? We may distinguish (a) a "language-ready brain" able to master language as the child matures within a language-using community from (b) a brain which "has" language (in the sense, e.g., of an *innate* principles and parameters Universal Grammar). I have argued that biological evolution endowed us with the former not the latter, and that the transition from protolanguage to full languages is a matter of human history rather than biological selection.

The question "A sentence is to speech as what is to action?" forced us to confront the prior question "What is a sentence to speech?" and see that the answer will be far more equivocal than "What is a sentence to well-punctuated writing?" But this legitimizes the view that we can characterize basic actions and then define a grammar of action whereby coordinated control programs can be combined to define "sentence-level actions." As is well known, compound actions may become so well practiced that they then become available to act as new "action words." In the same way, complex concepts which must be referred to by complex phrases may acquire single words if referred to frequently enough. Thus the line between word, phrase, and sentence can shift over time. On the other hand, just as sentences may be concatenated to yield the verbal equivalent of a paragraph, so too may "action sentences" be combined in less and less loosely structured ways.

Admitting this, I suggest that an *action sentence* is the interwoven structure of basic actions (recall that in general execution of actions is not strictly sequential) that results from combining a relatively small number of schemas to form a coordinated control program and then executing that program according to the contingencies of the current situation. I do not hold that action privileges the "action sentence" in the same way that written language privileges the sentence. But just as we ask what defines a constituent in language, and then ask how constituents aggregate to yield sentences that are relatively free standing, so can we do the same for action, and in this understanding of the relation of a complex behavior to the underlying coordinated control program that describes the interweaving of its components. It may be that construction grammar provides the best fit for the pragmatic skills of language use – with specific subskills corresponding to specific constructions.

Whereas many accounts of language are so structured as to hide the commonalities between action in general and language use in particular, I stress that much of what one does with language involves machinery that is available more generally. This does not deny that special machinery is needed to augment the general machinery to use special tools (words) to conduct communication in a relatively sequential manner. It is the task of neurolinguistics to understand how mechanisms supporting syntax and pragmatics compete and cooperate to make the perception and production of language a successful tool for communication and social coordination.

Acknowledgments

Preparation of the present paper was supported in part by a Fellowship from the Center for Interdisciplinary Research of the University of Southern California.

References

Adolphs, R., 1999. Social cognition and the human brain. *Trends Cogn. Sci.* **3**: 469–479

Allison, T., Puce, A., and McCarthy, G., 2000. Social perception from visual cues: role of the STS region. *Trends Cogn. Sci.* **4**: 267–278.

Arbib, M. A., 1981. Perceptual structures and distributed motor control. In V. B. Brooks (ed.) *Handbook of Physiology*, Section 2, *The Nervous System*, vol. 2, *Motor Control*, Part 1. Bethesda, MD: American Physiological Society, pp. 1449–1480.

1985. *In Search of the Person: Philosophical Explorations of Cognitive Science*. Amherst, MA: University of Massachusetts Press.

2001. Coevolution of human consciousness and language. *Ann. NY Acad. Sci.* **929**: 195–220.

2002. The mirror system, imitation, and the evolution of language. In C. Nehaniv and K. Dautenhahn (eds.) *Imitation in Animals and Artefacts*. Cambridge, MA: MIT Press, pp. 229–280.

2004. How far is language beyond our grasp? A response to Hurford. In D. K. Oller and U. Griebel (eds.) *Evolution of Communication Systems: A Comparative Approach*. Cambridge, MA: MIT Press, pp. 315–321.

2005a. From monkey-like action recognition to human language: an evolutionary framework for neurolinguistics. *Behav. Brain Sci.* **28**: 105–167.

2005b. Interweaving protosign and protospeech: further developments beyond the mirror. *Interaction Studies: Soc. Behav. Commun. Biol. Artif. Systems* **6**: 145–171.

2006. A sentence is to speech as what is to action? *Cortex*. (In press.)

Arbib, M. A., and Rizzolatti, G., 1997. Neural expectations: a possible evolutionary path from manual skills to language. *Commun. Cognit.* **29**: 393–423.

Arbib, M. A., Bischoff, A., Fagg, A. H., and Grafton, S. T., 1994. Synthetic PET: analyzing large-scale properties of neural networks. *Hum. Brain Map.* **2**: 225–233.

Arbib, M. A., Érdi, P. and Szentágothai, J. (1998) *Neural Organization: Structure, Function, and Dynamics*. Cambridge, MA: The MIT Press.

Arbib, M. A., Billard, A., Iacoboni, M., and Oztop, E., 2000. Synthetic brain imaging: grasping, mirror neurons and imitation. *Neur. Networks* **13**: 975–997.

Armstrong, D., Stokoe, W., and Wilcox, S., 1995. *Gesture and the Nature of Language*. Cambridge, UK: Cambridge University Press.

Barrett, A. M., Foundas, A. L., and Heilman, K. M., 2005. Speech and gesture are mediated by independent systems. *Behav. Brain Sci.* **28**: 125–126.

Bickerton, D. 1995. *Language and Human Behavior*. Seattle, WA: University of Washington Press.

2005. Beyond the mirror neuron: the smoke neuron? *Behav. Brain Sci.* **28**: 126.

Boesch, C., and Boesch, H., 1983. Optimization of nut-cracking with natural hammers by wild chimpanzees. *Behavior* **83**: 265–286.

Byrne, R. W., 2003. Imitation as behaviour parsing. *Phil. Trans. Roy. Soc. London B* **558**: 529–536.

Carey, D. P., Perrett, D. I., and Oram, M. W., 1997. Recognizing, understanding, and producing action. In M. Jeannerod and J. Grafman (eds.) *Handbook of Neuropsychology: Action and Cognition*, vol. 11. Amsterdam: Elsevier, pp. 111–130.

Cavada, C., and Goldman-Rakic, P. S., 1989. Posterior parietal cortex in rhesus macaque. II. Evidence for segregated corticocortical networks linking sensory and limbic areas with the frontal lobe. *J. Comp. Neurol.* **287**: 422–445.

Cobas, A., and Arbib, M., 1992. Prey-catching and predator-avoidance in frog and toad: defining the schemas. *J. Theor. Biol.* **157**: 271–304.

Corballis, M., 2002. *From Hand to Mouth: The Origins of Language*. Princeton, NJ: Princeton University Press.

Corina, D. P., Poizner H., Bellugi, U., *et al.* 1992. Dissociation between linguistic and nonlinguistic gestural systems: a case for compositionality. *Brain Lang.* **43**: 414–447.

Corina, D. P., Jose-Robertson, L. S., Guillemin, A., High, J., and Braun, A. R., 2003. Language lateralization in a bimanual language. *J. Cogn. Neurosci.* **15**: 718–730.

Coulmas, F., 2003. *Writing Systems: An Introduction to Their Linguistic Analysis*. Cambridge, UK: Cambridge University Press.

Croft, W., 2001. *Radical Construction Grammar: Syntactic Theory in Typological Perspective*. Oxford, UK: Oxford University Press.

de Jorio, A., 2000. *Gesture in Naples and Gesture in Classical Antiquity*. Translation of *La mimica degli antichi investigata nel gestire napoletano* (Gestural expression of the ancients in the light of Neapolitan gesturing), with an introduction and notes by Adam Kendon. Bloomington, IN: Indiana University Press.

Dixon, R. M. W., 1997. *The Rise and Fall of Languages*. Cambridge, UK: Cambridge University Press.

Donald, M., 1994. *Origin of the Modern Mind*. Cambridge, MA: Harvard University Press.

  1999. Precursors for the evolution of protolanguages. In M. C. Corballis and S. E. G. Lea (eds.) *Preconditions for the Evolution of Protolanguages*. Oxford, UK: Oxford University Press, pp. 138–154.

Emmorey, K., 2002. *Language, Cognition, and the Brain: Insights from Sign Language Research*. Mahwah, NJ: Lawrence Erlbaum.

Fagg, A. H., and Arbib, M. A., 1998. Modeling parietal–premotor interactions in primate control of grasping. *Neur. Networks* **11**: 1277–1303.

Falk, D., 2004. Prelinguistic evolution in early hominins: whence motherese. *Behav. Brain Sci.* **27**: 491–503; discussion 505–583.

Farnell, B., 1995. *Do You See What I Mean? Plains Indian Sign Talk and the Embodiment of Action*. Austin, TX: University of Texas Press.

Fellous, J.-M., and Arbib, M. A. (eds.), 2005. *Who Needs Emotions? The Brain Meets the Robot*. New York: Oxford University Press.

Ferrari, P. F., Gallese, V., Rizzolatti, G., and Fogassi, L., 2003. Mirror neurons responding to the observation of ingestive and communicative mouth actions in the monkey ventral premotor cortex. *Eur. J. Neurosci.* **17**: 1703–1714.

Fogassi, L., and Ferrari, P. F., 2004. Mirror neurons, gestures and language evolution. *Interaction Studies: Soc. Behav. Commun. Biol. Artific. Systems* **5**: 345–363.

Fogassi, L., Gallese, V., Fadiga, L., and Rizzolatti, G., 1998. Neurons responding to the sight of goal directed hand/arm actions in the parietal area PF (7b) of the macaque monkey. *Soc. Neurosci. Abstr.* **24**: 257.

Gallese, V., Fadiga, L., Fogassi, L., and Rizzolatti, G., 1996. Action recognition in the premotor cortex. *Brain* **119**: 593–609.

Gentilucci, M., 2003. Grasp observation influences speech production, *Eur. J. Neurosci.* **17**: 179–184.

Gentilucci, M., Santunione, P., Roy, A. C., and Stefanini, S., 2004a. Execution and observation of bringing a fruit to the mouth affect syllable pronunciation. *Eur. J. Neurosci.* **19**: 190–202.

Gentilucci, M., Stefanini, S., Roy, A. C., and Santunione, P., 2004b. Action observation and speech production: study on children and adults. *Neuropsychologia* **42**: 1554–1567.

Ghazanfar, A. A. (ed.), 2003. *Primate Audition: Ethology and Neurobiology*. Boca Raton, FL: CRC Press.

Ghazanfar, A. A., and Santos, L. R., 2004. Primate brains in the wild: the sensory bases for social interactions. *Nature Rev. Neurosci.* **5**: 603–616.

Gibson, J. J., 1979. *The Ecological Approach to Visual Perception*. Boston, MA: Houghton Mifflin.

Glover, S., and Dixon, P., 2002. Semantics affect the planning but not control of grasping. *Exp. Brain Res.* **146**: 383–387.

Glover, S., Rosenbaum, D. A., Graham, J., and Dixon, P., 2004. Grasping the meaning of words. *Exp. Brain Res.* **154**: 103–108.

Hauser, M. D., 1996. *The Evolution of Communication*. Cambridge, MA: MIT Press.

Hauser, M. D., Chomsky, N., and Fitch, W. T., 2002. The faculty of language: what is it, who has it, and how did it evolve? *Science* **298**: 1569–1579.

Hewes, G., 1973. Primate communication and the gestural origin of language. *Curr. Anthropol.* **14**: 5–24.

Hihara, S., Yamada, H., Iriki, A., and Okanoya, K., 2003. Spontaneous vocal differentiation of coo-calls for tools and food in Japanese monkeys. *Neurosci. Res.* **45**: 383–389.

Horwitz, B., Amunts, K., Bhattacharyya, R., *et al.*, 2003. Activation of Broca's area during the production of spoken and signed language: a combined cytoarchitectonic mapping and PET analysis. *Neuropsychologia* **41**: 1868–1876.

Hunt, G. R., and Gray, R. D., 2003. Diversification and cumulative evolution in New Caledonian crow tool manufacture. *Proc. Roy. Soc. London B* **270**: 867–874.

Hurford, J. R., 2004. Language beyond our grasp: what mirror neurons can, and cannot, do for language evolution. In D. Kimbrough OllerOller and U. Griebel (eds.) *Evolution of Communication Systems: A Comparative Approach*. Cambridge, MA: MIT Press, pp. 297–313.

Iacoboni, M., 2004. Understanding others: imitation, language, empathy. In S. Hurley and N. Chater (eds.) *Perspectives on Imitation: From Cognitive Neuroscience to Social Science*, vol. 1, *Mechanisms of Imitation and Imitation in Animals*. Cambridge, MA: MIT Press, pp. 77–99.

Iverson, J. M., and Goldin-Meadow, S. (eds.), 1998. *The Nature and Function of Gesture in Children's Communication*. New York: Jossey-Bass.

Jeannerod, M., 2005. How do we decipher others' minds? In J.-M. Fellous and M. A. Arbib (eds.) *Who Needs Emotions? The Brain Meets the Robot*. New York: Oxford University Press, pp. 147–169.

Jürgens, U., 1997. Primate communication: signaling, vocalization. In *Encyclopedia of Neuroscience*, 2 edn. Amsterdam: Elsevier, pp. 1694–1697.

2002. Neural pathways underlying vocal control. *Neurosci. Biobehav. Rev.* **26**: 235–258.

Kendon, A., 1988. *Sign Languages of Aboriginal Australia: Cultural, Semiotic, and Communicative Perspectives*. Cambridge, UK: Cambridge University Press.

Kimura, D., 1993. *Neuromotor Mechanisms in Human Communication*. Oxford, UK: Clarendon Press.

Kirby, S., 2000. Syntax without natural selection: how compositionality emerges from vocabulary in a population of learners. In C. Knight, M. Studdert-Kennedy and J. R. Hurford (eds.) *The Evolutionary Emergence of Language*. Cambridge, UK: Cambridge University Press.

Kohler, E., Keysers, C., Umiltà, M. A., *et al.*, 2002. Hearing sounds, understanding actions: action representation in mirror neurons. *Science* **297**: 846–848.

Lashley, K. S., 1951. The problem of serial order in behavior. In L. Jeffress (ed.) *Cerebral Mechanisms in Behavior: The Hixon Symposium*. New York: John Wiley, pp. 112–136.

Lieberman, P., 2000. *Human Language and my Reptilian Brain: The Subcortical Bases of Speech, Syntax, and Thought*. Cambridge, MA: Harvard University Press.

MacNeilage, P. F., 1998. The frame/content theory of evolution of speech production. *Behav. Brain Sci.* **21**: 499–546.

MacNeilage, P. F. and Davis, B. L., 2005. The frame/content theory of evolution of speech: comparison with a gestural origins theory. *Interaction Studies: Soc. Behav. Commun. Biol. Artif. Systems* **6**: 173–199.

Matelli, M., Camarda, R., Glickstein, M., and Rizzolatti, G., 1986. Afferent and efferent projections of the inferior area 6 in the macaque. *J. Comp. Neurol.* **251**: 281–298.

McNeill, D., 1992. *Hand and Mind: What Gestures Reveal about Thought*. Chicago, IL: University of Chicago Press.

Miller, G. A., Galanter, E., and Pribram, K. H., 1960. *Plans and the Structure of Behavior*. New York: Henry Holt.

Myowa-Yamakoshi, M., and Matsuzawa, T., 1999. Factors influencing imitation of manipulatory actions in chimpanzees (*P. troglodytes*). *J. Comp. Psychol.* **113**: 128–136.

Oztop, E., and Arbib, M. A., 2002. Schema design and implementation of the grasp-related mirror neuron system. *Biol. Cybernet.* **87**: 116–140.

Perrett, D. I., Mistlin, A. J., Harries, M. H., and Chitty, A. J., 1990. Understanding the visual appearance and consequence of hand actions. In M. A. Goodale (ed.) *Vision and Action: The Control of Grasping*. Norwood, NJ: Ablex, pp. 163–180.

Pizzuto, E., Capobianco, M., and Devescovi, A., 2004. Gestural–vocal deixis and representational skills in early language development. *Interaction Studies: Soc. Behav Commun. Biol. Artif. Systems* **6**: 223–252.

Rizzolatti, G., and Arbib, M. A., 1998. Language within our grasp. *Trends Neurosci.* **21**: 188–194.

Rizzolatti, G., and Luppino, G., 2001. The cortical motor system. *Neuron* **31**: 889–901.

2003. Grasping movements: visuomotor transformations. In M. A. Arbib (ed.) *The Handbook of Brain Theory and Neural Networks*, 2nd edn. Cambridge, MA: MIT Press, pp. 501–504.

Rizzolatti, G., Camarda, R., Fogassi L., *et al.*, (1998) Functional organization of inferior area 6 in the macaque monkey. II. Area F5 and the control of distal movements. *Exp. Brain Res.* **71**: 491–507.

Rizzolatti, G., Fadiga L., Gallese, V., and Fogassi, L., 1996. Premotor cortex and the recognition of motor actions. *Cogn. Brain Res.* **3**: 131–141.

Rizzolatti, R., Fogassi, L., and Gallese, V., 2001. Neurophysiological mechanisms underlying the understanding and imitation of action. *Nature Rev. Neurosci.* **2**: 661–670.

Seltzer, B., and Pandya, D. N., 1989. Frontal lobe connections of the superior temporal sulcus in the rhesus macaque. *J. Comp. Neurol.* **281**: 97–113.

Seyfarth, R. M., 2005. Continuities in vocal communication argue against a gestural origin of language. *Behav. Brain Sci.* **28**: 144–145.

Stokoe, W. C., 2001. *Language in Hand: Why Sign Came before Speech*. Washington, D. C.: Gallaudet University Press.

Studdert-Kennedy, M., 2000. Evolutionary implications of the particulate principle: imitation and the dissociation of phonetic form from semantic function. In C. Knight, M. Studdert-Kennedy and J. R. Hurford (eds.) *The Evolutionary Emergence of Language*. Cambridge, UK: Cambridge University Press, pp. 161–176.

Tagamets, M. A., and Horwitz, B., 1998. Integrating electrophysiological and anatomical data to create a large-scale model that simulates a delayed match-to-sample human brain imaging study. *Cereb. Cortex* **8**: 310–320.

Tomasello, M., 1999. The human adaptation for culture. *Annu. Rev. Anthropol.* **28**: 509–529.

Tomasello, M., and Call, J., 1997. *Primate Cognition*. Oxford, UK: Oxford University Press.

Turvey, M., Shaw, R., Reed, E., and Mace, W., 1981. Ecological laws of perceiving and acting. *Cognition* **9**: 237–304.

Umiltà, M. A., Kohler, E., Gallese, V., *et al.*, 2001. I know what you are doing: a neurophysiological study. *Neuron* **31**: 155–165.

Visalberghi, E., and Fragaszy, D., 2002. "Do monkeys ape?" Ten years after. In C. Nehaniv and K. Dautenhahn (eds.) *Imitation in Animals and Artifacts*. Cambridge, MA: MIT Press, pp. 471–499.

Wray, A., 1998. Protolanguage as a holistic system for social interaction. *Lang. Commun.* **18**: 47–67.

2000. Holistic utterances in protolanguage: the link from primates to humans. In C. Knight, M. Studdert-Kennedy and J. R. Hurford (eds.) *The Evolutionary Emergence of Language*. Cambridge, UK: Cambridge University Press, pp. 285–302.

2002. *Formulaic Language and the Lexicon*. Cambridge, UK: Cambridge University Press.

2005. The explanatory advantages of the holistic protolanguage model: the case of linguistic irregularity. *Behav. Brain Sci.* **28**: 147–148.

2

The origin and evolution of language: a plausible, strong-AI account

Jerry R. Hobbs

2.1 Framework

In this chapter I show in outline how human language as we know it could have evolved incrementally from mental capacities it is reasonable to attribute to lower primates and other mammals. I do so within the framework of a formal computational theory of language understanding (Hobbs *et al.*, 1993). In the first section I describe some of the key elements in the theory, especially as it relates to the evolution of linguistic capabilities. In the next two sections I describe plausible incremental paths to two key aspects of language – meaning and syntax. In the final section I discuss various considerations of the time course of these processes.

2.1.1 Strong AI

It is desirable for psychology to provide a reduction in principle of intelligent, or intentional, behavior to neurophysiology. Because of the extreme complexity of the human brain, more than the sketchiest account is not likely to be possible in the near future. Nevertheless, the central metaphor of cognitive science, "The brain is a computer," gives us hope. Prior to the computer metaphor, we had no idea of what could possibly be the bridge between beliefs and ion transport. Now we have an idea. In the long history of inquiry into the nature of mind, the computer metaphor gives us, for the first time, the promise of linking the entities and processes of intentional psychology to the underlying biological processes of neurons, and hence to physical processes. We could say that the computer metaphor is the first, best hope of materialism.

The jump between neurophysiology and intentional psychology is a huge one. We are more likely to succeed in linking the two if we can identify some intermediate levels. A view that is popular these days identifies two intermediate levels – the symbolic and the connectionist.

Action to Language via the Mirror Neuron System, ed. Michael A. Arbib. Published by Cambridge University Press. © Cambridge University Press 2006.

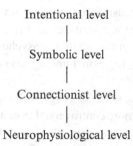

Intentional level

|

Symbolic level

|

Connectionist level

|

Neurophysiological level

The intentional level is implemented in the symbolic level, which is implemented in the connectionist level, which is implemented in the neurophysiological level.[1] From the "strong AI" perspective, the aim of cognitive science is to show how entities and processes at each level emerge from the entities and processes of the level below.[2] The reasons for this strategy are clear. We can observe intelligent activity and we can observe the firing of neurons, but there is no obvious way of linking these two together. So we decompose the problem into three smaller problems. We can formulate theories at the symbolic level that can, at least in a small way so far, explain some aspects of intelligent behavior; here we work from intelligent activity down. We can formulate theories at the connectionist level in terms of elements that are a simplified model of what we know of the neuron's behavior; here we work from the neuron up. Finally, efforts are being made to implement the key elements of symbolic processing in connectionist architecture. If each of these three efforts were to succeed, we would have the whole picture.

In my view, this picture looks very promising indeed. Mainstream AI and cognitive science have taken it to be their task to show how intentional phenomena can be implemented by symbolic processes. The elements in a connectionist network are modeled on certain properties of neurons. The principal problems in linking the symbolic and connectionist levels are representing predicate–argument relations in connectionist networks, implementing variable-binding or universal instantiation in connectionist networks, and defining the right notion of "defeasibility" or "non-monotonicity" in logic (see section 2.1.2) to reflect the "soft corners," or lack of rigidity, that make connectionist models so attractive. Progress is being made on all these problems (e.g., Shastri and Ajjanagadde, 1993; Shastri, 1999).

Although we do not know how each of these levels is implemented in the level below, nor indeed *whether* it is, we know that it *could* be, and that at least is something.

2.1.2 Logic as the language of thought

A very large body of work in AI begins with the assumptions that information and knowledge should be represented in first-order logic and that reasoning is theorem-proving.

---

[1] Variations on this view dispense with the symbolic or with the connectionist level.
[2] I take weak AI to be the effort to build smart machines, and strong AI to be the enterprise that seeks to understand human cognition on analogy with smart machines.

On the face of it, this seems implausible as a model for people. It certainly doesn't seem as if we are using logic when we are thinking, and if we are, why are so many of our thoughts and actions so illogical? In fact, there are psychological experiments that purport to show that people do not use logic in thinking about a problem (e.g., Wason and Johnson-Laird, 1972).

I believe that the claim that logic is the language of thought comes to less than one might think, however, and that thus it is more controversial than it ought to be. It is the claim that a broad range of cognitive processes are amenable to a high-level description in which six key features are present. The first three of these features characterize propositional logic and the next two first-order logic. I will express them in terms of "concepts," but one can just as easily substitute propositions, neural elements, or a number of other terms.

- **Conjunction** There is an additive effect ($P \wedge Q$) of two distinct concepts (P and Q) being activated at the same time.
- **Modus ponens** The activation of one concept (P) triggers the activation of another concept (Q) because of the existence of some structural relation between them ($P \supset Q$).
- **Recognition of obvious contradictions** It can be arbitrarily difficult to recognize contradictions in general, but we have no trouble with the easy ones, for example, that cats aren't dogs.
- **Predicate–argument relations** Concepts can be related to other concepts in several different ways. We can distinguish between a dog biting a man ($bite(D,M)$) and a man biting a dog ($bite(M,D)$).
- **Universal instantiation (or Variable Binding)** We can keep separate our knowledge of general (universal) principles ("All men are mortal") and our knowledge of their instantiations for particular individuals ("Socrates is a man" and "Socrates is mortal").

Any plausible proposal for a language of thought must have at least these features, and once you have these features you have first-order logic. Note that in this list there are no complex rules for double negations or for contrapositives (if P implies Q then not Q implies not P). In fact, most of the psychological experiments purporting to show that people don't use logic really show that they don't use the contrapositive rule or that they don't handle double negations well. If the tasks in those experiments were recast into problems involving the use of modus ponens, no one would think to do the experiments because it is obvious that people would have no trouble with the task.

There is one further property we need of the logic if we are to use it for representing and reasoning about commonsense world knowledge – defeasibility or non-monotonicity. Our knowledge is not certain. Different proofs of the same fact may have different consequences, and one proof can be "better" than another. Reasoning is *defeasible* if it is possible that facts may come to light which undercut or *defeat* the belief that results from the reasoning. This is contrary to traditional logic which is *monotonic* in the sense that if we once conclude something, it will always be true.

The mode of defeasible reasoning used here is "abduction,"[3] or inference to the best explanation. Briefly, one tries to prove something, but where there is insufficient knowledge,

---

[3] A term due to Peirce (1903).

one can make assumptions. One proof is better than another if it makes fewer, more plausible assumptions, and if the knowledge it uses is more plausible and more salient. This is spelled out in detail in Hobbs *et al.* (1993). The key idea is that intelligent agents understand their environment by coming up with the best underlying explanations for the observables in it. Generally not everything required for the explanation is known, and assumptions have to be made. Typically, abductive proofs have the following structure:

We want to prove R.
We know $P \wedge Q \supset R$.
We know P.
We assume Q.
We conclude R.

Abduction is "non-monotonic" because we could assume Q and thus conclude R, and later learn that Q is false.

There may be many Q's that could be assumed to result in a proof (including R itself), giving us alternative possible proofs, and thus alternative possible and possibly mutually inconsistent explanations or interpretations. So we need a kind of "cost function" for selecting the best proof. Among the factors that will make one proof better than another are the shortness of the proof, the plausibility and salience of the axioms used, a smaller number of assumptions, and the exploitation of the natural redundancy of discourse. A more complete description of the cost function is found in Hobbs *et al.* (1993).

2.1.3 Discourse interpretation: examples of definite reference

In the "Interpretation as abduction" framework, world knowledge is expressed as defeasible logical axioms. To interpret the content of a discourse is to find the best explanation for it, that is, to find a minimal-cost abductive proof of its logical form. To interpret a sentence is to deduce its syntactic structure and hence its logical form, and simultaneously to prove that logical form abductively. To interpret suprasentential discourse is to interpret individual segments, down to the sentential level, and to abduce relations among them.

Consider as an example the problem of resolving definite references. The following four examples are sometimes taken to illustrate four different kinds of definite reference.

I bought a new car last week. **The car** *is already giving me trouble.*
I bought a new car last week. **The vehicle** *is already giving me trouble.*
I bought a new car last week. **The engine** *is already giving me trouble.*
The engine *of my new car is already giving me trouble.*

In the first example, the same word is used in the definite noun phrase as in its antecedent. In the second example, a hypernym is used. In the third example, the reference is not to the "antecedent" but to an object that is related to it, requiring what

Clark (1975) called a "bridging inference." The fourth example is a determinative definite noun phrase, rather than an anaphoric one; all the information required for its resolution is found in the noun phrase itself.

These distinctions are insignificant in the abductive approach. In each case we need to prove the existence of the definite entity. In the first example it is immediate. In the second, we use the axiom

$$(\forall x)car(x) \supset vehicle(x)$$

In the third example, we use the axiom

$$(\forall x)car(x) \supset (\exists y)engine(y, x)$$

that is, cars have engines. In the fourth example, we use the same axiom, but after assuming the existence of the speaker's new car.

This last axiom is "defeasible" since it is not always true; some cars don't have engines. To indicate this formally in the abduction framework, we can add another proposition to the antecedent of this rule:

$$(\forall x)car(x) \wedge etc_i(x) \supset (\exists y)engine(y, x)$$

The proposition $etc_i(x)$ means something like "and other unspecified properties of x." This particular etc predicate would appear in no other axioms, and thus it could never be proved. But it could be assumed, at a cost, and could thus be a part of the least-cost abductive proof of the content of the sentence. This maneuver implements defeasibility in a set of first-order logical axioms operated on by an abductive theorem prover.

2.1.4 Syntax in the abduction framework

Syntax can be integrated into this framework in a thorough fashion, as described at length in Hobbs (1998). In this treatment, the predication

$$Syn(w, e, \ldots) \tag{2.1}$$

says that the string w is a grammatical, interpretable string of words describing the situation or entity e. For example, $Syn(\text{"John reads }Hamlet\text{"}, e, \ldots)$ says that the string "John reads *Hamlet*" (w) describes the event e (the reading by John of the play *Hamlet*). The arguments of Syn indicated by the dots include information about complements and various agreement features.

Composition is effected by axioms of the form

$$Syn(w_1, e, \ldots y, \ldots) \wedge Syn(w_2, y, \ldots) \supset Syn(w_1 w_2, e, \ldots) \tag{2.2}$$

A string w_1 whose head describes the eventuality e and which is missing an argument y can be concatenated with a string w_2 describing y, yielding a string describing e. For example, the string "reads" (w_1), describing a reading event e but missing the object y of the reading, can be concatenated with the string "*Hamlet*" (w_2) describing a book y, to yield a string "reads *Hamlet*" (w_1w_2), giving a richer description of the event e in that it does not lack the object of the reading.

Think of this in the following manner: let e represent the situation of John's reading *Hamlet*. We look at this situation e and describe it as "reading." That's a description of e, but not a very specific one. We then notice the book being read, and we now describe this same situation e in a more specific, more informative way as "reading *Hamlet*." Then we recognize the reader and describe it in an even more informative way – "John reads *Hamlet*." In each of the cases, the situation e being described is the same. The strings w used to describe it are different.

The interface between syntax and world knowledge is effected by "lexical axioms" of a form illustrated by

$$read'(e,x,y) \wedge text(y) \supset Syn(\text{"read"},e,\ldots,x,\ldots y,\ldots) \tag{2.3}$$

This says that if e is the eventuality of x reading y (the logical form fragment supplied by the word "read"), where y is a text (the selectional constraint imposed by the verb "read" on its object), then e can be described by a phrase headed by the word "read" provided it picks up, as subject and object, phrases of the right sort describing x and y, using composition axioms like (2.2).

To interpret a sentence w, one seeks to show it is a grammatical, interpretable string of words by proving there is an eventuality e that it describes, that is, by proving (2.1). One does so by decomposing it via composition axioms like (2.2) and bottoming out in lexical axioms like (2.3). This yields the logical form of the sentence, which then must be proved abductively, the characterization of interpretation we gave in Section 2.1.3.

A substantial fragment of English grammar is cast into this framework in Hobbs (1998), which closely follows Pollard and Sag (1994).

2.1.5 Discourse structure

When confronting an entire coherent discourse by one or more speakers, one must break it into interpretable segments and show that those segments themselves are coherently related. That is, one must use a rule like

$$Segment(w_1,e_1) \wedge Segment(w_2,e_2) \wedge rel(e,e_1,e_2) \supset Segment(w_1w_2,e)$$

That is, if w_1 and w_2 are interpretable segments describing situations e_1 and e_2 respectively, and e_1 and e_2 stand in some relation *rel* to each other, then the concatenation of w_1 and w_2 constitutes an interpretable segment, describing a situation e that is determined by the relation. The possible relations are discussed further in Section 2.4.

This rule applies recursively and bottoms out in sentences.

$$Syn(w,e,\ldots) \supset Segment(w,e)$$

A grammatical, interpretable sentence *w* describing eventuality *e* is a coherent segment of discourse describing *e*. This axiom effects the interface between syntax and discourse structure. *Syn* is the predicate whose axioms characterize syntactic structure; *Segment* is the predicate whose axioms characterize discourse structure; and they meet in this axiom. The predicate *Segment* says that string *w* is a *coherent* description of an eventuality *e*; the predicate *Syn* says that string *w* is a *grammatical and interpretable* description of eventuality *e*; and this axiom says that being grammatical and interpretable is one way of being coherent.

To interpret a discourse, we break it into coherently related successively smaller segments until we reach the level of sentences. Then we do a syntactic analysis of the sentences, bottoming out in their logical form, which we then prove abductively.[4]

2.1.6 Discourse as a purposeful activity

This view of discourse interpretation is embedded in a view of interpretation in general in which an agent, to interpret the environment, must find the best explanation for the observables in that environment, which includes other agents.

An intelligent agent is embedded in the world and must, at each instant, understand the current situation. The agent does so by finding an explanation for what is perceived. Put differently, the agent must explain why the complete set of observables encountered constitutes a coherent situation. Other agents in the environment are viewed as intentional, that is, as planning mechanisms, and this means that the best explanation of their observable actions is most likely to be that the actions are steps in a coherent plan. Thus, making sense of an environment that includes other agents entails making sense of the other agents' actions in terms of what they are intended to achieve. When those actions are utterances, the utterances must be understood as actions in a plan the agents are trying to effect. The speaker's plan must be recognized.

Generally, when a speaker says something, it is with the goal that the hearer believe the content of the utterance, or think about it, or consider it, or take some other cognitive stance toward it.[5] Let us subsume all these mental terms under the term "cognize." We can then say that to interpret a speaker *A*'s utterance to *B* of some content, we must explain the following:

$$goal(A, cognize(B, \text{content-of-discourse}))$$

Interpreting the content of the discourse is what we described above. In addition to this, one must explain in what way it serves the goals of the speaker to change the mental state

---

[4] This is an idealized, after-the-fact picture of the result of the process. In fact, interpretation, or the building up of this structure, proceeds word-by-word as we hear or read the discourse.

[5] Sometimes, on the other hand, the content of the utterance is less important than the nurturing of a social relationship by the mere act of speaking to.

of the hearer to include some mental stance toward the content of the discourse. We must fit the act of uttering that content into the speaker's presumed plan.

The defeasible axiom that encapsulates this is

$$(\forall s, h, e_1, e, w)[goal(s, cognize(h, e)) \land Segment(w, e) \supset utter(s, h, w)]^6$$

That is, normally if a speaker s has a goal of the hearer h cognizing a situation e and w is a string of words that conveys e, then s will utter w to h. So if I have the goal that you think about the existence of a fire, then since the word "fire" conveys the concept of fire, I say "Fire" to you. This axiom is only defeasible because there are multiple strings w that can convey e. I could have said, "Something's burning."

We appeal to this axiom to interpret the utterance as an intentional communicative act. That is, if A utters to B a string of words W, then to explain this observable event, we have to prove $utter(A,B,W)$. That is, just as interpreting an observed flash of light is finding an explanation for it, interpreting an observed utterance of a string W by one person A to another person B is to find an explanation for it. We begin to do this by backchaining on the above axiom. Reasoning about the speaker's plan is a matter of establishing the first conjunct in the antecedent of the axiom. Determining the informational content of the utterance is a matter of establishing the second. The two sides of the proof influence each other since they share variables and since a minimal proof will result when both are explained and when their explanations use much of the same knowledge.

2.1.7 A structured connectionist realization of abduction

Because of its elegance and very broad coverage, the abduction model is very appealing on the symbolic level. But to be a plausible candidate for how people understand language, there must be an account of how it could be implemented in neurons. In fact, the abduction framework can be realized in a structured connectionist model called SHRUTI developed by Lokendra Shastri (Shastri and Ajjanagadde, 1993; Shastri, 1999). The key idea is that nodes representing the same variable fire in synchrony. Substantial work must be done in neurophysiology to determine whether this kind of model is what actually exists in the human brain, although there is suggestive evidence. A good recent review of the evidence for the binding-via-synchrony hypothesis is given in Engel and Singer (2001). A related article by Fell *et al.* (2001) reports results on gamma band synchronization and desynchronization between parahippocampal regions and the hippocampus proper during episodic memorization.

By linking the symbolic and connectionist levels, one at least provides a proof of *possibility* for the abductive framework.

---

[6] This is not the real notation because it embeds propositions within predicates, but it is more convenient for this chapter and conveys the essential meaning. An adequate logical notation for beliefs, causal relations, and so on can be found in Hobbs (1985a).

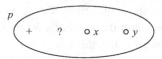

Figure 2.1 Predicate cluster for $p(x,y)$. The collecter node $(+)$ fires asynchronously in proportion to how plausible it is that $p(x,y)$ is part of the desired proof. The enabler node $(?)$ fires asynchronously in proportion to how much $p(x,y)$ is required in the proof. The argument nodes for x and y fire in synchrony with argument nodes in other predicate clusters that are bound to the same value.

There is a range of connectionist models. Among those that try to capture logical structure in the structure of the network, there has been good success in implementing defeasible *propositional* logic. Indeed, nearly all the applications to natural language processing in this tradition begin by setting up the problem so that it is a problem in propositional logic. But this is not adequate for natural language understanding in general. For example, the coreference problem, e.g., resolving pronouns to their antecedents, requires the expressivity of first-order logic even to state; it involves recognizing the equality of two variables or a constant and a variable presented in different places in the text. We need a way of expressing predicate–argument relations and a way of expressing different instantiations of the same general principle. We need a mechanism for universal instantiation, that is, the binding of variables to specific entities. In the connectionist literature, this has gone under the name of the *variable-binding problem*.

The essential idea behind the SHRUTI architecture is simple and elegant. A predication is represented as an assemblage or cluster of nodes, and axioms representing general knowledge are realized as connections among these clusters. Inference is accomplished by means of spreading activation through these structures.

In the cluster representing predications (Fig. 2.1), two nodes, a collector node and an enabler node, correspond to the predicate and fire asynchronously. That is, they don't need to fire synchronously, in contrast to the "argument nodes" described below; for the collector and enabler nodes, only the *level* of activation matters. The level of activation on the enabler node keeps track of the "utility" of this predication in the proof that is being searched for. That is, the activation is higher the greater the need to find a proof for this predication, and thus the more expensive it is to assume. For example, in interpreting "The curtains are on fire," it is very important to prove $curtains(x)$ and thereby identify which curtains are being talked about; the level of activation on the enabler node for that cluster would be high. The level of activation on the collector node is higher the greater the plausibility that this predication is part of the desired proof. Thus, if the speaker is standing in the living room, there might be a higher activation on the collector node for $curtains(c_1)$ where c_1 represents the curtains in the living room than on $curtains(c_2)$, where c_2 represents the curtains in the dining room.

We can think of the activations on the enabler nodes as prioritizing goal expressions, whereas the activations on the collector nodes indicate degree of belief in the predications, or more properly, degree of belief in the current relevance of the predications. The connections between nodes of different predication clusters have a strength of activation, or link weight, that corresponds to strength of association between the two concepts. This is one way we can capture the defeasibility of axioms in the SHRUTI model. The proof process then consists of activation spreading through enabler nodes, as we backchain through axioms, and spreading forward through collector nodes from something known or assumed. In addition, in the predication cluster, there are argument nodes, one for each argument of the predication. These fire synchronously with the argument nodes in other predication clusters to which they are connected. Thus, if the clusters for $p(x, y)$ and $q(z, x)$ are connected, with the two x nodes linked to each other, then the two x nodes will fire in synchrony, and the y and z nodes will fire at an offset with the x nodes and with each other. This synchronous firing indicates that the two x nodes represent variables bound to the same value. This constitutes the solution to the variable-binding problem. The role of variables in logic is to capture the identity of entities referred to in different places in a logical expression; in SHRUTI this identity is captured by the synchronous firing of linked nodes.

Proofs are searched for in parallel, and winner-takes-all circuitry suppresses all but the one whose collector nodes have the highest level of activation.

There are complications in this model for such things as managing different predications with the same predicate but different arguments. But the essential idea is as described. In brief, the view of relational information processing implied by SHRUTI is one where reasoning is a transient but systematic propagation of *rhythmic* activity over structured cell-ensembles, each active entity is a phase in the rhythmic activity, dynamic bindings are represented by the *synchronous* firing of appropriate nodes, and rules are high-efficacy links that cause the propagation of rhythmic activity between cell-ensembles. Reasoning is the spontaneous outcome of a SHRUTI network.

In the abduction framework, the typical axiom in the knowledge base is of the form

$$(\forall x, y)[p_1(x, y) \wedge p_2(x, y) \supset (\exists z)[q_1(x, z) \wedge q_2(x, z)]] \tag{2.4}$$

That is, the top-level logical connective will be implication. There may be multiple predications in the antecedent and in the consequent. There may be variables (x) that occur in both the antecedent and the consequent, variables (y) that occur only in the antecedent, and variables (z) that occur only in the consequent. Abduction backchains from predications in consequents of axioms to predications in antecedents. That is, to prove the consequent of such a rule, it attempts to find a proof of the antecedent. Every step in the search for a proof can be considered an abductive proof where all unproved predications are assumed for a cost. The best proof is the least cost proof.

The implementation of this axiom in SHRUTI requires predication clusters of nodes and axiom clusters of nodes (see Fig. 2.2). A predication cluster, as described above, has one collector node and one enabler node, both firing asynchronously, corresponding to the predicate and one synchronously firing node for each argument. An axiom cluster has one collector node and one enabler node, both firing asynchronously, recording the plausibility and the utility, respectively, of this axiom participating in the best proof. It also has one synchronously firing node for each variable in the axiom – in our example, nodes for

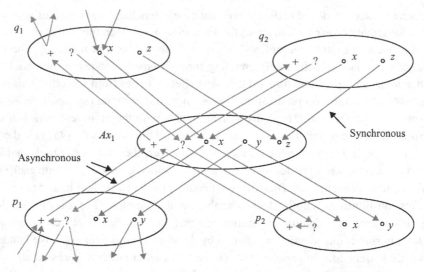

Figure 2.2 SHRUTI encoding of axiom $(\forall\ x,y)\ [p_1\ (x,y)\ \wedge\ p_2\ (x,y) \supset (\exists\ z)\ [q_1(x,z)\ \wedge\ q_2(x,z)]]$. Activation spreads backward from the enabler nodes (?) of the q_1 and q_2 clusters to that of the Ax_1 cluster and on to those of the p_1 and p_2 clusters, indicating the utility of this axiom in a possible proof. Activation spreads forward from the collector nodes (+) of the p_1 and p_2 clusters to that of the axiom cluster Ax_1 and on to those of the q_1 and q_2 clusters, indicating the plausibility of this axiom being used in the final proof. Links between the argument nodes cause them to fire in synchrony with other argument nodes representing the same variable.

x, y, and z. The collector and enabler nodes fire asynchronously and what is significant is their level of activation or rate of firing. The argument nodes fire synchronously with other nodes, and what is significant is whether two nodes are the same or different in their phases.

The axiom is then encoded in a structure like that shown in Fig. 2.2. There is a predication cluster for each of the predications in the axiom and one axiom cluster that links the predications of the consequent and antecedent. In general, the predication clusters will occur in many axioms; this is why their linkage in a particular axiom must be mediated by an axiom cluster.

Suppose (Fig. 2.2) the proof process is backchaining from the predication $q_1(x,z)$. The activation on the enabler node (?) of the cluster for $q_1(x,z)$ induces an activation on the enabler node for the axiom cluster. This in turn induces activation on the enabler nodes for predications $p_1(x,y)$ and $p_2(x,y)$. Meanwhile the firing of the x node in the q_1 cluster induces the x node of the axiom cluster to fire in synchrony with it, which in turn causes the x nodes of the p_1 and p_2 clusters to fire in synchrony as well. In addition, a link (not shown) from the enabler node of the axiom cluster to the y argument node of the same cluster causes the y argument node to fire, while links (not shown) from the x and z nodes cause that firing to be out of phase with the firing of the x and z nodes. This firing of the y node of the axiom cluster induces synchronous firing in the y nodes of the p_1 and p_2 clusters.

By this means we have backchained over axiom (2.4) while keeping distinct the variables that are bound to different values. We are then ready to backchain over axioms in which p_1 and p_2 are in the consequent. As mentioned above, the q_1 cluster is linked to other axioms as well, and in the course of backchaining, it induces activation in those axioms' clusters too. In this way, the search for a proof proceeds in parallel. Inhibitory links suppress contradictory inferences and will eventually force a winner-takes-all outcome.

2.1.8 Incremental changes to axioms

In this framework, incremental increases in linguistic competence, and other knowledge as well, can be achieved by means of a small set of simple operations on the axioms in the knowledge base:[7]

(1) The introduction of a new predicate, where the utility of that predicate can be argued for cognition in general, independent of language.

(2) The introduction of a new predicate p specializing an old predicate q:

$$(\forall x)p(x) \supset q(x)$$

For example, we learn that a beagle is a type of dog.

$$(\forall x)beagle(x) \supset dog(x)$$

(3) The introduction of a new predicate p generalizing one or more old predicates q_i:

$$(\forall x)q_1(x) \supset p(x), (\forall x)q_2(x) \supset p(x), \ldots$$

For example, we learn that dogs and cats are both mammals.

$$(\forall x)dog(x) \supset mammal(x), (\forall x)cat(x) \supset mammal(x)$$

(4) Increasing the arity of a predicate to allow more arguments.

$$p(x) \rightarrow p(x,y)$$

For example, we learn that "mother" is not a property but a relation.

$$mother(x) \rightarrow mother(x,y)$$

(5) Adding a proposition to the antecedent of an axiom.

$$p_1(x) \supset q(x) \rightarrow p_1(x) \wedge p_2(x) \supset q(x)$$

For example, we might first believe that a seat is a chair, then learn that a seat with a back is a chair.

$$seat(x) \supset chair(x) \rightarrow seat(x) \wedge has\text{-}back(x) \supset chair(x)$$

---

[7] Of course, a characterization of the data that would lead to these incremental changes merits deeper investigation.

(6) Adding a proposition to the consequent of an axiom.

$$p(x) \supset q_1(x) \rightarrow p(x) \supset q_1(x) \wedge q_2(x)$$

For example, a child might see snow for the first time and see that it's white, and then goes outside and realizes it's also cold.

$$snow(x) \supset white(x) \rightarrow snow(x) \supset white(x) \wedge cold(x)$$

It was shown in Section 2.1.7 that axioms such as these can be realized at the connectionist level in the SHRUTI model. To complete the picture, it must be shown that these incremental changes to axioms could also be implemented at the connectionist level. In fact, Shastri and his colleagues have demonstrated that incremental changes such as these can be implemented in the SHRUTI model via relatively simple means involving the recruitment of nodes, by strengthening latent connections as a response to frequent simultaneous activations (Shastri, 2001; Shastri and Wendelken, 2003; Wendelken and Shastri, 2003).

These incremental operations can be seen as constituting a plausible mechanism for both the development of cognitive capabilities in individuals and, whether directly or indirectly through developmental processes, their evolution in populations. In this chapter, I will show how the principal features of language could have resulted from a sequence of such incremental steps, starting from the cognitive capacity one could expect of ordinary primates.

Because of the equivalence of the logical and structured connectionist levels, we can proceed in the subsequent sections strictly in terms of the logical level.

2.1.9 Summary of background

To summarize, the framework assumed in this chapter has the following features:

A detailed, plausible, computational model for a large range of linguistic behavior.
A possible implementation in a connectionist model.
An incremental model of learning, development (physical maturation), and evolution.
An implementation of that in terms of node recruitment.

In the remainder of the chapter it is shown how two principal features of language – Gricean meaning and syntax – could have arisen from non-linguistic cognition through the action of three mechanisms:

incremental changes to axioms,
folk theories motivated independent of language,
compilation of proofs into axioms.

Gricean non-natural meaning and syntax are, in a sense, *the* two key features of language. The first tells how single words convey meaning in discourse. The second tells how multiple words combine to convey complex meanings.

2.2 The evolution of Gricean meaning

In Gricean non-natural meaning (Grice, 1957), what is conveyed is not merely the content of the utterance, but also the intention of the speaker to convey that meaning, and the intention of the speaker to convey that meaning *by means of* that specific utterance. When A shouts "Fire!" to B, A expects that

1. B will believe there is a fire
2. B will believe A wants B to believe there is a fire
3. (1) will happen because of (2)

Five steps take us from natural meaning, as in "Smoke means fire," to Gricean meaning. Each step depends on certain background theories being in place, theories that are motivated even in the absence of language. Each new step in the progression introduces a new element of defeasibility. The steps are as follows:

1. Smoke means fire
2. "Fire!" means fire
3. Mediation by belief
4. Mediation by intention
5. Full Gricean meaning

Once we get into theories of belief and intention, there is very little that is certain. Thus, virtually all the axioms used in this section are defeasible. That is, they are true most of the time, and they often participate in the best explanation produced by abductive reasoning, but they are sometimes wrong. They are nevertheless useful to intelligent agents.

The theories that will be discussed in this section – belief, mutual belief, intention, and collective action – are some of the key elements of a theory of mind (e.g., Premack and Woodruff, 1978; Heyes, 1998; Gordon, this volume). I discuss the possible courses of evolution of a theory of mind in Section 2.4.

2.2.1 Smoke means fire

The first required folk theory is a theory of causality (or rather, a number of theories *with* causality). There will be no definition of the predicate *cause*, that is, no set of necessary and sufficient conditions.

$$cause(e_1, e_2) \equiv \ldots$$

Rather there will be a number of domain-dependent theories saying what sorts of things cause what other sorts of things. There will be lots of necessary conditions

$$cause(e_1, e_2) \supset \ldots$$

and lots of sufficient conditions

$$\ldots \supset cause(e_1, e_2)$$

An example of the latter type of rule is

$$smoke(y) \supset (\exists x)[fire(x) \wedge cause(x, y)]$$

That is, if there's smoke, there's fire (that caused it).

This kind of causal knowledge enables prediction, and is required for the most rudimentary intelligent behavior.

Now suppose an agent B sees smoke. In the abductive account of intelligent behavior, an agent interprets the environment by telling the most plausible causal story. Here the story is that since fire causes smoke, there is a fire. B's seeing smoke causes B to believe there is fire, because B knows fire causes smoke.

2.2.2 *"Fire!" means fire*

Suppose seeing fire causes another agent A to emit a particular sound, say, "Fire!" and B knows this. Then we are in exactly the same situation as in Step 1. B's perceiving A making the sound "Fire!" causes B to believe there is a fire. B requires one new axiom about what causes what, but otherwise no new cognitive capabilities.

In this sense, sneezing means pollen, and "Ouch!" means pain. It has often been stated that one of the true innovations of language is its arbitrariness. The word "fire" is in no way iconic; its relation to fire is arbitrary and purely a matter of convention. The arbitrariness does not seem to me especially remarkable, however. A dog that has been trained to salivate when it hears a bell is responding to an association just as arbitrary as the relation between "fire" and fire.

I've analyzed this step in terms of comprehension, however, not production. Understanding a symbol–concept relation may require nothing more than causal associations. One can learn to perform certain simple behaviors because of causal regularities, as for example a baby crying to be fed and a dog sitting by the door to be taken out. But in general producing a new symbol for a concept with the intention of using it for communication probably requires more in an underlying theory of mind. A dog may associate a bell with being fed, but will it spontaneously ring the bell as a request to be fed? One normally at least has to have the notion of another individual's belief, since the aim of the new symbol is to create a belief in the other's mind.[8]

2.2.3 *Mediation by belief*

For the next step we require a folk theory of belief, that is, a set of axioms explicating, though not necessarily defining, the predicate *believe*. The principal elements of a folk theory of belief are the following:

---

[8] I am indebted to George Miller and Michael Arbib for discussions on this point.

(a) An event occurring in an agent's presence causes the agent to perceive the event.

$$cause(at(x, y, t), perceive(x, y, t))$$

This is only defeasible. Sometimes an individual doesn't know what's going on around him.

(b) Perceiving an event causes the agent to believe the event occurred. (Seeing is believing.)

$$cause(perceive(x, y, t), believe(x, y, t))$$

(c) Beliefs persist.

$$t_1 < t_2 \supset cause(believe(x, y, t_1), believe(x, y, t_2))$$

Again, this is defeasible, because people can change their minds and forget things.

(d) Certain beliefs of an agent can cause certain actions by the agent. (This is an axiom schema, that can be instantiated in many ways.)

$$cause(believe(x, P, t), ACT(x, t))$$

For example, an individual may have the rule that an agent's believing there is fire causes the agent to utter "Fire!"

$$fire(f) \supset cause(believe(x, f, t), utter(x, \text{``Fire!''}, t))$$

Such a theory would be useful to an agent even in the absence of language, for it provides an explanation of how agents can transmit causality, that is, how an event can happen at one place and time and cause an action that happens at another place and time. It enables an individual to draw inferences about unseen events from the behavior of another individual. Belief functions as a carrier of information.

Such a theory of belief allows a more sophisticated interpretation, or explanation, of an agent A's utterance, "Fire!" A fire occurred in A's presence. Thus, A believed there was a fire. Thus, A uttered "Fire!" The link between the event and the utterance is mediated by belief. In particular, the observable event that needs to be explained is that an agent A uttered "Fire!" and the explanation is as follows:

$$utter\ (A,\ \text{``Fire!''},\ t_2)$$
$$|$$
$$believe\ (A,\ f,\ t_2)\ \wedge fire\ (f)$$
$$|$$
$$believe\ (A,\ f,\ t_1) \wedge t_1 < t_2$$
$$|$$
$$perceive\ (A,\ f,\ t_1)$$
$$|$$
$$at\ (A,\ f,\ t_1)$$

There may well be other causes of a belief besides seeing. For example, communication with others might cause belief. Thus the above proof could have branched another way below the third line. This fact means that with this innovation, there is the possibility of "language" being cut loose from direct reference.

Jackendoff (1999) points out the distinction between two relics of one-word prelanguage in modern language. The word "ouch!", as pointed out above, falls under the case of Section 2.2.2; it is not necessarily communicative. The word "shh" by contrast has a necessary communicative function; it is uttered to induce a particular behavior on the part of the hearer. It could in principle be the result of having observed a causal regularity between the utterance and the effect on the people nearby, but it is more likely that the speaker has some sort of theory of others' beliefs and how those beliefs are created and what behaviors they induce.

Note that this theory of belief could in principle be strictly a theory of other individuals, and not a theory of one's self. There is no need in this analysis that the interpreter even *have* a concept of self.

2.2.4 Near-Gricean non-natural meaning

The next step is a close approximation of Gricean meaning. It requires a much richer cognitive model. In particular, three more background folk theories are needed, each again motivated independently of language. The first is a theory of goals, or intentionality. By adopting a theory that attributes agents' actions to their goals, one's ability to predict the actions of other agents is greatly enhanced. The principal elements of a theory of goals are the following:

(a) If an agent x has an action by x as a goal, that will, defeasibly, cause x to perform this action. This is an axiom schema, instantiated for many different actions.

$$cause(goal(x, ACT(x)), ACT(x)) \tag{2.5}$$

That is, wanting to do something causes an agent to do it. Using this rule in reverse amounts to the attribution of intention. We see someone doing something and we assume they did it because they wanted to do it.

(b) If an agent x has a goal g_1 and g_2 tends to cause g_1, then x may have g_2 as a goal.

$$cause(g_2, g_1) \supset cause(goal(x, g_1), goal(x, g_2)) \tag{2.6}$$

This is only a defeasible rule. There may be other ways to achieve the goal g_1, other than g_2. This rule corresponds to the body of a STRIPS planning operator as used in AI (Fikes and Nilsson, 1971). When we use this rule in the reverse direction, we are inferring an agent's ends from the means.

(c) If an agent A has a goal g_1 and g_2 enables g_1, then A has g_2 as a goal.

$$enables(g_2, g_1) \supset cause(goal(x, g_1), goal(x, g_2)) \tag{2.7}$$

This rule corresponds to the prerequisites in the STRIPS planning operators of Fikes and Nilsson (1971).

Many actions are enabled by the agent knowing something. These are knowledge prerequisites. For example, before picking something up, you first have to know where it is. The form of these rules is

$$enable(believe(x, P), ACT(x))$$

The structure of goals linked in these ways constitutes a plan. To achieve a goal, one must make all the enabling conditions true and find an action that will cause the goal to be true, and do that.

The second required theory is a theory of joint action or collective intentionality. The usual reason for me to inform you of a fact that will induce a certain action on your part is that this action will serve some goal that both of us share, or that we are somehow otherwise involved in each other's plans. A theory of collective intentionality is the same as a theory of individual intentionality, except that collectives of individuals can have goals and beliefs and can carry out actions. In addition, collective plans must bottom out in individual action. This is the point in the development of a theory of mind where a concept of self is probably required; one has to know that one is a member of the group like the rest of the community.

Agents can have as goals events that involve other agents. Thus, they can have in their plans knowledge prerequisites for other agents. A can have as a goal that B believe some fact. Communication is the satisfaction of such a goal.

The third theory is a theory of how agents understand. The essential content of this theory is that agents try to fit events into causal chains. The first rule is a kind of causal modus ponens. If an agent believes e_2 and believes e_2 causes e_3, that will cause the agent to believe e_3.

$$cause(believe(x, e_2) \wedge believe(x, cause(e_2, e_3)), believe(x, e_3))$$

This is defeasible since the individual may simply fail to draw the conclusion.

The second rule allows us to infer that agents backchain on enabling conditions. If an agent believes e_2 and believes e_1 enables e_2, then the agent will believe e_1

$$cause(believe(x, e_2) \wedge believe(x, enable(e_1, e_2)), believe(x, e_1))$$

The third rule allows us to infer that agents do causal abduction. That is, they look for causes of events that they know about. If an agent believes e_2 and believes e_1 causes e_2, then the agent may come to believe e_1.

$$cause(believe(x, e_2) \wedge believe(x, cause(e_1, e_2)), believe(x, e_1))$$

This is defeasible since the agent may have beliefs about other possible causes of e_2.

The final element of the folk theory of cognition is that all folk theories, including this one, are believed by every individual in the group. This is also defeasible. It is a corollary of this that A's uttering "Fire!" may cause B to believe there is a fire.

Now the near-Gricean explanation for the utterance is this: A uttered "Fire!" because A had the goal of uttering "Fire!", because A had as a goal that B believe there is a fire, because B's belief is a knowledge prerequisite in some joint action that A has as a goal (perhaps merely joint survival) and because A believes there is a fire, because there was a fire in A's presence.

2.2.5 Full Gricean non-natural meaning

Only one more step is needed for full Gricean meaning. It must be a part of B's explanation of A's utterance not only that A had as a goal that B believe there is a fire and that caused A to have the goal of uttering "Fire!", but also that A had as a goal that A's uttering "Fire!" would cause B to believe there is a fire. To accomplish this we must split the planning axiom (2.6) into two:

If an agent A has a goal g_1 and g_2 tends to cause g_1 then A may have as a goal that g_2 cause g_1 (2.6a)

If an agent A has as a goal that g_2 cause g_1, then A has the goal g_2 (2.6b)

The planning axioms (2.5), (2.6) and (2.7) implement means–end analysis. This elaboration captures the intentionality of the means–end relation.

The capacity for language evolved over a long period of time, after and at the same time as a number of other cognitive capacities were evolving. Among the other capacities were theories of causality, belief, intention, understanding, joint action, and (non-linguistic) communication. The elements of a theory of mind, in particular, probably evolved to make us more effective members of social groups. As the relevant elements of each of these capacities were acquired, they would have enabled the further development of language as well.

In Section 2.4 there is a discussion of possible evolutionary histories of these elements of a theory of mind.

2.3 The evolution of syntax

2.3.1 Protolanguage: the two-word stage

When agents encounter two objects in the world that are adjacent, they need to explain this adjacency by finding a relation between the objects. Usually, the explanation for why something is where it is is that that is its normal place. It is normal to see a chair at a desk, and we don't ask for further explanation. But if something is out of place, we do. If we walk into a room and see a chair on a table, or we walk into a lecture hall and see a dog in the aisle, we wonder why.

Similarly, when agents hear two adjacent utterances, they need to explain the adjacency by finding a relation between them. A variety of relations are possible. "Mommy sock" might mean "This is Mommy's sock" and it might mean "Mommy, put my sock on."

In general, the problem facing the agent can be characterized by the following pattern:

$$(\forall w_1, w_2, x, y, z)[B(w_1, y) \wedge C(w_2, z) \wedge rel(x, y, z) \supset A(w_1 w_2, x)] \qquad (2.8)$$

That is, to recognize two adjacent words or strings of words w_1 and w_2 as a composite utterance of type A meaning x, one must recognize w_1 as an object of type B meaning y, recognize w_2 as an object of type C meaning z, and find some relation between y and z, where x is determined by the relation that is found. There will normally be multiple possible relations, but abduction will choose the best.

This is the characterization of what Bickerton (1990) calls "protolanguage"[9] One utters meaningful elements sequentially and the interpretation of the combination is determined by context. The utterance "Lion. Tree." could mean there's a lion behind the tree or there's a lion nearby so let's climb that tree, or numerous other things. Bickerton gives several examples of protolanguage, including the language of children in the two-word phase and the language of apes. I'll offer another example: the language of panic. If a man runs out of his office shouting, "Help! Heart attack! John! My office! CPR! Just sitting there! 911! Help! Floor! Heart attack!" we don't need syntax to tell us that he was just sitting in his office with John when John had a heart attack, and John is now on the floor, and the man wants someone to call 911 and someone to apply CPR.

Most if not all rules of grammar can be seen as specializations and elaborations of pattern (2.8). The simplest example in English is compound nominals. To understand "turpentine jar" one must understand "turpentine" and "jar" and find the most plausible relation (in context) between turpentine and jars. In fact, compound nominals can be viewed as a relic of protolanguage in modern language.

Often with compound nominals the most plausible relation is a predicate–argument relation, where the head noun supplies the predicate and the prenominal noun supplies an argument. In "chemistry teacher," a teacher is a teacher of something, and the word "chemistry" tells us what that something is. In "language origin," something is originating, and the word "language" tells us what that something is.

The two-word utterance "Men work" can be viewed in the same way. We must find a relation between the two words to explain their adjacency. The relation we find is the predicate–argument relation, where "work" is the predicate and "men" is the argument.

2.3.2 From protolanguage to simple syntax

Let us now see how we can go from this type of protolanguage interpretation to a simple clausal syntax. The phrase structure rules

$$S \rightarrow NP\ VP;\ VP \rightarrow V\ NP$$

---

[9] See Arbib (this volume) for another view of protolanguage.

can be written in the abductive framework (Hobbs, 1998) as

$$(\forall w_1, w_2, x, e)[Syn(w_1, x) \wedge Syn(w_2, e) \wedge Lsubj(x, e) \supset Syn(w_1w_2, e)] \qquad (2.9)$$
$$(\forall w_3, w_4, y, e)[Syn(w_3, e) \wedge Syn(w_4, y) \wedge Lobj(y, e) \supset Syn(w_3w_4, e)] \qquad (2.10)$$

In the first rule, if w_1 is a string of words describing an entity x and w_2 is a string of words describing the eventuality e and x is the logical subject of e, then the concatenation w_1w_2 of the two strings can be used to describe e, in particular, a richer description of e specifying the logical subject. This means that to interpret w_1w_2 as describing some eventuality e, segment it into a string w_1 describing the logical subject of e and a string w_2 providing the rest of the information about e. The second rule is similar. These axioms instantiate pattern (2.8). The predicate *Syn*, which relates strings of words to the entities and situations they describe, plays the role of A, B, and C in pattern (2.8), and the relation *rel* in pattern (2.8) is instantiated by the *Lsubj* and *Lobj* relations.

Syntax, at a first cut, can be viewed as a set of constraints on the interpretation of adjacency, specifically, as predicate–argument relations.

Rule (2.9) is not sufficiently constrained, since w_2 could already contain the subject. We can prevent this by adding to the arity of *Syn*, one of the incremental evolutionary modifications in rules in Section 2.1.8, and giving *Syn* a further argument indicating that something is missing.

$$(\forall w_1, w_2, x, e)[Syn(w_1, x, -, -) \wedge Syn(w_2, e, x, -) \wedge Lsubj(x, e)$$
$$\supset Syn(w_1w_2, e, -, -)] \qquad (2.11)$$
$$(\forall w_3, w_4, y, e)[Syn(w_3, e, x, y) \wedge Syn(w_4, y, -, -) \wedge Lobj(y, e)$$
$$\supset Syn(w_3w_4, e, x, -)] \qquad (2.12)$$

Now the expression $Syn(w_3, e, x, y)$ says something like "String w_3 would describe situation e if strings of words describing x and y can be found in the right places."

But when we restructure the axioms like this, the *Lsubj* and *Lobj* are no longer needed where they are, because the x and y arguments are now available at the lexical level. The axioms become the following:

$$(\forall w_1, w_2, x, e)[Syn(w_1, x, -, -) \wedge Syn(w_2, e, x, -) \supset Syn(w_1w_2, e, -, -)] \qquad (2.13)$$
$$(\forall w_3, w_4, y, e)[Syn(w_3, e, x, y) \wedge Syn(w_4, y, -, -) \supset Syn(w_3w_4, e, x, -)] \qquad (2.14)$$

Rules (2.13) and (2.14) have the form of rule (2.2) shown in Section 2.1.4. Thus, the basic composition rules of syntax can be seen as direct developments from pattern (2.8). These rules describe the structure of syntactic knowledge; they do not presume any particular mode of processing that uses it.

In the full grammar of Hobbs (1998), the arity of *Syn* is increased to include arguments for category and feature information. This is used for expressing agreement and subcategorization constraints that specify what kinds of strings can concatenate with what other kinds of strings in grammatical sentences. For example, the subjects of most verbs have to be headed by nouns. We can equivalently capture these constraints by adding propositions to the antecedents of the composition axioms.

The lexical axioms (2.3) of Section 2.1.4, such as

$$(\forall\, e, x, y)[read'(e, x, y) \land text(y) \supset Syn(\text{"read"}, e, x, y)] \tag{2.15}$$

can be seen as specializations of causal rules that are mutually known in a community of language speakers. Hearing the word "reads" causes one to think of a reading situation. These axioms link predicates in the knowledge base with words in the language.

Metonymy is the device of referring to something by referring to something related to it. It is a pervasive characteristic of discourse. When we say

I've read Shakespeare.

we coerce "Shakespeare" into something that can be read, namely, the writings of Shakespeare. So syntax is a set of constraints on the interpretation of adjacency as predicate–argument relations plus metonymy. Metonymy can be realized formally by the axiom

$$(\forall w, e, x, z)[Syn(w, e, x, -) \land rel(x, z) \supset Syn(w, e, z, -)] \tag{2.16}$$

That is, if w is a string that would describe e providing the subject x of e is found, and rel is some metonymic "coercion" relation between x and z, then w can also be used as a string describing e if a subject describing z is found. Thus, z can stand in for x, as "Shakespeare" stands in for "the writings of Shakespeare." In this example, the metonymic relation rel would be *write*.

Metonymy is probably not a recent development in the evolution of language. Rather it is the most natural starting point for syntax. In many protolanguage utterances, the relation found between adjacent elements involves just such a metonymic coercion. If we combine axioms (2.10) and (2.16) into the single rule

$$(\forall w_3, w_4, y, z, e)[Syn(w_3, e) \land Syn(w_4, y) \land rel(y, z)$$
$$\land\, Lobj(z, e) \supset Syn(w_3 w_4, e)]$$

we can see this as an instantiation of axiom (2.8), where the first two conjuncts in the antecedent correspond to A and B in axiom (2.8), and the last two conjuncts in the antecedent correspond to $rel(x,y,z)$ in axiom (2.8).

In multi-word discourse, when a relation is found to link two words or larger segments into a composite unit, it too can be related to adjacent segments in various ways. The tree structure of sentences arises out of this recursion. Thus, "reads" and "*Hamlet*" concatenate into the segment "reads *Hamlet*," a verb phrase which can then concatenate with "John" to form the sentence "John reads *Hamlet*."

I have illustrated this advance – conveying predicate–argument relations by position – with the crucially important example of clause structure. But a similar story could be told about the equally important internal structure of noun phrases, which conveys a modification relation, a variety of the predicate–argument relation.

This account describes a way in which hominids *could have* progressed from proto-language to simple syntax. The competitive advantage this development confers is clear. When the relation between successive words or larger phrases is always the predicate–argument relation, there is less ambiguity in utterances and therefore more precision, and therefore more complex messages can be constructed. People can thereby engage in more complex joint action.

2.3.3 *Signaling predication and modification*

The languages of the world signal predication primarily by means of position and particles (or affixes). They signal modification primarily by means of adjacency and various concord phenomena. In what has been presented so far, we have seen how predicate–argument relations can be recovered from adjacency. Japanese is a language that conveys predicate–argument relations primarily through postpositional particles, so it will be useful to show how this could have arisen by incremental changes from pattern (2.8) as well. For the purposes of this example, a simplified view of Japanese syntax is sufficient: a verb at the end of the sentence conveys the predicate; the Japanese verb "iki" conveys the predicate *go*. The verb is preceded by some number of postpositional phrases, in any order, where the noun phrase is the argument and the postposition indicates which argument it is; "kara" is a postposition meaning "from," so "Tokyo kara" conveys the information that Tokyo is the *from* argument of the verb.

Signaling predication by postpositions, as does Japanese, can be captured in axioms, specializing and elaborating pattern (2.8) and similar to (2.11), as follows:

$$(\forall w_1, w_2, e, x)[Syn(w_1, x, \mathbf{n}, -) \wedge Syn(w_2, e, \mathbf{p}, x) \supset Syn(w_1 w_2, e, \mathbf{p}, -)]$$
$$(\forall w_3, w_4, e)[Syn(w_3, e, \mathbf{p}, -) \wedge Syn(w_4, e, \mathbf{v}, -) \supset Syn(w_3 w_4, e, \mathbf{v}, -)]$$
$$(\forall e, x)[from(e, x) \supset Syn(\text{"kara"}, e, \mathbf{p}, x)]$$
$$(\forall e)[go(e) \supset Syn(\text{"iki"}, e, \mathbf{v}, -)]$$

The first rule combines a noun phrase and a postposition into a postpositional phrase. The second rule combines a postpositional phrase and a verb into a clause, and permits multiple postpositional phrases to be combined with the verb. The two lexical axioms link Japanese words with underlying world-knowledge predicates.[10] The fourth rule generates a logical form for the verb specifying the type of event it describes. The third rule links that event with the arguments described by the noun phrases via the relation specified by the postposition.

The other means of signaling predication and modification, such as inflection and agreement, can be represented similarly.

Klein and Perdue (1997), cited by Jackendoff (1999), identify features of what they call the Basic Variety in second-language learning, one of the most important of which is

---

[10] Apologies for using English as the language of thought in this example.

the Agent First word order; word order follows causal flow. Once means other than position are developed for signaling predicate–argument relations, various alternations are possible, including passives and the discontinuous elements discussed next, enabling language users to move beyond the Basic Variety.

2.3.4 Discontinuous elements

Consider

John is likely to go.

This is an example of what is called "raising"; the subject "John" is "raised" out of its rightful position as the subject of "go" and made a subject of "is likely." To interpret this, a hearer must find a relation between "John" and "is likely." Syntax says that it should be a predicate–argument relation plus a possible metonymy. The predicate "is likely" requires a proposition or eventuality as its argument, so we must coerce "John" into one. The next phrase "to go" provides the required metonymic coercion function. That John will go is likely. This analysis can be represented formally by the following axiom:

$$(\forall w_3, w_4, e, e_1)[Syn(w_3, e, e_1, -) \wedge Syn(w_4, e_1, x, -) \supset Syn(w_3 w_4, e, x, -)]$$

In our example, $Syn(w_3, e, e_1, -)$ says that the string "likely" (w_3) describes the eventuality of John's going (e_1) being likely (e), provided we can find in the subject position something describing John's going, or at least something coercible into John's going. $Syn(w_4, e_1, x, -)$ says that the string "to go" (w_4) describes the eventuality of John's going (e_1), provided we can find something describing John (x). $Syn(w_4, e_1, x, -)$ is a relation between e_1 and x, and can thus be used to coerce e_1 into x as in axiom (2.16), thereby allowing the subject of "likely" to be John. $Syn(w_3 w_4, e, x, -)$ says that the string "likely to go" ($w_3 w_4$) describes John's going being likely (e) provided we find a subject describing John (x).

John (x) stands in for John's going (e_1) where the relation between the two is provided by the phrase "to go" (w_4). This axiom has the form of axiom (2.16), where the x of (2.16) is e_1 here, the z of (2.16) is x here, and the $rel(x, z)$ of (2.16) is the $Syn(w_4, e_1, x, -)$ in this axiom. (Hobbs (2001) provides numerous examples of phenomena in English that can be analyzed in terms of interactions between syntax and metonymy.)

This locution is then reinterpreted as a modified form of the VP rule (2.14), by altering the first conjunct of the above axiom, giving us the VP rule for "raising" constructions.

$$(\forall w_3, w_4, y, e)[Syn(w_3, e, x, e_1) \wedge Syn(w_4, e_1, x, -) \supset Syn(w_3 w_4, e, x, -)]$$

That is, suppose a string w_3 ("is likely") describes a situation e and is looking for a logical subject referring to x (John) and a logical object referring to e_1 (John's going). Concatenate it with a string w_2 ("to go") describing e_1 and looking for a subject x (John). Then the result describes the situation e provided we can find a logical subject describing x. This rule differs from rule (2.14) only in the second occurrence of x. Now the lexical axiom for "likely" will have e_1 available for its logical subject, and the lexical axiom for "go" will have x available for its logical subject.

Only two more rules like this suffice for all the complements of verbs in English.

This of course is only a plausible analysis of how discontinuous elements in the syntax could have arisen, but in my view the informal part of the analysis is very plausible, since it rests on the very pervasive interpretive move of metonymy. The formal part of the analysis is a direct translation of the informal part into the formal logical framework used here. When we do this translation, we see that the development is a matter of two simple incremental steps – specialization of a predicate (*rel* to *Syn*) and a modification of argument structure – that can be realized through the recruitment of nodes in a structured connectionist model.

2.3.5 Long-distance dependencies

One of the most "advanced" and probably one of the latest universal phenomena of language is long-distance dependencies, as illustrated by relative clauses and wh-questions. They are called *long-distance* dependencies because in principle the head noun can be an argument of a predication that is embedded arbitrarily deeply. In the noun phrase

> *the man John believes Mary said Bill saw*

the man is the logical object of the seeing event, at the third level of embedding.

In accounting for the evolution of long-distance dependencies, we will take our cue from the Japanese. (For the purposes of this example, all one needs to know about Japanese syntax is that relative clauses have the form of clauses placed before the head noun.) It has been argued that the Japanese relative clause is as free as the English compound nominal in its interpretation. All that is required is that there be *some* relation between the situation described by the relative clause and the entity described by the head noun (Akmajian and Kitagawa, 1974; Kameyama, 1994). They cite the following noun phrase as an example.

> *Hanako ga   iede   shita   Taroo*
> *Hanako Subj run-away-from-home did   Taroo*
> *Taroo such that Hanako ran away from home*

Here it is up to the interpreter to find some plausible relation between Taroo and Hanako's running away from home.

We may take Japanese as an example of the basic case. Any relation will explain the adjacency of the relative clause and the noun. In English, a further constraint is added, analogous to the constraint between subject and verb. The relation must be the predicate–argument relation, where the head noun is the argument and the predicate is provided, roughly, by the top-level assertion in the relative clause and its successive clausal complements. Thus, in "the man who John saw," the relation between the man and the seeing event is the predicate–argument relation – the man is the logical object of the seeing. The clause "John saw ( )" has a "gap" in it where the object should be, and that gap is filled by, loosely speaking, "the man." It is thus a specialization of pattern (2.8), and a constraint on the interpretation of the relation *rel* in pattern (2.8).

The constraints in French relative clauses lie somewhere between those of English and Japanese; it is much easier in French for the head to be an argument in an adjunct modifier of a noun in the relative clause. Other languages are more restrictive than English (Comrie, 1981, ch. 7). Russian does not permit the head to be an argument of a clausal complement in the relative clause, and in Malagasy the head must be in subject position in the relative clause.

The English case can be incorporated into the grammar by increasing the arity of the *Syn* predicate, relating strings of words to their meanings. Previously, we had arguments for the string, the entity or situation it described, and the missing logical subject and object. We will increase the arity by one, and add an argument for the entity that will fill the gap in the relative clause. The rules for relative clauses then become

$$(\forall w_1, e_1, x, y)[Syn(w_1, e_1, x, y, -) \land Syn("", y, -, -, -) \supset Syn(w_1 e_1, x, -, y)]$$
$$(\forall w_1, w_2, e_1, y)[Syn(w_1, y, -, -, -) \land Syn(w_2, e, -, -, y) \supset Syn(w_1 w_2, y, -, -, -)]$$

The first rule introduces the gap. It says a string w_1 describing an eventuality e_1 looking for its logical object y can concatenate with the empty string provided the gap is eventually matched with a head describing y. The second rule says, roughly, that a head noun w_1 describing y can concatenate with a relative clause w_2 describing e but having a gap y to form a string $w_1 w_2$ that describes y. The rare reader interested in seeing the details of this treatment should consult Hobbs (1998).

In conversational English one sometimes hears "which" used as a subordinate conjunction, as in *I did terrible on that test, which I don't know if I can recover from it.*

This can be seen as a relaxation of the constraint on English relative clauses, back to the protolanguage pattern of composition.

There are several ways of constructing relative clauses in the world's languages (Comrie, 1981, ch. 7). Some languages, like Japanese, provide no information about which argument in the relative clause, if any, is identical to the head. But in all of the languages that do, this information can be captured in fairly simple axioms similar to those above for English. Essentially, the final argument of *Syn* indicates which argument is to be taken as coreferential with the head, however the language encodes that information.

Relative clauses greatly enhance the possibilities in the modification of linguistic elements used for reference, and thereby enable more complex messages to be communicated. This in turn enhances the possibilities for mutual belief and joint action, conferring an advantage on groups whose language provides this resource.

Seeking a relation between adjacent or proximate words or larger segments in an utterance is simply an instance of seeking explanations for the observables in our environment, specifically, observable relations. Syntax can be seen largely as a set of constraints on such interpretations, primarily constraining the relation to be the predicate–argument relation. In this section I have shown how successively more complex syntax could arise by incremental steps, from protolanguage to simple clausal syntax, to discontinuous elements such as we see in raising constructions, to long-distance dependencies. These incremental changes taking us from the protolanguage pattern (2.8) to complex syntax are of three kinds.

- Specializing predicates that characterize strings of words, as the predicate *Syn* specializes the predicates in pattern (2.8).
- Increasing the arity of the *Syn* predicate, i.e., adding arguments, to transmit arguments from one part of a sentence to another.
- Adding predications to antecedents of rules, e.g., to capture agreement and subcategorization constraints.

The acquisition of syntax, whether in evolution or in development, can be seen as the incremental accumulation of such constraints.

As mentioned above, the particular treatment of syntax used here closely follows that of Pollard and Sag (1994). They go to great efforts to show the equivalence of their Head-driven Phrase Structure Grammar to the Government and Binding theories of Chomsky (1981) then current, and out of which the more recent Minimalist theory of Chomsky (1995) has grown. It is often difficult for a computational linguist to see how Chomsky's theories could be realized computationally, and a corollary of that is that it is difficult to see how one could construct an incremental, computational account of the evolution of the linguistic capacity. By contrast, the unification grammar used by Pollard and Sag is transparently computational, and, as I have shown in this section, one can construct a compelling plausible story about the incremental development of the capacity for syntax. Because of the work Pollard and Sag have done in relating Chomsky's theories to their own, the account given in this chapter can be seen as a counterargument to a position that "Universal Grammar" had to have evolved as a whole, rather than incrementally. Jackendoff (1999) also presents compelling arguments for the incremental evolution of the language capacity, from a linguist's perspective.

2.4 Remarks on the course of the evolution of language

Relevant dates in the time course of the evolution of language and language readiness are as follows:

1. Mammalian dominance: $c.65$–50 M years ago
2. Common ancestor of monkeys and great apes: $c.15$ M years ago
3. Common ancestor of hominids and chimpanzees: $c.5$ M years ago
4. Appearance of *Homo erectus*: $c.1.5$ M years ago
5. Appearance of *Homo sapiens sapiens*: $c.200$–150 K years ago
6. African/non-African split: $c.$ 90 K years ago

7. Appearance of preserved symbolic artifacts: *c*.70–40 K years ago
8. Time depth of language reconstruction: *c*.10 K years ago
9. Historical evidence: *c*.5 K years ago

In this section I will speculate about the times at which various components of language, as explicated in this chapter, evolved. I will then discuss two issues that have some prominence in this volume:

1. Was there a holophrastic stage before fully modern language? This is a question, probably, about the period just before the evolution of *Homo sapiens sapiens*.
2. When *Homo sapiens sapiens* evolved, did they have fully modern language or merely language readiness? This is a question about the period between the evolution of *Homo sapiens sapiens* and the appearance of preserved symbolic material culture.

2.4.1 Evolution of the components of language

Language is generally thought of as having three parts: phonology, syntax, and semantics. Language maps between sound (phonology) and meaning (semantics), and syntax provides the means for composing elementary mappings into complex mappings. The evolution of the components of language is illustrated in Fig. 2.3.

According to Arbib (Chapter 1, this volume), gestural communication led to vocal communication, which is phonology. As Emmorey (this volume) stresses, phonology can be defined in the manual domain as well. This arrow in the figure needs a bit of explication. Probably gesture and vocal communication have always both been there, as envisaged in Arbib's "expanding spiral." It is very hard to imagine that there was a stage in hominid evolution when individuals sat quietly and communicated to each other by gesture, or a stage when they sat with their arms inert at their sides and chattered with each other. Each modality, as Goldstein, Byrd, and Saltzman (this volume) point out, has its advantages. In some situations gestural communication would have been the most appropriate, and in others vocal communication. As Arbib points out, language developed in the region of the brain that had originally been associated with gesture and hand manipulations. A likely scenario is that there was a stage when manual gestures were the more expressive system. Articulatory gestures co-opted that region of the brain, and eventually became a more expressive system than the manual gestures.

In my view, the ability to understand composite event structure is a precursor to protolanguage, because protolanguage, in Bickerton's sense, requires one to recover the relation between two elements. In protolanguage two-word utterances *are* composite events. Protolanguage led to syntax, as I argue in Sections 2.3.1 and 2.3.2. For Arbib, composite event structure is part of the transition from protolanguage to language; for him this transition is post-biological. But he uses the term "protolanguage" to describe a holophrastic stage, which differs from the way Bickerton and I use the term.

According to the account in Section 2.2 of this chapter, the use of causal associations was the first requirement for the development of semantics. Causal association is possible

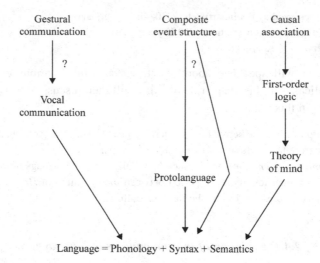

Figure 2.3 Evolution of the components of language.

in a brain that does the equivalent of propositional logic (such as most current neural models), but before one can have a theory of mind, one must have the equivalent of first-order logic. A creature must be able to distinguish different tokens of the same type. The last requirement is the development of a theory of mind, including models of belief, mutual belief, goals, and plans.

Arbib (2002, 2005) lists a number of features of language that have to evolve before we can say that fully modern language has evolved. It is useful to point out what elements in Fig. 2.3 are necessary to support each of these features. *Naming*, or rather the interpretation of names, requires only causal association. A causal link must be set up between a sound and a physical entity in the world. Dogs and parrots can do this. Production in naming, by contrast, requires a theory of others' beliefs. *Parity* between comprehension and production requires a theory of the mind of others. The *"Beyond here and now"* feature also requires a theory of mind; one function of belief, it was pointed out, is to transmit causality to other times and places. *Hierarchical structure* first appears with composite event structure. Once there is protolanguage, in the sense in which I am using the term, there is a *lexicon* in the true sense. The significance of *temporal order* of elements in a message begins somewhere between the development of protolanguage and real syntax. *Learnability*, that is, the ability of individuals to acquire capabilities that are not genetically hard-wired, is not necessary for causal association or first-order logic, but is very probable in all the other elements of Fig. 2.3.

Causal associations are possible from at least the earliest stages of multicellular life. A leech that moves up a heat gradient and attempts to bite when it encounters an object is responding to a causal regularity in the world. Of course, it does not know that it is responding to causal regularities; that would require a theory of mind. But the causal associations themselves are very early. The naming that this capability enables is quite within the capability of parrots, for example. Thus, in Fig. 2.3, we can say that causal association is pre-mammalian.

At what point are animals aware of different types of the same token? At what point do they behave as if their knowledge is encoded in a way that involves variables that can have multiple instantiations? That is, at what point are they first-order? My purely speculative guess would be that it happens early in mammalian evolution. Reptiles and birds have an automaton-like quality associated with propositional representations, but most mammals that I am at all familiar with, across a wide range of genera, exhibit a flexibility of behavior that would require different responses to different tokens of the same type. Jackendoff (1999) points out that in the ape language-training experiments, the animals are able to distinguish between "symbols for individuals (proper names) and symbols for categories (common nouns)" (p.273), an ability that would seem to require something like variable binding.

One reason to be excited about the discovery of the mirror neuron system (Rizzolatti and Arbib, 1998) is that it is evidence of an internal representation that abstracts away from a concept's role in perception or action, and thus is possibly an early solid indication of "first-order" features in the evolution of the brain.

Gestural communication, composite event structure, and a theory of mind probably appear somewhere between the separation of great apes and monkeys, and the first hominids, between 15 and 5 million years ago. Arbib discusses the recognition and repetition of composite events. There are numerous studies of the gestural communication that the great apes can perform. The evolution of the theory of mind is very controversial (e.g., Heyes, 1998; Stanford, this volume), but it has certainly been argued by some that chimpanzees have some form of a theory of mind. It is a clear advantage in a social animal to have a theory of others' behavior.

These three features can thus probably be assigned to the pre-hominid era. My, again purely speculative, guess would be that vocal communication (beyond alarm cries) emerged with *Homo erectus*, and I would furthermore guess that they were capable of protolanguage – that is, stringing together a few words or signals to convey novel though not very complex messages. The components of language readiness constitute a rich system and protolanguage would confer a substantial advantage; these propositions accord with the facts that *Homo erectus* was the dominant hominid for a million years, was apparently the first to spread beyond Africa, and was the stock out of which *Homo sapiens sapiens* was to evolve. It may also be possible to adduce genetic and anatomical evidence. It is impossible to say how large their lexicon would have been, although it might be possible to estimate on the basis of their life style.

Finally, fully modern language probably emerged simultaneously with *Homo sapiens sapiens*, and is what gave us a competitive advantage over our hominid cousins. We were able to construct more complex messages and therefore were able to carry out more complex joint action. As Dunbar (1996) has argued, fully modern language would have allowed us to maintain much larger social groups, a distinct evolutionary advantage.

A word on the adaptiveness of language: I have heard people debate whether language for hunting or language for social networks came first, and provided the impetus for language evolution. (We can think of these positions as the Mars and Venus theories.) This is a granularity mismatch. Language capabilities evolved over hundreds or thousands of generations, whereas hunting and currying social networks are daily activities. It thus seems highly implausible that there was a time when some form of language precursor was used for hunting but not for social networking, or vice versa. The obvious truth is that language is for establishing and otherwise manipulating mutual belief, enabling joint action, and that would be a distinct advantage for both hunting and for building social networks.

2.4.2 A holophrastic stage?

Wray (1998) proposes a picture of one stage of the evolution of language that is somewhat at odds with the position I espouse in Section 2.3, and it therefore merits examination here. She argues that there was a holophrastic stage in the evolution of language. First there were utterances – call them protowords – that denoted situations but were not broken down into words as we know them. These protowords became more and more complex as the lexicon expanded, and they described more and more complex situations. This is the holophrastic stage. Then these protowords were analyzed into parts, which became the constituents of phrases. One of her examples is this: Suppose by chance "mebita" is the protoword for "give her the food," and "kameti" is the protoword for "give her the stone." The occurrence of "me" in both is noted and is then taken to represent a singular female recipient.

Jackendoff (1999) points out that one of the important advances leading to language was the analysis of words into individual syllables and then into individual phonemes, providing an inventory out of which new words can be constructed. It is very likely that this happened by some sort of holophrastic process. We first have unanalyzed utterances "pig" and "pit" and we then analyze them into the sounds of /p/, /i/, /g/, and /t/, and realize that further words can be built out of these elements. This process, however, is much more plausible as an account of the evolution of phonology than it is of the evolution of syntax. The phonological system is much simpler, having many fewer elements, and phonemes have no semantics to overconstrain decompositions, as words do.

Wray says, "There is a world of difference between picking out the odd word, and forcing an entire inventory of arbitrary phonetic sequences representing utterances through a complete and successful analysis." (p.57) Indeed there is a world of difference. The latter problem is massively overconstrained, and a solution is surely mathematically impossible, as a synchronic process done all at once on an entire language. This is true even if the requirement of "complete and successful" is relaxed somewhat, as Wray goes on to do. The only way I could imagine such a development would be if the individuals were generating the protowords according to some implicit morphology, and the analysis was in fact a discovery of this morphology. If children go through such a process, this is the reason it is possible. They are discovering the syntax of adult language.

Kirby (2000) and Kirby and Christiansen (2003) consider the problem dynamically and argue that accidental regularities will be the most stable parts of a language as it is transmitted from one generation to the next, and that this stable, regular core of the language will gradually expand to encompass most of the language. Composition evolves because the learning system learns the composite structure of the underlying meanings. This is mathematically possible providing the right assumptions are made about the structure of meaning and about how partially novel meanings are encoded in language. But I think it is not very compelling, because such processes are so marginal in modern language and because the "composition via discourse" account, articulated in Section 2.3 and summarized below, provides a much more efficient route to composition.

Holophrases are of course a significant factor in modern adult language, for example, in idioms. But by and large, these have historical compositional origins (including "by and large"). In any specific example, words came first, then the composition, then the holophrase, the opposite of Wray's proposed course of language evolution. There is in language change the phenomenon of morphological reanalysis, as when we reanalyze the "-holic" in "alcoholic" to mean "addicted to" and coin words like "chocoholic." It is very much rarer to do this reanalysis because of an accidental co-occurrence of meaning. Thus, the co-occurrence of "ham" in the words "ham" and "hamburger" may have led to a reanalysis that results in words like "steakburger," "chickenburger," "soyburger," and so on, and the "-s" at the end of "pease" was reanalyzed into the plural morpheme. But this is simply not a very productive process, in contrast with "composition via discourse."

A holophrastic stage has sometimes been hypothesized in child language. Children go through a one-word stage followed by a two-word stage. The holophrastic stage would be between these two. The evidence is from "words" like "allgone," "whazzat," and "gimme." An alternative explanation for these holophrases is that the child has failed to segment the string, due to insufficient segmentation ability, insufficient vocabulary, insufficient contrastive data, and so on. For a holophrastic stage to exist, we would have to show that such holophrases don't occur in the one-word stage, and I know of no evidence in support of this.

In any case, children have models whose language is substantially in advance of their own. That was never the case in language evolution. Holophrasis in child language is a misanalysis. There was nothing for holophrasis in language evolution to be a misanalysis of.

A possible interpretation of Wray's position is that originally, in evolution and in development, protowords only describe situations. Thus, a baby's "milk" might always describe the situation "I want milk." At a later stage, situations are analyzed into objects and the actions performed on them; language is analyzed into its referential and predicational functions; the lexicon is analyzed into nouns and verbs. This then makes possible the two-word stage. I take Arbib (Chapter 1, this volume) to be arguing for something like this position. I do not find this implausible, although the evidence for it is unclear. The (controversial) predominance of nouns labeling objects in children's one-word stage would seem a counterindication, but perhaps those nouns originally denote situations for the child. But I read Wray as saying there is a further analysis of protowords describing situations into their protoword parts describing objects and actions, and this seems to me quite implausible for the reasons stated. One can argue that a holophrastic stage *precedes* the one-word stage, but then it plays no role in the development of a two-word stage or the subsequent development of syntax. That happens by composition of words.

I believe the coherence structure of discourse (e.g., Hobbs, 1985b) provides a more compelling account of the evolution of the sentence. Discourse and interaction precede language. Exchanges and other reciprocal behavior can be viewed as a kind of proto-discourse. Events in the world and in discourse cohere because they stand in coherence relations with each other. Among the relations are:

causality:
   "Smoke. Fire."
similarity:
   I signal that I go around to the right. I signal that you go around to the left.
ground-figure:
   "Bushes. Tiger."
occasion, or the next step in the process:
   You hand me grain. I grind it.
   "Approach antelope. Throw spear."
   "Scalpel. Sponge."[11]
and the predicate–argument or argument–predicate relation:
   "Sock. On."
   "Antelope. Kill."
   I point to myself. I point to the right.

While the evidence for a holophrastic stage in children's language development is scant, there *is* a stage that does often precede the two-word stage. Scollon (1979) and others have noted the existence of what have been called "vertical constructions". Children convey a two-concept message by successive one-word utterances, each with sentence intonation, and often with some time and some interaction between them. Hoff (2001, p.210) quotes a child near the end of the one-word stage saying, "Ow. Eye." Scollon reports a similar sequence: "Car. Go." In both of these examples, the adjacency conveys a predicate–argument relation.

It seems much more likely to me that the road to syntax was via coherence relations between successive one-word utterances, as described in Section 2.3, rather than via holo-phrasis. The coherence account requires no new mechanisms. It is just a matter of adding

[11] To pick a modern example.

constraints on the interpretation of temporal order as indicating predicate–argument relations. Construction is more plausible than deconstruction.

I think Wray exaggerates the importance of grammar in communication. She says, "Successful linguistic *comprehension* requires grammar, even if the *production* were to be grammarless. A language that lacks sufficient lexical items and grammatical relations can only hint at explicit meaning, once more than one word at a time is involved. " (pp.48–49) The problem with this statement is that discourse today has no strict syntax of the sort that a sentence has, and we do just fine in comprehending it. In a sense, discourse is still in the protolanguage stage. The adjacency of segments in discourse tells hearers to figure out a relation between the segments, and normally hearers do, using what they know of context.

Context has always been central in communication. The earliest utterances were one more bit of information added to the mass of information available in the environment. In the earliest discourse, understanding the relation between utterances was part of arriving at a coherent picture of the environment. The power of syntax in modern language, as Wray points out, is to constrain interpretations and thereby lessen the burden placed on context for interpretation and to enable the construction of more complex messages, culminating in communicative artifacts cut free from physical copresence and conveying very complex messages indeed, such as this book. But there was never a point at which situations involving more than one communicative act would have been uninterpretable.

Bickerton (2003) gives further persuasive arguments against a holophrastic stage in language evolution.

A succinct though perhaps crude formulation of my position is that it is more plausible that the sentence "Lions attack." derived from a discourse "Lions. Attack." than from a word "Lionsattack."

2.4.3 Language or language readiness?

Arbib (Chapter 1, this volume) expresses his belief that the first physically modern *Homo sapiens sapiens* did not have language, only language readiness. This is a not uncommon opinion. In most such accounts, language is a cultural development that happened with the appearance of preserved symbolic artifacts, and the date one most often hears is around 70 000 to 35 000 years ago. In one possible version of this account, anatomically modern humans of 150 000 years ago were language ready, but they did not yet have language. Language was a cultural achievement over the next 100 000 years, that somehow coincided with the species' spread over the globe.

Davidson (2003) presents a careful and sophisticated version of this argument. He argues, or at least suggests, that symbols are necessary before syntax can evolve, that surviving symbolic artifacts are the best evidence of a capacity for symbolism, and that there is no good evidence for symbolic artifacts or other symbolic behavior before 70 000 years ago in Africa and 40 000 years ago elsewhere, nor for symbolic behavior in any species other than *Homo sapiens sapiens*. (For example, he debunks reports of burials among Neanderthals.)

Although Davidson is careful about drawing it, the implication is if *Homo sapiens sapiens* evolved around 200 000 years ago and did not engage in symbolic behavior until 70 000 years ago, and if language is subsequent to that, then language must be a cultural rather than a biological development. (However, Davidson also casts doubt on the idea that we can tell very much about cognition from fossils.)

One problem with such arguments is that they are one bit of graffiti away from refutation. The discovery of one symbolic artifact could push our estimates of the origin of symbolic behavior substantially closer to the appearance of *Homo sapiens sapiens*, or before. Barber and Peters (1992) gave 40 000 to 35 000 years ago as the date at which humans had to have had syntax, on the basis of symbolic artifacts found up to that point. Davidson pushes that back to 70 000 years ago because of ochre found recently at a South African site and presumed to be used for bodily decoration. There have been a spate of recent discoveries of possible artifacts with possible symbolic significance. Two ochre plaques engraved with a criss-cross pattern, with no apparent non-symbolic utility, dated to 75 000 years ago, were found at Blombos Cave in South Africa (Henshilwood *et al.*, 2002). Pierced shells claimed to have been used as beads and dated to 75 000 years ago were found at the same site (Henshilwood *et al.*, 2004). Rocks stained with red ochre and believed to be used in burial practices were found in Qafzeh Cave in Israel (Hovers *et al.*, 2003); they were dated to 100 000 years ago. In northern Spain a single finely crafted pink stone axe was found in association with the fossilized bones of 27 *Homo heidelbergensis* individuals and is claimed as evidence for funeral rites; this site dates to 350 000 years ago (Carbonell *et al.*, 2003). A 400 000-year-old stone object which is claimed to have been sculpted into a crude human figurine was found in 1999 near the town of Tan-Tan in Morocco (Bednarik, 2003). All of these finds are controversial, and the older the objects are purported to be, the more controversial they are. Nevertheless, they illustrate the perils of drawing conclusions about language evolution from the surviving symbolic artifacts that we so far have found.

The reason for attempting to draw conclusions about language from symbolic artifacts is that they (along with skull size and shape) constitute the only archeological evidence that is remotely relevant. However, I believe it *is* only remotely relevant. *Homo sapiens sapiens* could have had language for a long time before producing symbolic artifacts. After all, children have language for a long long time before they are able to produce objects capable of lasting for tens of thousands of years. We know well that seemingly simple achievements are hard won. Corresponding to Arbib's concept of *language* readiness, we may hypothesize something called *culture* readiness (or more properly, *symbolic material culture* readiness). Symbolic material culture with some permanence may not have happened until 75 000 years ago, but from the beginning of our species we had *culture* readiness. The most reasonable position is that language is not identical to symbolic culture. Rather it is a component of culture readiness. As Bickerton (2003) puts it, "syntacticized language enables but it does not compel." (p.92)

One reservation should be stated here. It is possible that non-African humans today are not descendants of the *Homo sapiens sapiens* who occupied the Middle East 100 000 to

90 000 years ago. It is possible rather that some subsequent stress, such as glaciation or a massive volcanic eruption, created a demographic bottleneck that would enable further biological evolution, yielding an anatomically similar *Homo sapiens sapiens*, who however now had fully modern cognitive capacities, and that today's human population is all descended from that group. In that case, we would have to move the date for fully modern language forward, but the basic features of modern language would still be a biological rather than a cultural achievement.

I think the strongest argument for the position that fully modern language, rather than mere language readiness, was already in the possession of the earliest *Homo sapiens sapiens* comes from language universals. In some scholarly communities it is fashionable to emphasize how few language universals there are; Tomasello (2003), for example, begins his argument for the cultural evolution of language by emphasizing the diversity of languages and minimizing their common core. In other communities the opposite is the case; followers of Chomsky (e.g., 1975, 1981), for example, take it as one of the principal tasks of linguistics to elucidate Universal Grammar, that biologically based linguistic capability all modern humans have, including some very specific principles and constraints. Regardless of these differing perspectives, it is undeniable that the following features of language, among others, are universal:

- All languages encode predicate–argument relations and assertion-modification distinctions by means of word order and/or particles/inflection.
- All languages have verbs, nouns, and other words.
- All languages can convey multiple propositions in single clauses, some referential and some assertional.
- All languages have relative clauses (or other subordinate constructions that can function as relative clauses).
- Many words have associated, grammatically realized nuances of meaning, like tense, aspect, number, and gender, and in every language verbs are the most highly developed in this regard, followed by nouns, followed by the other words.
- All languages have anaphoric expressions.

These universal features of language may seem inevitable to us, but we know from formal language theory and logic that information can be conveyed in a very wide variety of ways. After the African/non-African split 100 000 to 90 000 years ago, uniform diffusion of features of language would have been impossible. It is unlikely that distant groups not in contact would have evolved language in precisely the same way. That means that the language universals were almost surely characteristic of the languages of early *Homo sapiens sapiens*, before the African/non-African split.

It may seem as if there are wildly different ways of realizing, for example, relative clauses. But from Comrie (1981) we can see that there are basically two types of relative clause – those that are adjacent to their heads and those that replace their heads (the internal-head type). The approach of Section 2.3.4 handles both with minor modifications of axioms using the same predicate *Syn*; at a deep level both types pose the problem of indicating what the head is and what role it plays in the relative clause, and the solutions rely on the same underlying machinery. In any case, there is no geographical coherence to the distribution of these two types that one would expect if relative clauses were a cultural development.

It is possible in principle that linguistic universals are the result of convergent evolution, perhaps with some diffusion, due to similar prelinguistic cognitive architecture and similar pressures. But to assess its plausibility, let's consider the case of technology. All cultures build their technologies with the same human brain, in response to very similar environmental challenges, using very similar materials. We know that technologies diffuse widely. Yet there have been huge differences in the level of technological development of various cultures in historical times. If the arguments for convergent evolution work anywhere, they should work for the evolution of technology. But they don't. Technological universals don't even begin to characterize the range of human technologies. It is clear that the original *Homo sapiens sapiens* were technology *ready* and that the development of fully modern technology was a subsequent cultural development. The situation with language is very different. We don't observe that level of variation.

There are some features of language that may indeed be a cultural development. These are features that, though widespread, are not universal, and tend to exhibit areal patterns. For example, I would be prepared to believe that such phenomena as gender, shape classifiers, and definiteness developed subsequently to the basic features of language, although I know of no evidence either way on this issue.

There are also areas of language that are quite clearly relatively recent cultural inventions. These include the grammar of numbers, of clock and calendar terms, and of personal names in various cultures, and the language of mathematics. These tend to have a very different character than we see in the older parts of language; they tend to be of a simpler, more regular structure.

If language were more recent than the African/non-African split, we would expect to see a great many features that only African languages have and a great many features that only non-African languages have. If, for example, only African languages had relative clauses, or if all African languages were VSO (i.e., the Verb precedes the Subject which precedes the Object) while all non-African languages were SVO (Subject–Verb–Object), then we could argue that they must have evolved separately, and more recently than 90 000 years ago. But in fact nothing of the sort is the case. There are very few phenomena that occur only in African languages, and they are not widespread even in Africa, and are rather peripheral features of language; among these very few features are clicks in the phonology (although these may also occur in some Australian mother-in-law languages) and logophoric pronouns, i.e., special forms of pronouns in complements to cognitive verbs that refer to the cognizer. There are also very few features that occur only in non-African languages. Object-initial word order may be one of them (although there may be such a language in Africa as well). These features are also not very widespread.[12]

---

[12] I am indebted to Chris Culy and Bill Croft for discussions about the material in this paragraph.

Finally, if language were a cultural achievement, rather than a biological achievement, we would expect to see significant developments in language in the era that we have more immediate access to, the last five or ten thousand years. For example, it might be that languages were becoming more efficient, more learnable, or more expressive in historical times. As a native English speaker, I might cite a trend from inflection and case markings to encode predicate–argument relations to word order for the same purpose. But in fact linguists detect no such trend. Moreover, we would expect to observe some unevenness in how advanced the various languages of the world are, as is the case with technology. Within the last century there have been numerous discoveries of relatively isolated groups with a more primitive material culture than ours. There have been no discoveries of isolated groups with a more primitive language.

I am not exactly appealing to monogenesis as an explanation. There may have been no time at which all *Homo sapiens sapiens* spoke the same language, although evolution generally happens in small populations. Rather I am arguing that language capacity and language use evolved in tandem, with the evolution of language capacity driven, through incremental stages like the ones proposed in this chapter, by language use. It is most likely that the apperance of fully modern language was contemporaneous with the appearance of anatomically modern humans, and that the basic features of language are not a cultural acquisition subsequent to the appearance and dispersion of *Homo sapiens sapiens*. On the contrary, fully modern language has very likely been, more than anything else, what made us human right from the beginning of the history of our species.

Acknowledgments

This chapter is an expansion of a talk I gave at the Meeting of the Language Origins Society in Berkeley, California, in July 1994. The original key ideas arose out of discussions I had with Jon Oberlander, Mark Johnson, Megumi Kameyama, and Ivan Sag. I have profited more recently from discussions with Lokendra Shastri, Bill Croft, Chris Culy, Cynthia Hagstrom, and Srini Narayanan, and with Michael Arbib, Dani Byrd, Andrew Gordon, and the other members of Michael Arbib's language evolution study group. Michael Arbib's comments on the original draft of this chapter have been especially valuable in strengthening its arguments. I have also profited from comments by Simon Kirby, Iain Davidson, and an anonymous reviewer of this chapter. None of these people would necessarily agree with anything I have said.

References

Akmajian, A., and Kitagawa, C. 1974. *Pronominalization, Relativization, and Thematization: Interrelated systems of Coreference in Japanese and English.* Bloomington, IN: Indiana University Linguistics Club.

Arbib, M. A., 2002. The mirror system, imitation, and the evolution of language. In C. Nehaniv and K. Dautenhahn (eds.) *Imitation in Animals and Artifacts*. Cambridge, MA: MIT Press, pp. 229–280.
  2005. From monkey-like action recognition to human language: an evolutionary framework for neurolinguistics. *Behavi. Brain Sci.* **28**: 105–167.
Barber, E. J. W., and Peters A. M. W., 1992. Ontogeny and phylogeny: what child language and archaeology have to say to each other. In J. A. Hawkins and M. Gell-Mann (eds.) *The Evolution of Human Languages*. Reading, MA: Addison-Wesley, pp. 305–352.
Bednarik, R. G., 2003. A figurine from the African Acheulian. *Curr. Anthropol.* **44**: 405–412.
Bickerton, D., 1990. *Language and Species*. Chicago, IL: University of Chicago Press.
  2003. Symbol and structure: a comprehensive framework for language evolution. In M. H. Christiansen and S. Kirby (eds.) *Language Evolution*. Oxford, UK: Oxford University Press, pp. 77–93.
Carbonell, E., Mosquera, M., Ollé, A., *et al.*, 2003. Les premiers comportements funéraires auraient-ils pris place à Atapuerca, il y a 350 000 ans? *L'Anthropologie* **107**: 1–14.
Chomsky, N., 1975. *Reflections on Language*. New York: Pantheon Books.
  1981. *Lectures on Government and Binding*. Dordrecht, Netherlands: Foris.
  1995. *The Minimalist Program*. Cambridge, MA: MIT Press.
Clark, H., 1975. Bridging. In *Proceedings of Conference on Theoretical Issues in Natural Language Processing*. Cambridge, MA: pp. 169–174.
Comrie, B., 1981. *Language Universals and Linguistic Typology*. Chicago, IL: University of Chicago Press.
Davidson, I., 2003. The archaeological evidence for language origins: states of art. In M. H. Christiansen and S. Kirby (eds.) *Language Evolution*. Oxford, UK: Oxford University Press, pp. 140–157.
Dunbar, R., 1996. *Grooming, Gossip and the Evolution of Language*. London: Faber and Faber.
Engel, A. K., and Singer, W., 2001. Temporal binding and the neural correlates of sensory awareness. *Trends Cogn. Sci.* **5**: 16–25.
Fell, J., Klaver, P., Lehnertz, K., *et al.*, 2001. Human memory formation is accompanied by rhinal-hippocampal coupling and decoupling. *Nature Neurosci.* **4**: pp. 1259–1264.
Fikes, R., and Nilsson, N. J., 1971. STRIPS: a new approach to the application of theorem proving to problem solving. *Artif. Intelli.* **2**: 189–208.
Grice, P., 1957. Meaning. *Philos. Rev.* **66**: 377–388.
Henshilwood, C., d'Errico, F., Yates, R., *et al.*, 2002. Emergence of modern human behavior: Middle Stone Age engravings from South Africa. *Science* **295**: 1278–1280.
Henshilwood, C., d'Errico F., Vanhaeren, M., van Niekirk, K., and Jacobs, Z., 2004. Middle Stone Age shell beads from South Africa. *Science* **304**: 404.
Heyes, C. M., 1998. Theory of mind in nonhuman primates. *Behav. Brain Sci.* **21**: 101–148.
Hobbs, J. R., 1985a. Ontological promiscuity. *Proceedings*, 25th *Annual Meeting of the Association for Computational Linguistics*, Chicago, IL, July 1985, pp. 61–69.
  1985b. *On the Coherence and Structure of Discourse*, Report No. CSLI-85–37. Stanford University, CA: Center for the Study of Language and Information.

1998. The syntax of English in an abductive framework. Available at http://www.isi. edu/~hobbs/discourse-inference/chapter4.pdf

2001. Syntax and metonymy. In P. Bouillon and F. Busa (eds.) *The Language of Word Meaning*. Cambridge, UK: Cambridge University Press, pp. 290–311.

Hobbs, J. R., Stickel, M., Appelt, D., and Martin, P., 1993. Interpretation as abduction. *Artif. Intell.* **63**: 69–142.

Hoff, E., 2001. *Language Development*. Belmont, CA: Wadsworth.

Hovers, E., Ilani, S., Bar-Yosef, O., and Vandermeersch, B., 2003. An early case of color symbolism: ochre use by modern humans in Qafzeh Cave. *Curr. Anthropol.* **44**: 491–522.

Jackendoff, R., 1999. Possible stages in the evolution of the language capacity. *Trends Cogn. Sci.* **3**: 272–279.

Kameyama, M., 1994. The syntax and semantics of the Japanese Language Engine. In R. Mazuka and N. Nagai (eds.) *Japanese Syntactic Processing*. Hillsdale, NJ: Lawrence Erlbaum.

Kirby, S., 2000. Syntax without natural selection: how compositionality emerges from vocabulary in a population of learners. In C. Knight, M. Studdert-Kennedy, and J. R. Hurford (eds.) *The Evolutionary Emergence of Language: Social Function and the Emergence of Linguistic Form*. Cambridge, UK: Cambridge University Press, pp. 303–323.

Kirby, S., and Christiansen, M. H., 2003. From language learning to language evolution. In M. H. Christiansen and S. Kirby (eds.) *Language Evolution*. Oxford, UK: Oxford University Press, pp. 272–294.

Klein, W., and Perdue, C., 1997. The Basic Variety, or couldn't language be much simpler? *Second Lang. Res.* **13**: 301–347.

Peirce, C. S., 1903. Abduction and induction. In J. Buchler (ed.) *Philosophical Writings of Pierce*. New York: Dover Publications, 1955, pp. 150–156.

Pollard, C., and Sag, I. A., 1994. *Head-Driven Phrase Structure Grammar*. Chicago, IL: University of Chicago Press.

Premack, D., and Woodruff, G., 1978. Does the chimpanzee have a theory of mind? *Behavi. Brain Sci.* **1**: 515–526.

Rizzolatti, G., and Arbib, M. A., 1998. Language within our grasp. *Trends Neurosci.* **21**: 188–194.

Scollon, R., 1979. A real early stage: an unzippered condensation of a dissertation on child language. In E. Ochs and B. B. Schiefielin (eds.) *Developmental Pragmatics* New York: Academic Press, pp. 215–227.

Shastri, L., 1999. Advances in SHRUTI: a neurally motivated model of relational knowledge representation and rapid inference using temporal synchrony. *Appl. Intell.* **11**: 79–108.

2001. Biological grounding of recruitment learning and vicinal algorithms in long-term potentiation. In J. Austin, S. Wermter, and D. Willshaw (eds.) *Emergent Neural Computational Architectures Based on Neuroscience*. Berlin: Springer-Verlag, pp. 348–367.

Shastri, L., and Ajjanagadde, V., 1993. From simple associations to systematic reasoning: a connectionist representation of rules, variables and dynamic bindings using temporal synchrony. *Behavi. Brain Sci.* **16**: 417–494.

Shastri, L., and Wendelken, C., 2003. Learning structured representations. *Neurocomputing* **52–54**: 363–370.

Tomasello, M., 2003. On the different origins of symbols and grammar. In M. H. Christiansen and S. Kilby (eds.) *Language Evolution*. Oxford, UK: Oxford University Press, pp. 94–110.

Wason, P. C., and Johnson-Laird, P., 1972. *Psychology of Reasoning: Structure and Content*. Cambridge, MA: Harvard University Press.

Wendelken, C., and Shastri, L., 2003. Acquisition of concepts and causal rules in SHRUTI. *Proceedings, 25th Annual Conference of the Cognitive Science Society*, Boston, MA, pp. 1224–1229.

Wray, A., 1998. Protolanguage as a holistic system for social interaction. *Lang. Commun.* **18**, 47–67.

Part II

Brain, evolution, and comparative analysis

3

Cognition, imitation, and culture in the great apes

Craig B. Stanford

3.1 Introduction: The problem of reconstructing cognition

Most psychologists and cognitive scientists and almost all linguists investigate the nature of the mind, intelligence, and language from a uniquely human perspective. But the origins of these attributes lie in the deep history of the primate brain, and all organic explanations for them must derive from non-human and human primate evolutionary history. There are a variety of approaches to reconstructing the origins of cognition and language. The fossil record is the only direct evidence of human evolution, but it is fragmentary. It can tell us little about the roots of cognition and language because the soft tissues involved do not fossilize.

Our understanding of human evolution is therefore greatly enhanced by evidence from living primates, especially those most closely related to us, the great apes – the chimpanzee (*Pan troglodytes*), bonobo (*P. paniscus*), gorilla (*Gorilla gorilla*), and orangutan (*Pongo pygmaeus*). While we must be careful not to consider apes as living fossils, they are exemplars of the way in which natural selection molds large-brained, social primates to the natural environment. They exhibit learned, interpopulational culturally varied behavior beyond that of any animal other than ourselves. They provide us with a sense of the likeliest range of options, behavioral and cognitive, that natural selection would have taken with ancestral forms of humans.

As an anthropologist, I study non-human primates because they inform us about the likely course that the evolution of human form and function has taken. Non-human primates share a high percentage of their DNA sequence with us, chimpanzees and humans having diverged from the same ancestral form only in the past 6 million or so years. Moreover, some of our closest non-human primate relatives live today in habitats very similar to those in which our own direct hominid ancestors evolved. These factors make non-human primates, and apes in particular, highly informative and important reference points for human evolution.

Action to Language via the Mirror Neuron System, ed. Michael A. Arbib. Published by Cambridge University Press.
© Cambridge University Press 2006.

In this chapter I consider aspects of the cognitive ecology of the great apes, with an emphasis on what the animals do in the natural habitat. In particular, I examine evidence for learned cultural behavior, including imitative behavior, and the nature of intelligence and culture in a human evolutionary framework. I also describe the way in which cognition is on display in the meat-eating and meat-sharing behavior of wild chimpanzees.

3.2 What is intelligence?

Intelligence is generally regarded as a prerequisite for learned, cultural behavior, including imitative behavior. Definitions of intelligence vary widely, but three schools of thought currently prevail.

3.2.1 Technical intelligence

Technical intelligence refers to the ability of some non-human primates to use cognitive skills to extract food and other resources from their natural environment by modifying that environment. Because we regard the advent of tool use to be a major hallmark of the rise of human intellect, anthropologists feel tool use by other primates offers intriguing clues about the evolution of intelligence and of culture. Darwin (1871) first postulated a connection between intelligence and technologically oriented cognition. He saw a feed-back loop involving bipedal posture, a freeing of the hands for tool manufacture and use, and a consequent evolutionary premium placed on neocortical reorganization and expansion. Darwin was wrong, because we now know these three events – the evolution of upright posture, the advent of stone tool use, and the expansion of the neocortex – were separated in time by millions of years. Bipedalism arose at least 5 million years ago; tool use 2.5 million years later, and the major increase in brain size only in the past 300 000 years. Technical intelligence, however, still has strong advocates (e.g. Byrne, 1995; Visalberghi and Trinca, 1989).

Tool use by wild chimpanzees has been documented in forests across equatorial Africa (Goodall, 1986; McGrew, 1979; Nishida, 1973; Yamagiwa *et al.*, 1988; Boesch and Boesch, 1982, 1989). As information on the distribution of tool use has accumulated, it has become clear that regions within the geographic range of chimpanzees differ markedly with respect to the types of tools used. Whiten et al. (1999) compiled evidence of systematic use of 38 tool types that cross-cut habitats, ecological settings, and gene pools, providing one of the strongest arguments that chimpanzee tool use variation is cultural, not genetic or ecologically determined.

3.2.2 Ecological intelligence

The ecological intelligence school of thought advocates that the key impetus for the expansion of the brain that occurred with hominids was the selective advantage of being able to navigate and find food in a highly complex environment. The extremely complex nature of a tropical forest, combined with the patchy and temporary availability and distribution of fruit placed a premium on the evolution of large brains, especially in frugivorous (fruit-eating) species, to remember food locations and be able to navigate among them. While many animals forage in ways that optimize their chances of stumbling onto good food patches – hummingbirds and bumblebees fly optimized route in fields of wild flowers, for example (Stephens and Krebs, 1986) – primates appear to possess elaborate mental maps of the landscape they live in. Evidence exists that highly frugivorous anthropoid primates may have larger brain-to-body-size ratios than folivorous (leaf-eating) species (Milton, 1981). In African forests, chimpanzees travel from tree to tree feeding all day long. Do they recall the locations of thousands of food trees, or do they travel a path that takes them randomly into food sources? Many fruit species, such as figs, ripen unpredictably, but a party of chimpanzees will usually be at the tree as soon as ripe fruits appear. This suggests that the apes are monitoring the fruiting status of trees as they forage, and remember which trees are worth waiting for. Some researchers have investigated this topic. Garber and Paciulli (1997) studied capuchin monkey food-finding abilities in South American rainforests, and found that even these little monkeys have the ability to recall the locations of hundreds of potential food trees.

If environmental complexity accounted for the increase in brain size that began with the emergence of the hominids, then the expansion of savanna that occurred in the late Miocene might have been the driving force. As forests became more fragmented, the patchiness of the habitat increased, and those emerging hominids with the capacity to navigate through it had a selective advantage.

The ecological intelligence school, which enjoyed much support in the 1980s, has become less accepted in recent years. First, some scholars have pointed out that many very small-brained animals navigate and forage in the same highly complex environment in which primates were thought to benefit from their large brains (Stanford, 1999). There is no evidence that other small mammals such as squirrels are less efficient foragers than primates. In addition, the premise of the ecological complexity argument for hominization, that hominids arose in fragmented forest with patches of open grass, has been questioned in recent years (Kingston *et al.*, 1994).

3.2.3 Social intelligence

The prevailing view among scientists today is that the brain size increase that occurred in great apes and was extended into the hominids was due to the premium natural selection placed on individuals that were socially clever. This theory, often called social intelligence or Machiavellian intelligence (Byrne and Whiten, 1988a), argues that the primary evolutionary benefit to large brain size was that it allowed apes and hominids to cope with and even exploit increasingly complex social relations (see Section 3.4.1 for further discussion). In large social groups, each individual must remember the network of alliances, rivalries, debts, and credits that exist among group members. This is not so different from the politics of our own day-to-day lives. De Waal (1987) pointed out that chimpanzees seem to engage in a "service economy" in which they barter alliances and other forms of support with one another.

Those individuals best able to exploit this web of social relationships would be likely to have reaped more mating success than their group mates. Whiten and Byrne (1997) point out that the ability to subtly manipulate others is a fundamental aspect of group life. Dunbar (1992) argues that as average group size increased, the cerebrum, or neocortex, of the primate brain increased in size to handle the additional input of social information, in much the same way that a switchboard would be enlarged to handle added telephone traffic. This effect holds true even when the evolved patterns of social grouping are taken in account. Dunbar observes that small-brained primates, such as prosimians, typically live solitarily or in smaller groups than do most monkeys and apes. Other cognitive skills also indicate a high level of social intelligence in anthropoid primates. Cheney and Seyfarth (1991) studied vervet monkey social behavior in Amboseli National Park, Kenya. They wanted to know the extent to which the monkeys understand the nature of social relationships within their group, and how that knowledge might be used by group members to manipulate and deceive one another. Vervet monkeys very clearly understand kinship patterns in the group. Using tape-recorded playbacks of baby vervet monkeys' distress calls issuing from cleverly placed loudspeakers, Cheney and Seyfarth showed that vervet monkeys are keenly aware of which baby belongs to which mother. When the call of a given baby was played by the researchers, the other animals in the group looked at the mother of the missing infant. Female vervet monkeys, in other words, understand patterns of maternity in their groups.

Cheney and Seyfarth also found that vervet monkeys engage in *tactical deception*, or lying. In their study, a vervet monkey would give a predator alarm call as the group fed in a desired fruit tree. As other group members fled from the "predator," the call-giver would capitalize on its lie by feeding aggressively in their absence. Byrne and Whiten (1988b) collected examples of potential lying in non-human primates, and concluded that this behavior showed an evolutionary trend, being more widespread in social primates. Great apes seem to be skilled at deceiving one another, while lemurs rarely if ever engage in tactical deception. I once watched a low-ranking Gombe male chimpanzee named Beethoven mate with a female despite the presence of the alpha male Wilkie by using tactical deception. As a party of chimpanzees sat in a forest clearing, Beethoven did a charging display through the middle of the group. As a low-ranking male, this was taken by the alpha Wilkie as an act of insubordination. As Beethoven charged past Wilkie and into dense thickets, Wilkie pursued and launched into his own display. With Wilkie absorbed in his display of dominance Beethoven furtively made his way back to the clearing and mated with an eagerly awaiting female.

Primate researchers believe deception is at the heart of understanding the roots of human cognition because in order to lie to someone, you must possess a theory of mind (see Section 3.4.1). That is, you must be able to place yourself in the mind of another, to be able to attribute mental states to others. The ability to lie, to imitate, and to teach all rely on the assumption that the object of your actions thinks as you do. Whether non-human primates possess a human-like theory of mind is a subject of intense debate, especially among great ape researchers (Povinelli, 2000; Tomasello and Call, 1997).

3.3 Cultural traditions and communication in great apes

"What is culture?" is the central question of the discipline of anthropology. It has, however, deeper roots in the animal kingdom. Culture is the learned set of behaviors that characterize any animal population, as opposed to those behaviors that are believed to be instinctual. The definition of culture as a human-specific concept has long been debated within anthropology, but its application to non-human animals has generated enormous controversy. The concept of culture was devised for humans only, just as the concept of language is anthropocentric. Gaining acceptance for a more expansive definition has been a long struggle for animal behaviorists.

We find cultural variation in traits to be far more common in some great apes than is seen in any other primate. This is presumably because great apes have larger, more encephalized brains that go along with longer lifespans, more extended ontogeny, and more complex sociality than other non-human mammals exhibit.

Among the four great apes species cultural diversity is most widespread in chimpanzees. In the wild, chimpanzees display regional variation mainly in functional behaviors, such as tool-use practices and grooming styles. A few symbolic behaviors vary cross-culturally as well, such as leaf-clipping (clipping leaves with the front teeth to indicate frustration (Bossou, Guinea), or sexual interest (Mahale, Tanzania), depending on the local culture) (Whiten *et al.*, 1999). Chimpanzees display by far the most varied and widespread degree of cultural variation (Table 3.1). Whiten et al. (1999) presented a systematic analysis of tools and other cultural aspects of chimpanzee societies in seven long-term study sites across Africa. They distinguished among anecdotal, habitual, and customary use of such tools, and showed that in at least 39 cases, cultural behaviors including tool use show a pattern of customary use, more reasonably attributed to learned traditions than to ecological influences on the pattern.

Chimpanzee tool use varies widely across Africa, although the mosaic of occurrence suggests that appearances and extinctions of local tool technologies may occur frequently. Certainly there is no evidence that the occurrence of particular tools is related to either ecological (i.e., forest resource availability) or genetic differences among chimpanzee populations (McGrew, 1992). Instead, chimpanzee tool-use patterns appear to reflect local cultural traditions that arise in individuals and spread within, and perhaps among, breeding populations.

Some generalizations about tool use can be made, however. Three broad categories of tools are found: stick tools that are used to increase either arm reach or lever arm strength; sponges, made of chewed leaves, to absorb fluids; and hammers, made of wood or stone and used to crack open hard-shelled food objects. In eastern Africa, stick tools are used to fish for termites and ants, as honey probes, and occasionally to brandish against

Table 3.1 *Evidence of imitative behavior in wild great apes*

	Chimpanzees	Bonobos	Gorillas	Orangutans
Tool use	Widespread and varied; many sources	Very limited	No evidence	Very limited
Teaching	Boesch and Boesch, 1982, 1989	No evidence	No evidence	No evidence
Vocal dialects ?	Yes; Mitani *et al.*, 1999	No evidence	No evidence	No evidence
Cultural variation in social behaviors	Widespread	No evidence	Little evidence	Little evidence

intracommunity rivals. These traditions themselves vary among nearby communities. In Gombe National Park in western Tanzania, chimpanzee fish for termites from earthen mounds using simple twigs stripped of leaves (Goodall, 1986). In Mahale National Park, only 100 km away, chimpanzees fish for ants from tree trunks using the same methods (Nishida, 1990), but never apply this tool to termites in mounds, although they are abundantly available. In Taï National Park in Ivory Coast, western Africa, chimpanzees use stones or wooden clubs to break open nuts and hard-shelled fruits (Boesch and Boesch, 1982). In Gombe, despite an abundance of rocks of suitable size and widespread nuts and fruits of species related to those in Ivory Coast, chimpanzees have never been reported to use stones as hammers.

The ontogeny of tool use is still poorly understood. There are only a few reports of "teaching" by mothers of tool-use skills to their offspring (Boesch and Tomasello, 1998). As young chimpanzees mature, they become more skilled tool-users (McGrew, 1992); simple tool-using ability appears to come from observational learning, imitation, and trial-and-error experiment. This stands in contrast to the role of caregivers in teaching tool-using skills in humans (see, e.g., Zukow-Goldring, this volume).

The closely related bonobo occurs in a far more ecologically restricted habitat than the versatile chimpanzee. Although much attention has been given to perceived sophistication in bonobo social dynamics, sexuality, and male–female relationships, bonobos are demonstrably poor tool-users in the wild (Hohmann and Fruth, 2003). High-ranking bonobos are known to drag branches across the forest floor as a means of signaling others in the foraging party of their desire to travel. But there are no reports of the use of either stick or stone tools in any of the contexts in which chimpanzee tool use is commonplace.

Gorillas are even more limited tool-users than bonobos. Although these largest of the primates sometimes eat termites and ants, they obtain them with their fingers, and no reports of tool use exist. In Bwindi Impenetrable National Park, Uganda, chimpanzees forage for honey and bee larvae using a variety of tool types (apparently selected according to the species, and perhaps temperament, of the bees; Stanford *et al.*, 2000). Byrne and Byrne (1993) have described the cognitive ecology of mountain gorilla foraging in the Virunga Volcanoes of Rwanda; a complex sequence of steps is used to feed on plants that contain defensive bristles. Some evidence exists that other nearby populations of mountain gorillas, such as those in Bwindi Impenetrable National Park, Uganda, process the same plant foods in different ways.

Orangutans are the most arboreal of the great apes, living in Indonesian rainforests. They have only recently been reported to use tools in a limited way, primarily as probes to extract small animals and plant food from tree cavities. Van Schaik *et al.* (1996) documented interpopulational variation in the use of stick tools by Sumatran orangutans. Apart from this limited context, however, orangutans do not use tools or exhibit cross-populational variation in behavioral traditions.

Chimpanzee cultural variation is an outgrowth of the high degree of learned behavior in the chimpanzee ethological repertoire, which requires a long period of socialization and relatively few purely instinctual behavior patterns. Vocalizations are one area of chimpanzee behavior that seems fairly hard-wired. Although dialectical variants are reported across Africa (Mitani *et al.*, 1999), the vocal repertoire is uniform in any chimpanzee culture one observes. The overall social structure in which chimpanzees live is also quite uniform across the species' geographic distribution. In general, however, within-group social behaviors in chimpanzees are highly labile and believed to be acquired through observational learning.

A primary difference between human and ape culture is, of course, language use. Language allows for the indirect transmission of culture across geographic and generational boundaries (Boesch and Tomasello, 1998). The well-documented acquisition and use of simple linguistic skills by chimpanzees reared in human home settings is not easily related to a "natural" ape context. Chimpanzees in the wild do not use the language skills they so easily acquire when raised in settings similar to those in which human infants are reared. When captive reared in language-rich environments featuring the same sort of intensive adult–infant bonds that are natural for all apes and humans, chimpanzees can acquire an understood language set of over 1000 spoken words (Savage-Rumbaugh and Lewin, 1994; Fouts and Mills, 1997) and a generated vocabulary using signs or symbol board of several hundred words. The chimpanzees in Fout's study spontaneously sign short sentences to one another in the absence of trainers or observers (captured by remote video monitoring).

Wild chimpanzees also employ a variety of hand gestures in communication. Hand signals indicate, for example, a desire for food (the begging gesture common to Western culture), an intention signal to a female expressing a desire to mate (Goodall, 1986). In some chimpanzee study populations, symbolic gestures occur; the leaf-grooming behavior, in which animals indicate apparent mating interest by carefully grooming the surface of a plucked leaf (Goodall, 1986). These hand gestures are part of the natural communication repertoire of wild chimpanzee. In the wild, chimpanzee use a wide range of vocalizations, some of them apparently relying heavily on context to provide meaning (Mitani *et al.*, 1999). Some studies have shown that in captivity chimpanzees taught sign language use it spontaneously with one another outside experimental regimens (Fouts and

Mills, 1997). Chimpanzees are as adapted to gestural communication as to complex vocal communication; while the known ethogram of calls by wild chimpanzees is about 50 (Goodall, 1986).

The key question about ape linguistic abilities should not be "Do apes have language?", but rather "Do apes and humans share the same foundations for language?" Apes in the wild do not acquire anything resembling human language. In captivity, foster-reared apes can learn something that closely resembles a simple version of human language, of the sort seen in 2-year-old children. Since the concept of language was invented for human use only, it is not surprising that linguists and cognitive scientists have long been reluctant to apply it more expansively to other linguistically skilled animals. If a chimpanzee shares with a human child the basic "software" for basic language acquisition (the ape lacks the vocal "hardware" needed to produce intelligible speech sounds), this would be a profound statement about the deep history of language in the protohominid lineage.

Perhaps the most persuasive research done on language acquisition in an ape is Savage-Rumbaugh's (1999) work with Kanzi, a male bonobo. Kanzi communicates by touching symbols on a lexicon board; he receives information through his understanding of spoken English. Savage-Rumbaugh estimates his production vocabulary as 300 words and his understanding of spoken English as over 1000 words. Work by Savage-Rumbaugh and many other researchers has conclusively settled at least two arguments over ape language. First, she demonstrates that apes understand and employ the concept of reference; symbolically using words to represent things in their environment. Second, these words and phrases are spontaneously combined to communicate, request, and give information, as well as to comment on or describe the world around them (Byrne, 1995). Savage-Rumbaugh *et al.* (1998) report that Kanzi and a 2.5-year-old girl were tested on their comprehension of 660 sentences phrased as simple requests (presented once). Kanzi was able to carry out the request correctly 72% of the time, whereas the girl scored 66% on the same sentences and task. This seemed to mark the limits of Kanzi's abilities but was just the beginning for the human child. No non-human primate has exhibited any of the richness of human language that distinguishes the adult human from the 2-year-old, suggesting a biological difference in the brain's "language-readiness" between humans and other primates.

3.4 Imitation

This focus on culture is essential to understanding the imitative and emulative abilities of great apes. Great ape cognition is demonstrably social in nature; the context in which their cognitive skills evolved was, like our own, a complex web of interindividual interactions. Learning through observation and interaction in a social group is far different from that in isolation, in that it relies more heavily on socialization. This makes the experimental study of learned behaviors, including imitation, problematic to conduct and the results difficult to interpret under captive laboratory conditions.

By imitation, I mean behavior performed by one individual modeled upon that of another. As Byrne (1995) points out, animal behaviorists have tended to define imitation in the negative; by what it is not. Only after one can exclude lower-order forms of behavioral modeling can one refer to a behavior as imitation. Many bird species, for instance, are accomplished mimics and may have extensive, highly accurate repertoires of songs of other species. But there is no evidence that birds have any understanding of the content they mimic (although see Pepperberg (1999) for "intelligent imitation" by an African gray parrot).

3.4.1 Imitation and emulation

Determining whether an ape understands what it appears to be copying is at the heart of the debate over the role of imitation in ape intelligence. Skeptics point to two barriers in ape cognition. First, the ape must be able to place itself in the mind of another; this is the theory-of-mind issue (Byrne, 1995). By theory of mind, I mean the ability for an individual to place himself in the mind of others, to ascribe their mental state, and to act on that insight.

Second, the ape must be able to copy an entire, sometimes elaborate, sequence of behaviors that serve little function individually, into a "whole" behavior sequence that is processually the equivalent of the model for the behavior. This involves not simply the ability to repeat a fixed sequence of behaviors but also the ability to recognize subgoals and thus "repeat x until subgoal g is achieved." There is no question that many animals, and certainly many primates, can imitate simple actions they observe performed by humans or by members of their own species. Attempts at imitation may be facilitated by familiarity of the identity and context of the modeler of the behavior. Sumita *et al.* (1985) showed that captive chimpanzees readily copied tool-use behavior they observed performed by other chimpanzees, even though the same tool use performed by a human modeler elicited little copying. Japanese researchers have observed the spread of the symbolic leaf-clipping behavior from one chimpanzee community to an adjacent one (Nishida, 1980; Sugiyama, 1981).

To adopt the mental perspective of another human is one thing; to recognize it in a non-verbal animal is quite another. How do we know when an ape copies a physical action that it understands what it is copying? Because of the difficulty of studying the mental perspective of the ape, researchers usually make this judgement by distinguishing apparent imitation from trial and error. Some researchers have attempted to distinguish between imitation, which they regard as a human cognitive attribute (e.g., Tomasello, 1996; Povinelli, 2000), and emulation, which they regard as an ape's approximation of human imitative abilities. As an emulator, an ape would be able to duplicate the results of other individuals' behavior without using their methods. Tomasello (1999) has argued that many claimed instances of chimpanzee observational learning of tool skills are really examples of emulation; the chimpanzee has seen the goal accomplished by a group mate and attempts to achieve the same goal, but is not necessarily able to replicate the intermediate steps toward achieving it. To these researchers, imitative behavior is copying with an understanding of both the goal and each step toward it. When an ape sees a person place a bandage on a cut, and then copies this behavior by placing a bandage on its own arm, is this imitation? Povinelli and Tomasello argue it is emulation, not imitation, because there is no evidence the ape understood the meaning of the act of placing the bandage. For example, while Boesch and Tomasello (1998) consider whether chimpanzees learn to use tools through imitation or even through maternal teaching, Povinelli (2000) argues that the generalizations chimpanzees make from an observed tool-use task are entirely superficial. That is, while a human child would watch a tool-use task and gain an understanding of deeper causation, the chimpanzee lacks the ability to understand the causational aspects of the process. Wohlschlager and Bekkering (2000) have also argued that imitation is based on a mirror-neuron system similar to that proposed by Arbib (2005).

In an often-cited study, Tomasello (1996) set out to determine if a chimpanzee could imitate the way a person performs a task requiring forethought and planning. He set up an experiment in which a ball could be reached at the end of a table only by inserting a rake through a grate from the opposite end, by which the ball could be dragged toward the person holding the rake. When a small child observed this, it took the child only one trial to learn how to get the ball using the rake, and the child perfectly imitated the researcher's demonstration. The chimpanzee was also able to get the ball after only watching a single trial. However, the chimpanzee devised its own method of using the rake to obtain the ball, not the style the researcher had demonstrated. Tomasello concluded from this that the chimpanzee failed to imitate the process – even though it could achieve the same result – because it lacked a theory of mind that is necessary for true imitation. Tomasello labeled what the chimpanzee did as emulation; achieving the goal without understanding the importance of imitating the process.

Tomasello's experiment and others like it have been carried out by researchers who doubt that chimpanzees possess a theory of mind. But critics (e.g., Fouts and Mills, 1997) point out that the rake–ball test is conducted in a context highly familiar to many children, and utterly unfamiliar and unnatural to most chimpanzees. This is a persistent problem for laboratory studies of great apes, which evolved cognitive abilities in response to the ecological and social pressures of tropical forest environments, not captive settings. Most laboratories provide severely impoverished social learning environments for their study subjects relative to what children growing up in families experience. No human orphan who had been traumatically separated from his or her mother as a toddler and then reared in a socially impoverished environment would be deemed an appropriate model for studying normal childhood cognitive development. Apes, psychologically and emotionally similar to young children of similar age, may be no more suitable as models.

Povinelli and colleagues' (Povinelli *et al.*, 1994; Povinelli, 2000) skepticism about an ape's ability to imitate is based on the question whether a chimpanzee possesses a theory of mind. For Povinelli, an understanding of knowledge, belief, and other advanced mental states is the fundamental aspect of a human mind, one which chimpanzees lack. He cites studies in which both chimpanzee and orangutan subjects were unable to hold a false belief such as required to attract the attention of an observer for a reward that the ape knew was not actually present (Call and Tomasello, 1999), the critical test in many scholars' view for a theory of mind. Critics of Call and Tomasello might point out that the circumstances of such a training regimen – social deprivation and captivity itself – are artificial enough to make negative results dubious. Povinelli therefore argues that social cognition, at which great apes demonstrably excel, must have evolved independently from the capacity to reason about other individuals' mental states.

Though this may seem paradoxical, Povinelli (2000) reasons that earlier forms of social primates evolved cognitive abilities such as tactical deception, reconciliation, and empathy long before the emergence of a theory of mind, which in his view occurred only after the evolutionary split between apes and hominids. He sees natural selection acting to mold these traits in the absence of representation of minds of group mates. In other words, Povinelli rejects the social or Machiavellian intelligence model of the origins of human intelligence (Byrne and Whiten, 1988a) that is the prevailing view among researchers today. Only after the emergence of a hominid lineage does Povinelli see the rise of mental state attribution. In other non-human primate lineages, the package of pre-existing cognitive traits would therefore have been unaffected by the protohuman cognitive innovations. Although Povinelli argues that social primates would be as effective, or even more effective, at group living if they lacked a theory of mind, he offers no evidence, direct or even circumstantial, to back up this claim. Povinelli's views therefore seem almost creationist with respect to cognition. It may be more sensible to view theories of mind as existing at multiple levels; perhaps an ape theory of mind can exist separately from a human theory of mind without implying the binary logic of some researchers.

3.4.2 Program-level imitation

Byrne (1995) and Byrne and Byrne (1993) have done extensive research on program-level imitation, in which entire sequences of behaviors must be learned, in proper sequence, to achieve some goal (cf. the notion of complex imitation used by Arbib, this volume). They postulate that for complex behavior sequences, what appears to be an assortment of separately conceived actions engaged in to reach a goal may be a well-defined hierarchy. In theory, smart animals should be able to copy either individual behaviors, segments of behaviors, or entire behavioral sequences. In practice, copying entire hierarchical programs of behavior should be most common when survival and reproduction are at stake. For instance, Byrne and Byrne (1993) documented how mountain gorillas employ a logical chain of behavior when foraging for difficult-to-eat plant foods. *Gallium* is a tough, bristly plant commonly eaten by gorillas. To do so, the animals employ a hierarchy of fine motor movements with their hands and teeth in a sequence that rarely varies.

Program-level imitation involves both the imitation of a hierarchical routine and the imitation of individual subroutines. Success in the former is not possible without achieving the latter. A gorilla might successfully imitate a subroutine but not master the entire logical sequence, thereby rendering the entire behavior pattern useless. At the same time, subroutines or individual pieces of behavior might be imitated by an ape with a perfect understanding, while the entire hierarchy might be copied without such a deep understanding of the process. Previous studies (e.g., Bauer and Mandler, 1989) have found that subjects (in this case children) could imitate behavioral sequences more accurately and with less experiential learning if the steps in the sequence are causally connected.

3.5 Meat-sharing in wild chimpanzees

One interesting form of learned behavior in wild apes is hunting. Most species of non-human primates eat animal protein, mainly in the form of insects and other invertebrates. Only a few of the higher primates eat other mammals on a frequent basis. In the New World, capuchin monkeys of the genus *Cebus* are voracious predators on a variety of smaller animals including squirrels and immature coatis. Baboons can also be avid hunters of small mammals such as hares and antelope fawns, and in at least one site meat-eating by baboons was as frequent as for any non-human primate population recorded. Only in one great ape, however, do we see the sort of systematic predatory behavior that we believe began to be a part of our early hominid ancestors' behavior between 3 and 2 million years ago. Chimpanzees occur across equatorial Africa, and in all forests in which they have been studied intensively, they prey on a variety of vertebrates including other mammals. At least 35 species of vertebrate prey have been recorded in the diet of wild chimpanzees, some of the prey weighing as much as 20 kg. One prey species, the red colobus monkey, is the most frequently eaten prey in all forests in which chimpanzees and the colobus occur together (Fig. 3.1). In some years, chimpanzees in Gombe National Park, Tanzania, kill more than 800 kg of prey biomass per community, most of it in the form of red colobus. Individual hunters may kill up to 10% of the entire prey population within the community's hunting range. After four decades of research on the hunting behavior of chimpanzees at Gombe, we know a great deal about their predatory patterns (Stanford, 1995, 1998). At Gombe, red colobus account for more than 80% of the prey items eaten. But Gombe chimpanzees do not select the colobus they will kill randomly; infant and juvenile colobus are caught in greater proportion than their availability; 75% of all colobus killed are immature (Stanford *et al.*, 1994). Chimpanzees are largely fruit-eaters, and meat-eating composes only about 3% of the time they spent eating overall, less than in nearly all human societies. Adult and adolescent males do most of the hunting, making about 90% of the kills recorded at Gombe over the past decade. Females also hunt, though more often they receive a share of meat from the male who either captured the meat or stole it from the captor. Although lone chimpanzees, both male and female, sometimes hunt by themselves, most hunts are social. In other species of hunting animals, cooperation among hunters is positively correlated with

Figure 3.1 An adult male chimpanzee with a colobus monkey he has caught. Wild chimpanzees are avid meat-eaters, and hunting is clearly a learned skill requiring years of practice.

greater success rates, thus promoting the evolution of cooperative behavior. In both Gombe and in the Taï Forest in the Ivory Coast (Boesch and Boesch, 1989), there is a strong positive correlation between the number of hunters and the odds of a successful hunt, although researchers differ on the extent to which cooperative hunting is pervasive at Taï.

During a hunt, there probably are communication signals given that remain for researchers to understand. In at least one site, Kibale National Park in Uganda, a hunting call has been noted (Mitani and Watts, 1999). Other calls may be communicating information to other local hunters based on context rather than content (personal observation). No gestural communication has been noted, although the post-hunt meat-sharing involves extensive begging for meat, using a human-like outstretch supplicating hand (Fig. 3.2).

The amount of meat eaten, even though it comprises a small percentage of the chimpanzee diet, is substantial. I estimated that in some years, the 45 chimpanzees of the Kasakela community at Gombe kill and consume a sum of hundreds of kilograms of prey biomass of all species (Stanford, 1998). This is far more than most previous estimates of the weight of live animals eaten by chimpanzees. During the peak dry season months, the estimated per capita meat intake is about 65 g of meat per day for each adult chimpanzee. This approaches the meat intake by the members of some human foraging societies in the lean months of the year. Chimpanzee dietary strategies may thus approximate those of human hunter–gatherers to a greater degree than previously imagined.

In the early years of her research, Jane Goodall (1986) noted that the Gombe chimpanzees tend to go on "hunting crazes," during which they would hunt almost daily and kill large numbers of monkeys and other prey. The most intense hunting binge we have seen occurred in the dry season of 1990. From late June through early September, a period of

Figure 3.2 A female chimpanzee begs meat from a male who has captured a monkey. Meat transfer is a highly ritualized social activity in which meat-possessors manipulate those who want to receive meat.

68 days, the chimpanzees were observed to kill 71 colobus monkeys in 47 hunts. It is important to note that this is the observed total, and the actual total of kills that includes hunts at which no human observer was present may be one-third greater.

Hunting by wild chimpanzees appears to have both a nutritional and social basis. Understanding when and why chimpanzees should choose to undertake a hunt of colobus monkeys rather than simply continue to forage for fruits and leaves, even though the hunt involves risk of injury from colobus canine teeth and a substantial risk of failure to catch anything, has been a major goal of my research.

In his pioneering study of Gombe chimpanzee predatory behavior in the 1960s, Teleki (1973) considered hunting to have a strong social basis. Some early researchers had said that hunting by chimpanzees might be a form of social display, in which a male chimpanzee tries to show his prowess to other members of the community. Wrangham (1975) conducted the first systematic study of chimpanzee behavioral ecology at Gombe and concluded that predation by chimpanzees was nutritionally based, but that some aspects of the behavior were not well explained by nutritional needs alone. Nishida (1990) and his colleagues in the Mahale Mountains chimpanzee research project reported that the alpha there, Ntologi, used captured meat as a political tool to withhold from rivals and dole out to allies. McGrew (1992) has shown that those female Gombe chimpanzees who receive generous shares of meat after a kill have more surviving offspring, suggesting a reproductive benefit tied to meat-eating.

My own preconception was that hunting must be nutritionally based. After all, meat from monkeys and other prey would be a package of protein, fat, and calories hard to equal from any plant food. I therefore examined the relationship between the odds of success and the amount of meat available with different numbers of hunters in relation to each hunter's expected pay-off in meat obtained. That is, when is the time, energy, and risk (the costs) involved in hunting worth the potential benefits, and therefore when should a chimpanzee decide to join or not join a hunting party? And how does it compare to the costs and benefits of foraging for plant foods? These analyses are still under way because of the difficulty in learning the nutritional components of the many plant foods in the chimpanzees' diverse diet, but the preliminary results have been surprising. I expected that as the number of hunters increased, the amount of meat available for each hunter would also increase. This would have explained the social nature of hunting by Gombe chimpanzees. If the amount of meat available per hunter declined with increasing hunting party size (because each hunter got smaller portions as party size increased), then it would be a better investment of time and energy to hunt alone rather than join a party. The hunting success rate of lone hunters is only about 30%, while that of parties with ten or more hunters is nearly 100%. As it turned out, there was no relationship, either positive or negative, between the number of hunters and the amount of meat available per capita. This may be because even though the likelihood of success increases with more hunters in the party, the most frequently caught prey animal is a 1-kg baby colobus monkey. Whether shared among four hunters or 14, such a small package of meat does not provide anyone with much food.

Sharing is common among social animals, as recently reviewed by Winterhalder (2001). Vampire bats, for instance, engage in extensive reciprocal sharing which is crucial to their survival. Wilkinson (1988) showed that these bats often fail to find their normal meal of mammalian blood, and mortality would be extremely high if not for blood they receive as shared meals regurgitated by other, often unrelated, bats in the colony. Blood-sharers form alliances, the partners in which regurgitate food back and forth. This is a classically reciprocal altruistic relationship, predicated upon long term relationships, mutual need, and ease of resource transfer. Similarly, Heinrich and Marzluff (1995) studied meat-sharing among ravens These birds share a great deal of information about the location of the carcasses of deer and moose that are their primary wintertime food sources. Ravens call other, unrelated ravens to inform them of the presence of a carcass, which in the winter represents a key resource without which starvation might rapidly occur. They appear to share information readily, rather than hoard both the knowledge of the meat and the meat itself, both because the cost of sharing is extremely low and because the benefits of a reciprocal system in which one reaps the reward of information later in exchange for giving it now is great.

These two examples of sharing behavior reflect the idea that cooperation occurs mainly when there are major selfish interests at stake. Packer and Ruttan (1988) showed that when a lion pride's food (savanna ungulates) is abundant, meat consumption is unrelated to group size. But when food is scarce, those lions that either forage alone or else hunt cooperatively in groups of five or six achieve the greatest per capita consumption of meat. In-between hunting party sizes of two to four reduces intake. This same relationship between the number of hunters and their meat consumption was shown by Hill *et al*. (1987) to occur as in the Aché, a society of foraging people in Paraguay, and I found a similar relationship in my own work on chimpanzee predatory behavior. It suggests that individuals may adjust their pattern of association in order to maximize their odds of obtaining meat for themselves.

Darwinian explanations for sharing therefore focus on the selfish incentive for what often appears to be generous altruism. Blurton-Jones (1984) first proposed that when a hunter–gatherer controls a carcass, he may allow others to take meat from it simply because the energy expended in preventing the theft would be greater than the loss of the bit of meat itself. This model, known as tolerated theft, can occur when the carcass is big enough to be divisible. If so, then it is subject to diminishing marginal value as its pieces are taken and the consumer becomes sated. A second model for sharing, risk-sensitive sharing, is the hedge against shortfalls that we examined earlier. Altruistically reciprocal sharing of a tit-for-tat kind occurs when resources are swapped directly for each other.

3.6 Conclusion: ape cognition, culture, and imitation

The extent to which imitation, action, and language are interrelated clearly depends very much on one's perspective on cognition and language. The dichotomy drawn by many cognitive scientists between human language and the linguistic abilities of great apes, whether false or not, has stymied a better understanding of the evolution of human language. Human language use must have evolved in much the same way as other cognitive traits: through the selective retention by natural selection of facilities allowing either gestural or verbal communication, or both. Given the signing dexterity of great apes (Fouts and Mills, 1997) gestural communication would seem likely as a natural extension of what apes do in the wild. In this context, gestural communication might have served as a template upon which verbal language then evolved. This might be the "language-ready brain" to which Arbib refers (Chapter 1, this volume).

Going from a non-verbal protolanguage to full-blown verbal language may have happened in an organic way through natural selection, or in an epigenetic, culture-driven way. If the latter, we are unable to test hypotheses about the origin and early evolution of language, since it would have been not only a unique occurrence but the only known human trait that arose through non-evolutionary means without natural selection as the driving force. If full-blown syntactical language was an epigenetic phenomenon that emerged *de novo* from protosigns and protospeech, then investigations into language origins will always remain largely speculation. If, however, language was an outgrowth of a long process of cognitive evolution, increasingly sophisticated imitative abilities, and increasingly complex social interaction networks, then some key aspects of its origins should be discernible with future human evolutionary research.

References

Arbib, M. A., 2005. From monkey-like action recognition to human language: an evolutionary framework for neurolinguistics. *Behav. Brain Sci.* **28**: 105–124.

Bauer, P., and Mandler, J., 1989. One thing follows another: effects of temporal structure on 1- and 2-year-olds' recall of events. *Devel. Psychol.* **25**: 197–206.

Blurton-Jones, N., 1984. A selfish origin for human food sharing: tolerated theft. *Ethol. Sociobiol.* **4**: 145–147.

Boesch, C., and Boesch, H., 1982. Optimization of nut-cracking with natural hammers by wild chimpanzees. *Behaviour* **3**: 265–286.

1989. Hunting behavior of wild chimpanzees in the Taï National Park. *Am. J. Phys. Anthropol.* **78**: 547–573.

Boesch, C. and Tomasello, M., 1998. Chimpanzee and human culture. *Curr. Anthropol.* **39**: 591–614.

Byrne, R. W., 1995. *The Thinking Ape*. Oxford, UK: Oxford University Press.

Byrne, R. W., and Byrne, J. M. E., 1993. Complex leaf-gathering skills of mountain gorillas (*Gorilla g. beringei*). *Am. J. Primatol.* **31**: 241–261.

Byrne, R. W., and Whiten, A. (eds.), 1988a. *Machiavellian Intelligence*. Oxford, UK: Clarendon Press

1988b. Towards the next generation in data quality: a new survey of primate tactical deception. *Behav. Brain Sci.* **11**: 267–273.

Call, J., and Tomasello, M., 1999. A nonverbal false belief task: the performances of children and great apes. *Child Devel.* **70**: 381–395.

Cheney, D. L., and Seyfarth, R. M., 1991. *How Monkeys See the World*. Chicago, IL: University of Chicago Press.

Darwin, C., 1871. *The Descent of Man and Selection in Relation to Sex*. London: John Murray.

de Waal, F. B. M., 1987. Tension regulation and nonreproductive functions of sex in captive bonobos (*Pan paniscus*). *Natl Geogr. Res. Rep.* **3**: 318–335.

Dunbar, R. I. M., 1992. Neocortex size as a constraint on group size in primates. *J. Hum. Evol.* **20**: 469–493.

Fouts, R., and Mills, S. T., 1997. *Next of Kin*. New York: William Morrow.

Garber, P. A., and Paciulli, L. M., 1997. Experimental field study of spatial memory and learning in wild capuchin monkeys (*Cebus capucinus*). *Folia Primatol.* **68**: 236–253.

Gergely, G., Bekkering, H., and Kiraly, I., 2002. Rational imitation in preverbal infants. *Nature* **415**: 755.

Goodall, J., 1986. *The Chimpanzees of Gombe: Patterns of Behavior*. Cambridge, MA: Harvard University Press.

Heinrich, B., and Marzluff, J., 1995. Why ravens share. *Am. Scientist* **83**: 342–349.

Hill, K., Kaplan, H., Hawkes, K., and Hurtado, A. M., 1987. Foraging decisions among Aché hunter–gatherers: new data and implications for optimal foraging models. *Ethol. Sociobiol.* **8**: 1–36.

Hohmann, G., and Fruth, B., 2003. Culture in bonobos? Between-species and with-species variation in behavior. *Curr. Anthropol.* **44**: 563–571.

Kingston, J. D., Marino, B. D., and Hill, A., 1994. Isotopic evidence for Neogene hominid paleoenvironments in the Kenya rift valley. *Science* **264**: 955–959.

McGrew, W. C., 1979. Evolutionary implications of sex differences in chimpanzee predation and tool use. In D. A. Hamburg and E. R. McCown (eds.) *The Great Apes*. Mento Park, CA: Benjamin/Cummings, pp. 441–469.

1992. *Chimpanzee Material Culture*. Cambridge, UK: Cambridge University Press.

Milton, K., 1981. Distribution pattern of tropical food plants as a stimulus to primate mental development. *Am. Anthropologist* **83**: 534–548.

Mitani, J. C., and Watts, D., 1999. Demographic influences on the hunting behavior of chimpanzees. *Am. J. Phys. Anthropol.* **109**: 439–454.

Mitani, J. C., Hunley, K. L., and Murdoch, M. E., 1999. Geographic variation in the calls of wild chimpanzees: a reassessment. *Am. J. Primatol.* **47**: 133–151.

Nishida, T., 1973. The ant-gathering behavior by the use of tools among wild chimpanzees of the Mahali Mountains. *J. Hum. Evol.* **2**: 357–370.

1980. The leaf-clipping display: a newly discovered expressive gesture in wild chimpanzees. *J. Hum. Evol.* **9**: 117–128.

(ed.), 1990. *The Chimpanzees of the Mahale Mountains.* Tokyo: University of Tokyo Press.

Packer, C., and Ruttan, L., 1988. The evolution of cooperative hunting. *Am. Naturalist* **132**: 159–198.

Pepperberg, I. M., 1999. *The Alex Studies: Cognitive and Communicative Abilities of Grey Parrots.* Cambridge, MA: Harvard University Press.

Povinelli, D. J., 2000. *Folk Physics for Apes: The Chimpanzee's Theory of How the World Works.* New York: Oxford University Press.

Povinelli, D. J., Rulf, A. B., and Biershwale, D., 1994. Absence of knowledge attribution and self-recognition in young chimpanzees (*Pan troglodytes*). *J. Comp. Psychol.* **180**: 74–80.

Savage-Rumbaugh, S., 1999. Ape language: between a rock and a hard place. In B. J. King (ed.) *The Origins of Language: What Nonhuman Primates Can Tell Us.* Santa Fe, NM: School of American Research Press, pp. 115–189.

Savage-Rumbaugh, S., and Lewin, R., 1994. *Kanzi: The Ape at the Brint of the Human Mind.* New York: John Wiley.

Savage-Rambaugh, S., Shankar, S. G., and Taylor, T. J., 1998. *Apes, Language, and the Human Mind.* Oxford, UK: Oxford University Press.

Stanford, C. B., 1995. The influence of chimpanzee predation on group size and anti-predator behaviour in red colobus monkeys. *Anim. Behav.,* **49**: 577–587.

1998. *Chimpanzee and Red Colobus: The Ecology of Predator and Prey.* Cambridge, MA: Harvard University Press.

1999. *The Hunting Apes.* Princeton, NJ: Princeton University Press.

Stanford, C. B., Wallis, J., Matama, H., and Goodall, J., 1994. Patterns of predation by chimpanzees on red colobus monkeys in Gombe National Park, Tanzania, 1982–1991. *Am. J. Phys. Anthropol.* **94**: 213–228.

Stanford, C. B., Gambaneza, C., Nkurunungi, J. B., and Goldsmith, M., 2000. Chimpanzees in Bwindi Impenetrable National Park, Uganda, use different tools to obtain different types of honey. *Primates* **41**: 335–339.

Stephens, D. W., and Krebs, J. R., 1986. *Foraging Theory.* Princeton, NJ: Princeton University Press

Sugiyama, Y., 1981. Observations on the population dynamics and behavior of wild chimpanzees of Bossou, Guinea, 1979–1980. *Primates* **22**: 435–444.

Sumita, K., Kitahara-Frisch, J., and Norikoshi, K., 1985. The acquisition of stone-tool use in captive chimpanzees. *Primates* **26**: 168–181.

Teleki, G., 1973. *The Predatory Behavior of Wild Chimpanzees.* Lewisburg, PA: Bucknell University Press.

Tomasello, M., 1996. Do apes ape? In C. Heyes and B. Galef Jr. (eds.) *Social Learning in Animals: The Roots of Culture.* New York: Academic Press, pp. 319–346.

1999. Emulation learning and cultural learning. *Behav. Brain Sci.* **21**: 703–704.

Tomsello, M., and Call, J., 1997. *Primate Cognition*. Oxford, UK: Oxford University Press.

van Schaik, C. P., Fox, E. A., and Sitompul, A. F., 1996. Manufacture and use of tools in wild Sumatran orangutans: implications for human evolution. *Naturwissenschaften* **83**: 186–188.

Visalberghi, E., and Trinca, L., 1989. Tool use in capuchin monkeys, or distinguish between performing and understanding. *Primates* **30**: 511–521.

Whiten, A., and Byrne, R. W., 1997. *Machiavellian Intelligence, vol. 2, Extensions and Evaluations*. Cambridge, UK: Cambridge University Press.

Whiten, A., Goodall, J., McGrew, W. C., *et al.*, 1999. Cultures in chimpanzees. *Nature* **399**: 682–685.

Wilkinson, G. S., 1988. Reciprocal altruism in bats and other mammals. *Ethol. Sociobiol.* **9**: 85–100.

Winterhalder, B., 2001. Intragroup resource transfers: comparative evidence, models, and implications for human evolution. In C. B. Stanford and H. T. Bunn (eds.) *Meat-Eating and Human Evolution*. New York: Oxford University Press, pp. 279–301.

Wolschlager, A., and Bekkering, H., 2000. Is human imitation based on a mirror-neuron system ? Some behavioral evidence. *Exp. Brain Res.* **143**: 335–341.

Wrangham, R. W., 1975. Behavioural ecology of chimpanzees in Gombe National Park, Tanzania. Ph.D. dissertation, Cambridge University.

Yamagiwa, J., Yamoto, T., Mwanza, N., and Maruhashi, T., 1988. Evidence of the tool-use by chimpanzees (*Pan troglodytes Schwenfurthii*) for digging out a bee-nest in the Kahuzi-Biega National Park, Zaire. *Primates* **29**: 405–411.

4

The signer as an embodied mirror neuron system: neural mechanisms underlying sign language and action

Karen Emmorey

4.1 Introduction

"Mirror" neurons are found in area F5 of the monkey brain, and they fire both when a monkey grasps an object and when the monkey observes another individual grasping the object (e.g., Rizzolatti *et al.*, 1996; see Arbib, Chapter 1, this volume, for further discussion). Mirror neurons have also been found in the rostral part of the monkey inferior parietal lobule (Gallese *et al.*, 2002). Like mirror neurons, signers must associate the visually perceived manual actions of another signer with self-generated actions of the same form. Sign language comprehension and production requires a direct coupling between action observation and action execution. However, unlike mirror neurons for hand movements recorded in monkey, signing is not tied to object manipulation. Mirror neurons for grasping in monkey fire only when an object is present or understood to be present and do not fire when just the grasping movement is presented (Umiltà *et al.*, 2001). Furthermore, unlike grasping and reaching movements, sign articulations are structured within a phonological system of contrasts. The hand configuration for a sign is determined by a phonological specification stored in the lexicon, not by the properties of an object to be grasped. These facts have interesting implications for the evolution of language and for the neural systems that underlie sign language and action.

The fact that sign language exhibits form-based patterning of meaningless elements (i.e., phonology) distinguishes signs from actions, even when the two appear quite similar on the surface; for example, the American Sign Language (ASL) sign TO-HAMMER resembles the act of hammering. In this chapter, I first discuss the ramifications of the existence of sign phonology for theories of language evolution. Specifically, sign languages escape their pantomimic and iconic roots and develop duality of patterning, which argues against the superiority of vocal gestures in this regard. Next, I explore the relationship between production and perception for sign, speech, and action. The mirror neuron system must recognize that an observed action is the same (in some relevant sense) as a self-generated action – how does this happen? Finally, the similarities and differences between the neural systems that support speech, sign, and action are

Action to Language via the Mirror Neuron System, ed. Michael A. Arbib. Published by Cambridge University Press.

discussed, focusing on production (execution) and perception (observation). The chapter ends with a discussion of the possible role of the mirror neuron system for sign language.

4.2 How signs differ from actions: manual phonology, duality of patterning, and the evolution of language

The discovery that sign languages have a phonology (traditionally defined as the *sound* patterns of language) was ground-breaking and crucial to our understanding of the nature of human language (for reviews see Brentari, 1998; Corina and Sandler, 1993). In spoken languages, words are constructed out of sounds that, in and of themselves, have no meaning. The words "bat" and "cat" differ only in the initial sounds that convey no inherent meanings of their own. As discussed by Goldstein, Byrd, and Saltzman (this volume), the atomic units of speech (traditionally, segments) are arbitrary, discrete, and are combinable into larger units (e.g., syllables and words). The ability to create an unlimited set of meaningful elements (words) from a closed set of recombinable mean-ing*less* elements is referred to as *duality of patterning*. Duality of patterning also refers to the fact that the patterning of phonological units within a word is distinct from the patterning of syntactic units (words and morphemes) within a clause. Like spoken languages, signed languages also exhibit duality of patterning. Individual signs are constructed out of meaningless components that are combined to create morphemes and words, and the patterning of signs at the level of the syntactic clause is distinct from the patterning of components of form within a sign.

Signs can be minimally distinguished by hand configuration, place of articulation, movement, and orientation with respect to the body. Figure 4.1 provides an illustration of four *minimal pairs* from ASL. The signs in a minimal pair are identical except for one component, and if you substitute that component for another, it changes the meaning of the sign. These components of form are arbitrary and not dependent on iconicity. For example, the movement of the sign TRAIN (Fig. 4.1C) is unrelated to the movement of a train and does not contribute to the semantics of the sign. The place of articulation for the sign ONION (Fig. 4.1B) is non-meaningful, although folk etymology traces its origin to the location of tears when cutting an onion. Historically, the form of a sign may be iconically motivated (see Taub, 2001), but such iconicity plays little role in the phono-logical representation of signs. The form of signs is constrained by the set of available phonological units within a sign language and by language-specific constraints on their combination – not by properties of iconicity.

Not all hand configurations and all places of articulation are distinctive in any given sign language. For example, the T handshape (thumb inserted between index and middle fingers) appears in ASL, but does not occur in European sign languages. Chinese Sign Language contains a handshape formed with an open hand with all fingers extended except for the ring finger, which is bent; this hand configuration does not occur in ASL. Although the upper cheek and chin are distinctive places of articulation in ASL (see

Figure 4.1 Examples of minimal pairs in American sign language. (A) Signs that contrast in hand configuration, (B) signs that contrast in place of articulation, (C) signs that contrast in movement, (D) signs that contrast in orientation.

Fig. 4.1B), the back jaw is not a phonologically contrastive location (e.g., there are no minimal pairs involving this place of articulation). Thus, the set of phonological units is closed for a given sign language, and the members of the set may differ cross-linguistically. Further, borrowing signs from another signed language does not extend the phonological repertoire of a language; rather, the form of the borrowed sign is altered to conform to the phonological constraints of the adoptive language (Quinto-Pozos, 2003).

Spoken languages represent sequential structure in terms of the linear ordering of segments (or complex gestures within Articulatory Phonology). Similarly, signs are not holistic expressions, and linear structure is represented in terms of sequences of phonological units. The number of sequential units within a word tends to be less for signed languages (e.g., canonical signs have only three sequential segments; Sandler, 1989), and more phonological features are represented simultaneously (akin to the representation of tone for spoken languages). Of course, the relevant articulators are manual rather than vocal, and phonological patterns are described in terms of manual features (e.g., selected fingers, body contact) rather than oral features (e.g., wide glottis, bilabial constriction).

In addition, there is fairly strong evidence for the existence of syllables in sign languages – the syllable is a unit of structure that is below the level of the word but above the level of the segment. Syllables are defined as dynamic units, such as path movements or changes in handshape or in orientation. Certain phonological constraints must refer to the sign syllable; for example, handshape changes must be temporally coordinated with respect to movement at the level of the syllable, not at the level of the lexical word or morpheme (Brentari, 1998). However, there is little evidence for onset–rhyme distinctions within a sign syllable or resyllabification process. Finally, sign languages exhibit phonological rules such as assimilation, constraints on well formedness that refer to segmental structure, and form patterns can be accounted for using (manual) feature geometry. Thus, both signed and spoken languages exhibit a linguistically significant, yet meaningless, level of structure that can be analyzed as *phonology* for both language types.

Goldstein, Byrd, and Saltzman (this volume) argue that the evolution of phonology requires the emergence of discrete combinatorial units and that the non-iconic nature of vocal gestures is advantageous in this regard. It has been suggested that manual actions do not provide a ready source of discreteness to differentiate similar actions with different symbolic meaning. However, the existence and nature of modern sign languages argues against such a claim. Sign languages exhibit both iconicity and phonological structure (i.e., discrete combinatorial units that are non-meaningful). Therefore, it is possible that early gestural communication could evolve phonology.

Interestingly, the fact that sign languages exhibit phonological structure actually presents a problem for the theory of language evolution put forth by Arbib (2005; Chapter 1, this volume). Arbib (along with Corballis, 2002) hypothesizes that there was an early stage in the evolution of language in which communication was predominantly gestural. Arbib proposes that the transition from gesture to speech was not abrupt, and he suggests that "protosign" and "protospeech" developed in an expanding spiral until protospeech became dominant. The problem for this hypothesis is that signed languages are just as complex, just as efficient, and just as useful as spoken languages. As noted above, the manual modality is not deficient with respect to duality of patterning and affords the evolution of phonological structure. In addition, visual–gestural languages easily express abstract concepts, have recursive hierarchical syntactic structure, are used for artistic expression, are similarly acquired by children, etc. (see Emmorey (2002) for review). Thus, in principle, there is no linguistic reason why the expanding spiral between protosign and protospeech could not have resulted in the evolutionary dominance of sign over speech. A gestural origins theory of language evolution must explain why speech evolved at all, particularly when choking to death is a potential by-product of speech evolution due to the repositioning of the larynx (but see Clegg and Aiello, 2000).

Corballis (2002) presents several specific hypotheses for why speech might have won out over gesture, but none is satisfactory. Corballis suggests that speech may have an advantage because more arbitrary symbols are used, but sign languages also consist of arbitrary symbols, and there is no evidence that the iconicity of some signs limits expression or processing. The problem of signing in the dark is another oft-cited disadvantage for sign language. However, early signers/gesturers could sign in moonlight or firelight, and a tactile version of sign language could even be used if it were pitch black (i.e., gestures/signs are felt). Furthermore, speech has the disadvantage of attracting predators with sound at night or alerting prey during a hunt. Corballis argues that speech would allow for communication simultaneously with manual activities, such as tool construction or demonstration. However, signers routinely sign with one hand, while the other hand holds or manipulates an object (e.g., turning the steering wheel while driving and signing to a passenger). It is true that operation of a tool that requires two hands would necessitate serial manual activity, interspersing gesturing with object manipulation. But no deaths have been attributed to serial manual activity.

Everyone agrees that the emergence of language had clear and compelling evolutionary advantages. Presumably, it was these advantages that outweighed the dangerous change in the vocal tract that allowed for human speech but increased the likelihood of choking. If communicative pantomime and protosign preceded protospeech, it is not clear why protosign simply did not evolve into sign language. The evolutionary advantage of language would already be within the grasp of early humans.

4.3 Language and mirror neurons: mapping between perception and production

The existence of mirror neurons has been taken by some as supporting the motor theory of speech perception (Liberman, 1996) or the *direct realist* theory of speech perception (Fowler, 1986; Best, 1995). Under these theories, listeners perceive speech as gestures; that is, they parse the acoustic stream using gestural information (see Goldstein, Byrd, and Saltzman, this volume). However, as pointed out by Goldstein and Fowler (2003), the existence of mirror neurons is not particularly explanatory. For example, it is not at all clear how a mirror neuron "recognizes" the correspondence between a visually observed action and a self-executed action of the same form (but see Oztop, Bradey, and Arbib, this volume, for modeling that attempts to address this question). Similarly, we are far from understanding how a signer recognizes and represents the correspondence between a visually perceived sign and a self-produced sign. In fact, understanding how signer–viewers (or speaker–listeners) map production to perception and vice versa may have implications for understanding how mirror neurons come to achieve their function. For example, the potential role of mirror neurons for language is clear: to help make communication between individuals possible by establishing parity between self-produced forms and perceived forms. However, the functional role of mirror neurons in the context of behavior for non-human primates is currently unclear, particularly given that imitation is not a frequently observed behavior in the wild (Tomasello *et al.*, 1993). Understanding how parity of linguistic form emerges during development might provide some insight into the possible function(s) of mirror neurons in non-human primates.

A critical distinction between speech perception and sign perception is that the articulators are entirely visible for sign, but not for speech. For sign language, "what-you-see" is "what-you-produce" in terms of the relation between perception and production. In contrast, a major challenge for speech perception research is to understand the relation between the acoustic signal perceived by a listener and the movements of the vocal articulators. This is a complex problem because the same segment can have different acoustic realizations depending upon the surrounding context, speaking rate, and the individual speaker. Furthermore, different articulatory gestures can give rise to the same acoustic perception, and a simple invariant mapping between acoustic features and perceived phonemes has been extremely difficult to find, despite intense research efforts (Stevens and Blumstein, 1981; see Skipper, Nusbaum, and Small, this volume, for further discussion). For the visual signal, the problem of invariance takes the form of a problem in high-level vision: how are objects (e.g., the hands) recognized as "the same" when seen from different views and distances or in different configurations? Furthermore, vision researchers are struggling with the problem of how information about object motion, object shape, and object location are integrated (the "binding problem"). However, even when this problem is solved, the mapping between such recognition and patterns of motoneuron activity remains a challenge. Thus, although mapping between the perceived visual signal and mental representations of sign form may appear straightforward, how such perceptual mapping is achieved is not well understood.

Given that sign gestures are not hidden and do not have to be inferred from acoustic information, there might be no need for a distinction between perceptual and articulatory phonological features. That is, perceptual targets may not exist for sign. Only articulatory targets may be relevant. For speech, many sounds can be made in more than one way. For example, the same vowel can be produced with different jaw positions or a change in pitch can be produced either by extra respiratory effort or by altering the tension of the vocal cords (Ladefoged, 2000). What is critical is the perceptual target, not the specific articulatory means of achieving it. Crasborn (2001) argues that, despite the visibility of the sign articulators, perceptual targets exist for sign as well. The argument is based on whispering data from Sign Language of the Netherlands, but examples can be found in American Sign Language as well. Signers whisper by making reduced movements in a constricted volume when they wish to be visually quiet or do not want their signing to be viewed by others. The relevant observation here is that when some signs are whispered, a change in location (path movement) is realized as a change in orientation. Figure 4.2 provides an example from ASL in which the citation form of the sign NOT is produced with outward movement from the chin, whereas the whispered form is produced with a change in hand orientation and no path movement. Crasborn (2001) argues that change in location and change in orientation are abstract perceptual categories that can be articulated either by proximal or distal joints. That is, there is an abstract perceptual target that does not refer to the articulatory means employed to make it visible. Thus, despite directly observable articulations, perceptual factors nonetheless may characterize aspects of phonological form for sign language. However, it remains unclear whether such effects must be represented with perceptual (visual) features or whether they can be accounted for by sufficiently abstract articulatory features or specifications.

Finally, investigations of the link between perception and production for sign language may provide unique insight into the link between action observation and action execution in non-linguistic domains and in non-human primates. Some investigators have hypothesized that action perception may automatically and unconsciously engender an internal simulation of that action (a forward model) which can serve to anticipate observed actions (e.g., Blakemore and Decety, 2001). Sign language provides an outstanding tool for addressing questions about how the human action system might contribute to action perception. For example, sign perception might be accompanied by an unconscious internal simulation of sign production that might aid in the visual parsing of sign input – this is a hypothesis that we are currently exploring. The distinct biological basis of sign language results in a unique interface between vision and language and between action systems and language production. Investigation of this interface will not only inform us about relations between linguistic perception and production, but will also provide insight into broader issues within cognitive neuroscience.

4.4 Neural mechanisms underlying the production of speech, sign, and action

It has been hypothesized that the human homologue of area F5 in monkey is Broca's area, the classic speech production region for humans (Rizzolatti and Arbib, 1998; Iacoboni *et al.*, 1999; Arbib and Bota, this volume). Given that Broca's area is anterior to primary motor cortex controlling mouth and lip movements, it is reasonable that an area involved in speech production would be anatomically located near regions involved in control of the speech articulators. Is the same area involved in sign language production? Or is the functional equivalent of Broca's area shifted superiorly so that it is next to the motor representation for the hand and arm? Using cortical stimulation mapping, Corina *et al.* (1999) were able to identify the areas involved in mouth and lip movements in a deaf ASL signer. While the subject produced manual signs in a picture-naming task, they stimulated

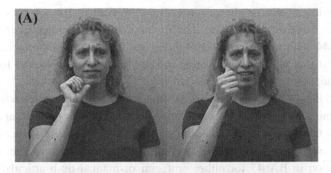

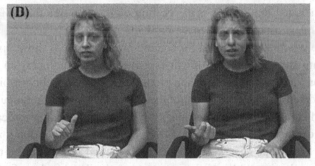

Figure 4.2 Illustration of (A) the citation form of the ASL sign NOT and (B) its whispered form. Whispering is used when signers wish to be visually quiet and/or when they do not want their signing to be viewed by others.

the posterior aspect of Broca's area (BA 44). This site was just anterior to the sites that evoked facial motor responses. Stimulation resulted in signing errors best characterized as phonetic or motor execution errors. Signs were articulated with non-specific movements and laxed handshape configurations. Semantic naming errors did not occur with stimulation to BA 44, nor did phonological errors (e.g., producing an erroneous, but clearly articulated, handshape substitution). In contrast, both phonological substitution errors and semantic errors occurred during sign production when stimulation was applied to the left supramarginal gyrus in the same deaf signer.

The cortical stimulation mapping results with this deaf ASL signer suggest that BA 44 is involved in phonetic aspects of linguistic expression, regardless of the anatomy of the language articulators. This hypothesis is supported by the results of Horwitz *et al.* (2003), who used a combination of cytoarchitectonic mapping and positron emission tomography (PET) data to investigate activation of Broca's area during the production of spoken and signed language by hearing ASL–English bilinguals. Subjects were asked to produce autobiographical narratives in either English or ASL, and the baseline comparison tasks were laryngeal and orofacial articulatory movements or bilateral non-routinized hand and arm movements (from Braun *et al.*, 1997, 2001). When contrasted with the baseline articulation conditions, there was little or no activation within BA 44 for sign production or for speech production. However, when the articulatory baseline tasks were contrasted with a rest condition, significant activation was observed in BA 44 for both language types (more for orofacial articulation than for manual articulation). Interestingly, activation in BA 44 within the right hemisphere was observed for orofacial articulation compared to rest, but very little activation in right BA 44 was observed for (bimanual) limb articulation.

In contrast to BA 44, Horwitz *et al.* (2003) found only left hemisphere activation in BA 45 for both signing and speaking. Activation in left BA 45 was only observed when the linguistic tasks (producing an English or an ASL narrative) were contrasted with the articulatory baseline tasks. The contrast between the articulation tasks and rest did not produce activation in BA 45 for either orofacial or manual-limb articulation. Horwitz *et al.* (2003) concluded that BA 45 is the portion of Broca's area that is fundamental to modality-independent aspects of language production.

Given that the articulators required for signing are the same as those involved in non-linguistic reaching and grasping, one might expect Broca's area to play an important role in both signing and in non-linguistic action production. However, for grasping tasks, hand configuration is determined by the nature of the object to be held or manipulated. For sign production, hand configuration is determined by the phonological specification stored in the lexicon. For example, as shown in Fig. 4.1B, the hand configuration for the ASL sign APPLE is an X handshape (fist with extended and curved index finger), rather than the hand configuration that would be used to grasp an apple. The hand configuration used to grasp objects is functionally determined, but the hand configuration for signs is dependent on the lexicon and phonology of a particular sign language, e.g., the sign for APPLE in British Sign Language is formationally distinct from ASL.

Nonetheless, ASL contains some signs that appear to mimic grasping motions and the hand configuration in these signs depicts how the human hand holds or manipulates an instrument or the hand depicts the instrument itself. Specifically, ASL *handling classifier verbs* denote actions performed with an implement. For example, the sign BRUSH-HAIR is produced with a grasping handshape and a brushing motion at the head (see Fig. 4.3). For *instrument classifier verbs*, the instrument itself is represented by the articulator, and the movement of the sign reflects the stylized movement of the tool or implement. Such verbs are referred to as *classifier verbs* because the handshape is morphemic and refers to a property of the referent object (e.g., the handle of a brush; the shape of a screwdriver); see papers in Emmorey (2003) for a discussion of classifier constructions in signed languages. As can be seen in Fig. 4.3A, the form of classifier verbs is quite iconic.

In addition, ASL nouns denoting tools or manipulable objects are often derived from instrument classifier verbs. For example, the sign SCREWDRIVER shown in Fig. 4.3B is made with a twisting motion, and the H handshape (fist with index and middle fingers extended) depicts the screwdriver itself, rather than how the hand would hold a screwdriver. In general, the movement of a noun in ASL reduplicates and shortens the movement of the related verb (Supalla and Newport, 1978). Thus, the twisting motion of the sign SCREWDRIVER is repeated and relatively short compared to that employed for the verb SCREW.

(A) BRUSH-HAIR   (B) SCREWDRIVER   (C) YELL

Figure 4.3 Illustration of an ASL verb and an ASL noun that resemble pantomime (A and B) and an ASL verb that does not exhibit such sensory–motor iconicity (C).

Given the motoric iconicity of handling classifier verbs and of many ASL nouns referring to manipulable objects, Emmorey *et al.* (2004) investigated whether such iconicity impacts the neural systems that underlie tool and action naming for deaf ASL signers. In this PET study, signers were asked to name actions performed with an implement (producing handling classifier verbs), actions performed without a tool (e.g., YELL in Fig. 4.3C), and manipulable objects (nouns referring to tools or utensils). The results showed that the sensory–motoric iconicity of ASL signs denoting tools (e.g., SCREWDRIVER) and of handling classifier verbs denoting actions performed with a tool (e.g., BRUSH-HAIR) does not alter the neural systems that underlie lexical retrieval or sign production.

Specifically, naming tools or tool-based actions engaged a left premotor–parietal cortical network for both signers and speakers. The neural activation maximum observed within left premotor cortex for ASL handling classifier verbs was similar to the premotor activation observed when English speakers named the function or use of tools (Grafton *et al.*, 1997). Naming tools with iconic ASL signs also engaged left premotor cortex, and this activation maximum was similar to that found when English speakers named tools (Grabowski *et al.*, 1998; Chao and Martin, 2000). These premotor sites were generally superior to Broca's area. Activation maxima within the left inferior parietal cortex when naming tools or tool-based actions were also similar for signers and speakers. When signers or speakers named actions that did not involve tools (e.g., yelling, sleeping) or non-manipulable objects (e.g., animals that are not pets) activation within this left premotor–parietal network was not observed (Damasio *et al.*, 1996, 2001; Emmorey *et al.*, 2004, 2003a). Thus, activation within this network may represent retrieval of knowledge about the sensory- and motor-based attributes of grasping movements associated with tools and commonly manipulated objects.

In addition, Broca's area (BA 45) was activated when signers named tools, tool-based actions, and actions without tools (Emmorey *et al.*, 2004). This activation is hypothesized to reflect lexical retrieval and/or selection processes during naming (Emmorey, 2006). Activation in Broca's area is not reported when subjects are asked to pantomime tool use (Moll *et al.*, 2000; Choi *et al.*, 2001). The fact that the production of ASL verbs resembling pantomime (e.g., BRUSH-HAIR) and non-pantomimic verbs (e.g., YELL) both engaged Broca's area suggests that handling classifier verbs are lexical forms, rather than non-linguistic gestures. This result is complemented by two case studies of aphasic signers who exhibit a dissociation between the ability to sign and to pantomime (Corina *et al.*, 1992; Marshall *et al.*, 2004). Corina *et al.* (1992) describe the case of WL who had a large frontotemporoparietal lesion in the left hemisphere. The lesion included Broca's area, the arcuate fasciculus, a small portion of inferior parietal lobule (BA 40) and considerable damage to the white matter deep to the inferior parietal lobule. WL exhibited poor sign comprehension, and his signing was characterized by phonological and semantic errors with reduced grammatical structure. An example of a phonological error by WL was his production of the sign SCREWDRIVER. He substituted an A-bar handshape (fist with thumb extended) for the required H handshape (see Fig. 4.3B). In contrast to his sign production, WL was unimpaired in his ability to produce pantomime. For example, instead of signing DRINK (a C handshape (fingers extended and curved as if holding a cup) moves toward the mouth, with wrist rotation – as if drinking), WL cupped his hands together to form a small bowl. WL was able to produce stretches of pantomime and tended to substitute pantomimes for signs, even when pantomime required more complex movements. Such pantomimes were not evident before his brain injury.

Marshall *et al.* (2004) report a second case of a deaf aphasic signer who demonstrated a striking dissociation between pantomime and sign (in this case, British Sign Language). "Charles" had a left temporoparietal lesion and exhibited sign anomia that was parallel to speech anomia. For example, his sign-finding difficulties were sensitive to sign frequency and to cueing, and he produced both semantic and phonological errors. However, his pantomime production was intact and superior to his sign production even when the forms of the signs and pantomimes were similar. Furthermore, this dissociation was impervious to the iconicity of signs. His production of iconic signs was as impaired as his production of non-iconic signs. Thus, the lesion data complement the neuroimaging results. The neural systems supporting sign language production and pantomimic expression are non-identical. Specifically, Broca's area is engaged during the lexical retrieval of signs denoting tools and tool-based actions, but not during tool-use pantomime (Moll *et al.*, 2000; Choi *et al.*, 2001).

Although Broca's area (specifically, BA 44) is activated when subjects imitate non-linguistic finger movements (Iacoboni *et al.*, 1999; Tanaka and Inui, 2002), activation in Broca's area has not been found when signers imitate (repeat) visually presented signs, in contrast to visual fixation (Petitto *et al.*, 2000, supplemental material). Activation in the posterior inferior frontal gyrus (including Broca's area) is rarely reported when speakers repeat (i.e., imitate) spoken words (Indefrey and Levelt, 2000). Repetition of manual signs (or oral words) does not require lexical retrieval/selection and is cognitively quite distinct from imitating novel manual movements because mental representations specifying form exist only for lexical items, not for finger/hand movements. Thus, although Broca's area may be engaged during the imitation of manual actions, it appears to play a different role in sign language production.

Finally, Fig. 4.4 illustrates cortical sites that are consistently engaged during the production of sign language. All six studies included in this summary figure found strong left hemisphere lateralization for sign production. In addition, the same perisylvian regions implicated in spoken language production are also implicated in sign language production: left inferior frontal gyrus, left posterior temporal cortices, and left inferior parietal lobule (Indefrey and Levelt, 2004). However, the left superior parietal lobule is more involved in sign production than in speech production. In Fig. 4.4, the white circles (with letters) indicate sites that showed greater activation for signing than for speaking, specifically the left superior parietal lobule and left supramarginal gyrus (Braun *et al.*, 2001; Emmorey and Grabowski, 2004). In their meta-analysis of imaging studies of spoken word production, Indefrey and Levelt (2004) found that the superior parietal lobule was rarely engaged during speech production (but see Watanabe *et al.*, 2004). In contrast, several studies have found activation in the superior parietal lobule during action imitation and pantomime production (Iacoboni *et al.*, 1999; Moll *et al.*, 2000; Choi *et al.*, 2001; Tanaka *et al.*, 2001; Tanaka and Inui, 2002).

Emmorey *et al.* (2004) hypothesize that for signing, the left superior parietal lobule may play a role in the proprioceptive monitoring of sign production. Studies with both non-human primates (e.g., Mountcastle *et al.*, 1975) and humans (e.g., Wolpert *et al.*, 1998; McDonald and Paus, 2003) demonstrate an important role for the superior parietal lobule in proprioception and the assessment of self-generated movements. Proprioceptive monitoring may play a more important role in sign production because visual monitoring of signed output presents an unusual signal for language perception. For spoken language, speakers can monitor their speech output by listening to their own voice — a perceptual loop feeds back to the speech comprehension mechanism. In contrast, signers do not look directly at their hands and cannot see their own faces. The visual input from their own signing is quite distinct from the visual input of another's signing. Therefore, a simple perceptual loop that feeds back to the sign comprehension mechanisms is problematic. Sign production may crucially involve the proprioceptive monitoring of movement, particularly since sign production is not visually guided.

In sum, the evidence to date suggests that action production (e.g., pantomime), sign production, and speech production are subserved by similar, but non-identical neural systems. Pantomime production can be dissociated from sign language production (Corina *et al.*, 1992; Marshall *et al.*, 2004), and the pantomimic properties of (certain) signs do not alter the neural systems that underlie their production (Emmorey *et al.*, 2004). Thus, sign and pantomime production depend upon partially segregated neural systems. Sign production parallels speech production in that both engage left perisylvian regions. These regions appear to support similar language production functions, such as lexical selection and phonological encoding (see San José-Robertson *et al.* (2004), for further evidence and discussion). However, sign production parallels non-linguistic action production in that both engage the superior parietal lobule, perhaps due to the need to internally monitor limb movements.

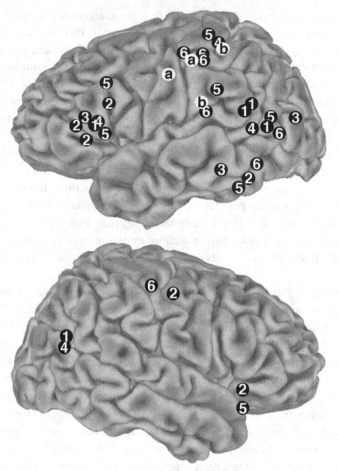

Figure 4.4 Peak activation sites from six studies of *sign production* are overlaid onto an individual brain, using *Y* and *Z* Talairach coordinates. Talairach coordinates specify locations within a standard three-dimensional brain space, where *X* specifies the lateral–medial dimension, *Y* specifies the anterior–posterior dimension, and *Z* specifies the superior–inferior dimension (Talairach and Tournoux, 1988). Medial, basal, and subcortical activations are not shown. **1**: Braun *et al.* (2001): ASL – limb motor comparison (p.2036). **2**: Corina *et al.* (2003): ASL right-hand verb generation – ASL right-hand Noun Repetition (p.722). **3**: Emmorey *et al.* (2003a): naming animals – yes/no judgment of upright/inverted faces (p.90–91). **4**: Emmorey *et al.* (2004): naming actions performed without an implement – yes/no judgment of upright/inverted faces (p.34). **5**: Petitto *et al.* (2000): generate ASL verb – view ASL signs (p.13963 and supplemental data). **6**: San José-Robertson *et al.* (2004); Repetition of nouns – passive viewing of nouns (p.160). Isolated peak activation sites (farther than 10 mm from an activation peak reported by another study) were not included in the figure in order to provide a summary illustration of sites consistently activated during sign production. **White circles: a**: Braun *et al.* (2001): activation sites that were more engaged for the production of ASL narratives than English narratives in hearing ASL-English bilinguals, using individual motor contrasts in the interaction (p.107). **b**: Emmorey and Grabowski (2004): activation sites that were more engaged during the production of ASL nouns by deaf signers compared to the production of English nouns by hearing speakers, using a yes/no judgement (either signed or spoken response) of upright/inverted faces in the interaction.

4.5 Neural mechanisms underlying the perception of speech, sign, and action

Given that Wernicke's area is adjacent to auditory cortex, deficits in comprehension following temporal lobe damage might be expected for spoken language but perhaps not for signed language, which does not depend upon auditory input for perception. However, Hickok *et al.* (2002) reported that deaf signers with damage to the left temporal lobe performed significantly worse on tests of ASL comprehension than signers with left frontal and/or parietal damage and worse than signers with right temporal lobe damage. The comprehension deficit associated with left temporal lobe damage was most severe for complex sentences (e.g., multi-clausal commands) but was also observed for simple sentences (single clause commands) and single sign comprehension. Hickok *et al.* (2002) concluded that language comprehension for either spoken or signed language depends upon the integrity of the left temporal lobe.

Functional imaging data support this conclusion. Several studies using a variety of techniques have reported activation within left temporal cortices during sign language comprehension (Söderfeldt *et al.*,1994, 1997; Neville *et al.*, 1998; Nishimura *et al.*, 1999; Petitto *et al.*, 2000; Levänen *et al.*, 2001; MacSweeney *et al.*, 2002, 2004). In addition, several studies have found activation within auditory cortices, including the planum temporale, during the visual perception of sign language (Nishimura *et al.*, 1999; Petitto *et al.*, 2000; MacSweeney *et al.*, 2002). Thus, temporal lobe structures, including regions generally considered to be dedicated to auditory processing are recruited during sign language comprehension.

In contrast, deficits in the perception of actions or the recognition of pantomime are not typically associated with damage to superior temporal cortex. Rather, deficits in action and pantomime recognition are most often associated with apraxia and parietal damage (e.g., Heilman *et al.*, 1982; Rothi *et al.*, 1985). Furthermore, Corina *et al.* (1992) and Marshall *et al.* (2004) reported dissociations not only in production (see above) but also between pantomime comprehension and sign language comprehension. Specifically, single sign comprehension can be impaired, while the ability to interpret action pantomimes is spared. The available evidence suggests that the neural systems that underlie sign language comprehension and action recognition are non-identical and that superior temporal cortices may be more involved in sign language comprehension.

Nonetheless, a recent functional magnetic resonance imaging (fMRI) study by MacSweeney *et al.* (2004) found a surprisingly high degree of similarity between the neural systems supporting sign language perception and the perception of non-meaningful gestures that resemble sign language (Tic Tac, a gestural system used by bookies to communicate odds, unknown to the study participants). In particular, activation was observed in the superior temporal gyrus and the planum temporale for all study participants, regardless of hearing status or knowledge of British Sign Language (BSL). In contrast, Petitto *et al.* (2000) found no activation within the planum temporale when hearing non-signers passively viewed either lexical signs or phonologically legal pseudo-signs, but planum temporale activation was observed for deaf signers for both signs and

Karen Emmorey

pseudosigns. Task differences between the two studies may account for the conflicting results. Participants in the Petitto *et al.* (2000) study passively viewed signs, whereas in the MacSweeney *et al.* (2004) study, both deaf and hearing participants were asked to indicate which BSL sentence or Tic Tac "sentence" was semantically anomalous. In an attempt to find meaning in the gestural stimuli, hearing participants may have tried to label the gestures with English. Activation in the planum temporale has been observed for internally generated speech (e.g., McGuire *et al.*, 1996). Thus, activation in the planum temporale for hearing non-signers may have been due to self-generated internal speech, rather than to the perception of gestural stimuli.

The role of the superior temporal gyrus and the planum temporale in the perception of sign language is not clear. One possibility suggested by Petitto *et al.* (2000) is that the planum temporale "may be dedicated to processing specific distributions of complex, low-level units in rapid temporal alternation, rather than to sound, *per se.*" (p. 13966) However, the rapid temporal alternations of sign do not approach the 40 ms rate found for the sound alternations of speech, arguing against an explanation based on rapid temporal patterns. Nonetheless, the planum temporale may be recruited during the visual perception of gestural phonetic units for both spoken and signed language. Several studies have reported activation within the planum temporale when hearing people lip-read silent speech (e.g., Calvert *et al.*, 1997; MacSweeney *et al.*, 2000; Campbell *et al.*, 2001). The superior temporal gyrus may be critical to interpreting gestures that are organized within a linguistic system of contrasts, i.e., phonological units. Finally, hearing ability modulates activation within the superior temporal gyrus (including the planum temporale), this activation being greater for deaf than hearing signers (MacSweeney *et al.*, 2004; Sadato *et al.*, 2004). Further, non-linguistic visual stimuli (e.g., coherent dot-motion) produce activation within the superior temporal gyrus for deaf individuals (Finney *et al.*, 2001; Sadato *et al.*, 2004). Thus, sign language perception may engage auditory cortices to a greater extent for deaf signers because auditory deprivation from birth induces cross-modal plasticity.

Fig. 4.5 illustrates cortical sites that are consistently engaged during sign language comprehension. In this summary illustration, activation sites are displayed on the lateral surface of the brain; thus, sites within the sylvian fissure (BA 42) and superior temporal sulcus appear on the surface. Most studies report greater left than right hemisphere involvement for sign comprehension (Neville *et al.* (1998) is the exception). Some controversy surrounds whether sign comprehension recruits the right hemisphere to a greater extent than speech comprehension (Bavelier *et al.*, 1998; Hickok *et al.*, 1998). However, when sign language perception is directly compared to audiovisual speech (rather than to reading), bilateral activation is observed for both sign and speech, with more extensive activation in the left hemisphere (MacSweeney *et al.*, 2002; Sakai *et al.*, 2005).

As illustrated in Fig. 4.5, left perisylvian regions are engaged during sign language comprehension, as they are for speech. Specifically, activation is found in left superior temporal cortex, left parietal cortex, and left inferior frontal cortex (Broca's area). In

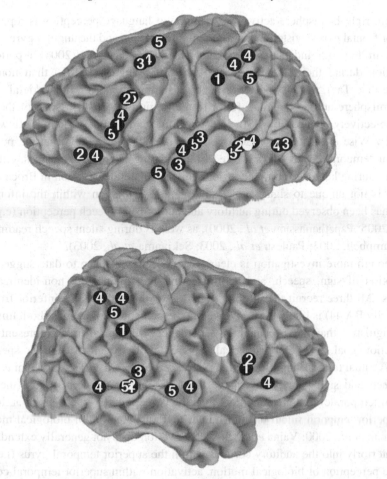

Figure 4.5 Peak activation sites from six studies of *sign perception* are overlaid onto an individual brain, using *Y* and *Z* Talairach coordinates. Medial, basal, and subcortical activations are not shown. **1**: MacSweeney *et al*. (2002): perception of British Sign Language (BSL) sentences and monitoring for a semantic anomaly – viewing the signer at rest while monitoring the display for a change (a small dot superimposed on the chin of the signer changed color); deaf native signers (p.1587). **2**: MacSweeney *et al*. (2004): same tasks as MacSweeney *et al*. (2002) but with a different set of BSL sentences; deaf native signers (p.1609): **3**: Petitto *et al*. (2000): view ASL signs – fixation (p.13965; supplemental data). **4**: Sadato *et al*. (2004): Viewing animated signs (Japanese Sign Language) and making a lexical decision – viewing an upside-down signer; deaf non-native signers (p.4). **5**: Newman *et al*. (2002: viewing ASL sentences – viewing meaningless sign-like gestures (stimuli recognition was tested after scanning); hearing native ASL signers (note: International Consortium for Brain Mapping (ICBM) coordinates were reported (using Montreal Neurological Institute (MNI)'s average T1 MRI) and do not necessarily represent the true centroid of activation). (p.77). Isolated peak activation sites (farther than 10 mm from an activation peak reported by another study) were not included in the figure in order to provide a summary illustration of sites consistently activated during sign perception. **White circles:** Activation sites that were more active during the perception of BSL than during the perception of Tic Tac, a gestural system used to communicate betting odds (MacSweeney *et al*., 2004; pg. 1612). A basal activation site is not shown.

addition, right hemisphere activation during sign language perception is reported in the inferior frontal gyrus, posterior superior temporal cortex, and the angular gyrus. The white circles in Fig. 4.5 indicate sites where MacSweeney *et al.* (2004) reported greater activation during the perception of British Sign Language (BSL) than non-linguistic gesture (Tic Tac), namely, posterior superior temporal and inferior parietal sites in the left hemisphere and a bilateral inferior frontal region (BA 6/44). None of these regions were selectively engaged by BSL perception for hearing non-signers. Following a proposal by Wise *et al.* (2001), MacSweeney *et al.* (2004) conclude that the posterior left superior temporal cortex plays a role in representing phonetic sequences, either oral or manual, during language comprehension. Finally, bilateral activation in Broca's area (BA 44/45) is not unique to sign perception. Bilateral activation within the inferior frontal gyrus has been observed during auditory and audiovisual speech perception (e.g., Skipper *et al.*, 2005; Papathanassiou *et al.*, 2000), as well as during silent speech reading (Calvert and Campbell, 2003; Paulesu *et al.*, 2003; Sekiyama *et al.*, 2003).

Although more investigation is clearly needed, the evidence to date suggests that the perception of sign, speech, and action recruit overlapping, but non-identical, cortical regions. All three recruit the inferior frontal gyrus bilaterally. The inferior frontal gyrus (primarily BA 44) is likely to be involved in the perception of actions (both linguistic and non-linguistic) that are executed in accordance with an abstract representation of a production goal, i.e., a phonological representation for either sign or speech and a cognitive goal for non-linguistic actions (see Rizzolatti *et al.*, 2002; Iacoboni *et al.*, 2005).

Speech and sign perception engage the left superior temporal gyrus (including the planum temporale) to a greater extent than the perception of other human actions. While the superior temporal sulcus is engaged during the perception of biological motion (e.g., Grossman *et al.*, 2000; Vaina *et al.*, 2001), activation does not generally extend superiorly and anteriorly into the auditory cortices within the superior temporal gyrus (i.e., BA 42). For the perception of biological motion, activation within superior temporal cortex tends to be bilateral or right lateralized, while for language perception activation is greater and more extensive within left superior temporal cortex (see also McCullough *et al.*, 2005).

Action and sign perception appear to engage parietal cortex to a greater extent than speech perception. Although left inferior parietal cortex is involved in the phonological processing of speech (particularly near the temporal–parietal–occipital junction; see Hickok and Poeppel, 2000), parietal activation appears to be more extensive and more superior and posterior for sign perception. The activation peaks reported for sign perception often lie within the superior parietal lobule (BA 7) or near the border between the inferior and superior parietal lobules. In addition, Sakai *et al.* (Sekiyama *et al.*, 2005) reported greater activation within the inferior parietal lobule bilaterally for sign perception compared to speech perception. Differential parietal activation during sign perception may simply reflect the biology of sign production. Namely, signing involves bilateral hand and arm movements, as well as facial movements (distinct facial expressions convey linguistic contrasts). Within parietal cortex, action observation exhibits a dorsal to ventral somatotopic organization for foot, hand, and mouth actions (Buccino *et al.*, 2001). More dorsal activation in parietal cortex during sign observation may reflect involvement of the hands and arms in sign production.

In sum, sign perception, like speech perception but unlike action observation, engages the superior temporal gyri bilaterally ("auditory" cortices). Sign perception, like action observation but unlike speech perception, engages left posterior parietal cortex. Finally, the perception of sign, speech, and action all engage the inferior frontal gyrus (BA 44) in both the left and right hemisphere.

4.6 Discussion

To date, no study has explicitly investigated the neural regions that are engaged during both sign production and sign perception, i.e., the mirror neuron system for sign language. However, a comparison of separate studies suggests that frontoparietal regions (and possibly the left posterior superior temporal sulcus) are engaged during both perception and production (see Fig. 4.4 and 4.5). These regions are also engaged during the observation and execution of non-linguistic actions such as grasping, finger and hand movements, and even dancing (Calvo-Merino *et al.*, 2004). Regions that appear to be engaged during sign perception, but not production, include the left superior temporal gyrus and the left middle temporal gyrus. Regions that may be engaged only during sign production include the left anterior superior parietal lobule and possibly the left inferior temporal cortex (at least for single sign production). Clearly, more research is needed to map out the neural systems that support sign language.

As hypothesized by Arbib (Chapter 1, this volume; Rizzolatti and Arbib, 1998), the mirror neuron system for language appears to be strongly related to the mirror neuron system for action. The fact that modern sign languages in some sense emerge from action and pantomime provides interesting insight into changes that may occur when the mirror neuron system supporting action observation/execution comes to subserve language perception/production. Evidence from emerging sign languages such as Nicaraguan Sign Language (Senghas, 2003), as well as historical analysis of established sign languages, indicate that pantomimic-like communication can evolve into language. For example, diachronic linguistic analyses have traced grammaticalization pathways in ASL that originate with gesture (Janzen and Shaffer, 2002). For instance, grammatical markers of modality in ASL (e.g., CAN, MUST) are derived from lexical signs (STRONG, OWE), and these lexical signs are in turn derived from non-linguistic communicative gestures (clenching the fists and flexing muscles to indicate strength and a deictic pointing gesture indicating monetary debt). Of course, ASL is a modern sign language acquired and created by modern human brains, but the evidence illustrates the potential for communicative gestures to evolve into language.

As language emerges from pantomime and a community of users arises, duality of patterning and phonological structure appear. Linguistic actions (signs) not only have symbolic meaning and syntactic properties (e.g., they belong to syntactic classes), they also have form features that are not governed by meaning or syntax. One role of the mirror neuron system for language may be to establish (and maintain) parity between a self-generated sign and a visually perceived sign produced by another. The system may exploit an internalized phonological system of contrasts (i.e., phonology) to determine what counts as "the same" sign. It is notable that some mirror neurons identify similar actions as belonging to the same category (e.g., "grasping"); for example some mirror neurons fire during observation of different kinds of grasping actions (Gallese *et al.*, 1996). Similarly, ASL signers exhibit categorical perception effects (better discrimination across phonological categories than within a category), particularly for hand configuration (Emmorey *et al.*, 2003b). Thus, the mirror neuron system may code actions and signs with respect to abstract representations of a production goal that do not necessarily specify a particular muscle or joint configuration to be used in the articulation of the action or sign.

However, the production goals for language (either sign or speech) may in part be perceptually motivated. Here the mirror neuron system for action may differ from that for sign due to the distinctive interplay between perception and production for language. For example, visibility can affect the form of articulation for sign but probably has little influence on the form of grasping. During whispering (see Fig. 4.2), the form of a sign is reduced and altered. Conversely, when signing to a large audience, the form of signs is perceptually enhanced (e.g., more proximal joints are used; Brentari, 1998). For both sign and speech, there is a trade off between ease of articulation and ease of perception (for discussion see Emmorey, 2005). Thus, sign articulations are not like other types of manual activities, such as reaching or grasping. Speakers alter their production to suit listeners – "speaking for listening" (Cutler, 1987), and signers alter their production to suit viewers – "signing for viewing" (Emmorey, 2005). But there does not appear to be "grasping for viewing" (although there may be "dancing for viewing"!).

The mirror neuron system may play a role during sign perception that parallels the role proposed by Skipper, Nusbaum and Small (this volume) for speech. Skipper *et al.* argue that the mirror neuron system creates an inverse model of speech production by mapping audiovisual speech to abstract representations of the actions that the perceiver herself would have produced. These action representations are mapped to the relevant pre- and primary motor cortices, and the resulting motoric activation results in a forward model (a motor hypothesis), which creates sensory feedback that can be compared with the incoming information. In this way, inverse-forward models predict upcoming sensory information thus reducing the number of alternative interpretations for acoustic patterns and thereby improving speech perception. Skipper *et al.* suggest that this process allows cognitive resources to be shifted to other aspects of the linguistic signal, such as semantic interpretation.

It is possible that the mirror neuron system performs a similar function for sign perception. Viewing signing may activate abstract representations of sign actions via the mirror neuron system, which results in inverse-forward models that can be used to predict upcoming sensory information thereby facilitating sign perception. However, Skipper *et al.* argue that for speech the mirror neuron system is often not engaged during auditory-only speech perception because "it would be quite costly to initiate inverse-forward models if the predictive value were low and the redundant information high as in auditory speech perception alone." Is there any evidence that inverse-forward models might facilitate sign perception? One bit of evidence is the fact that late learners of ASL take significantly longer to identify ASL signs than native signers in a visual gating task (a sign is presented repeatedly, with the length of each presentation increased by one videoframe, and subjects attempt to identify the sign after each presentation; Emmorey and Corina, 1990). If native signers have efficient inverse-forward models, they could better predict the trajectory of a sign action and thus identify the sign earlier. Late-learners of sign language have been argued to have a "phonological bottleneck" that causes them to expend additional cognitive effort deciphering phonological patterns, which prevents them from quickly accessing the meanings of lexical signs (Mayberry and Fischer, 1989; Mayberry, 1994, 1995). This pattern of results suggests that the mirror neuron system might play a role in facilitating early visual–phonetic processing.

Finally, for modern humans the mirror neuron system is "ready" for language and need not depend upon symbolic action or pantomime precursors. For example, deaf signing children do not start with gestures or pantomimes and then move to language. Although there may be an early communicative advantage for gesture for both hearing and deaf children (see Volterra and Iverson, 1995), communicative gesture does not drive sign language acquisition. Deaf children's first signs are not iconic or pantomimic (Newport and Meier, 1985); signing children sometimes make errors when producing personal pronouns which are identical to pointing gestures (e.g., a child points to herself to mean "you"; Petitto, 1987); and signing children have difficulty acquiring headshake as a grammatical marker for negation, despite its early use to communicate "no!" (Anderson and Reilly, 1997). Signs (and non-manual grammatical markers) are acquired as part of a linguistic system, and children do not automatically adopt communicative actions into this linguistic system.

The acquisition facts are consistent with the neuropsychological and neuroimaging data indicating that gesture and sign language are subserved by partially segregated neural systems. Pantomime comprehension and production can be preserved, while the comprehension and production of signs (even iconic signs) is impaired. However, imaging studies that directly compare sign and gesture production or compare action observation and sign/gesture comprehension have not yet been conducted. Such studies will help to determine how the mirror neuron system responds when similar human actions constitute meaningful symbols, transitive actions, or meaningful forms that are part of a linguistic system.

Acknowledgments

Preparation of this chapter was supported by National Institutes of Health grants RO1 HD13249 from the National Institute of Child Health and Human Development and RO1 DC006708 from the National Institute of Deafness and other Communicative Disorders and National Science Foundation grant BCS 0132291 (Linguistics program).

References

Anderson, D., and Reilly, J. S., 1997. The puzzle of negation: how children move from communicative to grammatical negation in ASL. *Appl. Psycholing.* **18**: 411–429.

Arbib, M. A., 2005. From monkey-like action recognition to human language: an evolutionary framework for neurolinguistics. *Behav Brain Sci.* **28**: 105–167.

Bavelier, D., Corina, D., Jezzard, P., *et al.*, 1998. Hemispheric specialization for English and ASL: left invariance-right variability. *Neuroreport* **9**: 1537–1542.

Best, C., 1995. A direct realist perspective on cross-language speech perception. In W. Strange and J. J. Jenkins (eds.) *Cross-Language Speech Perception*. Timonium, MD: York Press, pp. 171–204.

Blakemore, S.-J., and Decety, J., 2001. From the perception of action to the understanding of intention. *Nature Rev.* **2**: 561–567.

Braun, A. R., Varga, M., Stager, S., *et al.*, 1997. Altered patterns of cerebral activity during speech and language production in developmental stuttering: an H_2 (15)O positron emission tomography study. *Brain* **120**: 761–784.

Braun, A. R., Guillemin, A., Hosey, L., and Varga, V., 2001. The neural organization of discourse: an $H_2^{15}O$ PET study of narrative production in English and American Sign Language. *Brain* **124**: 2028–2044.

Brentari, D., 1998. *A Prosodic Model of Sign Language Phonology*. Cambridge, MA: MIT Press

Buccino, G., Binkofski, F., Fink, G. R., *et al.*, 2001. Action observation activates premotor and parietal areas in a somatotopic manner: an fMRI study. *Eur. J. Neurosci.* **13**: 400–404.

Calvert, G. A., and Campbell, R., 2003. Reading speech from still and moving faces: the neural substrates of visible speech. *J. Cogn. Neurosci.* **15**: 57–70.

Calvert, G. A., Bullmore, E. T., Brammer, M. J., *et al.*, 1997. Activation of auditory cortex during silent lipreading. *Science* **276**: 593–596

Calvo-Merino, B., Glaser, D. E., Grezes, J., Passingham, R. E., and Haggard, P., 2004. Action observation and acquired motor skills: an fMRI study with expert dancers. *Cereb Cortex.* **15**: 1243–1249.

Campbell, R., MacSweeney, M., Sugurladze, S., *et al.*, 2001. The specificity of activation for seen speech and form meaningless lower-face acts (gurning). *Brain Res. Cog. Res.* **12**: 233–243.

Chao, L., and Martin, A., 2000. Representation of manipulable man-made objects in the dorsal stream. *Neuroimage* **12**: 478–484.

Choi, S. H., Na, D. L., Kang, E., *et al.*, 2001. Functional magnetic resonance imaging during pantomiming tool-use gestures. *Exp. Brain Res.* **139**: 311–317.

Clegg, M., and Aiello, L. C., 2000. Paying the price for speech? An analysis of mortality statistics for choking on food. *Am. J. Phys. Anthropol.* (Suppl. 30), **126**: 9482–9483.

Corballis, M., 2002. *From Hand to Mouth: The Origins of Language*. Princeton, NJ: Princeton University Press.

Corina, D. P., and Sandler, W., 1993. On the nature of phonological structure in sign language. *Phonology* **10**: 165–207.

Corina, D. P., Poizner, H., Bellugi, U., *et al.*, 1992. Dissociation between linguistic and non-linguistic gestural systems: a case for compositionality. *Brain Lang.* **43**: 414–447.

Corina, D. P., McBurney, S. L., Dodrill, C., *et al.*, 1999. Functional roles of Broca's area and supramarginal gyrus: evidence from cortical stimulation mapping in a deaf signer. *Neuroimage* **10**: 570–581.

Crasborn, O., 2001. *Phonetic Implementation of Phonological Categories in Sign Language of the Netherlands*. LOT: Utrecht, Netherlands.

Cutler, A., 1987. Speaking for listening. In A. Allport and D. MacKay (eds.), *Language Perception and Production: Relationship between Listening, Speaking, Reading, and Writing*. San Diego, CA: Academic Press, pp. 23–40.

Damasio, H., Grabowski, T. J., Tranel, D., Hichwa, R., and Damasio, A. R., 1996. A neural basis for lexical retrieval. *Nature* **380**: 499–505.

Damasio, H., Grabowski, T. J., Tranel, D., *et al.*, 2001. Neural correlates of naming actions and of naming spatial relations. *Neuroimage* **13**: 1053–1064.

Emmorey, K., 2002. *Language, Cognition, and the Brain: Insights from Sign Language Research*. Mahwah, NJ: Lawrence Erlbaum.

(ed.), 2003. *Perspectives on Classifier Constructions in Signed Languages*. Mahwah, NJ: Lawrence Erlbaum.

2005. Signing for viewing: some relations between the production and comprehension of sign language. In A. Cutler (ed.) *Twenty-First Century Psycholinguistics: Four Cornerstones*. Mahwah, NJ: Lawrence Erlbaum, pp. 293–309.

2006. The role of Broca's area in sign language. In Y. Grodzinsky and K. Amunts (eds.) *Broca's Region*. Oxford, UK: Oxford University Press, pp. 167–182.

Emmorey, K., and Corina, D., 1990. Lexical recognition in sign language: effects of phonetic structure and morphology. *Percept. Motor Skills* **71**: 1227–1252.

Emmorey, K., and Grabowski, T., 2004. Neural organization for sign versus speech production. *J. Cogn. Neurosci.* (suppl.): 205.

Emmorey, K., Grabowski, T., McCullough, S., *et al.*, 2003a. Neural systems underlying lexical retrieval for sign language. *Neuropsychologia* **41**: 85–95.

Emmorey, K., McCullough, S., and Brentari, D., 2003b. Categorical perception in American Sign Language. *Lang. Cogn. Processes* **18**: 21–45.

Emmorey, K., Grabowski, T., McCullough, S., *et al.*, 2004. Motor-iconicity of sign language does not alter the neural systems underlying tool and action naming. *Brain Lang.* **89**: 27–37.

Finney, E. M., Fine, I., and Dobkins, K. R., 2001. Visual stimuli activate auditory cortex in the deaf. *Nature Neurosci.* **4**: 1171–1173.

Fowler, C., 1986. An event approach to the study of speech perception: a direct-realist perspective. *J. Phonet.* **14**: 3–28.

Gallese, V., Fadiga, L., Fogassi, L., and Rizzolatti, G., 1996. Action recognition in the premotor cortex. *Brain* **119**: 593–609.

Gallese, V., Fadiga, L., Fogassi, L., and Rizzolatti, G., 2002. Action representation and the inferior parietal lobule. In W. Prinz and B. Hommel (eds.) *Common Mechanisms in Perception and Action*. Oxford, UK: Oxford University Press, pp. 334–355.

Goldstein, L., and Fowler, C., 2003. Articulatory phonology: a phonology for public language use. In N. Schiler and A. Meyer (eds.) *Phonetics and Phonology in Language Comprehension and Production*. New York: Mouton de Gruyter, pp. 159–208.

Grabowski, T. J., Damasio, H., and Damasio, A., 1998. Premotor and prefrontal correlates of category-related lexical retrieval. *Neuroimage* **7**: 232–243.

Grafton, S., Fadiga, L., Arbib, M., and Rizzolatti, G., 1997. Premotor cortex activation during observation and naming of familiar tools. *Neuroimage* **6**: 231–236.

Grossman, E., Donnelly, M., Price, R., *et al.*, 2000. Brain areas involved in perception of biological motion. *J. Cogn. Neurosci.* **12**: 711–720.

Heilman, K. M., Rothi, I. J., and Valenstein, E., 1982. Two forms of ideomotor apraxia. *Neurology* **32**: 342–346.

Hickok, G., and Poeppel, D., 2000. Towards a functional neuroanatomy of speech perception. *Trends Cogn. Sci.* **4**: 131–138.

Hickok, G., Bellugi, U., and Klima, E. S. (1998). What's right about the neural organization of sign language? A perspective on recent neuroimaging results. *Trends Cogn. Sci.* **2**: 465–468.

Hickok, G., Love-Geffen, T., and Klima, E. S., 2002. Role of the left hemisphere in sign language comprehension. *Brain Lang.* **82**: 167–178.

Horwitz, B., Amunts, K., Bhattacharyya, R., *et al.*, 2003. Activation of Broca's area during the production of spoken and signed language: a combined cytoarchitectonic mapping and PET analysis. *Neuropsychologia* **41**: 1868–1876.

Iacoboni, M., Woods, R. P., Brass, M., *et al.*, 1999. Cortical mechanisms of human imitation. *Science* **286**: 2526–2528.

Iacoboni, M., Molnar-Szakacs, I., Gallese, V., *et al.*, 2005. Grasping the intentions of others with one's own mirror neuron system. *PLoS Biol.* **3** (3): e79.

Indefrey, P., and Levelt, W. J., 2004. The spatial and temporal signatures of word production components. *Cognition* **92**: 101–144.

Indefrey, P., and Levelt, W., 2000. The neural correlates of language production. In M. Gazzaniga (ed.) *The New Cognitive Neurosciences*. Cambridge, MA: MIT Press, pp. 845–865.

Janzen, T., and Shaffer, B., 2002. Gesture as the substrate in the process of ASL grammaticalization. In R. P. Meier, K. Cormier, and D. Quinto-Pozos (eds.) *Modality and Structure in Signed and Spoken Languages*. Cambridge, UK: Cambridge University Press, pp. 199–223.

Ladefoged, P., 2000. *A Course in Phonetics*, 3rd edn. Zurich, Switzerland: Heinle.

Levänen, S., Uutela, K., Salenius, S., and Hari, R., 2001. Cortical representation of sign language: comparison of deaf signers and hearing non-signers. *Cereb. Cortex* **11**: 506–512.

Liberman, A., 1996. *Speech: A Special Code*. Cambridge, MA: Bradford Books.

MacSweeney, M., Amaro, E., Calvert, G. A., *et al.*, 2000. Activation of auditory cortex by silent speechreading in the absence of scanner noise: an event-related fMRI study. *Neuroreport* **11**: 1729–1734.

MacSweeney, M., Woll, B., Campbell, R., *et al.*, 2002. Neural systems underlying British Sign Language and audio-visual English processing in native users. *Brain* **125**: 1583–1593.

MacSweeney, M., Campbell, R., Woll, B., *et al.*, 2004. Dissociating linguistic and nonlinguistic gestural communication in the brain. *Neuroimage* **22**: 1605–1618.

Marshall, J., Atkinson, J., Smulovitch, E., Thacker, A., and Woll, B., 2004. Aphasia in a user of British Sign Language: dissociation between sign and gesture. *Cogn. Neuropsychol.* **21**: 537–554.

Mayberry, R., 1994. The importance of childhood to language acquisition: insights from American Sign Language. In J. C. Goodman and H. C. Nusbaum (eds.) *The Development of Speech Perception: The Transition from Speech Sounds to Words*. Cambridge, MA: MIT Press, pp. 57–90.

1995. Mental phonology and language comprehension or What does that sign mistake mean? In K. Emmorey and J. Reilly (eds.) *Language, Gesture, and Space*. Hillsdale, NJ: Lawrence Erlbaum, pp. 355–370.

Mayberry, R. I., and Fischer, S. D., 1989. Looking through phonological shape to lexical meaning: the bottleneck of non-native sign language processing. *Mem. Cogn.* **17**: 740–754.

McCullough, S., Emmorey, K., and Sereno, M., 2005. Neural organization for recognition of grammatical and emotional facial expressions in deaf ASL signers and hearing nonsigners. *Brain Res. Cogn. Brain Res.* **22**: 193–203.

McDonald, P. A., and Paus, T., 2003. The role of parietal cortex in awareness of self-generated movements: a transcranial magnetic stimulation study. *Cereb. Cortex* **13**: 962–967.

McGuire, P., Silbersweig, D. A., and Frith, C. D., 1996. Functional neuroanatomy of verbal self-monitoring. *Brain* **119**: 907–917.

Moll, J., de Oliveira-Souza, R., Passman, L. J., *et al.*, 2000. Functional MRI correlates of real and imagined tool-use pantomimes. *Neurology* **54**: 1331–1336.

Mountcastle, V. B., Lynch, J. C., Georgopoulos, A., Sakata, H., and Acuna, C., 1975. Posterior parietal association cortex of the monkey: command functions for operations within extrapersonal space. *J. Neurophysiol.* **38**: 871–908.

Neville, H. J., Bavelier, D., Corina, D., *et al.*, 1998. Cerebral organization for language in deaf and hearing subjects: biological constraints and effects of experience. *Proc. Natl Acad. Sci. USA* **95**: 922–929.

Newman, A., Bavelier, D., Corina, D., Jezzard, P., and Neville, H. J., 2002. A critical period for right hemisphere recruitment in American Sign Language processing. *Nature Neurosci.* **5**: 76–80.

Newport, E. L., and Meier, R. P., 1985. The acquisition of American Sign Language. In D. Slobin (ed.) *The Cross-Linguistic Study of Language Acquisition*. Hillsdale, NJ: Lawrence Erlbaum, pp. 881–938.

Nishimura, H., Hashikawa, K., Doi, K., *et al.*, 1999. Sign language "heard" in the auditory cortex. *Nature* **397**: 116.

Papathanassiou, D., Etard, O., Mellet, E., *et al.*, 2000. A common language network for comprehension and production: a contribution to the definition of language epicenters with PET. *Neuroimage* **11**: 347–357.

Paulesu, E., Perani, D., Blasi, V., *et al.*, 2003. A functional–anatomical model for lipreading. *J. Neurophysiol.* **90**: 2005–2013.

Petitto, L. A., 1987. On the autonomy of language and gesture: evidence from the acquisition of personal pronouns in American Sign Language. *Cognition* **27**: 1–52.

Petitto, L. A., Zatorre, R. J., Gauna, K., *et al.*, 2000. Speech-like cerebral activity in profoundly deaf people processing signed languages: implications for the neural basis of human language. *Proc. Natl Acad. Sci. USA* **97**: 13961–13966.

Quinto-Pozos, D., 2003. Contact between Mexican Sign Language and American Sign Language in two Texas border areas. Ph. D. dissertation, University of Texas, Austin, TX.

Rizzolatti, G. and Arbib, M., 1998. Language within our grasp. *Trends Neurosci.* **21**: 188–194.

Rizzolatti, G., Fadiga, L., Matelli, M., *et al.*, 1996. Premotor cortex and the recognition of motor actions. *Cogn. Brain Res.* **71**: 491–507.

Rizzolatti, G., Fogassi, L., and Gallese, V., 2002. Motor and cognitive functions of the ventral premotor cortex. *Curr. Opin. Neurobiol.* **12**: 149–154.

Rothi, L. J., Heilman, K. M., and Watson, R. T., 1985. Pantomime comprehension and ideomotor apraxia. *J. Neurol. Neurosurg. Psychiat.* **48**: 207–210.

Sadato, N., Yamada, H., Okada, T., *et al.*, 2004. Age-dependent plasticity in the superior temporal sulcus in deaf humans: a functional MRI study. *BMC Neurosci.* **5**: 56–61.

Sakai, K. L., Tatsuno, Y., Suzuki, K., Kimura, H., and Ichida, Y., 2005. Sign and speech: amodal commonality in left hemisphere dominance for comprehension of sentences. *Brain* **128**: 1407–1417.

Sandler, W., 1989. *Phonological Representation of the Sign: Linearity and Nonlinearity in American Sign Language*. Dordrecht, Netherlands: Foris.

San José-Robertson, L., Corina, D. P., Ackerman, D., Guillemin, A., and Braun, A. R., 2004. Neural systems for sign language production: mechanisms supporting lexical selection, phonological encoding, and articulation. *Hum. Brain Map.* **23**: 156–167.

Sekiyama, K., Kanno, I., Miura, S., and Sugita, Y., 2003. Auditory–visual speech perception examined by fMRI and PET. *Neurosci. Res.* **47**: 277–287.

Senghas, A., 2003. Intergenerational influence and ontogenetic development in the emergence of spatial grammar in Nicaraguan Sign Language. *Cogn. Devel.* **18**: 511–531.

Skipper, J. I., Nusbaum, H. C., and Small, S. L., 2005. Listening to talking faces: motor cortical activation during speech perception. *Neuroimage* **25**: 76–89.

Söderfeldt, B., Rönnberg, J., and Risberg, J., 1994. Regional cerebral blood flow in sign language users. *Brain Lang.* **46**: 59–68.

Söderfeldt, B., Ingvar, M., Rönnberg, J., *et al.*, 1997. Signed and spoken language perception studied by positron emission tomography. *Neurology* **49**: 82–87.

Stevens, K., and Blumstein, S., 1981. The search for invariant acoustic correlates of phonetic features. In P. D. Eimas and J. L. Miller (eds.) *Perspectives on the Study of Speech*. Hillsdale, NJ: Lawrence Erlbaum, pp. 1–38.

Supalla, T., and Newport, E., 1978. How many seats in a chair? The derivation of nouns and verbs in American Sign Language. In P. Siple (ed). *Understanding Language through Sign Language Research*. New York: Academic Press.

Talairach, J., and Tournoux, P., 1988. *Co-Planar Stereotaxic Atlas of the Human Brain*. New York: Thieme.

Tanaka, S., and Inui, T., 2002. Cortical involvement for action imitation of hand/arm versus finger configurations: an fMRI study. *Neuroreport* **13**: 1599–1602.

Tanaka, S., Inui, T., Iwaki, S., Konishi, J., and Nakai, T., 2001. Neural substrates involved in imitating finger configurations: an fMRI study. *Neuroreport* **12**: 1171–1174.

Taub, S., 2001. *Language from the Body: Iconicity and Metaphor in American Sign Language*. New York: Cambridge University Press.

Tomasello, M., Savage-Rumbaugh, S., and Kruger, A. C., 1993. Imitative learning of actions on objects by children, chimpanzees, and enculturated chimpanzees. *Child Devel.* **64**: 1688–705.

Umiltá, M. A., Kohler, E., Gallese, V., *et al.*, 2001. I know what you are doing: a neurophysiological study. *Neuron* **31**: 155–165.

Vaina, L. M., Solomon, J., Chowdhury, S., Sinha, P., and Belliveau, J. W., 2001. Functional neuroanatomy of biological motion perception in humans. *Proc. Natl Acad. Sci. USA* **98**: 11656–11661.

Volterra, V., and Iverson, J., 1995. When do modality factors affect the course of language acquisition? In K. Emmorey and J. Reilly (eds.) *Language, Gesture, and Space*. Hillsdale, NJ: Lawrence Erlbaum, pp. 371–390.

Watanabe, J., Sugiura, M., Miura, N., *et al.*, 2004. The human parietal cortex is involved in spatial processing of tongue movement: an fMRI study. *Neuroimage* 21: 1289–1299.

Wise, R. J. S., Scott, S. K., Blank, S. C., *et al.*, 2001. Separate neural subsystems within 'Wernicke's' area. *Brain* 124: 83–95.

Wolpert, D. M., Goodbody, S. J., and Husain, M., 1998. Maintaining internal representations: the role of the human superior parietal lobe. *Nature Neurosci.* 1: 529–533.

5

Neural homologies and the grounding
of neurolinguistics

Michael A. Arbib and Mihail Bota

5.1 Introduction

5.1.1 Homologies and language evolution

We are interested here in *homologous* brain structures in the human and the macaque, those which may be characterized as descended from the same structure of the brain of the common ancestor. However, quite different paths of evolution, responsive to the need of different organisms for similar functions, may yield organs with similar functions yet divergent evolutionary histories – these are called *homoplasic*, rather than homologous. Arbib and Bota (2003) forward the view that homology is not usefully treated as an all-or-none concept except at the grossest level, such as identifying visual cortex across mammalian species. Even if genetic analysis were to establish that two brain regions were homologous in that they were related to a common ancestral form it would still be important to have access to a measure of similarity to constrain too facile an assumption that homology guarantees similarity across all criteria.

Indeed, from the perspective of computational and comparative neuroscience, declared homologies may be the start, rather than the end, of our search for similarities that will guide our understanding of brain mechanisms across diverse species. When two species have diverged significantly from their common ancestors, a single organ x in ancestor A might differentiate into two organs y and y' in modern species B and three organs z, z', and z'' in modern species C. For example, a region involved in hand movements in monkey might be homologous to regions involved in both hand movements and speech in humans. Thus there may be no absolute homology between these five modern organs since they are differentially modified from the common ancestor. We thus use the terms *degree of homology* and *degree of similarity*, so that rather than asserting that two structures are homologous as an all-or-none-concept, we should weigh the evidence for homology, using such varied criteria as relative position, cytoarchitecture, and hodology (the sets of afferent and efferent connections). Other useful criteria for comparison are myeloarchitecture, chemoarchitecture, functionality, and continuity through intermediate species.

Action to Language via the Mirror Neuron System, ed. Michael A. Arbib. Published by Cambridge University Press.
© Cambridge University Press 2006.

What then are the human brain structures we wish to homologize? Figure 5.1A presents a lateral view of the left hemisphere of the human brain showing various sulci as well as a number of areas (numbered according to Brodmann) relevant to language performance. The generally accepted view of the human cortical areas involved in language gives special prominence to Broca's area and Wernicke's area, both lateralized in the left hemisphere for most humans. However, many other areas are also implicated in language and there is no one-to-one correlation between cortical areas and the damage caused by strokes, tumors, or surgery. Thus future research must not only seek a firmer basis for (degrees of) homology between cortical areas of macaque and human but also a firmer understanding of the functional contribution of each area. The discussion of homologies for Broca's area and Wernicke's area in Section 5.2 marks the beginning of the quest, not its end. Indeed, we would claim that we need a far fuller computational account of the brain mechanisms involved in language performance, one that is informed by a deeper understanding of diverse systems of the macaque brain, to provide a more refined account of relevant regions of the human brain that will be the target of neurohomological analysis. The results will then lead to the refinement of models, and so on to a better characterization of degrees of homology, in a virtuous circle.

Broca's area is located on the inferior frontal gyrus (pars triangularis and opercularis), and comprises BA (Brodmann area) 44 and BA 45. Some publications use the term Broca's area to cover BA 44, 45, and 47, others use it for BA 44 only. In any case, the frontal language networks are more widespread than areas 44 and 45. Amunts *et al.* (1999) conducted a three-dimensional reconstruction of ten human brains and found that area 44 but not area 45 was left-over-right asymmetrical in all brains and they see this as evidence for viewing area 44 as the anatomical correlate of the functional lateralization of speech production.

Wernicke's area is usually considered to be in the posterior part of the superior temporal gyrus and in the floor of the sylvian sulcus (Aboitiz and García, 1997). It corresponds to the posterior part of BA 22, or area Tpt (temporoparietal) as defined by Galaburda and Sanides (1980). Lesion-based views of Wernicke's area may include not only the posterior part of BA 22 but also (in whole or in part) areas 42, 39, 40, and perhaps 37.

We are concerned with homologies of the macaque and human brain which may help ground our hypotheses on what changes occurred in the primate brain from the brain of the putative macaque–human common ancestor of perhaps 20 million years ago to the brain of *Homo sapiens* today. We should thus note the striking difference in the parcellations of the macaque brain in Figs. 5.1B and 5.1C. This poses challenges both for study of the macaque and for comparative analysis of macaque and human. Not only are the labels different in Figs. 5.1B and 5.1C, but so too are the parcellations of cerebral cortex (i.e., its division into different regions) to which they are attached. Arbib and Bota (2003) offer a detailed analysis of competing parcellations of the macaque brain and the problems solved in deciding upon the most plausible homologies; the present chapter will use the results of their analysis without replicating the details.

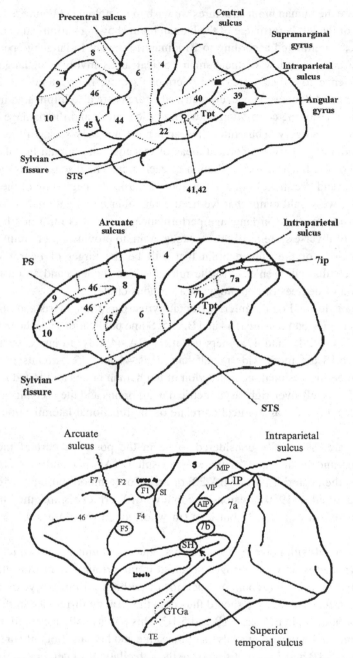

Figure 5.1 (A, B). Architectonic areas considered by Aboitiz and García (1997) in comparing the human (A) and the macaque (B). Numbers correspond to Brodmann's classification. AG, angular gyrus; AIP, anterior intraparietal cortex; AS, arcuate sulcus; CS, central sulcus; IPS, intraparietal sulcus; LIP, lateral intraparietal cortex; MIP, medial intraparietal cortex; PCS, precentral sulcus; PS, principal sulcus; SF, sylvian fissure; SMG, supramarginal gyrus; STS, superior temporal sulcus;

Semendeferi *et al.* (2002) argue that magnetic resonance imaging (MRI) shows that human frontal cortices are not disproportionately large in comparison to those of the great apes; though Deacon (2004) offers a critique of their methodology (as well as a response to Arbib and Bota (2003) – see also Arbib and Bota, 2004). They suggest that the special cognitive abilities of humans may be due to differences in individual cortical areas and to a richer interconnectivity, rather than an increase in the overall relative size of the frontal lobe during human evolution. However, restricting connectivity may also have been important as it would focus the types of structured associations that could be learned as the brain develops.

5.1.2 The structure of the chapter

Section 5.2.1 briefly recalls key aspects of the Mirror System Hypothesis (MSH) from Arbib (Chapter 1, this volume) [1]. MSH outlines the macaque basis for the evolution of the language-ready brain, but does so with an emphasis on the relation between visual input and manual output. These relationships were captured in the FARS model of visually directed grasping (involving premotor area F5 and the anterior intraparietal sulcus (AIP) and, less centrally, inferotemporal cortex (IT) and prefrontal cortex (PFC), as well as the role of the basal ganglia (BG) in controlling sequences) and the MNS model of recognition of visually present hand–object trajectories (involving the division of F5 into mirror as well as canonical neurons, parietal area PF ($\approx$ 7b) as well as AIP, and superior temporal sulcus (STS). Complementing this, the Aboitiz–García hypothesis (AGH; Section 5.2.2) attends not only to the lexicon but also to the working memory structures that support language. After a brief discussion of dorsal and ventral streams in the visual system (Section 5.2.3), we provide in Section 5.2.4 a *preliminary* synthesis of MSH and AGH that frames the rest of the chapter. Section 5.3 follows Arbib and Bota (2003) and others in exploring macaque homologues of Broca's and Wernicke's areas, foreshadowing some of the issues addressed in the next three sections:

1. Although minor attention to sequencing is included in the FARS model, it was not addressed in the MNS model. We thus exclude BG from the MSH core, and return to the possible roles of SMA and BG in "Sequences and Hierarchy" as the first supplement to the MSH core

Caption for Figure 5.1 (cont.)
TPT, temporoparietal; VIP, vental intraparietal cortex. (C). A side view of the left hemisphere of the macaque brain. Area 7b is also known as area PF. (A and B adapted from Aboitiz and García, 1997; C adapted from Jeannerod et al., 1995.)

[1] The present chapter has evolved from Arbib and Bota (2003), but does not address one of the major themes of that paper – the development of neuroinformatics tools to support the kind of work presented here. The NeuroHomology Database (NHDB), outlined in Arbib and Bota (2003), is presented in more detail in Bota and Arbib (2004) and has been built upon in current work on the Brain Architecture Knowledge Management System (BAMS; Bota *et al.*, 2003).

(Section 5.4). This topic must also include attention to "phonological working memory" that keeps track of the perception and production of structured utterances or actions.

2. Recent data on mirror neurons show us the relevance of auditory input and orofacial output. This requires analyzing the relation of the MSH core to other circuitry for audition and vocalization, including data reviewed by Aboitiz *et al.* (2005) in updating AGH. Section 5.5.1 analyzes the auditory system in this spirit.

3. Complementing the circuitry required to generate or recognize an action is the circuitry that merges emotional and context information in organizing vocalization and other output. We thus address the role of anterior cingulate cortex (ACC) in Section 5.5.2.

Finally, Section 5.7, "Towards an action-oriented neurolinguistics," draws these strands together, adding considerably to the complexity of the schematic offered in the overview of Section 5.2.4.

5.2 Hypotheses on evolving the language-ready brain

5.2.1 The Mirror System Hypothesis briefly recalled

The *parity requirement* for language in humans – that what counts for the speaker must count approximately the same for the hearer – roots language (and communication more generally) in a social context in which speaker and hearer are motivated to communicate with each other. Since normal face-to-face speech involves manual and facial as well as vocal gestures, and because signed languages are fully developed human languages, the "speaker" and "hearer" may use hand and face gestures to complement or instead of vocal gestures. This motivates a study of the evolution of the language-ready brain not restricted to speech mechanisms.

As we saw in Chapter 1, the system of the macaque brain for visuomotor control of grasping has its premotor outpost in an area called F5 (see Fig. 5.1C) which contains a set of neurons, called *mirror neurons*, such that each mirror neuron is active not only when the monkey executes a specific grasp but also when the monkey observes a human or other monkey execute a more-or-less similar grasp (Rizzolatti *et al.*, 1996). Thus macaque F5 contains a *mirror system for grasping* which employs a common neural code for *executed* and *observed* manual actions. The homologous region of the human brain is in or near Broca's area (the posterior part of the inferior frontal gyrus, including at least areas 44 and 45; Fig. 5.1A), traditionally thought of as a speech area, but which has been shown by brain imaging studies to be active when humans both execute and observe grasps. It is posited that the mirror system for grasping was also present in the common ancestor of humans and monkeys (perhaps 20 million years ago) and that of humans and chimpanzees (perhaps 5 million years ago). These findings ground:

The Mirror System Hypothesis (henceforth MSH) (Arbib and Rizzolatti, 1997; Rizzolatti and Arbib, 1998): The mechanisms which support language in the human brain evolved atop a basic mechanism *not* originally related to communication, the *mirror system for grasping* with its capacity to generate *and* recognize a set of actions, and this provides the evolutionary basis for *language parity* – i.e., an utterance means roughly the same for both speaker and hearer. In particular, human Broca's area contains a mirror system for grasping which is homologous to the F5 mirror system of macaque.

The current version of the Mirror System Hypothesis (Arbib, 2002, 2005) postulates a progression of mirror systems for:

1. grasping and manual praxic actions,
2. imitation of grasping and manual praxic actions,
3. pantomime of grasping and manual praxic actions,
4. pantomime of actions outside the panto-mimic's own behavioral repertoire (e.g., flapping the arms to mime a flying bird),
5. conventional gestures used to formalize and disambiguate pantomime (e.g., to distinguish "bird" from "flying"),
6. conventionalized manual, facial, and vocal communicative gestures ("protowords") separate from pantomime.

We saw in Chapter 1 that neurophysiological studies of the macaque ventral agranular cortex allowed us to distinguish *mirror neurons* in F5 (these discharge not only when the macaque grasped or manipulated objects in a specific way, but also when the macaque observed the experimenter make a similar gesture) from *canonical neurons* in F5 (which are active only when the macaque itself performs the relevant actions). Canonical neurons lie in the region of F5 buried in the dorsal bank of the arcuate sulcus (AS), F5ab, while mirror F5 neurons lie in the convexity located caudal to the AS (Rizzolatti and Luppino, 2001). Both sectors receive a strong input from the secondary somatosensory area (SII) and parietal area (PF) (Brodmann area 7b) encoding observations of arm and hand movements (Rizzolatti and Luppino, 2001; Rizzolatti *et al.*, 2002). Canonical neurons also receive object-related input from AIP.

Computational analysis has explored some interactions observed experimentally between these regions and hypothesized others in constructing working models of the mechanisms of the macaque brain related to grasping and the mirror system for grasping. The FARS model (Fagg and Arbib, 1998; Section 1.2.1, this volume) shows how prefrontal cortex may modulate AIP–F5 interactions in action selection through task-related analysis of object identity by the "what" path of inferotemporal cortex, through working memory, and through selection cues in conditional tasks. The Mirror Neuron System (MNS) model (Oztop and Arbib, 2002; Section 1.2.2, this volume) shows how STS and PF inputs may be associated with activity in F5 canonical neurons to train F5 mirror neurons so that they may respond to data on the actions of others and not just during action execution. These models challenge us to further anatomical and neurophysiological analysis of the macaque, to "lift" these models to the human brain, to extend the resulting models to models of language performance, and then test the result in part through more rigorous analysis of macaque–human homologies (Arbib and Bota, 2003).

Looking in more detail at items (4) and (5): the transition from the use of hand movements for praxis to their use for pantomime, with communication being the primary and intended effect of an action rather than a side effect, does seem to involve a genuine neurological change. A grasping movement that is not directed toward a suitable object will not elicit mirror neuron firing in macaque (Umiltà *et al.*, 2001). By contrast, in pantomime, the observer sees the movement in isolation and *infers* what non-hand movement is being mimicked by the hand movement, and thus the goal of the action.

Item (6) combines the results of two of the stages reviewed in Chapter 1:

S5 *Protosign*, a manual-based communication system, breaks through the fixed repertoire of primate vocalizations to yield an open repertoire.

S6 *Protospeech* results as protosign mechanisms evolved to control a vocal apparatus of increasing flexibility.

It is not claimed that stage S5 was completed before stage S6 was initiated. Rather, the notion is that protosign provided the initial scaffolding but that protosign and protospeech evolved thereafter in an expanding spiral. To begin thinking about the "auditory outreach" of the mirror system for grasping, recall from Chapter 1 the data showing that macaque F5 mirror neurons are not limited to visual recognition of hand movements. Kohler *et al.* (2002) studied mirror neurons that were activated by the characteristic sound of an action as well as sight of the action. Ferrari *et al.* (2003) found orofacial motor neurons in F5 that discharge when the monkey observes another individual performing mouth actions. Whereas the majority of these "mouth mirror neurons" become active during the execution and observation of ingestive actions, there are some neurons active during the execution of ingestive actions for which the most effective visual stimuli in triggering them are communicative mouth gestures such as lip-smacking, lip protrusion, tongue protrusion, teeth-chatter, and lips/tongue protrusion – all a long way from the sort of vocalizations that occur in speech.

5.2.2 The Aboitiz–García hypothesis

Aboitiz and García (1997) propose the development of brain mechanisms supporting the lexicon and syntax as the keys to the evolution of the human brain to support language:

1. The capacity to give names to yield the *lexicon*. This underlies the ability to refer to objects or events in the external world. They associate this with the elaboration of a precursor of Wernicke's area in the superior temporal lobe as a zone for cross-modal associations which include a phonological correlate.

2. *Syntax* arose to express regularities in the ways in which different elements are combined to form linguistic utterances. They associate this with the differentiation of an inferoparietal–frontal (Broca's) area with its connections to the incipient Wernicke's region developing as a phonological rehearsal device that eventually differentiated into the language areas, this phonological-rehearsal apparatus providing some basic syntactic rules at the levels of phonemes and morphemes. The coordinated operation of networks involving granular frontal cortex and

the semantic system represented in the temporoparietal lobes, together with the phonological-rehearsal loop just mentioned, generated higher levels of syntax and discourse.

We now review these hypotheses in somewhat more detail.

In search of the lexicon

Aboitiz and García (1997) adopt the hypothesis (Geschwind, 1964) that cross-modal sensory associations in the monkey need an intact limbic system to develop, while in the human non-limbic cortex, cross-modal cortico-cortical interactions facilitated establishment of associations between the sound of a vocalization and the image of an object. This, Geschwind argued, permitted the generation of a lexicon in which arbitrary sounds (vocalizations) represented objects identified through the visual or the tactile system. This account has also been espoused by Wilkins and Wakefield (1995), but remains controversial, as can be seen from the commentaries on that article.

Indeed, mere association of sight and sound cannot be the secret of the lexicon. We have noted that Kohler *et al.* (2002) show that for certain manual actions (e.g., the breaking of a peanut), characteristic sounds associated with the action can activate some of the monkey mirror neurons associated with that action. Chimpanzees and gorillas trained appropriately from an early age are capable of learning hand signs or the use of "lexigrams" to denote a hundred or so objects and actions, and can combine two or three of these in "protosentences." However, there is still no evidence that apes can move beyond the capability of a 2-year-old human infant, never reaching the "naming explosion" that occurs in humans in their third year. What is it about the human brain that supports it? Arbib (Chapter 1, this volume) notes the dissociation between pantomime and sign language in the human brain to stress the special nature of (proto)linguistic representations. This stresses that Geschwind's hypothesis leaves open the question of how multimodal concepts can obtain a linguistic dimension.

Vervet monkeys have different cries specifying distinct predators (Cheney and Seyfarth, 1990). The acoustic structure of each alarm call is in great part genetically preprogrammed but its innate meaning seems to be refinable through learning (cf. Winter *et al.*, 1973). Vervet monkey infants, for instance, react to eagles with warning calls but, in contrast to adults, react in the same way to pigeons and geese. Further, infants react to hearing these warning calls with alertness but do not show the orientation toward the sky typical for adults. Thus in this case it seems that the relevant "phonological sequences" are innate but that their perceptual and motor associations are subject to change.

Aboitiz and García (1997) thus concede that the neural substrate for the development for a lexicon exists in an incipient form in higher primates. But this concession weakens Geschwind's hypothesis, and forces us to be more specific about the changing functionality and the neural changes that made it possible. Moreover, a key element of MSH has been to explain why Broca's area is not the homologue of the anterior cingulate area which is the area of cerebral cortex primarily involved in monkey vocalization (see Jürgens, 1997, 2002, and Section 5.5.2 below for a review).

Aboitiz and García (1997) hypothesize that multimodal concepts in Wernicke's area are mapped into phonological sequences as follows (compare Figs. 5.1A and 5.1B): the system of long temporoparietal–prefrontal connections serves to integrate sensory and mnemonic information from the temporoparietal lobes with the organization of behavior, both short- and long-term, by the frontal systems. They further stress that language processing is closely linked to working memory (a) in terms of the anatomical arrangement of the neural networks involved, and (b) because it operates in the context of an efficient working memory system. One of their main suggestions is that selective pressure for the capacity to learn complex vocalizations through imitation and repeated practice was a key aspect in establishing a phonological working memory system that allowed temporary storage of phonological representations in order to rehearse them internally. Through the action of natural selection favoring good learners, this system eventually differentiated into primordial language regions. Concomitantly, a prefrontal system in which information from other sensory modalities was integrated and coordinated with the representation of complex vocalizations was also being developed.

Phonological rehearsal

Baddeley and Hitch (1974) proposed a now classic working memory model which posits a *central executive* (an attentional controller) coordinating two subsidiary systems, the *phonological loop*, capable of holding speech-based information, and the *visuospatial sketchpad* – the emphasis here being on the role of working memory in sentence processing. Baddeley (2003) added an episodic long-term memory (LTM) to the Baddeley–Hitch model, with the ability to hold language information complementing the phonological loop and (the idea is less well developed) an LTM for visual semantics complementing the visuospatial sketchpad. He further adds an *episodic buffer*, controlled by the central executive, which is assumed to provide a temporary interface between the phonological loop and the visuospatial sketchpad and LTM. (See Itti and Arbib (this volume) for a more general view of an active visuospatial working memory which could include sign language "phonology" as well as a working memory for recent events and actions.)

More generally, *working memory* holds information about, e.g., objects or events (whether recently perceived, recalled from LTM, or inferred), for some period prior to executing some action for which this information may be relevant. Specific neurons have been observed that hold the encoding of some stimulus for the "delay period" from the time the stimulus is observed to the time when the action is initiated (Goldman-Rakic, 1987; Fuster, 1995). We – like Aboitiz and García – will follow Goldman-Rakic (1995) in appealing to a multiplicity of special-purpose working memory systems organized in parallel across the prefrontal cortex. Working memory for the spatial location of objects involves connections between parietal area 7 and dorsolateral prefrontal BA 46 and BA 8; working memory for object characteristics depends on connections between area TE (von Bonin and Bailey, 1947) of the inferior temporal lobe and the inferior convexity of the

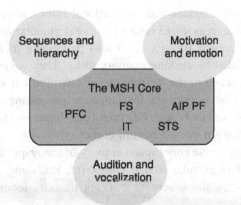

Figure 5.2 A schematic view of brain regions considered in extending the Mirror System Hypothesis (MSH). At the center are the core areas of the macaque brain considered in grounding MSH (Arbib, chapter 1, this volume). The FARS model of visually-directed grasping involves premotor area F5 and the anterior intraparietal sulcus AIP, inferotemporal cortex IT and prefrontal cortex PFC; while the MNS model of recognition of visually present hand–object trajectories involves the division of F5 into mirror as well as canonical neurons and considers parietal area PF ($\approx$ 7b) as well as AIP, and superior temporal sulcus STS. We add three systems which must be considered in extending MSH: (1) the extension of MSH to "Sequences and Hierarchy" which also includes "phonological working memory"; (2) recent data on mirror neurons related to auditory input and orofacial output must be combined with relevant data on "Audition and Vocalization"; (3) complementing the circuitry required to generate or recognize an action is the role of emotion and motivation both in shading communicative expression and in determining whether or not the creature is motivated to act. We focus on (1) and (2) rather than (3) in this chapter.

prefrontal cortex, BA 45 and BA 12 (Wilson *et al.* 1993). Rainer *et al.* (1998) enrich this account by noting that many neurons of PFC convey information about both an object's identity (*what*) and its location (*where*), and then mapping the "memory fields" of these neurons in monkeys required to remember both an object and its location at varied positions throughout central vision. They found that many PFC neurons which conveyed object information had highly localized memory fields that emphasized the contralateral, but not necessarily foveal, visual field. These results indicate that PF neurons can simultaneously convey precise location and object information and thus may play a role in constructing a unified representation of a visual scene. This helps us reiterate our point that the phonological loop is but one of many working memory resources that we should consider in linking language to vision and action.

Aboitiz and García (1997) argue that linguistic working memory involves connections between inferoparietal areas BA 39–40 and frontal areas BA 44–47 and provide the anatomical basis for their phonological-rehearsal loop (Fig. 5.2). Granular frontal areas (BA 9 and especially BA 46) relate with more general aspects of working memory in humans.

Based on data from brain lesions and connectional information, Aboitiz and García (1997) propose that beside areas 44/45 (and 47), frontal granular areas such as 9 and 46 (forming together an extended Broca's area) also participate in language processing, especially in aspects related to working memory tasks. They suggest that these frontal granular areas not only relate to the distribution of attention but also handle cognitive (semantic) information that is relevant for language processing. For example, when recalling the objects observed in a room, one might say "there is a lamp with a red shade (object/feature information) in the left bottom corner (visuospatial information)." Tasks such as this probably require the coordinated activity of the respective working memory circuits that are located in granular frontal cortex (cf. Rolls and Arbib, 2003 and Itti and Arbib (this volume) on the structured working memory required for visual scene perception).

Aboitiz and García (1997) take pains to relate these human areas to the macaque brain (Fig. 5.1). In the macaque the equivalent of Broca's area (BA 45) receives major projections from the inferior parietal and the inferior temporal lobes. Aboitiz and García propose that the inferoparietal areas from which some of these projections arise in the monkey (areas 7b and 7ip) are homologous to areas 40 (supramarginal gyrus) and perhaps 39 (angular gyrus) in the human. Aboitiz and García (1997) equate human Wernicke's area with temporoparietal (Tpt) and see it as feeding (directly or indirectly) areas 40, 39, etc., that project to Broca's area – but we might also read this as further support for the view that Wernicke's area may include (parts of) Brodmann areas 22, 42, 39, 40, and perhaps 37.

In humans, Tpt also projects directly to prefrontal cortex (Broca's area in an extended version), thus participating in language working memory. Since frontal projections from area Tpt ($\approx$ human Wernicke's region) do not terminate massively in areas 44–45 ($\approx$ human Broca's area) in monkey, Aboitiz and García (1997) propose that in human evolution, area Tpt may have become increasingly connected with inferoparietal regions such as the supramarginal gyrus (area 40) thus feeding the latter with auditory information to be used in the phonological loop (Fig. 5.3). They then suggest that in the human the posterior superior temporal region represents a transitional zone in which concepts progressively acquire a phonological correlate while the supramarginal gyrus (area 40, and perhaps also the angular gyrus, area 39) in the parietal lobe stores this phonological representation for a brief time. Some neurons in human area 40 (and perhaps in area 39) may then project to Broca's region (areas 44 and 45), thus establishing a neuronal circuit for the phonological-rehearsal system of linguistic working memory. If they exist, direct projections from Tpt or neighboring areas to Broca's region may participate in language processing in at least two possible ways: (a) generating a shortcut between Wernicke's and Broca's regions for some automatic routines, and (b) participating in higher levels of language processing. More recently, however, Aboitiz *et al.* (2005) have extended AGH by taking account of the post-1997 data on temporo-frontal interactions in the macaque auditory system which we summarize in Section 5.5.1.

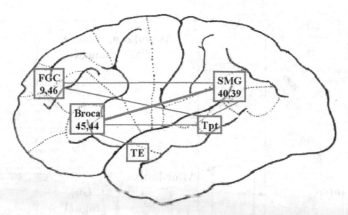

Figure 5.3 Network of connectivity for language in the human brain proposed by Aboitiz and García (1997), emphasizing the connection, which may largely correspond to the arcuate fasciculus, between SMG (supramarginal gyrus) and Broca's area. Area TE, which projects to area 45, may also participate in language processing. Connections between the temporoparietal area Tpt and Broca's area, and between Tpt and SMG (direct or indirect) have not been substantially confirmed in the monkey but (especially the latter) are proposed by Aboitiz and García to have developed in the hominid line. FGC, frontal granular cortex.

5.2.3 Dorsal and ventral streams

Given our concern with hand use and language, it is particularly striking that the ability to use the size of an object to preshape the hand while reaching to grasp it can be dissociated by brain lesions from the ability to consciously recognize and describe that size. This involves lesions that affect either

- *the dorsal stream*: the pathway from primary visual cortex, area 17 (V1), towards posterior parietal cortex (PP) – the so-called "where/how" pathway; or
- *the ventral stream*: from V1 to inferotemporal cortex (IT), the so-called "what" pathway (see Ungerleider and Mishkin (1982) for the "where/what" distinction in the macaque visual system).

Goodale and Milner (1992) studied a patient (DF) who developed a profound visual form of agnosia following carbon monoxide poisoning in which most of the damage to cortical visual areas was apparent not in primary visual cortex, area 17 (V1), but bilaterally in the adjacent areas 18 and 19 of visual cortex to disconnect the "dorsal stream" from V1 to IT. When asked to indicate the width of a single block by means of her index finger and thumb, her finger separation bore no relationship to the dimensions of the object and showed considerable trial to trial variability. Yet when DF was asked simply to reach out and pick up the block, the peak aperture (well before contact with the object) between her index finger and thumb changed systematically with the width of the object, as in normal controls. A similar dissociation was seen in her responses to the orientation of stimuli. In other words, DF could preshape accurately, even though she appeared to

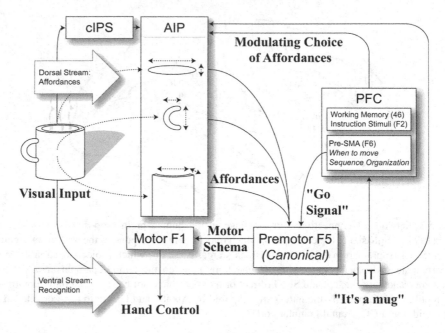

Figure 5.4 "FARS Modificato" (Fig. 1.4a). The key point for the present discussion is that AIP does not "know" the identity of the object, but can only extract affordances (opportunities for grasping for the object consider as an unidentified solid). Prefrontal cortex (PFC) uses the identification of the object by inferotemporal cortex (IT), in concert with task analysis and working memory, to help AIP select the appropriate action from its "menu."

have no conscious appreciation (expressible either verbally or in pantomime) of the visual parameters that guided the preshape.

Jeannerod *et al.* (1994) report the impairment of grasping in a patient (AT) with a bilateral posterior parietal lesion of vascular origin that left IT and the ventral pathway V1→IT relatively intact, but grossly impaired the dorsal pathway V1→PP. This patient is the "opposite" of DF – she can use her hand to pantomime the size of a cylinder, and can reach without deficit toward the location of such an object, but cannot preshape appropriately when asked to grasp it. Instead of an adaptive preshape, AT will open her hand to its fullest, and only begin to close her hand when the cylinder hits the "web" between index finger and thumb. But, interestingly, when the stimulus used for the grasp was a familiar object – such as a reel of thread, or a lipstick – for which the "usual" size is part of the subject's knowledge, AT showed a relatively adaptive preshape.

We thus see in the AT and DF data a dissociation between parietal and inferotemporal pathways, respectively, for the praxic use of size information and the "declaration" of that information either verbally or through pantomime. A corresponding distinction in the role of the dorsal and ventral pathways in the monkey is crucial to the FARS model (Fig. 5.4) discussed in Chapter 1. The FARS model provides a computational analysis of how the "canonical system," centered on generation of affordances and setting up of related motor schemas by the dorsal AIP → F5 pathway, may be modulated by IT (inferotemporal cortex) and PFC (prefrontal cortex) in selection of an affordance. The dorsal stream (from primary visual cortex to parietal cortex) carries amongst other things the information needed for AIP to recognize that different parts of the object can be grasped in different ways, thus extracting affordances for the grasp system which (according to the model) are then passed on to F5. The dorsal stream does not know "what" the object is, it can only see the object as a set of possible affordances. The ventral stream (from primary visual cortex to inferotemporal cortex), by contrast, is able to recognize what the object is. This information is passed to prefrontal cortex which can then, on the basis of the current goals of the organism and the recognition of the nature of the object, bias AIP to choose the affordance appropriate to the task at hand. The visual system is not the only system that may be functionally segregated into "what" and "where" pathways. The hypothesis of separate channels for processing the identity and location of an auditory object is discussed in Section 5.5.1.

5.2.4 Integrating the hypotheses: a first pass

The role of IT and AIP in the FARS model (Fig. 5.4) would appear to ground the human dissociation between parietal and inferotemporal pathways, respectively, for the praxic use of size information and the "declaration" of that information either verbally or through pantomime. The "minimal mirror system" for grasping in monkeys includes mirror neurons in the parietal area PF (7b) as well as F5, and some not-quite-mirror neurons in the region STSa in the superior temporal sulcus (STS). The "F5 mirror system" is thus shorthand for "F5–PF–STSa mirror system for manual and orofacial actions." Correspondingly, human data show that grasp observation significantly activates the superior temporal sulcus), the inferior parietal lobule, and the inferior frontal gyrus (area 45, part of Broca's area) in the left hemisphere. The full neural representation of the Cognitive Form $Grasp_A$ (Agent, Object) requires not only the regions AIP, STSa, 7a, 7b, and F5 mirror neurons but also inferotemporal cortex (IT) which holds the identity of the object and other regions of the STS which extract the identity of the agent (e.g., through face recognition). STSa is also part of a circuit that includes the amygdala and the orbitofrontal cortex and so may be involved in the elaboration of affective aspects of social behavior (Amaral *et al.*, 1992). The same act can be perceived in different ways. If attention is focused on the agent's hand, then the appropriate case structure would be $Grasp_A$ (Hand, Object) as a special case of $Grasp_A$ (Instrument, Object).

Arbib and Bota (2003) developed a diagram (Fig. 5.5) to synthesize lessons about the language mechanisms of the human brain they had gleaned from a comparative analysis of Aboitiz and García (1997) and the Mirror System Hypothesis, extending an earlier sketch for a "Mirror Neurolinguistics" (Arbib, 2001). The "model" of Fig. 5.5 is designed to elicit further modeling; it does not have the status of fully implemented models such as the FARS and MNS and Infant Learning to Grasp (ILGM) models (see Oztop *et al.*, 2004;

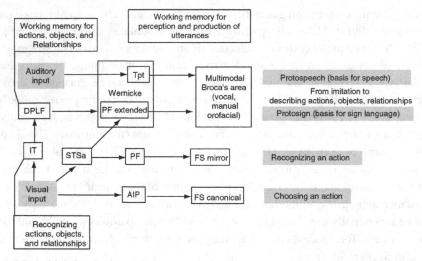

Figure 5.5 A high-level view of the cumulative emergence of three fronto-parietal systems: choosing an action → recognizing an action → describing an action (in multiple modalities). Note that this simple figure does not address issues of lateralization. DPLF, dorsolateral prefrontal cortex. (From Arbib and Bota (2003), expanding upon Arbib (2001).)

and Chapter 1, and Oztop, Bradley, and Arbib, this volume) whose relation to, and prediction of, empirical results has been probed through computer simulation. We will introduce the main ideas in Fig. 5.5 now, then produce a more elaborate version (Fig. 5.7) in Section 5.6, characterized especially by that attempt to address the data on the auditory system that we review in Section 5.5.

An over-simple analysis of praxis, action understanding, and language production might focus on the following parallel parieto-frontal interactions:

I. object → AIP →$F5_{canonical}$ praxis
II. action → PF → $F5_{mirror}$ action understanding
III. scene → Wernicke's → Broca's   language production

However, the patient AT described in the previous section could use her hand to pantomime the size of a cylinder, but could not preshape appropriately when asked to grasp it. This suggests the following scheme:

IV. Parietal "affordances"   → preshape
V. IT "perception of object" → pantomime or verbally describe size, i.e., one cannot pantomime or verbalize an affordance; but rather one needs a "recognition of the object" (IT) to which attributes can be attributed before one can express them.

Recall now the path shown in Fig. 5.4 from IT to AIP both directly and via PFC. We postulate that similar pathways link IT and PF. We show neither of these pathways in

Fig. 5.5, but rather show how this pathway might in the human brain not only take the form needed for praxic actions but also be "lifted" into a pathway that supports the recognition of communicative manual actions. We suggest that the IT "size-signal" has a diffuse effect on grasp programming – it is enough to bias a choice between two alternatives, or provide a default value when PP cannot offer a value itself, but not strong enough to perturb a single sharply defined value when it has been established in PP by other means (cf. Bridgeman *et al.* (1997) and Bridgeman (1999) for related psychophysical data). In any case, the crucial point for our discussion is that communication must be based on the size estimate generated by IT, not that generated by PP. We would then see the "extended PF" of this pathway as functionally integrated with the posterior part of Brodmann's area 22, or area Tpt. Indeed, lesion-based views of Wernicke's area may include not only the posterior part of Tpt but also (in whole or in part) areas in human cortex that correspond to macaque PF (see Arbib and Bota (2003) for further details). In this way, we see Wernicke's area as combining capabilities for recognizing protosign and protospeech to support a language-ready brain that is capable of learning signed languages as readily as spoken languages. Finally, we note that Arbib and Bota (2003) responded to the analysis of Aboitiz and García (1997) by including a number of working memories crucial to the linkage of visual scene perception, motor planning, and the production and recognition of language. However, they did not provide data on the integration of these diverse working memory systems into their anatomical scheme.

Figure 5.5 is a hybrid, mixing macaque and human regions and gives the impression that three fronto-parietal systems end in three distinct brain structures. In fact, current research has not really settled the question. In any case, the possibility that one monkey area may be homologous to different human regions implicated in one or more of praxic hand movements, pantomime, protosign, protospeech, signed language, and speech offers challenge to all-or-none views of homology (Section 5.1). Much needs to be done to delineate subareas of Broca's area that can be distinguished on this basis and their connections with other regions – while noting that differences that are found (and the variations in the pattern of such differences from individual to individual) may reflect the self-organization of the brain as the child grows within a language community rather than any innate "fate map" for these differences. DeRenzi *et al.* (1966) found that the majority of patients with apraxia of speech had oral apraxia and a high coexistence of oral and limb apraxia while Marquardt and Sussman (1984) found that all 12 of their 15 patients with Broca's aphasia had apraxia of speech while five had limb apraxia. Double dissociations occur in individual cases. Thus, either separate networks of neurons are engaged in the generation of speech and non-speech movement of the same muscles, or the same general network underlies speech and non-speech movements but these have separate control mechanisms which can be differentially damaged (Code, 1998).

Barrett *et al.* (2005) summarize functional and structural evidence supporting differential localization of mechanisms for limb praxis, speech and language, and emotional communication, showing further that in most humans, the left hemisphere may be dominant in the control of vocalization associated with propositional speech, but the right hemisphere often controls vocalization associated with emotional prosody. Such data must be taken into account in refining MSH, but in no way contradict it. As general principles of cortical evolution, one may state that increasing complexity of behavior is paralleled by increases in the overall size and number of functional subdivisions of neocortex and the complexity of internal organization of the subdivisions, and reduplication of circuitry may form the basis for differential evolution of copies of a given system, with differing connectivities to serve a variety of functions (Kaas, 1993; Striedter, 2004).

With this background we next review certain ideas on macaque homologues of Broca's and Wernicke's areas, and then in turn consider neural mechanisms for sequences and hierarchies, audition and vocalization. Section 5.6 will then present an expanded version of Fig. 5.5 that reflects certain aspects of this review. The new figure (Fig. 5.7) offers an elaboration of the anatomical context for MSH, summarizing mechanisms for the generation and recognition of *single* actions and how these "lift" to communicative actions. Complementing this, Itti and Arbib (this volume) set forth an overall framework for the functional analysis of how attention guides vision in the analysis of a visual scene, and how the structure of "minimal subscenes" (linking the recognition of agents, actions, and objects) may be linked with language production both in describing a scene or in answering questions about the scene.

5.3 Macaque homologues of Broca's and Wernicke's areas

The relative positions of Broca's area on the inferior part of the frontal cortex and Wernicke's area on the superior part of the temporal lobe suggest that candidates for homologous structures of these language-related areas may be located in corresponding locations in the macaque cortex. Accordingly, the homologous structures of the human Broca's area may be found on the inferior part of the macaque agranular frontal cortex, in the vicinity of the AS (considered to be the macaque homologue of the human precentral and prefrontal sulci), and the macaque homologues of Wernicke's area may be located at the junction between the temporal and parietal cortices.

5.3.1 Broca's area

Broca's area in humans includes Brodmann areas 44 and 45. One might hope to relate human areas 44 and 45 to the corresponding structures in the cortex of macaques but (as spelled out by Arbib and Bota, 2003) different neuroanatomists have different views on whether and how to use areas 44 and 45 to parcellate macaque cortex. Clearly, shared notation does not automatically ensure homology between human and macaque cortical areas!

Matelli (in Rizzolatti and Arbib, 1998) argued that areas 44 and 45 in the left hemisphere of the human brain are homologous with area F5 in the macaque. The distribution of the sulci play the key role in this analysis. However, the great variability in anatomy of individual human brains means homologies based on the pattern of sulci in "typical" brains provide only a first approximation to the analysis of the individual human brain shaped by genetic and experiential particularities. Indeed, we present below evidence in support of the view that F5 is homologous to area 44 alone, rather than to the combination of areas 44 and 45.

Area 44 in humans has large neurons in layer IV (Pandya and Yeterian, 1996). Layer II is densely packed, layer III contains small and medium pyramidal cells, especially in the upper part, and layer V is divided in two sublayers, Va and Vb, with medium-sized pyramidal cells present in VIa (Petrides and Pandya, 1994). Von Bonin and Bailey (1947) provided a description of area 44 in the macaque which generally resembles the structure of the human area 44 and is identical with area FCBm and F5 as defined by Matelli *et al.* (1985). Pandya and Yeterian (1996) identify an area with architectonic features similar to human area 44 in the caudal bank of the lower ramus of the AS, which appears to be identical with that part of F5 which is contained in the caudal bank of the inferior AS. Therefore, the minimal extent of the macaque homologue of the human area 44, according to position relative to the AS and cytoarchitectonic criteria is a subpart of the agranular premotor cortex, area FCBm ($\approx$ F5).

We saw earlier that canonical neurons lie in the region of F5 buried in the dorsal bank of the AS, F5ab, while mirror F5 neurons lie in the convexity located caudal to the AS. Arbib and Bota (2003) argue that the minimal extent of Brodmann's area 44 in macaque at least overlaps F5c and therefore may contain the mirror neurons. It is possible that the macaque homologue of area 44 also includes the F5 canonical neurons, since these two populations of neurons are not totally segregated (Rizzolatti and Luppino, 2001).

Recent neurophysiological evidence suggests that the mirror neurons may be further dissociated. Kohler *et al.* (2002) report that some of the mirror neurons of F5 respond to auditory stimuli, and are called audiovisual mirror neurons. Presumably the existence of such neurons can be explained by the auditory inputs received from the auditory cortices, including Tpt, which we will discuss below. Area 45 also appears to contain auditory-responsive neurons (Romanski and Goldman-Rakic, 2002) but, unlike audiovisual mirror neurons of area 44, these cells may be involved in non-spatial acoustic processing. More on this below.

We now briefly summarize further data supporting the homology between F5 and area 44. In both F5 and human 44 there is a representation of hand and mouth actions. Perhaps the strongest motor activation of area 44 was obtained in a task in which participants had to continuously change finger grip (Binkofski *et al.*, 1999). Moreover, the somatotopy of macaque area F5 is markedly different from that in primary motor cortex (F1 $\approx$ M1). Matelli *et al.* (1986) found that while there are virtually no connections between hand and mouth areas in F1, the two representations are heavily connected in F5. Turning to humans, fMRI experiments (e.g., Buccino *et al.*, 2001) show activations during action observation located in the dorsal sector of area 44, extending into area 6. Why should these activations include both dorsal 44 and ventral area 6? Giacomo Rizzolatti (personal communication) suggests that when there are actions including arm movements, area 6 (not just area 44) is activated. F5 is strictly connected with F1 where individual finger movements are specifically coded. The independent control of fingers is even more developed in human M1 than in monkey F1. It is thus likely that actions where individual control of fingers is the fundamental motor aspect are represented in 44/F5, while more global actions (perhaps even hand actions) are represented in area 6.

Area 45 in humans is granular, with clusters of large pyramidal cells in the lower part of layer III and a well-developed layer IV. The description of area 45 provided by Walker (1940) suggests that the possible homologue in the macaque cortex is located ventrally to area 8A and in the anterior part of the inferior limb of the AS. The macaque area 45 is characterized by large pyramidal cells in layers III and V (Walker, 1940). Petrides and Pandya (2002) partially agree with the description provided by Walker, locating area 45 on the rostral part of the inferior AS, but include in it the frontal bank of the inferior part of the AS, as well as the part of the inferior prefrontal convexity between the AS and infraprincipalis dimple (Petrides and Pandya, 1994, 2002; Pandya and Yeterian, 1996). Based on such data, Arbib and Bota (2003) conclude that the human area 45 may have as macaque counterpart the subarea 45B of Petrides and Pandya which has the cytoarchitectonical characteristics of human area 45: large neurons in the deeper part of layer III, a well-developed layer IV, and medium-sized neurons in layer V. This area is ventral to that part of the rostral bank of the AS which is included in the macaque frontal eyefield (FEF), and contains an orofacial representation (Cadoret *et al.*, 2000). However, the macaque counterpart of the human area 45 may extend into the inferior prefrontal convexity. A comprehensive evaluation of similarities between the human area 45 and the macaque areas 45A and 45B will have to include not only position relative to landmarks and cytoarchitecture, but also a complete hodological analysis and single neurons recordings.

5.3.2 Wernicke's area

We saw that Wernicke's area, in the most limited definition, corresponds to the posterior part of BA 22, or area Tpt (temporoparietal) as defined by Galaburda and Sanides (1980), whereas lesion-based views of Wernicke's area may include not only the posterior part of BA 22 but also (in whole or in part) areas 42, 39, 40, and perhaps 37 (Wise *et al.*, 2001). In this section, we focus on the narrow definition, and start by characterizing the macaque area Tpt. This is part of the lateral belt line of auditory-related areas in the superior temporal gyrus (Preuss and Goldman-Rakic, 1991). It is distinguished from the adjacent auditory structure by its cytoarchitectonics profile which resembles more that of the neighboring posterior parietal cortex than those of temporal cortices (Galaburda and Pandya, 1983). Area Tpt appears to be the macaque homologue of BA 22, based on criteria of relative position and cytoarchitecture (Galaburda and Sanides, 1980; Preuss and Goldman-Rakic, 1991). Both BA 22 in humans and Tpt in the macaque are located in the posterior part of the superior temporal gyrus. Both structures present a layer IV which is not as strong as in the anteriorly located auditory structures and fuses with layer V, a sublayer IIIc, a layer V which is split, and a densely populated layer VI (Galaburda and Sanides, 1980; Galaburda and Pandya, 1983). The differences between Wernicke's area and the macaque Tpt are in their relative sizes, the human area 22 being more extended than Tpt (Aboitiz and García, 1997), and highly asymmetric towards the left hemisphere (Aboitiz and García, 1997).

In the macaque there is no fiber tract corresponding to the human arcuate fasciculus, which is claimed to link Wernicke's area to Broca's area, even though there are fibers projecting from the superior temporal gyrus toward the frontal and prefrontal cortices (Seltzer and Pandya, 1989). Petrides and Pandya (1984) showed that areas located around the inferior ramus of the AS receive connections from the auditory cortices TS2 and TS3, while the output of Tpt is mainly directed towards the dorsal parts of areas 6 and 8.

Areas TS2 and TS3, together with area TS1, make up the rostral part of the macaque superior temporal gyrus (STG), and are distinguished according to topographic and cytoarchitectural criteria (Galaburda and Pandya, 1983). Area TS1 occupies the most rostroventral position in the STG, and it is dorsally bordered by TS2. TS3 is located dorsal to TS2 and ventral to Tpt. Area TS3, as defined by Galaburda and Pandya (1983), is identical, or at least overlaps with the rostral parabelt region (RP) as defined by Hackett *et al.* (1998). Area Tpt as defined in nomenclature of Galaburda and Pandya (1983) may be identical or overlaps with the caudal parabelt region (CP) of Hackett *et al.* (1998)

However, retrograde tracing experiments performed by Deacon (1992) show that the ventrocaudal part of Tpt (i.e., that part buried inside the superior temporal sulcus) sends projections to the caudal part of the inferior ramus of the AS, which corresponds to area 44. When Petrides and Pandya (2002) injected retrograde tracers in area 45 they found strong inputs from the auditory cortices and from the association areas from the superior temporal gyrus, including Tpt. Other relevant inputs originated from the inferior parietal lobule areas PG and POa (Petrides and Pandya, 2002). These experiments show that at least those parts of Tpt which are neighboring or are inside the superior temporal gyrus send projections to areas 45 and 44 as defined by Petrides and Pandya.

An additional set of connections that terminate in the inferior part of the AS and in those structures which are considered the homologues of Broca's area originate from areas of the inferior parietal lobule. Area 45 (as defined by Goldman-Rakic, 1987) which corresponds to 45A (as defined by Petrides and Pandya, 2002) receives connections from inferior parietal areas 7a and 7b (Cavada and Goldman-Rakic, 1989). Area 7b as defined here corresponds to areas PF and PFG of Rizzolatti and Luppino (2001), and area 7a overlaps with area PG, while area POa partially corresponds to LIP, situated on the lateral bank of the intraparietal sulcus (Seltzer and Pandya, 1986; Lewis and Van Essen, 2000). Macaque area 44 (area F5) receives connections from the anterior intraparietal area (AIP), PF, PFG, and PG (Matelli *et al.*, 1986; Luppino *et al.*, 1999; Luppino and Rizzolatti, 2000, 2001).

5.4 Sequences and hierarchies

Lieberman (2000) emphasizes that the roles of Broca's and Wernicke's areas cannot be considered apart from a large range of neocortical and subcortical circuits, citing data from studies of Broca's aphasia, Parkinson's disease, focal brain damage, etc., to demonstrate the importance of basal ganglia in sequencing the discrete elements that constitute a complete motor act, syntactic process, or thought process. Similarly, in building upon Fig. 5.5, we need to bear in mind the definition of "complex imitation" (Arbib, Chapter 1, this volume) as the ability to recognize another's performance as a set of familiar movements and then repeat them, but also to recognize that such a performance combines novel actions which can be approximated by (i.e., more or less crudely be imitated by) variants of actions already in the repertoire. Moreover, in the FARS model, the interactions shown in Fig. 5.4 are supplemented in the computer implementation of the model by code representing the role of the basal ganglia in administering sequences of actions; and Bischoff-Grethe *et al.* (2003) model the possible role of the basal ganglia in interactions with the pre-supplementary motor area (pre-SMA) in sequence learning. Thus we agree with Visalberghi and Fragaszy's (2002, p.495) suggestion that "[mirror] neurons provide a neural substrate for segmenting a stream of action into discrete elements matching those in the observer's repertoire, as Byrne (2003) has suggested in connection with his string-parsing theory of imitation" (cf. Stanford, this volume), while adding that the success of complex imitation requires that the appropriate motor system be linked to appropriate working memories (as in Fig. 5.5) as well as to pre-SMA and basal ganglia (not shown in Fig. 5.5) to extract and execute the overall structure of the compound action, which may be sequential, or a more general coordinated control program (Arbib, 1981).

In Section 5.5.2, we review the role of the anterior cingulate cortex (ACC) in vocalization and emotional behavior. Here we note (following Jürgens, 2002) that both ACC and the supplementary motor area (SMA) have been reported to produce vocalization when electrically stimulated – but with crucial differences between humans and non-human primates: the SMA has been found to produce vocalization only in humans, not in other mammals. Lesions in the SMA do not affect spontaneous vocal behavior in the squirrel monkey (Kirzinger and Jürgens, 1982) and macaques (Sutton *et al.*, 1985). Lesions invading the pre-SMA, an area bordering the SMA rostrally, however, decrease the number of spontaneous vocalizations. The anterior cingulate gyrus, in contrast, produces vocalization in non-human mammals, such as the rhesus monkey, squirrel monkey, cat, and bat, but not in humans. Human speech needs input from the ventral premotor and prefrontal cortex, including Broca's area, for motor planning of longer purposeful utterances, as well as input from the SMA and pre-SMA which give rise to the motor commands executed by the motor cortex. In humans, lesions invading anterior cingulate gyrus as well as SMA and/or pre-SMA usually are characterized in the beginning by an akinetic mutism which, after some time, changes into transcortical motor aphasia (Rubens, 1975). In this state, the patient has recovered the ability to produce well-articulated, grammatically correct sentences. There is a severe reduction in the number of speech utterances, however. Most of the utterances represent responses to questions of the partner; there are only very few spontaneous utterances. If the patient is asked to produce short sentences with different emotional intonations, the sentences are all spoken in the same monotonous manner (Jürgens and von Cramon, 1982).

Jürgens (2002) argues that the differential role of SMA and ACC in human and non-human vocalization might reflect the far greater role that motor learning plays in the vocal behavior of humans as compared to other mammals. In this regard, we may note Tanji's (2002) review of the issue of motor selection to arrange multiple movements in an appropriate temporal order. He reviews studies on human subjects and non-human primates showing that the medial motor areas in the frontal cortex and the basal ganglia play particularly important roles in the temporal sequencing of multiple movements. In particular, SMA and pre SMA take part in constructing the time structure for the sequential execution of multiple movements. The involvement of SMA in, e.g., sequential hand movements in the macaque whereas stimulation of the SMA does not produce vocalization only in non-human mammals is certainly consistent with the hypothesis that protosign may well have provided the scaffolding for the emergence of protospeech.

It is beyond the scope of the present chapter to do justice to the theme of "sequences and hierarchies." However, it is important to stress that our interest – whether in praxic action or language – combines two very different kinds of sequence:

a. Fixed sequences such as the sequence of phonemes that constitutes a word (though even here, the detailed control of articulators will vary from occasion to occasion).
b. Sequences generated on-line to meet some specific goal. Whether a sentence or a goal-directed action, the "sequence" will in fact be structured as the output of a nested hierarchy of constituents.

Two examples (from Arbib, 2006, discussed in Chapter 1 of this volume) illustrated option (b):

i. The plan for opening a child-proof aspirin bottle was given as a hierarchical structure which unpacks to different sequences of action on different occasions, with subsequences conditioned on the achievement of goals and subgoals.
ii. The sentence "Serve the handsome old man on the left." can be parsed to yield its hierarchical constituent structure by applying syntactic rules. However, we saw the utility of viewing this sequence as the result of the speaker endeavoring to reach a *communicative goal*: to get the waiter to serve the intended customer. The *sentence planning strategy* repeats the "loop" <add adjective or prepositional phrase> until the ambiguity in specifying the correct customer is (thought to be) resolved.

Bringing in the role of prefrontal cortex in the "on-line" planning of the sequential expression of hierarchical structures of goals and subgoals thus remains a major challenge both for expanding our understanding of the neural basis of motor control and for building a bridge from action to language.

5.5 Audition and vocalization

5.5.1 Further thoughts on the auditory system

We noted in Section 5.2.2 that Aboitiz and García (1997) presented macaque area Tpt as the homologue of human Wernicke's region but that Aboitiz *et al.* (2005) have taken account of the post-1997 literature to update their view of the relevant auditory areas of the macaque. We now review the relevant data, and more.

The starting point is that the functional organization of the macaque auditory system may be similar to that of the visual system in that it is dissociated in a ventral and a dorsal stream (recall Section 5.2.3). The processing of sound is performed and conveyed by three hierarchically organized groups of auditory regions: the core, belt, and parabelt, and the dissociation of the auditory "what" pathway from the "where" pathway follows a rostrocaudal gradient (Wise, 2003).

Analysis of the patterns of connections of macaque areas 44 and 45 show that the homologous structures of Broca's area receive connections from the auditory cortices, as well as from Tpt. Romanski *et al.* (1999a, 1999b) have studied auditory input to the inferior frontal lobe in macaques, including area 45. Romanski and Goldman-Rakic (2002) identified an auditory domain in the monkey prefrontal cortex, located in areas 12lat, 12orb, and the portion of area 45 rostral to the ventral part of the arcuate cortex, in which most neurons preferred vocalizations to other acoustic stimuli, while some neurons were also responsive to visual stimuli. The anterior superior temporal (aST) region projects to the inferior frontal gyrus (IFG) and other parts of the ventrolateral prefrontal cortex (VLPFC). Together, aST and IFG seem to form a "what" stream for the recognition of auditory objects (Rauschecker, 1998; Romanski *et al.*, 1999b), perhaps related to the role of IT in visual object identification mentioned earlier for the FARS model (Fig. 5.4). Moreover, neurons in macaque aST are quite selective for species-specific vocalizations.

Rauschecker (2005) argues that aST in non-human primates is a precursor of the same region in humans and that non-human primate vocalizations are an evolutionary precursor to human speech sounds. However, recognizing speech sounds is just one small component of being language-ready. It is quite consistent with MSH that as protospeech "spirally evolves" with protosign (Arbib, Chapter 1 this volume, Section 1.4), it would co-opt available resources rather than invent a new periphery. Following Rauschecker (2005), one might view the projection from aST to IF via the uncinate fascicle as serving a similar role for "auditory objects" to that served by inferotemporal cortex for the visual system. This hypothesis is also supported by Wise (2003), who views the information conveyed by the uncinate fasciculus as related to the identity of the auditory object. Thus, the "what" auditory pathway in monkey includes the rostral areas of the lateral belt and of the parabelt, and the higher-order areas TS1 and TS2 (Wise, 2003). One should also note that the rostral and caudal streams do overlap, with the middle belt regions sending connections to both rostral and caudal prefrontal areas (Fig. 5.6) (Wise, 2003).

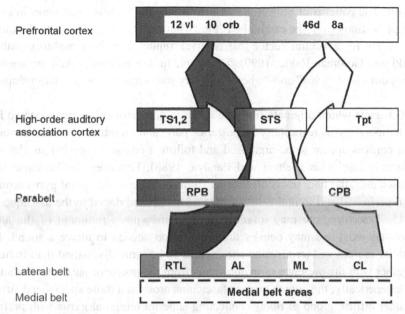

Figure 5.6 The "what" and "where" auditory streams preserve a rostrocaudal cortical topography. The streams separation is not absolute since the middle regions of the auditory lateral belt and parabelt project to both rostral and caudal areas of the prefrontal cortex. AL, anterolateral belt area; CL, caudolateral belt area; CPB, caudal parabelt area; ML, middle lateral belt area; RPB, rostral parabelt area; RTL, rostrotemporal lateral belt area. (Adapted from Wise, 2003.)

Wernicke's area might then be seen as providing an input stage to parietal cortex and being part of an auditory dorsal pathway which is involved in processing the location of an auditory object. The "where" auditory pathway is hypothesized to include the caudal regions of the auditory lateral belt and parabelt area, Tpt area, and it projects to dorsal frontal and prefrontal areas (Wise, 2003).

The auditory "what" and "where" pathways hypothesized by Romanski do not explain the tract tracing experiments of Petrides and Pandya (2002), which show that areas 45A and 45B receive connections from association cortices TS1 and TS2, the lateral belt and parabelt areas, and area Tpt (Romanski and Goldman-Rakic, 2002; Petrides and Pandya, 2002). The difference between the findings is possibly due to the techniques employed. The tract tracing results of Petrides and Pandya (2002) and Galaburda and Pandya (1983) were obtained using retrograde tracers and the tracer injection covered most of area 45 and extended into the anterior wall of the AS, while Romanski used anterograde tracers, and the injections were small and did not cover the extent of the lateral belt auditory areas. Moreover, the injections performed by Romanski were made in the more medial parts of the lateral belt areas, while the experiments of Petrides and Pandya show neurons that project to area 45 tend to cluster in the more lateral sectors of the lateral auditory belt, and the parabelt. The patterns of projections of the "what" and "where" pathways in area 45 show the projections from the caudal belt (the "where" pathway) cover area 45A almost entirely, except for a ventral sector that receives inputs from both auditory pathways (Romanski and Goldman-Rakic, 1999). Therefore, further empirical data are needed to test the hypothesis of "what" and "where" auditory pathways and their relationships with area 45.

The fiber tracts which originate from the auditory and association areas located in the superior temporal gyrus (especially its posterior part) and targeting different frontal and prefrontal cortices appear to be organized and follow a course somewhat similar to the human arcuate fasciculus (Seltzer and Pandya, 1988). However, unlike those of the arcuate fasciculus, the fiber tracts originating from the superior temporal gyrus terminate on several prefrontal and frontal cortices, located rostral and dorsal to the macaque areas 44 and 45. Therefore, one may infer that in the macaque a rudiment of the arcuate fasciculus may exist and may convey the information needed to locate a sound. However, both its sources and termination sites appear to be more diversified than in human. This, together with the presence of major inputs from the posterior parietal areas suggest that, phylogenetically, the human arcuate fasciculus arose as a more specialized structure from a rather diffuse group of fibers connecting superior temporal gyrus with prefrontal and frontal areas in the macaques, possibly incorporating the posterior parietal inputs. Functionally, the human arcuate fasciculus may be partially analogous to the projections from the posterior part of STG in the macaque, and therefore participate to the "where" auditory stream in humans.

5.5.2 Vocalization, motivation, and the anterior cingulate cortex

The macaque ACC plays a crucial role in primate vocalization. ACC occupies two-thirds of the medial part of the cerebral hemispheres and is composed of the ventrally located anterior cingulate gyrus (ACG) and of the more dorsal anterior cingulate sulcus. ACC is made of several areas that are distinguished on cytoarchitectonical and hodological grounds: Brodmann areas 24, 25, and 32 (Paus, 2001). Area 25 is located in the rostroventral part of ACC and is characterized by a deep layer V–VI with the highest cell density of all cingulate regions, no layer IV, and a superficial layer II–III. Area 32 is located dorsal to area 25, and is characterized by homogeneous layers II–III and Vb–VI, with a distinct layer Va. Area 24 is located dorsal to the corpus callosum and is further subdivided into areas 24a–c. Area 24a is located on the ACG and does not present layers II–III, layers Va and VB are difficult to differentiate, and layer VI is distinct. Area 24b is also located on ACG and has distinct layers II, III, and Va. Finally, area 24c is agranular, is located in the cingulate sulcus, and is characterized by large pyramidal neurons in layer III (Vogt *et al.* 1987; Morecraft and van Hoesen, 1998).

Matelli and his colleagues divide the cingulate area 24 into four subparts: 24a and 24b that are identical to those defined by Vogt, and 24c and 24d. The differences between areas 24c and 24d are in the thickness of layer V, the radial organization of the deep cortical layers, and in the density of the myelinated fibers and their orientation (Matelli *et al.*, 1991). Carmichael and Price (1994) proposed a parcellation scheme of the ACC based on cytoarchitectonic, myeloarchitectonic, and chemoarchitectonic criteria, and which is similar to that proposed by Vogt and colleagues.

The cortical areas making up the ACC have overlapping but distinct afferent connection patterns. Area 25 receives cortical connections preferentially from prefrontal cortices, association auditory areas TS2 and TS3, the subiculum and the CA1 field of the hippocampus and from the lateral and accessory basal nuclei of the amygdala. Areas 24a and 24b receive projections mostly from the prefrontal areas, the parahippocampus, and from the lateral and accessory basal nuclei of the amygdala. Area 24c receives input preferentially from the anterior cingulate areas 24a, 24b, and 25, posterior parietal areas PGm and PG, insula, and frontal areas 12 and 46 (Vogt and Pandya, 1987; Morecraft and van Hoesen, 1998; Barbas *et al.*, 1999).

The ACC also receives major input from the mesencephalic dopaminergic system (Paus, 2001) that originates in the A8 and A9 fields (Williams and Goldman-Rakic, 1998). The dopamine projections to the ACC are also specific: area 24 appears to be the major recipient of the dopamine fibers (Williams and Goldman-Rakic, 1998; Berger *et al.*, 1988). While the role of dopamine inputs to the ACC is not unequivocally established, it may be related to error detection and action monitoring (Gehring and Knight, 2000; Paus, 2001).

The ACC projects to different parts of the monkey central nervous system, with the component areas each having specific projection patterns. Areas 24a and 24b project to the dorsal and medial premotor areas (pre-SMA, SMA and the ventrorostral part of F2 that includes a hand representation), and area 24c projects to the SMA, the ventral premotor areas F4 and F5, and the primary motor cortex (Morecraft and van Hoesen, 1992; Rizzolatti and Luppino, 2001). Areas 24b and 24c provide a strong input to the motor cortical larynx area (Simonyan and Jürgens, 2005), and area 24c includes neurons that project directly to the spinal cord (Dum and Strick, 1996). Dum and Strick (1996) define three regions of the cingulate cortex, which contain neurons projecting to the spinal cord: a more rostral region, the rostral cingulate motor area (CMAr) that is included in area 24c, and two areas located caudal to the genu of the corpus callosum.

Different regions of the ACC also project to nuclei of amygdala and hypothalamus. Areas 25 and 32 project strongly to the basolateral and basomedial nuclei of amygdala (Barbas and de Olmos, 1990). The hypothalamic targets of areas 32 and 25 include the lateral and the medial preoptic areas, the anterior hypothalamic and the lateral hypothalamic area, and area 24b projects preferentially to the ventromedial hypothalamic nucleus, and the posterior parts of the periventricular nucleus (Ongür *et al.*, 1998).

Lesion studies and recording experiments showed that the ACC is involved in monkey vocalizations. Lesions of the ACG showed that this region is necessary for mastering vocal operant conditioning tasks, but the lesioned animals can still react vocally to unconditioned stimuli (Jürgens, 2002). More extensive bilateral lesions of the ACC that extend in the subcallosal cortex abolish spontaneously produced long-distance calls in socially isolated animals, but do not impair animals' response to calls uttered by other conspecifics (Jürgens, 2002).

Single cell recordings in macaques have shown that ACC contains neurons that modulate their activity with vocalization, opening of the jaw, and with vocalization and jaw opening together (West and Larson, 1995). The neurons with activity related to vocalization have been found to be distributed in two regions of the ACC: one located rostral to the genu of the corpus callosum and overlapping area 25, area 32, and the rostral sector of area 24, and a second region located more caudally and that overlaps with areas 24c and 24b and extends into pre-SMA (West and Larson, 1995). The more rostral vocalization region, rostral cingulate vocalization area (CVAr), was previously described in the literature, and the second region, caudal cingulate vocalization area (CVAc), overlaps with area CMAr of Dum and Strick.

The existence of two vocalization regions in the ACC, each located in areas with specific patterns of afferent and efferent connections, suggests specific roles for each of them. The more caudal region, CVAc, may be part of a motor network of brain regions involved in vocalization. The presence of the dopaminergic input in area 24 indicates a possible error detection and motor correction mechanism in area CVAc. The more rostral region, CVAr, may be involved in integration of auditory information relayed by the higher order auditory areas TS1 and TS2 with multimodal information provided by the hippocampus, and with emotional and context-related information provided by the amygdala. In its turn, CVAr may be involved in regulation of the internal states of the animal, through the projections to different nuclei of the hypothalamus, and of the amygdala. However, the specific roles of CVAr and CVAc in monkey vocalization remain to be unequivocally established by lesion experiments and single neuron recordings.

The human ACC is made of the agranular Brodmann areas 33, 24, 25, and 32. Each of these cortical areas can be further subdivided in several subparts, based on cytoarchitectonic and chemoarchitectonic grounds (Vogt et al., 2004). Similar to the macaque ACC, the human ACC areas receive connections from the amygdalar nuclei and project to brainstem autonomic nuclei, thought to be involved in the control of autonomic functions (Vogt et al., 2004). The pattern of connections of the amygdala with the anterior cingulate regions is similar to that in the macaque in that it projects primarily to the areas 25 and 32, and to the rostral sector of area 24 (Vogt et al., 2004).

The human ACC can be divided functionally in three regions: a visceromotor control region that is located in area 25, a region involved in autonomic regulation of emotional behaviors, located in areas 32 and the rostroventral part of area 24, and a region involved in skeletomotor control that is located in dorsocaudal sectors of areas 32 and 24 (Vogt et al., 2004). As noted in Section 5.4, lesions that extend into the human ACC and the SMA and/or pre-SMA initially yield akinetic mutism, after which the ability of the patient to produce speech is partially restored, but the patient does not utter spontaneously, and the ability to produce sentences with different emotional intonations is lost (Jürgens and von Cramon, 1982; Jürgens, 2002). This reminds us that MSH addresses the parity problem – how meaning may be shared between "speaker" and "hearer" – but not what motivates the speaker to "say" something, and the hearer to "listen."

The neuroanatomical and functional data reviewed in this section indicate that the various regions of the ACC have different roles in motor, cognitive, and affective processing (Paus, 2001). The more ventral and rostral regions of the ACC may be involved more in processing emotional input conveyed from the amygdalar nuclei and in the subsequent alteration of the internal state of the animal, through direct output towards the hypothalamic nuclei. The more dorsal and caudal regions of the ACC appear to be involved in motor output that is related to the information processed by the more rostral and ventral cingulate regions, and in a possible error-matching and motor correction function. In any case, the roles of ACC appear to be complementary to that of the F3 SMA–pre-SMA system and its parietal extension.

Positron emission tomography (PET) measurements of the activity of ACC in speech behavioral paradigms revealed two foci located in the rostral and in the dorsal part of the cingulate cortex, respectively (Paus *et al.*, 1993). The more rostral speech related region overlaps area 24 and in the intermediary area 32, and the dorsal speech related region is located in the dorsal sector of area 24. Comparing the macaque and human maps of the ACC, one may assume a possible homology between the human speech-related areas and the monkey vocalization regions, based on relative position. Functionally, the speech-related human cingulate areas also may be similar to those of macaques. Thus, the rostral speech-related region may be involved in emotional and contextual aspects, while the caudal speech-related region is involved in motor aspects of the utterances.

5.6 Towards an action-oriented neurolinguistics

We now build upon Fig. 5.5 and the data reviewed in the previous sections to offer a more complete high-level view diagram (Fig. 5.7) of the fronto-parieto-temporal cortices involved in choosing an action, recognizing an action and describing an action. As noted earlier, Itti and Arbib (this volume) complement Fig. 5.7 by providing an overall functional framework for how the structure of "minimal subscenes" may be linked with language production both in describing a visual scene or in answering questions about the scene. Their account links the recognition of agents, actions, and objects – thus immensely expanding the simple box at bottom left of Fig. 5.7, and expanding the network of its relations with other parts of the system.

The main extension of the cortical network shown in Fig. 5.5 is the specification of the auditory pathway and the inclusion of the "what" and "where" auditory channels. The auditory "what" pathway includes at least two stations, the anterior auditory lateral belt (AL) and the prefrontal cortices 9/46 and possibly the most ventral sector of area 45A (Romanski *et al.*, 1999b, 2004; Romanski and Goldman-Rakic, 2002). This pathway may also convey information about the sound source identity ("who"). Recordings performed in the macaque ventrolateral prefrontal cortex located rostral to the AS showed that some

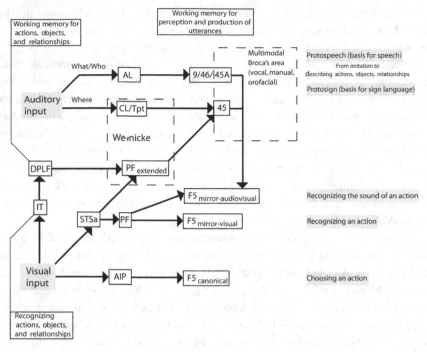

Figure 5.7 The extension of the fronto-parietal network shown in Fig. 5.5 to a fronto-parieto-temporal network that specifies the "what/who" and "where" auditory pathways, and views area 45 as mediating a convergence of these pathways. The network allows us to add one more step from action recognition to language: recognition of the sound of an action.

neurons increase their firing rates with monkey or human vocalizations, which may indicate a processing of the identity of the auditory stimulus (Romanski and Goldman-Rakic, 2002). We however do not exclude the hypothesis of a dissociation of the "what" and "who" channels in terms of processing stations and neuroanatomical pathways. Thus, the network shown in Fig. 5.7 may be further extended with the parabelt areas, the association cortices TS2 and TS3, and possibly with more prefrontal regions (Rauschecker and Tian, 2000). The "where pathway" includes the caudal lateral belt (CL) and possibly Tpt (the extended PF), and it is thought to project to area 45 (Petrides and Pandya, 2002). As for the "what/who" pathway, more cortical regions may be included in this channel, and therefore the network shown in Fig. 5.7 can be further extended and refined. Area 45 may be seen as a frontal cortical region where the two functionally segregated auditory pathways may converge, as the results of Petrides and Pandya (2002) indicate. This, together with the projections from area 45 to area 44, as described by Petrides and Pandya (2002), may explain the existence of the audiovisual neurons, described by Kohler.

Because of the data of Kohler *et al.* (2002) demonstrating mirror neurons which were activated by the characteristic sound of an action as well as sight of the action, Fig. 5.7 not only adds the specification of the auditory pathways to Fig. 5.5, but also makes explicit the role of audiovisual mirror neurons in the recognition of the sound of an action. The auditory regions and pathways shown in Fig. 5.7 are based mostly on macaque data. Therefore, the Figure may need to be changed as new human neuroanatomical and functional data become available. The existence of two auditory channels in the macaque, which are to a great extent spatially segregated, does not eliminate the possibility that in humans the lexicon and word meaning are also processed in the posterior part of the superior temporal cortex. In any case, as we turn from macaque to human, more experimental data are needed to dissociate the neural correlates of "hearing an action" from those of "hearing a word."

Figure 5.7 preserves "Wernicke's area" as an entity made of the extended PF and the CL/Tpt. However, the "what/who pathway" shown in Fig. 5.7, and thus the neural pathway associated to the lexicon, includes AL and not CL. This may imply that the concept of "Wernicke's area" has now become meaningless as other than a historical term, as suggested by Wise *et al.* (2001), and the discussion on the neural correlates of hearing of spoken language should be centered on the auditory cortical areas that make up the superior temporal gyrus and sulcus. Therefore, the discussion on the neural correlates of language should be shifted from the analysis of a single region, to a distributed system that is involved in lexicon and word meaning. However, multi-modality of the language system remains crucial. Skipper, Nusbaum, and Small (this volume) make specific reference to the integration of auditory cues and visual (e.g., lip movement) cues in understanding speech.

Aboitiz and García (1997) stress the role of the lexicon of spoken words and syntax to bind them into sentences in language. They then look at the features of the human brain, seek the homologous areas of the macaque brain, note what has changed (some areas enlarge, some connections are strengthened) and then suggest how these changes could support a lexicon and a syntax. By contrast, Arbib and Rizzolatti (1997; Rizzolatti and Arbib, 1998) start from an analysis of the monkey's capabilities, especially the facts that species-specific vocalizations have their cortical outpost in the anterior cingulate but that a different area, involved in hand movements, is homologous to Broca's area. The version of MSH summarized in Section 5.2.1 assumes that the human brain evolved (in part) to support protosign and protospeech, with the richness of human languages being a "post-biological accumulation of inventions," and offers hypotheses on how intermediate stages from the mirror system for grasping led via imitation and protosign to protospeech. However, Rizzolatti and Arbib are relatively silent on the phonological loop and other working memory systems whose emphasis is an important feature of the AGH.

Whether we are talking about the words of a language or the "protowords" of a protolanguage, we must distinguish the "sign" from the "signified" (Fig. 1.7). The neural representation of the "sign" inherits mirror properties from item (6) above – linking the production of vocal, manual, and/or facial gestures for a word to their recognition. This shared neural representation for the word must then be linked to activity in the neural networks representing schemas for related concepts. In discussing the "pre-grammar" for

F5 canonical neurons, we drew the implication that 'if the same principle holds for linguistic commands as for motor commands, Broca's area would code "verb phrases" and constraints on the noun phrases that can fill the slots, but not details of the noun phrases themselves. This knowledge (objects/noun phrases) could be completely outside F5/Broca's area. In keeping all this straight it is necessary to distinguish verbs and nouns as words in the sense of perceivable and reproducible sequences of articulatory gestures (which we see as being "in the mirror") from verbs (some but not all of which are "in the mirror") and nouns and other words (which are not "in the mirror"). Reflecting the MNS model up to the level of language, the notion is Broca's area must be linked into PFC planning (and its administration by the basal ganglia) to assemble verb–argument and more complex hierarchical structures, finding the words and binding them correctly.

Neininger and Pulvermüller (2003) show that patients with lesions in the right frontal lobe showed most severe deficits in processing action verbs, whereas those with lesions in their right temporo-occipital areas showed most severe deficits in processing visually related nouns. This suggests that humans possess a bilateral naming system but that the left hemisphere has enough initial extra complexity that it has an initial advantage in handling not only the action recognition and naming tasks but also the translation of that recognition and the binding of actions to objects into the appropriate word order and use of function words. Other studies tell us more about the dominant hemisphere.[2] Several studies (e.g., Miceli *et al.*, 1984; McCarthy and Warrington, 1985; Damasio and Tranel, 1993) show that motion verbs are represented in the frontal lobe, essentially in Broca's area whereas object naming is located posteriorly. Miceli *et al.* (1984) hypothesized that the omission of main verbs in agrammatic speech is caused, at least in part, by a lexical as opposed to a syntactic deficit. They found that agrammatic patients showed a marked impairment in naming actions in contrast to anomic aphasics and normal controls who named actions better than objects. However, we have some concern about the disjunction of "a lexical as opposed to a syntactic deficit" given the notion that a noun can be considered in isolation, whereas a verb, we would suggest, comes as a lexical item with links to its "slot-fillers" and the conventions to express them with word order and function words or case markings. McCarthy and Warrington (1985) also studied a severely agrammatic patient whose extremely abnormal verb phrase constructions in spontaneous speech rested on an impairment both in the retrieval and in the comprehension of action names and verbs, rather than of nouns. They argue that his semantic representation of verbs is impaired, and that this lexical deficit may well be at the core of his type of agrammatism. This would seem to be in agreement with the analysis we gave above once one accepts that syntax is not autonomous, and that the "lexical item" that is a verb has the semantics of its slot-fillers inextricably intertwined with their syntactic expression. Damasio and Tranel (1993) studied three patients to demonstrate a double dissociation between verb retrieval (impaired by a lesion of the left frontal region) and the retrieval of concrete nouns (impaired by lesions of the left anterior and middle temporal lobe).

---

[2] Our thanks to Giacomo Rizzolatti for an earlier version of this material.

The results of human brain imaging will be best understood when they can be grounded in analysis of detailed circuitry. Such grounding can be of two kinds: (1) relatively direct, where a human system may be posited to be "directly homologous" (a high degree of homology) with a corresponding system in the macaque or other species (as in certain working memory systems for AGH; and the mirror system for grasping for MSH); and (2) relatively indirect when a human system is "somewhat related" (a low degree of homology) to some system in the macaque or, indeed, some system elsewhere in the human brain (consider parallels between different loops linking basal ganglia and cerebral cortex). In the former case, computational models of the neural networks of the human system can be based rather directly on models of the homologous macaque system; in the latter case, partial homologies can be used to define a search space of models which can then be tested by synthetic brain imaging. In each case, we need neuroinformation to develop in a fashion which more tightly integrates modeling with databases (Bischoff-Grethe *et al.*, 2001) to allow the more effective integration of data from neuroanatomy, neurophysiology, brain imaging, and all the other modalities we have seen as helpful in providing criteria for the establishment of degrees of homology. The resultant cross-species framework will allow progress in understanding the neural mechanisms of language (and diverse other cognitive processes) that would be impossible with too narrow a focus on the data of human brain imaging alone.

References

Aboitiz, F., and García, V. R., 1997. The evolutionary origin of the language areas in the human brain: a neuroanatomical perspective. *Brain Res. Rev.* **25**: 381–396.

Aboitiz, F., García, R., Brunetti, E., and Bosman, C., 2005. The origin of Broca's area and its connections from an ancestral working memory network. In Y. Grodzinsky and K. Amunts (eds.) *Broca's Region* Oxford, UK: Oxford University Press.

Amaral, D. G., Price, J. L., Pitkänen, A., and Carmichael, S. T., 1992. Anatomical organization of the primate amygdaloid complex. In J. P. Aggleton (ed.) *The Amygdala: Neurobiological Aspects of Emotion, Memory, and Mental Dysfunction.* New York: Wiley-Liss, pp. 1–66.

Amunts, K., Schleicher, A., Burgel, U., *et al.*, 1999. Broca's region revisited: cytoarchitecture and intersubject variability. *J. Comp. Neurol.* **412**: 319–341.

Arbib, M. A., 1981. Perceptual structures and distributed motor control. In V. B. Brooks (ed.) *Handbook of Physiology*, Section 2, *The Nervous System*, vol. 2, *Motor Control*, Part 1. Bethesda, MD: American Physiological Society, pp. 1449–1480.

  2001. The Mirror System Hypothesis for the language-ready brain. In A. Cangelosi and D. Parisi (eds.) *Computational Approaches to the Evolution of Language and Communication.* London: Springer-Verlag, pp. 229–254.

  2002. The mirror system, imitation, and the evolution of language. In C. Nehaniv and K. Dautenhahn (eds.) *Imitation in Animals and Artefacts.* Cambridge, MA: MIT Press, pp. 229–280.

  2005. From monkey-like action recognition to human language: an evolutionary framework for neurolinguistics. *Behavi. Brain Sci.* **28**: 145–167.

  2006. A sentence is to speech as what is to action? *Cortex.* (In press.)

Arbib, M. A., and Bota, M., 2003. Language evolution: neural homologies and neuroinformatics. *Neur. Networks* **16**: 1237–1260.
    2004. Response to Deacon: evolving mirror systems – homologies and the nature of neuroinformatics. *Trends Cogn. Sci.* **8**: 290–291.
Arbib, M., and Rizzolatti, G., 1997. Neural expectations: a possible evolutionary path from manual skills to language. *Commun. Cogn.* **29**: 393–424.
Baddeley, A. D., 2003. Working memory: looking back and looking forward. *Nature Rev. Neurosci.* **4**: 829–839.
Baddeley, A. D., and Hitch, G. J., 1974. Working memory. In G. A. Bower (ed.) *The Psychology of Learning and Motivation*. New York: Academic Press, pp. 47–89.
Barbas, H., and de Olmos, J., 1990. Projections from the amygdala to basoventral and mediodorsal prefrontal regions in the rhesus monkey. *J. Comp. Neurol.* **300**: 549–571.
Barbas, H., Ghashghaei, H., Dombrowski, S. M., and Rempel-Clower, N. L., 1999. Medial prefrontal cortices are unified by common connections with superior temporal cortices and distinguished by input from memory-related areas in the rhesus monkey. *J. Comp. Neurol.* **410**: 343–367.
Barrett, A. M., Foundas, A. L., and Heilman, K. M., 2005. Speech and gesture are mediated by independent systems. *Behav. Brain Sci.* **28**: 125–126.
Berger, B., Trottier, S., Verney, C., Gaspar, P., and Alvarez, C., 1988. Regional and laminar distribution of the dopamine and serotonin innervation in the macaque cerebral cortex: a radioautographic study. *J. Comp. Neurol.* **273**: 99–119.
Binkofski, F., Buccino, G., Posse, S., *et al.*, 1999. A fronto-parietal circuit for object manipulation in man: evidence from an fMRI study. *Eur. J. Neurosci.* **11**: 3276–3286.
Bischoff-Grethe, A., Spoelstra, J., and Arbib, M. A., 2001. Brain models on the web and the need for summary data. In M. A. Arbib and J. Grethe (eds.) *Computing the Brain: A Guide to Neuroinformatics*. San Diego, CA: Academic Press, pp. 287–296.
Bischoff-Grethe, A., Crowley, M. G., and Arbib, M. A., 2003. Movement inhibition and next sensory state predictions in the basal ganglia. In A. M. Graybiel, M. R. Delong, and S. T. Kitai (eds.) *The Basal Ganglia*, vol. 6. New York: Kluwer Academic, pp. 267–277.
Bota, M., and Arbib, M. A., 2004. Integrating databases and expert systems for the analysis of brain structures: connections, similarities, and homologies. *Neuroinformatics* **2**: 19–58.
Bota, M., Dong, H. W., and Swanson, L. W., 2003. From gene networks to brain networks. *Nature Neurosci.* **6**: 795–799.
Bridgeman, B., 1999. Separate representations of visual space for perception and visually guided behavior. In G. Aschersleben, T. Bachmann, and J. Müsseler (eds.) *Cognitive Contributions to the Perception of Spatial and Temporal Events*. Amsterdam: Elsevier Science, pp. 3–13.
Bridgeman, B., Peery, S., and Anand, S., 1997. Interaction of cognitive and sensorimotor maps of visual space. *Percept. Psychophys.* **59**: 456–469.
Buccino, G., Binkofski, F., Fink, G. R., *et al.*, 2001. Action observation activates premotor and parietal areas in a somatotopic manner: an fMRI study. *Eur. J. Neurosci.* **13**: 400–404.
Byrne, R. W., 2003. Imitation as behaviour parsing. *Phil. Trans. Roy. Soc. Lond. B* **358**: 529–536.

Cadoret, G., Bouchard, M., and Petrides, M., 2000. Orofacial representation in the rostral bank of the inferior ramus of the arcuate sulcus of the monkey. *Soc. Neurosci. Abstr.* **26**: 680 abstr. no. 253.13.

Carmichael, S. T., and Price, J. L., 1994. Connectional networks within the orbital and prefrontal cortex of macaque monkeys. *J. Comp. Neurol.* **371**: 179–207.

Cavada, C., and Goldman-Rakic, P. S., 1989. Posterior parietal cortex in rhesus macaque. II. Evidence for segregated corticocortical networks linking sensory and limbic areas with the frontal lobe. *J. Comp. Neurol.* **287**: 422–445.

Cheney, D. L., and Seyfarth, R. M., 1990. *How Monkeys See the World: Inside the Mind of Another Species.* Chicago, IL: University of Chicago Press.

Code, C., 1998. Models, theories and heuristics in apraxia of speech. *Clin. Linguist. Phonet.* **12**: 47–65.

Damasio, A. R., and Tranel, D., 1993. Nouns and verbs are retrieved with differently distributed neural systems. *Proc. Natl. Acad. Sci. USA* **90**: 4957–4960.

Deacon, T. W., 1992. Cortical connections of the inferior arcuate sulcus cortex in the macaque brain. *Brain Res.* **573**: 8–26.

2004. Monkey homologues of language areas: computing the ambiguities. *Trends Cogn. Sci.* **8**: 288–290.

DeRenzi, E., Pieczuro, A., and Vignolo, L. A., 1966. Oral apraxia and aphasia. *Cortex* **2**: 50–73.

Dum, R. P., and Strick, P. L., 1996. Spinal cord termination of the medial wall motor areas in macaque monkeys. *J. Neurosci.* **16**: 6513–6525.

Fagg, A. H., and Arbib, M. A., 1998. Modeling parietal–premotor interactions in primate control of grasping. *Neur. Networks* **11**: 1277–1303.

Ferrari, P. F., Gallese, V., Rizzolatti, G., and Fogassi, L., 2003. Mirror neurons responding to the observation of ingestive and communicative mouth actions in the monkey ventral premotor cortex. *Eur. J. Neurosci.* **17**: 1703–1714.

Fuster, J. M., 1995. *Memory in the Cerebral Cortex.* Cambridge, MA: MIT Press.

Galaburda, A. M., and Pandya, D. N., 1983. The intrinsic architectonic and connectional organization of the superior temporal region of the rhesus monkey. *J. Comp. Neurol.* **221**: 169–184.

Galaburda, A. M., and Sanides, F., 1980. Cytoarchitectonic organization of the human auditory cortex. *J. Comp. Neurol.* **190**: 597–610.

Gehring, W. J., and Knight, R. T., 2000. Prefrontal–cingulate interactions in action monitoring, *Nature Neurosci.* **3**: 516–520.

Geschwind, N., 1964. The development of the brain and the evolution of language. *Monogr Ser. Lang. Ling.* **1**: 155–169.

Goldman-Rakic, P. S., 1987. Circuitry of the prefrontal cortex and the region of behavior by representational knowledge. In F. Blum and V. Mountcastle (eds.) *Handbook of Physiology*, Section 1, *The Nervous System*, vol. 5, *Higher Functions of the Brian*, Part 1. Bethesda, MD: American Physiological Society, pp. 373–417.

1995. Architecture of the prefrontal cortex and the central executive. *Ann. N. Y. Acad. Sci.* **769**: 71–83.

Goodale, M. A., and Milner, A. D., 1992. Separate visual pathways for perception and action. *Trends Neurosci.* **15**: 20–25.

Hackett, T. A., Stepniewska, I., and Kaas, J. H., 1998. Subdivisions of auditory cortex and ipsilateral cortical connections of the parabelt auditory cortex in macaque monkeys. *J. Comp. Neurol.* **394**: 375–395.

Jeannerod, M., Decety, J., and Michel, F., 1994. Impairment of grasping movements following a bilateral posterior parietal lesion. *Neuropsychologia* **32**: 369–380.

Jeannerod, M., Arbib, A., Rizzolatti, G., and Sakata, H., 1995. Grasping objects: the cortical mechanisms of visuomotor transformation. *Trends Neurosci.* **18**: 314–320.

Jürgens, U., 1997. Primate communication: signaling, vocalization. In *Encyclopedia of Neuroscience*, 2nd edn. Amsterdam: Elsevier.

2002. Neural pathways underlying vocal control. *Neurosci. Biobehav. Rev.* **26**: 235–258.

Jürgens, U., and von Cramon, D., 1982. On the role of the anterior cingulate cortex in phonation: a case report. *Brain.* **15**: 234–248.

Kaas, J., 1993. Evolution of multiple areas and modules within the cortex. *Persp. Devel. Neurobiol.* **1**: 101–107.

Kohler, E., Keysers, C., Umiltá, M. A., *et al.*, 2002. Hearing sounds, understanding actions: action representation in mirror neurons. *Science* **297**: 846–848.

Kirzinger, A., and Jürgens, U., 1982. Cortical lesion effects and vocalization in the squirrel monkey. *Brain Res.* **233**: 299–315.

Lewis, J. W., and Van Essen, D. C., 2000. Corticocortical connections of visual, sensorimotor, and multimodal processing areas in the parietal lobe of the macaque. *J. Comp. Neurol.* **428**: 112–137.

Lieberman, P., 2000. *Human Language and my Reptilian Brain: The Subcortical Bases of Speech, Syntax, and Thought.* Cambridge, MA: Harvard University Press.

Luppino, G., and Rizzolatti, G., 2000. The organization of the frontal motor cortex. *News Physiol. Sci.* **15**: 219–224.

2001. The cortical motor system. *Neuron* **31**: 889–901.

Luppino, G., Murata, A., Govoni, P., and Matelli, M., 1999. Largely segregated parietofrontal connections linking rostral intraparietal cortex (areas AIP and VIP) and the ventral premotor cortex (areas F5 and F4). *Exp. Brain Res.* **128**: 181–187.

Marquardt, T. P., and Sussman, H., 1984. The elusive lesion: apraxia of speech link in Broca's aphasia. In J. C. Rosenbek, M. R. McNeil, and A. E. Aronson (eds.) *Apraxia of Speech: Physiology, Acoustics, Linguistics, Management.* San Diego, CA: College-Hill Press, pp. 99–112.

Matelli, M., Luppino, G., and Rizzolatti, G., 1985. Patterns of cytochrome oxidase activity in the frontal agranular cortex of the macaque. *Behav. Brain Res.* **18**: 125–137

Matelli, M., Camarda, R., Glickstein, M., and Rizzolatti, G., 1986. Afferent and efferent projections of the inferior area 6 in the macaque. *J. Comp. Neurol.* **251**: 281–298.

Matelli, M., Luppino, G., and Rizzolatti, G., 1991. Architecture of superior and mesial area 6 and the adjacent cingulate cortex in the macaque monkey. *J. Comp. Neurol.* **311**: 445–462.

McCarthy, R., and Warrington, E. K., 1985. Category specificity in an agrammatic patient: the relative impairment of verb retrieval and comprehension. *Neuropsychologia* **23**: 709–727.

Miceli, G., Silveri, M. C., Villa, G., and Caramazza, A., 1984. On the basis for the agrammatic's difficulty in producing main verbs. *Cortex* **20**: 207–220.

Morecraft, R. J., and van Hoesen, G. W., 1992. Cingulate input to the primary and supplementary motor cortices in the rhesus monkey: evidence for somatotopy in areas 24c and 23c. *J. Comp. Neurol.* **322**: 471–489.

1998. Convergence of limbic input to the cingulated motor cortex in the rhesus monkey. *Brain Res. Bull.* **45**: 209–232.

Neininger, B., and Pulvermüller, F., 2003. Word-category specific deficits after lesions in the right hemisphere. *Neuropsychologia* **41**: 53–70.

Ongür, D., An, X., and Price, J. L., 1998. Prefrontal cortical projections to the hypothalamus in macaque monkeys. *J. Comp. Neurol.* **401**: 489–505.

Oztop, E., and Arbib, M. A., 2002. Scheme design and implementation of the group-related mirror system. *Biol. Cybernet.* **87**: 116–140.

Oztop, E., Bradley, N. S., and Arbib, M. A., 2004. Infant grasp learning: a computational model. *Exp. Brain Res.* **158**: 480–503.

Pandya, D. N., and Yeterian, E. H., 1996. Comparison of prefrontal architecture and connections. *Phil. Trans. Roy. Soc. London. B* **351**: 1423–1432.

Paus, T., 2001. Primate anterior cingulate cortex: where motor control, drive and cognition interface. *Nature Rev. Neurosci.* **2**: 417–424.

Paus, T., Petrides, M., Evans, A. C., and Meyer, E., 1993. Role of the human anterior cingulate cortex in the control of oculomotor, manual and speech responses: a positron emission tomography study. *J. Neurophysiol.* **70**: 453–469.

Petrides, M., and Pandya, D. N., 1984. Projections to the frontal cortex from the posterior parietal region in the rhesus macaque. *J. Comp. Neurol.* **228** : 105–116.

1994. Comparative architectonic analysis of the human and macaque frontal cortex. In F. Boller and J. Graham (eds.) *Handbook of Neuropsychology*, vol. 9. New York: Elsevier, pp. 17–58.

2002. Comparative cytoarchitectonic analysis of the human and the macaque ventrolateral prefrontal cortex and corticocortical connection patterns in the macaque. *Eur. J. Neurosci.* **16**: 291–310.

Preuss, T. M., and Goldman-Rakic, P. S., 1991. Myelo- and cytoarchitecture of the granular frontal cortex and surrounding regions in the strepsirhine primate *Galago* and the anthropoid primate *Macaca*. *J. Comp. Neurol.* **310**: 429–474.

Rainer, G., Asaad, W. F., and Miller, E. K., 1998. Memory fields of neurons in the primate prefrontal cortex. *Proc. Natl Acad. Sci. USA* **95**: 15008–15013.

Rauschecker, J. P., 1998. Cortical processing of complex sounds. *Curr. Opin. Neurobiol.* **8**: 516–521.

2005. Vocal gestures and auditory objects. *Brain Behav. Sci.* **28**: 143.

Rauschecker, J. P., and Tian, B., 2000. Mechanisms and streams of "what" and "where" in auditory cortex. *Proc. Natl Acad. Sci. USA* **97**: 11800–11806.

Rizzolatti, G., and Arbib, M. A., 1998. Language within our grasp. *Trends Neurosci.* **21**: 188–194.

Rizzolatti, G., and Luppino, G., 2001. The cortical motor system. *Neuron* **31**: 889–901.

Rizzolatti, G., Luppino, G., and Matelli, M., 1996. The classic supplementary motor area is formed by two independent areas. *Adv. Neurol.* **70**: 45–56.

Rizzolatti, G., Fogassi, L., and Gallese V., 2002. Motor and cognitive functions of the ventral premotor cortex. *Curr. Opin. Neurobiol.* **12**: 149–154.

Rolls, E. T., and Arbib, M. A., 2003. Visual scene perception: neurophysiology principles. In M. A. Arbib (ed.) *The Handbook of Brain Theory and Neural Networks*, 2nd edn. Cambridge, MA: MIT Press, pp. 1210–1215.

Romanski, L. M., and Goldman-Rakic, P. S., 2002. An auditory domain in primate prefrontal cortex. *Nature Neurosci.* **5**: 15–16.

Romanski, L. M., Bates, J. F., and Goldman-Rakic, P. S., 1999a. Auditory belt and parabelt projections to the prefrontal cortex in the vhesus monkey. *J. Comp. Neurol.* **403**: 141–157.

Romanski, L. M., Tian, B., Fritz, J., *et al.*, 1999b. Dual streams of auditory afferents target multiple domains in the primate prefrontal cortex. *Nature Neurosci.* **12**: 1131–1136.

Romanski, L. M., Averbeck, B. B., and Diltz, M., 2004. Neural representation of vocalizations in the primate ventrolateral prefrontal cortex. *J. Neurophysiol.* **93**: 743–747.

Rubens, A. B., 1975. Aphasia with infarction in the territory of the anterior cerebral artery. *Cortex* **11**: 239–250.

Seltzer, B., and Pandya, D. N., 1986. Posterior parietal projections to the intraparietal sulcus of the rhesus monkey. *Exp. Brain Res.* **62**: 459–469.

    1988. Frontal lobe connections of the superior temporal sulcus in the rhesus monkey. *J. Comp. Neurol.* **281**: 97–113.

    1989. Frontal lobe connections of the superior temporal sulcus in the rhesus macaque. *J. Comp. Neurol.* **281**: 97–113.

Semendeferi, K., Lu, A., Schenker, N., and Damasio, H., 2002. Humans and great apes share a large frontal cortex. *Nature Neurosci.* **5**: 272–276.

Simonyan, K., and Jürgens, U., 2005. Afferent cortical connections of the motor cortical larynx area in the rhesus monkey. *Neuroscience* **130**: 133–149.

Striedter, G. F., 2004. *Principles of Brain Evolution.* Sunderland, MA: Sinauer Associates.

Sutton, D., Trachy, R. E., and Lindeman, R. C., 1985. Discriminative phonation in macaques: effects of anterior mesial cortex damage. *Exp. Brain Res.* **59**: 410–413.

Tanji, J., 2002. Sequential organization of multiple movements: involvement of cortical motor areas. *Annu. Rev. Neurosci.* **24**: 631–651.

Umiltà, M. A., Kohler, E., Gallese, V., *et al.*, 2001. I know what you are doing: a neurophysiological study. *Neuron* **31**: 155–165.

Ungerleider, L. G., and Mishkin, M., 1982. Two cortical visual systems. In D. J. Ingle, M. A. Goodale, and R. J. W. Mansfield (eds.) *Analysis of Visual Behavior.* Cambridge, MA: MIT Press, pp. 549–586.

Visalberghi, E., and Fragaszy, D., 2002. "Do monkeys ape?" Ten years after. In C. Nehaniv and K. Dautenhahn (eds.) *Imitation in Animals and Artifacts* Cambridge, MA: MIT Press, pp. 471–499.

Vogt, B. A., and Pandya, D. N., 1987. Cingulate cortex of the rhesus monkey. II. Cortical afferents. *J. Comp. Neurol.* **262**: 256–270.

Vogt, B. A., Pandya, D. N., and Rosene, D. L., 1987. Cingulate cortex of the rhesus monkey. I. Cytoarchitecture and thalamic afferents. *J. Comp. Neurol.* **262**: 271–289.

Vogt, B. A., Hof, P. R., and Vogt, L. J., 2004. Cingulate gyrus. In G. Paxinos and J. K. Mai (eds.) *The Human Nervous System.* New York: Academic Press, pp. 915–946.

von Bonin, G., 1949. Architecture of the precentral motor cortex and some adjacent areas. In P. C. Bucy (ed.) *The Precentral Cortex.* Urbana, IL: University of Illinois Press, pp. 83–110.

von Bonin, G., and Bailey, P., 1947. *The Neocortex of* Macaca mulatta. Urbana, IL: University of Illinois Press.

Walker, A. E., 1940. A cytoarchitectural study of the prefrontal area of the macaque monkey. *J. Comp. Neurol.* **73**: 59–86.

West, R. A., and Larson, C. R., 1995. Neurons of the anterior mesial cortex related to faciovocal activity in the awake monkey. *J. Neurophysiol.* **74**: 1856–1869.

Williams, S. M., and Goldman-Rakic, P. S., 1998. Widespread origin of the primate mesofrontal dopamine system. *Cereb. Cortex* **8**: 321–345.

Wilkins, W. K., and Wakefield, J., 1995. Brain evolution and neurolinguistic preconditions. *Behav. Brain Sci.* **18**: 161–181.

Wilson, F. A. W., Ó Scalaide, S. P., and Goldman-Rakic, P. S., 1993. Dissociation of object and spatial processing domains in primate prefrontal cortex. *Science* **260**: 1955–1958.

Wise, R. J. S., 2003. Language systems in normal and aphasic human subjects: functional imaging studies and inferences from animal studies. *Br. Med. Bull.* **65**: 95–119.

Wise, R. J. S., Scott, S. K., Blank, S. C., *et al.*, 2001. Separate neural subsystems within 'Wernicke's area'. *Brain* **124**: 83–95.

Winter, P., Handley, P., Ploog, D., and Schott, D., 1973. Ontogeny of squirrel monkey calls under normal conditions and under acoustic isolation. *Behaviour* **47**: 230–239.

Part III

Dynamic systems in action and language

6

Dynamic systems: brain, body, and imitation

Stefan Schaal

6.1 Introduction

The Mirror System Hypothesis (MSH) (see Arbib, Chapter 1, this volume) postulates that higher communication skills and language may be grounded in the basic skill of action recognition, enhanced by the ability of movement pantomime and imitation. This chapter will examine from a computational point of view how such recognition and imitation skills can be realized. Our goal is to take our formal knowledge of how to produce simple actions, also called movement primitives, and explore how a library of such movement primitives can be built and used to compose an increasingly large repertoire of complex actions. An important constraint in the development of this material comes from the need to account for movement imitation and movement recognition in one coherent framework, which naturally leads to an interplay between perception and action.

The existence of motor primitives (a.k.a. synergies, units of actions, basis behaviors, motor schemas, etc.) (Bernstein, 1967, Arbib, 1981, Viviani, 1986, Mataric, 1998, Miyamoto and Kawato, 1998; Schaal, 1999; Sternad and Schaal, 1999; Dautenhahn and Nehaniv, 2002) seems, so far, the only possibility for how one could conceive that biological and artificial motor systems are able to cope with the complexity of motor control and motor learning (Arbib, 1981; Schaal, 1999, 2002a, 2002b; Byrne 2003). This is because learning based on low-level motor commands, e.g., individual muscle activations, becomes computationally intractable for even moderately complex movement systems. Our computational approach to motor control with movement primitives is sketched as an abstract flowchart in Fig. 6.1 (Schaal, 1999). In the following sections, we will first sketch our idea of dynamic movement primitives (DMPs), originally introduced in Ijspeert *et al.* (2001, 2002, 2003) and related to the early ideas of Bullock and Grossberg (1988), and then illustrate their potential for imitation learning and movement recognition. We will thus lay down a basic methodology that tentatively could be the computational basis of an action generation/action recognition system as required by the Mirror System Hypothesis (MSH).

Action to Language via the Mirror Neuron System, ed. Michael A. Arbib. Published by Cambridge University Press.
© Cambridge University Press 2006.

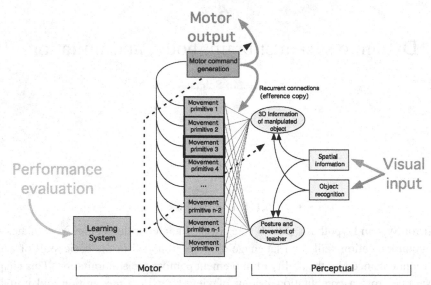

Figure 6.1 Conceptual computational sketch of motor control and motor learning with movement primitives. The right side of the figure contains primarily perceptual elements and indicates how visual information is transformed into spatial and object information, as needed for supervised learning from demonstration. The left side focuses on motor elements, illustrating how a set of movement primitives competes for a demonstrated behavior, and finally selects the most appropriate one for execution and further refinement through trial and error. Motor commands are generated from input of the most appropriate primitive. Learning can adjust both movement primitives and the motor command generator.

6.1.1 From control policies to motor primitives

From a computational point of view, the goal of motor learning can generally be formalized in terms of finding a task-specific control policy:

$$\mathbf{u} = \pi(\mathbf{x}, t, \alpha) \qquad (6.1)$$

that maps the continuous state vector $\mathbf{x}$ of a control system and its environment, possibly in a time t dependent way, to a continuous control vector $\mathbf{u}$ (Bellman, 1957; Dyer and McReynolds, 1970). The parameter vector α denotes the problem-specific adjustable parameters in the policy π, e.g., the weights in neural network or a generic statistical function approximator.[1] In simple words, all motor commands for all actuators (e.g., muscles or torque motors) at every moment of time depend (potentially) on all sensory and perceptual information available at this moment of time, and possibly even past information. Given some cost criterion that can evaluate the quality of an action $\mathbf{u}$ in a particular state $\mathbf{x}$, dynamic programming (DP), and especially its modern relative,

---

[1] Note that different parameters may actually have different functionality in the policy: some may be more low level and just store a learned pattern, while others may be higher level, e.g., as the position of a goal, that may change every time the policy is used. See, for instance, Barto and Mahadevan (2003), or the following sections of this chapter.

reinforcement learning (RL), provide a well-founded set of algorithms of how to compute the policy π for complex non-linear control problems. In essence, both RL and DP derive an optimal policy by optimizing the accumulated reward (in statistic expectation) over a long-term horizon (Sutton and Barto, 1998) – the reward is given from the cost criterion above at every time step. Unfortunately, as already noted in Bellman's original work, learning of π becomes computationally intractable for even moderately high-dimensional state–action spaces, e.g., starting from about 6–10 continuous dimensions, as the search space for an optimal policy becomes too large or too non-linear to explore empirically. Although recent developments in reinforcement learning increased the range of complexity that can be dealt with (e.g., Tesauro, 1992; Bertsekas and Tsitsiklis, 1996; Sutton and Barto, 1998), it still seems that there is a long way to go to apply general policy learning to complex control problems like human or humanoid movement.

In many theories of biological motor control and most robotics applications, the full complexity of learning a control policy is strongly reduced by assuming prior information about the policy. The most common priors are in terms of a desired trajectory, $[\mathbf{x}_d(t), \dot{\mathbf{x}}_d(t)]$, that is computed from some optimization criterion (Kawato and Wolpert, 1998). For instance, by using a tracking error driven feedback (e.g., proportional-derivative (PD)) controller, a (explicitly time-dependent) control policy can be written as:

$$\mathbf{u} = \pi\left(\mathbf{x}, \alpha(t), t\right) = \pi\left(\mathbf{x}, [\mathbf{x}_d(t), \dot{\mathbf{x}}_d(t)], t\right) = \mathbf{K}_x\left(\mathbf{x}_d(t) - \mathbf{x}\right) + \mathbf{K}_{\dot{x}}\left(\dot{\mathbf{x}}_d(t) - \dot{\mathbf{x}}\right) \quad (6.2)$$

For problems in which the desired trajectory is easily generated and in which the environment is static or fully predictable, such a shortcut through the problem of policy generation is highly successful. However, since policies like those in (6.2) are usually valid only in a local vicinity of the time course of the desired trajectory $[\mathbf{x}_d(t), \dot{\mathbf{x}}_d(t)]$, they are not very flexible. A typical toy example for this problem is the tracking of the surface of a ball with the fingertip. Assume the fingertip movement was planned as a desired trajectory that moves every second 1 cm forward in tracing the surface. Now imagine that someone comes and holds the fingertip for 10 seconds, i.e., no movement can take place. In these 10 seconds, however, the trajectory plan has progressed 10 cm, and upon the release of your finger, the error-driven control law in (6.2) would create a strong motor command to catch up. The bad part, however, is that (6.2) will try to take the shortest path to catch up with the desired trajectory, which, due to the convex surface in our example, will actually try to traverse through the inside of the ball. Obviously, this behavior is inappropriate and would hurt the human and potentially destroy the ball. Many daily-life motor behaviors have similar properties. Thus, when dealing with a dynamically changing environment in which substantial and reactive modifications of control commands are required, one needs to adjust desired trajectories appropriately, or even generate entirely new trajectories by generalizing from previously learned knowledge. In certain cases, it is possible to apply scaling laws in time and space to desired trajectories (Hollerbach, 1984; Kawamura and Fukao, 1994), but those can provide only limited flexibility. For the time being, the "desired trajectory" approach seems to be too restricted for general-purpose motor control and planning in dynamically changing environments, as needed in every biological motor system.

From the viewpoint of learning from empirical data, (6.1) constitutes a non-linear function-approximation problem. A typical approach to learning complex non-linear functions is to compose them out of a superposition of elements of reduced complexity, usually discussed in terms of "basis function" or "kernel" approaches. The same line of thinking generalizes to learning policies: a complicated policy could be learned from the combination of (a not necessarily fixed number of) simpler policies, i.e., policy primitives or movement primitives, as for instance:

$$\mathbf{u} = \pi(\mathbf{x}, \alpha, t) = \sum_{k=1}^{K} \pi_k(\mathbf{x}, \alpha_k, t) \tag{6.3}$$

Indeed, related ideas have been suggested in various fields of research, for instance in computational neuroscience as Schema Theory (Arbib, 1981), reinforcement learning as macro action or options (Sutton *et al.*, 1999; Barto and Mahadevan, 2003), and mobile robotics as behavior-based or reactive robotics (Brooks, 1986; Mataric, 1998). In particular, the latter approach also emphasized removing the explicit time dependency of π. Explicit timing is cumbersome as it requires maintaining a clocking signal, e.g., a time counter that increments every 1 ms (as typically done in robotics). Besides the fact that it is disputed whether biological system have access to such clocks (e.g., Keating and Thach, 1997; Roberts and Bell, 2000; Ivry *et al.*, 2002), there is an additional level of complexity needed for aborting, halting, or resetting the clock when unforeseen disturbances happen during movement execution – in robot-control software, a large number of heuristic if-then statements are usually required for this purpose. Without direct time dependencies, policies are more flexible and the combination of policy primitives becomes simplified.[2] Despite the successful application of policy primitives in the mobile robotics domain, so far, it remains a topic of ongoing research (Lohmiller and Slotine, 1998; Burridge *et al.*, 1999; Inamura *et al.*, 2002) how to generate and combine primitives in a principled and autonomous way, and how such an approach generalizes to complex movement systems, like human hand, arms, and legs.

Thus, a key research topic, both in biological and artificial motor control, keeps on revolving around the question of movement primitives:

1. what is a good set of primitives,
2. how can they be formalized,
3. how can they interact with perceptual input,
4. how can they be adjusted autonomously,
5. how can they be combined for specific tasks,
6. and what is the origin of primitives?

---

[2] As a technical remark, an explicit time dependency in policies allows learning actions that are more complex and can be shown to be optimal under certain circumstances (Bagnell *et al.*, 2003).

In order to address the first four of these questions (the others we will consider towards the end of the paper), we resort to some of the most basic ideas of dynamic systems theory. A dynamic system can generally be written as a differential equation:

$$\dot{\mathbf{x}} = f(\mathbf{x}, \alpha, t) \tag{6.4}$$

which is almost identical to (6.1), except that the left-hand side denotes a change of state, not a motor command. Such a kinematic formulation is, however, quite suitable for motor control if we conceive of this dynamic system as a kinematic planning policy, whose outputs are subsequently converted to motor commands by an appropriate controller (Wolpert, 1997). Planning in kinematic space is often more suitable for motor control since kinematic plans generalize over a large part of the workspace – non-linearities due to gravity and inertial forces are taken care off by the controller at the motor execution stage (cf. Fig. 6.1 and Fig. 6.2). Kinematic plans can also theoretically be cleanly superimposed to use multiple movement primitives to form more complex behaviors, as indicated in (6.3). It should be noted, however, that a kinematic representation of movement primitives is not necessarily independent of the dynamic properties of the limb. Proprioceptive feedback can be used on-line to modify the attractor landscape of a dynamic movement primitive in the same way as perceptual information (Rizzi and Koditschek, 1994; Schaal and Sternad, 1998; Williamson, 1998; Nakanishi *et al.*, 2004). Figure 6.2 indicates this property with the "perceptual coupling" arrow.

Adopting the framework of dynamics systems theory for movement primitives allows us to draw from a large body of well-established knowledge to formalize this idea (Strogatz, 1994). The two most elementary behaviors of a non-linear dynamic system are point attractive and limit cycle behaviors, in analogy to discrete (e.g., reaching, grasping) and rhythmic (e.g., locomotion, scratching, dancing) movement in motor control. The idea of DMPs is to exploit well-known simple formulations of such attractor equations to code the basic behavioral pattern (i.e., rhythmic or discrete) as the attractor behavior – later sections will also demonstrate some evidence from brain imaging studies to support this separation in discrete and rhythmic movement. Statistical learning will be used to adjust the shape of the attractor of the DMP for the detailed needs of the task. As will be outlined in the next section, several appealing properties, such as perception–action coupling and reusability of the primitives, can be accomplished in this framework.

6.1.2 Dynamic movement primitives

Motor command generation with DMPs

We assume that the variables of a DMP represent the desired kinematic state of a limb, i.e., desired positions, velocities, and accelerations for each joint. Alternatively, the DMP could also be defined in task space, and we would use appropriate task variables (e.g., the distance of the hand from an object to be grasped) as variables for the DMP – for the discussions in this chapter, this distinction is, however, of subordinate importance, and,

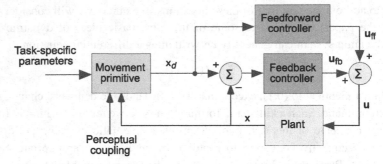

Figure 6.2 Sketch of a control diagram with dynamic movement primitives, showing in particular how the primitive is inserted into a controller with feedback (i.e., error-driven) and feedforward (i.e., anticipatory or model-based) components.

for the ease of presentation, we will focus on formulations in joint space. As shown in Fig. 6.2, kinematic variables are converted to motor commands through a feedforward controller – usually by employing an inverse dynamics model – and stabilized by low-gain[3] feedback control.[4] The example of Fig. 6.2 corresponds to a classical computed torque controller (Craig, 1986) which has also been suggested for biological motor control (Kawato, 1999), but any other control scheme could be inserted here. Thus, the motor execution of DMPs can incorporate any control technique that takes as input kinematic trajectory plans, and in particular, it is compatible with current theories of model-based control in computational motor control.

Motor planning with DMPs

In order to accommodate discrete and rhythmic movement plans, two kinds of DMPs are needed: point-attractive systems and limit-cycle systems. The key question of DMPs is how to formalize non-linear dynamic equations such that they can be flexibly adjusted to represent complex motor behaviors without the need for manual parameter tuning and the danger of instability of the equations. We will sketch our approach in the example of a discrete dynamic system for reaching movements – an analogous development holds for rhythmic systems.

Assume we have a basic point-attractive system, instantiated by the second-order dynamics

$$\tau \dot{z} = \alpha_z \Big(\beta_z (g - y) - z \Big), \tau \dot{y} = z + f \tag{6.5}$$

---

[3] The emphasis of low-gain feedback control is motivated by the desire to have a movement system that is compliant when interacting with external objects or unforeseen perturbation, similar to humans, but quite unlike traditional high-gain industrial robotics.

[4] The concepts of feedforward control and feedback control are among the most elementary of control theory. Feedback control usually compares a desired state with the current state and computes a motor command based on this error. Feedforward control denotes an anticipatory motor command, e.g., in the amount of biceps flexion as needed to overcome gravity when the elbow is 90° flexed in the sagittal plane.

where g is a known goal state, α_z and β_z are time constants, τ is a temporal scaling factor (see below) and y, $\dot{y}$ correspond to the desired position and velocity generated by (6.5), interpreted as a movement plan as used in Fig. 6.2. For instance, y, $\dot{y}$ could be the desired states for a 1-degree-of-freedom motor system, e.g., the elbow flexion–extension. Without the function f, (6.5) is nothing but the first-order formulation of a linear spring-damper, and, after some reformulation, the time constants α_z and β_z have an interpretation in terms of spring stiffness and damping. For appropriate parameter settings and $f = 0$, these equations form a globally stable linear dynamic system with g as a unique point attractor, which means that for any start position the limb would reach g after a transient, just like a stretched spring, upon release, will return to its equilibrium point. Our key goal, however, is to instantiate the non-linear function f in (6.5) to change the rather trivial exponential and monotonic convergence of y towards g to allow trajectories that are more complex on the way to the goal. As such a change of (6.5) enters the domain of non-linear dynamics, an arbitrary complexity of the resulting equations might be expected. To the best of our knowledge, this problem has prevented research from employing generic learning in non-linear dynamic systems so far. We will address this problem by first introducing a bit more formalism, and then by analyzing the resulting system equations.

The easiest way to force (6.5) to become more complex would be to create a function f as an explicit function of time. For instance, $f(t) = \sin(\omega t)$ would create an oscillating trajectory y, or $f(t) = \exp(-t)$ would create a speed-up of the initial part of the trajectory y. But, as mentioned before, we would like to avoid explicit time dependencies. To achieve this goal, we need an additional dynamic system (x, v)

$$\tau \dot{v} = \alpha_v \left(\beta_v (g - x) - v \right), \tau \dot{x} = v \qquad (6.6)$$

and the non-linear function f in form of

$$f(x, v, g) = \frac{\sum_{i=1}^{N} \psi_i w_i v}{\sum_{i=1}^{N} \psi_i}, \text{ where } \psi_i = \exp\left(-h_i \left(\frac{x}{g} - c_i \right)^2 \right) \qquad (6.7)$$

Equation (6.6) is a linear spring-damper system similar to (6.5), however, not modulated by a non-linear function – we will call this equation the *canonical* system from now on, as it is among the most basic dynamic system available to create a point attractor. The monotonic global convergence of (6.6) to g can be guaranteed with a proper choice of α_v and β_v, e.g., in the same manner that a damped spring returns to its equilibrium point no matter from which stretched position it is released. Equation (6.7) is a standard representation of a non-linear function in terms of basis function, as commonly employed in modeling population coding in the primate brain (e.g., Mussa-Ivaldi, 1988; Georgopoulos, 1991). The monotonic convergence of x to the goal g in (6.6) becomes a substitute for time: all that time does is that it monotonically increases, similar to the time course of x. If

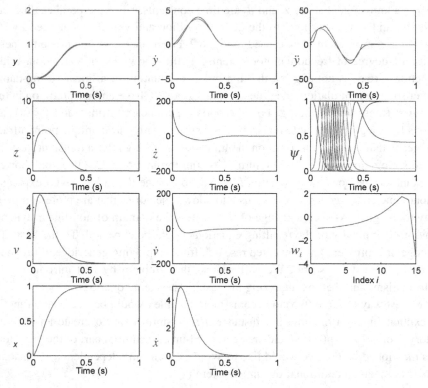

Figure 6.3 Example of all variables of a discrete movement dynamic primitive as realized in a minimum jerk movement from zero initial conditions to goal state $g = 1$.

we assume that all initial conditions of the state variables x, v, y, z are zero, the quotient $x / g \in [0,1]$ will be a time substitute that starts at zero and converges to 1. Such a variable is called a "phase" variable as one can read out from the value of this variable in which phase of the movement we are, where "0" is the start, and "1" is the end. The non-linear function f is now generated by anchoring its Gaussian basis functions ψ_i (characterized by a center c_i and bandwidth h_i) in terms of the phase variable x. The phase velocity v appears multiplicative in (6.7) such that the influence of f vanishes at the end of the movement (see below). It can be shown that the combined system in (6.5), (6.6), and (6.7) asymptotically converge to the unique point attractor g.

The example in Fig. 6.3 clarifies the ingredients of the discrete DMP. The top row of Fig. 6.3 illustrates the position, velocity, and acceleration trajectories that serve as desired inputs to the motor command generation stage (cf. Fig. 6.2) – acceleration was derived from numerical differentiation of $\dot{y}$. In this example, the trajectories realize a minimum jerk trajectory (Hogan, 1984), a smooth trajectory as typically observed in human behavior (the ideal minimum jerk trajectory, which minimizes the integral of the squared jerk along the trajectory, is superimposed to the top three plots of Fig. 6.3, but the difference to the DMP output is hardly visible). The remaining plots of Fig. 6.3 show the time course of all internal variables of the DMP, as given by (6.5), (6.6), and (6.7). Note that the trajectory of x is just a strictly monotonically increasing curve, i.e., it can be interpreted as a phase variable that codes the progress of the movement behavior, where "0" indicates the beginning, and "1" the end of the behavior. The trajectory for v, in contrast, is just a phasic signal that starts at zero, creates a positive pulse, and then decays back to zero. As v multiplies the non-linearity in (6.7), the non-linearity only acts in a transient way, one of the main reasons that these non-linear differential equations remain relatively easy to analyze. The basis function activations ψ_i are graphed as a function of time, and demonstrate how they essentially partition time into shorter intervals in which the function value of f can vary.

It is not the particular instantiation in (6.5), (6.6), and (6.7) that is the most important idea of DMPs, but rather it is the design principle that matters. A DMP consists of two sets of differential equations: a *canonical* system

$$\tau\dot{\mathbf{x}} = \mathbf{h}(\mathbf{x}, \theta) \tag{6.8}$$

and an *output* system

$$\tau\dot{\mathbf{y}} = \mathbf{g}(\mathbf{y}, f, \theta) \tag{6.9}$$

where we just inserted θ as a place-holder for all parameters of these systems, like goal, time constants, etc. The canonical system needs to generate two quantities: a phase variable[5] x, and a phase velocity v, i.e., $\mathbf{x} = [x\ v]^T$. The phase x is a substitute for time and allows us to anchor our spatially localized basis functions (6.7). The appealing property of using a phase variable instead of an explicit time representation is that we can now manipulate the time evolution of phase, e.g., by speeding up or slowing down a movement as appropriate by means of additive coupling terms or phase resetting techniques (Nakanishi *et al.*, 2004) – in contrast, an explicit time representation cannot be manipulated as easily. For instance, (6.6) could be augmented to be

$$\tau\dot{v} = \alpha_v\left(\beta_v(g - x) - v\right), \tau\dot{x} = v - \alpha_c(y_{\text{actual}} - y)^2 \tag{6.10}$$

The term $(y_{\text{actual}} - y)$ is the tracking error of the motor system; if this error is large, the time development of the canonical system comes to a stop, until the error is reduced – this is exactly what one would want if a motor act gets suddenly perturbed.

As already mentioned above, the phase velocity v is a multiplicative term in the non-linearity in (6.7). Since v becomes zero at the end of a movement, the influence of the non-linearity vanishes in the output system, and the dynamics of the output system with $f = 0$ dominate its time evolution. In the design of a DMP, we usually choose a structure

---

[5] A phase variable, in our notation, monotonically increases (or decreases) its value from movement start to end under unperturbed conditions. For periodic motion, it may reset after one period to its initial value.

for canonical and outputs systems that are analytically easy to understand, such that the stability properties of the DMP can be guaranteed (Schaal *et al.*, 2003b).

An especially useful feature of this general formalism is that it can be applied to rhythmic movements as well, simply by replacing the point attractor in the canonical system with a limit-cycle oscillator (Ijspeert *et al.*, 2003). Among the simplest limit-cycle oscillators is a phase-amplitude representation (i.e., constant speed rotation around a circle with radius r):

$$\tau\dot{r} = \alpha_r(A - r), \tau\dot{\phi} = 1 \qquad (6.11)$$

where r is the amplitude of the oscillator, A the desired amplitude, and ϕ its phase. For this case, (6.7) is modified to

$$f(r, \phi) = \frac{\sum_{i=1}^{N} \psi_i \mathbf{w}_i^T \mathbf{v}}{\sum_{i=1}^{N} \psi_i}, \text{ where } \mathbf{v} = [r\cos\phi, r\sin\phi]^T \qquad (6.12)$$

$$\text{and } \psi_i = \exp\left(h_i\left(\cos(\phi - c_i) - 1\right)\right)$$

The changes in (6.12) are motivated by the need to make the function f a function that lives on a circle, i.e., $\mathbf{v}$ is a vector of first-order Fourier terms, and ψ_i are computed from a Gaussian function that lives on a circle (called van Mises function). The output system in (6.5) remains the same, except that we now identify the goal state g with a setpoint around which the oscillation takes place. Thus, by means of A, τ, and g, we can control amplitude, frequency, and setpoint of an oscillation independently.

At this point, it will be useful to re-evaluate what we actually call a movement primitive. We consider our DMP approach as a general method to create primitives. The primitive is either discrete or rhythmic in nature, and the parameters w_i in the function f determine the family of trajectories that the primitive can create. Thus, a multidimensional DMP for reach-and-grasp would form the reach-and-grasp primitive, a tennis forehand primitive could exist, or we may have primitives for writing the letters of the alphabet. From an algorithmic point of view, all we do is to store the parameters w_i in some memory, which becomes thus our primitive library. If desirable, the parameterization of a primitive could be made more complex, e.g., by including the locations c_i of the basis functions, and their bandwidths h_i. From the viewpoint of learning, adding more basis functions will also create more powerful learning systems, e.g., with 100 basis function we could learn more complex trajectories than with 10 basis functions. Thus, from a technical point of view, DMPs could represent very complex movements, so complex that the word "primitive" may not be so suitable any more. We assume that in biology, there is a limit to the complexity that a single primitive can represent, such that it will be easier to create complex movement by sequencing multiple simpler primitives instead of learning one very complex one. This topic remains for future work and discussions.

Summary

In summary, by anchoring a linear learning system with non-linear basis functions in the *phase space* of a *canonical dynamic system with guaranteed attractor properties*, we are able to define complex movement behaviors as an attractor of non-linear differential equations without endangering the asymptotic convergence to the goal state. Both discrete and rhythmic movements can be coded in the DMPs, and almost arbitrarily complex (but smooth) trajectory profiles are possible. By modifying the goal parameter (or amplitude parameter in rhythmic movement), and the overall time constant of the equation, variants of the same movement can be generated, i.e., the movement primitive can be reused for temporal spatial task variation. This strategy opens a large range of possibilities to create movement primitives, e.g., for reaching, grasping, object manipulation, and locomotion. In particular, imitation learning as well as trial and error learning methods can easily be built around this approach, as outlined in the next section.

6.2 Movement primitives for imitation

After developing the formal idea of DMPs in the previous sections, we will now turn to imitation learning. In fact, imitation learning with DMPs is technically rather straightforward, which is not surprising, as the theory of DMPs was built with imitation learning in mind. It is worth contrasting this technical simplicity with MSH, which postulates that imitation is beyond the grasp of the monkey, and available in only simple form in the chimpanzee – but what is easily done nowadays in algorithms may require much evolution to find its way into a neural system. In a second step of the next sections, movement recognition based on the representations of DMPs will be discussed. The idea of using the generative representations of motor control for recognition opens the discussion of generative-recognition models, a concept that has found increasing attention in theoretical neuroscience, and which will be discussed at the end of this section.

6.2.1 Imitation learning with DMPs

If we assume, for simplicity, that the number of basis functions in (6.7) or (6.12) is known, the crucial parameters of a movement primitive are the weights w_i in the non-linear function f, as they define the spatiotemporal path of a DMP. Given that f is a normalized basis function representation, linear in the coefficients of interest (i.e., w_i) (e.g., Bishop, 1995), a variety of learning algorithms exist to find w_i. In an imitation learning scenario, at an advanced level of preprocessing of sensory data about the teacher's demonstration (Fig. 6.1), we can suppose that we are given a sample trajectory $y_{demo}(t)$, $\dot{y}_{demo}(t)$, $\ddot{y}_{demo}(t)$ with duration T. Based on this information, a supervised learning problem results with the following target for f:

$$f_{\text{target}} = \tau \dot{y}_{\text{demo}} - z_{\text{demo}}$$

where

$$\tau \dot{z}_{\text{demo}} = \alpha_z \left(\beta_z (g - y_{\text{demo}}) - z_{\text{demo}} \right)$$

(6.13)

In order to obtain a matching input for f_{target}, the canonical system needs to be integrated. For this purpose, in (6.6), the initial state of the canonical system is set to $v = 0$, $x = y_{\text{demo}}$ (0) before integration. An analogous procedure is performed for the rhythmic DMPs. The time constant τ is chosen such that the DMP with $f = 0$ achieves 95% convergence at $t = T$. With this procedure, a clean supervised learning problem is obtained over the time course of the movement to be approximated with training samples ($\mathbf{v}, f_{\text{target}}$).

For solving the function approximation problem, we chose a non-parametric regression technique from locally weighted learning (LWPR) (Vijayakumar and Schaal, 2000) as it allows us to determine the necessary number of basis functions N, their centers c_i, and bandwidth h_i automatically. In essence, for every basis function ψ_i, LWPR performs a locally weighted regression of the training data to obtain an estimate of the tangent of the function to be approximated within the scope of each basis function. Predictions for a query point are generated by a ψ_i-weighted average of the predictions of all local models. Given that the parameters $\mathbf{w}_i$ learned by LWPR are independent of the number of basis functions, they can be used robustly for categorization of DMPs (see below). In simple words, we create a piecewise linear approximation of f_{target}, where each linear function piece belongs to one of the basis functions.

As evaluations of the suggested approach to movement primitives, in Ijspeert *et al.* (2002) we demonstrated how a complex tennis forehand and tennis backhand swing can be learned from a human teacher, whose movements were captured at the joint level with an exoskeleton. Figure 6.4 illustrates imitation learning for a rhythmic trajectory using the phase oscillator DMP from (6.11) and (6.12). The images in the top of Fig. 6.4 show four frames of the motion capture of a figure-8 pattern and its repetition on the humanoid robot after imitation learning of the trajectory. The plots in Fig. 6.5 demonstrate the motion captured and fitted trajectory of a bimanual drumming pattern, using 6 DOFs (degrees of freedom) per arm. All DMPs referred to the same canonical system (Schaal *et al.*, 2003b). Note that rather complex phase relationships between the individual DOFs can be realized. For one joint angle, the right elbow joint (R_EB), Fig. 6.5b exemplifies the effect of various changes of parameter settings of the DMP. Here it is noteworthy how quickly the pattern converges to the new limit-cycle attractor, and that parameter changes do not change the movement pattern qualitatively, an effect that can be predicted theoretically (Schaal *et al.*, 2003). The non-linear function of each DMP employed 15 basis functions.

While this tennis example is obviously an idealized technical realization of imitation learning, all the ingredients could be realized by biological systems as well. Our exoskeleton movement recording could be replaced by visual posture extraction, which seems to be solved in humans in the temporal lobe (Perrett *et al.*, 1989a). The imitation learning problem in (6.13) could be accomplished by any form of supervised learning mechanism

Figure 6.4 Humanoid robot learning a figure-8 movement from a human demonstration.

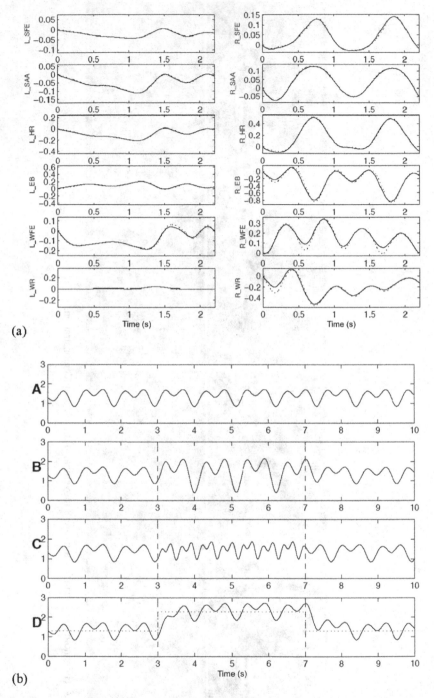

Figure 6.5 (a) Recorded drumming movement performed with both arms (6 DOFs per arm). The dotted lines and continuous lines correspond to one period of the demonstrated and learned trajectories, respectively – due to rather precise overlap, they are hardly distinguishable.

in the brain, e.g., as often assumed to take place in the cerebellum (Thach, 1998; Wolpert *et al.*, 1998; Schweighofer *et al.*, 1998). For hitting the tennis ball, its location in three-dimensional space needs to be perceived and transformed into appropriate target coordinates in joint space, an inverse kinematics problem that can be acquired easily by neural network architectures. (Jordan and Rumelhart, 1992). The joint space targets are subsequently the goals for the DMP of each degree of freedom of the motor system. Of course, true imitation has additional complexity if teacher and student have different body kinematics (Alissandrakis *et al.*, 2002), and imitation may often be more suitable in task space than in joint space (Schaal *et al.*, 2003a). But, at least in principle, the given methodology will be capable to model basic goal-directed and generalizable motor behaviors in all these scenarios, while obviously additional problems remain to be addressed for particular problems. It should also be mentioned that reach and grasp movements, which are more in the spotlight of MSII, are equally in the scope of our DMP approach – finger movements are just additional degrees of freedom of the movement system, such that movement planning for fingers proceeds exactly in the same way as for arm movements. An interesting component of grasping are contact forces, which, however, we would address more on the level of force control than on the planning level with DMP (Nakanishi *et al.*, 2005).

6.2.2 Movement recognition with DMPs

Using our formal model of movement primitives, we can address methods of movement recognition by making use of a generative model of motor control in terms of DMPs. The goal of movement recognition is that of statistical movement classification, i.e., given an observed movement, is there a similar one in the existing movement repertoire. In our modeling approach, a movement is described completely by its parameters w_i – as detailed in Ijspeert *et al.* (2003); the parameters w_i are invariant to the distance to the goal, or movement amplitude for rhythmic movement, and the movement speed, and provide thus a robust description of a movement. For instance, the figure-8 in Fig. 6.4 would have the same parameters if translated, scaled, or executed at different speeds, but it would have different parameters when rotated to be a lying figure-8. We assume that a movement repertoire exists, which is a library of parameter setting w_i for different movement primitives, which are additionally classified as either discrete or rhythmic. Starting from this assumption, there are at least two possibilities for movement

Caption for Figure 6.5 (cont.)
(b) Modification of the learned rhythmic pattern (flexion/extension of the right elbow, R_EB). **A**: trajectory learned by the rhythmic DMP; **B**: temporary modification with $A \leftarrow 2A$; **C**: temporary modification with $\tau \leftarrow \tau/2$; **D**: temporary modification with $g \leftarrow g+1$ (dotted line). Modified parameters were applied between $t = 3$ s and $t = 7$ s. Note that in all modifications, the movement patterns does not change qualitatively, and convergence to the new attractor under changed parameters is very rapid.

recognition, one based on prediction, and one based on comparisons in parameter space (Schaal, 1999).

The predictive approach assumes that there are, in parallel, mental simulations of all relevant movement primitives that try to predict the currently observed movement (cf. Fig. 6.1) (Demiris and Hayes, 1999; Schaal, 1999). Again, we assume that the observed movement has already been preprocessed by the perceptual system such that it is represented in the same coordinate frame as used for the movement primitives, e.g., joint angles in our simplified approach. Thus, we arrive at a similar initial setting as in the imitation learning section above, i.e., we have an observed trajectory $y_{demo}(t)$, $\dot{y}_{demo}(t)$, $\ddot{y}_{demo}(t)$, just that this time we need to find out which of our primitives in the motor library can predict this trajectory the best. As we can extract the start state, target (or amplitude), and duration (or period) from the demonstrated trajectory, we can initialize all existing primitives with this information, and, subsequently, run them in a mental simulation. Each primitive will generate a predicted trajectory, and the one that matches the demonstrated trajectory the best (e.g., in a least-squares difference sense), will be the winning motor primitive.

As an alternative to the predictive approach, we can simply repeat the parameter fitting of the imitation learning method above and obtain the optimal parameters $\mathbf{w}_i$ for the observed movement. Afterwards, a simple comparison of these parameters against all other parameters in our motor library allows picking the best fitting primitive. In Ijspeert *et al.* (2003), we employed for classification the correlation coefficient between the parameter vectors of the observed movement $\mathbf{w}_{demo}$ and every primitive in the library $\mathbf{w}_j$, i.e.,

$$r_j^2 = \frac{\mathbf{w}_{demo}^T \mathbf{w}_j}{|\mathbf{w}_{demo}||\mathbf{w}_j|} \tag{6.14}$$

but more sophisticated methods could be used to compare the weight vectors.

In order to illustrate this approach to movement recognition, we tested it in a character classification test, using the graffiti characters of personal digital assistant (PDA) computers. Trajectories were recorded by digitizing movement with the help of a computer mouse. Note that in this example, DMPs are used in task space, and not in joint space, as this task is inherently defined in a plane. Five repetitions of all 26 characters of the alphabet were collected from one subject. Figure 6.6 depicts the results for four characters, i.e., N, I, P, S. Fig. 6.6a–d shows that the trajectories can be well approximated with our movement primitives, which employed two simultaneous dynamic systems, one for each of the two spatial dimensions X and Y. Figure 6.6e provides the correlation matrix between all five repetitions of the characters. As can be seen from the block-diagonal dark entries of this matrix (dark entries correspond to high correlation), the comparison of parameter vectors using (6.14) was reliable in finding the highest correlation between similar characters. Over the entire set of characters, more than 80% correct classification could be achieved with this simple correlation measure without any special tuning – determining the start and end of a trajectory is source of the most variance from our experience. More sophisticated classifiers would undoubtedly achieve better results. It should be noted that, in contrast to standard character classification, our approach employs both temporal *and* spatial information for classification, which makes it more sensitive to temporal characteristics of a movement, but which is normally not necessary in character recognition, due to the purely spatial nature of this task.

Some issues in this approach to movement recognition are worth further discussions. First, given that both target (or amplitude) and the duration (or period) of the movement need to be known for either the prediction or parameter comparison method to work, it seems to be necessary that an observed movement be completed before classification can begin. It is, however, known that humans can infer a movement class from partial information (Oztop and Arbib, 2002), e.g., as in tennis, where professional players can anticipate the type of a serve from only the initial racket movement. But one could argue that professional athletes develop specialized means to accomplish this skill, for instance, by pure visual pattern discrimination. Oztop and others (Oztop *et al.*, 2005) examined computational models for movement recognition with partial information. In essence, they employed a target state estimator that was updated continuously as more movement information became available, and they demonstrated that deceptive movements could lead the observer astray in his/her target estimation. It currently seems that this area of research in movement recognition is not well explored.

A second point of interest is the flexibility and robustness of predictive vs. parameter estimation methods for movement recognition. The evaluation criterion in predictive approaches is the prediction error, which can be obtained even if the parameters of a motor primitive are hard to fit to an observed movement. The evaluation criterion in parameter comparison is the error in parameter space, which is inherently more abstract and a more artificial method of comparison. It can also be that a small error in parameter space may actually lead to large differences in the actual movement. Thus, despite the appealing success of our toy example above, predictive comparisons may be more general for movement recognition, an insight that resonates with various other models in the literature (e.g., Wolpert and Kawato, 1998; Demiris and Hayes, 1999; Haruno *et al.*, 2001).

6.2.3 Summary

Sharing the same representations for movement recognition and movement generation is one of the key findings that resulted from research on mirror neurons (Rizzolatti *et al.*, 1996). Although ideas about a "motor theory of perception" have been around for some while (e.g., Liberman, 1996) a formalization of such reciprocal interaction between a generative model and a recognition model had been emphasized before only rarely, with the exception of a few pieces of work, including Dayan *et al.*'s Helmholtz machine (Dayan *et al.*, 1995; Hinton *et al.*, 1995) and Kawato's bidirectional theory of motor control (Kawato, 1996; Miyamoto and Kawato, 1998). In neural network literature, the motivation of generative-recognition models came originally from theoretical considerations of unsupervised learning: in order to avoid that unsupervised learning finds useless structure in data, a useful "reality check" is to try to re-generate the original training data

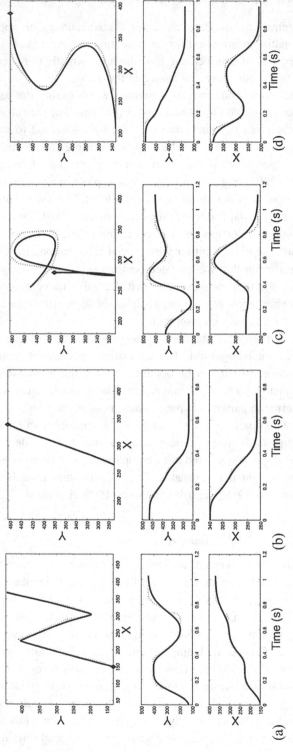

Figure 6.6 Example of movement recognition with DMPs, using a correlation index between parameter vectors of two movements for classification. (a)–(d) Four two-dimensional sample trajectories from the graffiti alphabet of PDA computers; the dotted line is the desired trajectory, the solid line is the fitted trajectory. Both spatial and spatiotemporal appearance of the trajectories is provided.

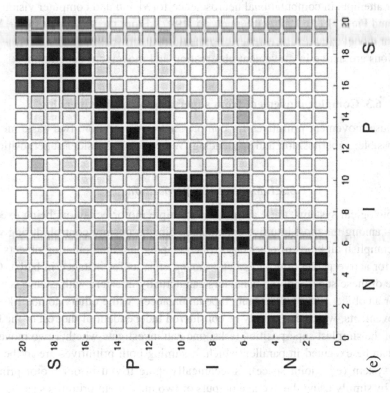

(e)

Figure 6.6 (e) Correlation matrix for each of the four letters in (a) to (d). The weight vector of each trajectory was correlated with each other trajectory. Dark entries in the correlation matrix show high correlation coefficients, light or white illustrates low correlation.

from the higher-order representation that unsupervised learning created (Hinton *et al.*, 1995). The process of mapping observed data to a higher-order representation is thus termed a recognition model, and the inverse operation is the generative model. Bottom–up and top–down processing, as often seen crucial in brain processing, have interesting parallels to the generative-recognition approach.

Our suggested approach to imitation learning and movement recognition with DMPs follows the flavor of generative-recognition models, although we predetermined the higher-level representations in forms of dynamic systems, i.e., we did not employ unsupervised learning to acquire these representations. While an unsupervised formation of motor primitives would be highly desirable, it has so far proven to be more complicated than similar attempts in computational neuroscience for vision and computer vision (e.g., Olshausen and Field, 1996; Bell and Sejnowski, 1997). From the viewpoint of the MSH, however, our model captures the essence of some of the neuroscientific data, i.e., the same representations are active in movement observation and movement generation.

6.3 Complex movement from superposition and sequencing

In order to use movement primitives for complex action generation, two basic mechanisms are possible: superposition and sequencing. We will first consider superposition.

6.3.1 Movement superposition

In mature biological systems, there are usually multiple motor behaviors being executed in parallel – among the most intuitive examples is bipedal posture control, during which humans accomplish all kinds of other tasks, like grasping and manipulating objects, head movement for active perception, or esthetic movements, as for instance in ballet. Obviously, some of these superimposed behaviors can actually interfere with each other (e.g., reaching for an object while maintaining bipedal balance), while others do not, like head and eye movements, which can largely be considered independent of the rest of the body.

As one of the simplest superposition tasks, one can investigate whether two movement primitives can be executed in parallel, which, assuming both primitives are in the same coordinate system (e.g., joint space), is technically quite trivial in our motor primitive framework by simply using the average outputs of two movement primitives as the target signal. There is also some evidence that humans can accomplish similar superposition performance. For instance, the "target-switching paradigm" investigated how movement trajectories change when a target in a reaching task is suddenly displaced during the ongoing movement. Flash and collaborators (Flash and Henis, 1991; Henis and Flash, 1995) concluded from their behavioral data that the target displacement triggered a second movement towards this target, that was simply superimposed onto the ongoing movement to the original target. However, similar effects have been modeled without the use of superposition. If we assume that a reaching task is performed on a table-top, i.e., it is planar in the transversal plane, the reaching movement can be modeled by a two-dimensional discrete motor primitive in hand space – desired hand trajectories can be converted to joint space trajectories by traditional inverse kinematics algorithms (Bullock *et al.*, 1993; Tevatia and Schaal, 2000). A target switch simply changes the goal state g in (6.5), (6.6), and (6.7), and the attractor dynamics automatically takes care of reaching the new goal. Similar to Hoff and Arbib (1992), who investigated target switching from the view of an on-line realization of the minimum jerk planning model, we can obtain a very good model of human behavior without the need of superposition of motor primitives. This line of research thus awaits additional behavioral and modeling studies to conclude whether a supposition of primitives exists or not.

In other projects, our two fundamental movement primitive categories, i.e., discrete and rhythmic movements, were examined as to whether they can exist in superposition or not. For example, take the task of bouncing a ball on a racket: intuitively, the vertical sinusoidal movement of the racket is a rhythmic movement, superimposed by a "tracking" movement that keeps the racket under the ball in the horizontal plane. Such a decomposition strategy was successfully implemented on a humanoid robot (Schaal *et al.*, 1992; Schaal and Atkeson, 1993; Schaal, 1997) with the help of a preliminary version of the motor primitive framework of the previous sections. Behavioral investigations of this topic also exist, but are not fully conclusive so far. Adamovich *et al.* (1994) conducted an experiment where subjects performed a rhythmic elbow movement that, at a trigger signal, had to shift its mean position from a more flexed to a more extended point, or vice versa, but without aborting the rhythmic movement. As these authors observed some phase resetting effects in the rhythmic movement after the discrete shift, they concluded that the rhythmic movement was terminated during the discrete shift, and after the shift, it was restarted. In a richer behavioral design, Sternad and co-workers (Sternad *et al.*, 1999, 2000, 2002; de Rugy and Sternad, 2003; Sternad and Dean, 2003; Wei *et al.*, 2003) and other studies (Staude and Wolf, 1997) accumulated increasing evidence against the theory of Adamovich *et al.*, 1994, and rather advocated a superposition theory of discrete and rhythmic movement. Phase resetting and other behavioral phenomena were attributed to a cerebral and/or spinal interaction of the two motor primitives. In a recent study, Mohajerian *et al.* (2004) provided evidence that the observed interaction effects are most likely all on the spinal level, i.e., due to reciprocal inhibition of flexor and extensor muscles.

From a technical point of view, superposition of motor behaviors can be treated in a hierarchical constraint approach. Essentially, behaviors are ranked according to their importance, e.g., posture control is the highest objective, reaching the goal is the second highest objective, esthetics the third highest objective, etc. Given the highly redundant human motor system, it is possible to select the contribution of different degrees of freedom for each task objective such that interference is avoided. In the control theory literature, such hierarchical constraints can be formalized in terms of the range space and null space of a task, i.e., ideas from linear algebra, which can be nested hierarchically until the lowest level null space has zero dimensionality. Sentis and Khatib (2004) suggested one of the most elegant solutions to this problem, and demonstrated its success on humanoid robot simulation studies, which accomplished manipulation tasks under postural constraint.

6.3.2 Movement sequencing

Creating complex actions out of a sequence of motor primitives is, from a computational viewpoint, a field of research with a long history, particularly in the domain of Markov decision processes (MDP). The MDP framework is formalized by states, actions, and rewards, i.e., when the movement system is in a particular state, it can take an action, and will end up in another state. Taking an action in a state creates a particular reward, which is used to judge the quality of the selected action. The objective of the movement system is to maximize the accumulated reward over time. In the 1950s, MDPs were primarily discussed in the field of dynamic programming (Bellman, 1957; Dyer and McReynolds, 1970), and in recent years, research in reinforcement learning (RL) (e.g., Sutton and Barto, 1998) has augmented theory and algorithms for solving MDPs significantly, particularly for scenarios where an analytical model of the system behavior is not available. For our discussion of movement primitives, the most recent advances in RL in terms of abstract actions are of particular interest. Originally, RL research considered only basic actions as motor commands, like a torque command to the motor of a robot joint. However, planning complex movements based on such atomic actions quickly becomes infeasible in high-dimensional motor systems due to the exponential explosion of the possible solutions to a complex behavior. Instead of using atomic action, motor primitives can be used instead, also called macros, options, schemas, basis behaviors, etc., in the literature (Barto and Mahadevan, 2003). Figure 6.7 depicts a generic sketch of motor primitives in a MDP[6] setting, which is essentially a state machine with each motor primitive as a node, and state-transition probabilities between the nodes. The dimmed heavy line arrows indicate the preferred sequence through the motor primitives, e.g., as might have been extracted from watching a teacher. Each motor primitive must have a subgoal, again, potentially provided through a movement demonstration. Learning could be used to refine execution of the sequence of motor primitives in terms of (1) refining the subgoals, (2) refining the motor primitive itself by adjusting its weights, (3) refining the transition to the next motor primitive, and (4) finding new paths through this motor primitive graph in order to achieve improved behavior. Points (1) and (2) are natural elements in reinforcement learning for motor primitives, e.g., as developed by Peters *et al.* (2003). From the viewpoint of dynamic movement primitives, point (3), i.e., transitions to another motor primitive, could be based on the behavioral phase variable, such that learning needs to determine at which value of the phase to trigger the next primitive – this is a typical component of learning with abstract actions in MDPs (Barto and Mahadevan, 2003). For example, if time optimality were required and it is not important to fully achieve the subgoal of each primitive, one can easily image some form of "co-articulation" in the execution of the sequence that smoothes the primitives more and more together. Such a process would not be unlike human learning of complex motor skills. Figure 6.8 illustrates an example of the smoothing together of motor primitives in

---

[6] Formally, an MDP with temporally extended actions is a semi-Markov decision process.

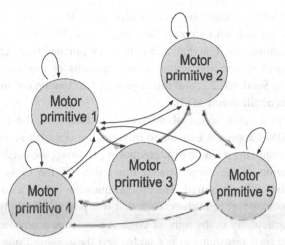

Figure 6.7 State machine based on motor primitives. Nodes denote a particular primitive, and arrows indicate possible state transitions; each of them would be associated with a transition probability. The dimmed heavy arrows indicate the preferred sequence through the state machine.

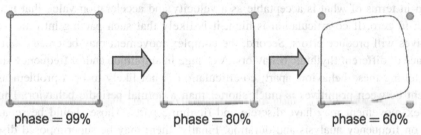

Figure 6.8 "Co-articulation" in executing a sequence of motor primitives. The moment starts at the bottom-left dot, then moves up, right, down, and left. The three different realizations demonstrate increasing co-articulation between the motor primitives due to changing at which phase value the transition to the next movement occurs.

the task of drawing a rectangle. Sosnik *et al.* (2004) presented data that is similar to this example, although their interpretation was in a different modeling framework of optimal control. Burridge *et al.* (1999) explored related ideas in dynamic manipulation task in robotics. Future work will be needed to fully flesh out how to apply the framework of reinforcement learning to the learning of motor sequences with movement primitives, and also how to address point (4) above, which is the most challenging.

6.3.3 Action recognition

The task of recognizing complex action is challenging. There are two different approaches in the computational literature. One primarily focuses on classifying action from observations, without the intention to reproduce the action – such research is conducted more in the domain of computer vision (Gavrila, 1997; Aggarwal and Cai, 1999) and human–computer interaction, and will not be pursued here. The other approach emphasizes movement recognition based on mechanisms of movement generation, as already mentioned in Section 6.2.2, and this approach will remain our main interest in the context of action recognition and MSH.

Before looking into some select alternative approaches in the literature, we will examine how our DMP approach extends to recognizing complex actions. Section 6.2.2 outlined that for recognizing a single discrete motor primitive, we need the start and goal state, and the duration of the demonstrated movement. If the movement is periodic, amplitude and period will be required. The crucial question is how a complex movement can be parsed into constituent primitives, such that it is possible to apply the mechanisms of Section 6.2.2, particularly in the light of co-articulation of primitives as illustrated in Fig. 6.8. While our own research has not addressed these issues, one can nevertheless examine the possible scenarios that need to be dealt with.

First, the complex movement may be decomposed into a series of concatenated discrete primitives. Without co-articulation, primitives can be separated by simultaneous zero velocity and zero acceleration crossings. With co-articulation, some thresholding will be needed in terms of what is acceptable as a velocity and acceleration value that is close enough to zero. If co-articulation is high, it is likely that such parsing into movement primitives will produce errors. Second, the complex movement may be created out of a sequence of different rhythmic behaviors. A change in amplitude and/or frequency should allow telling these behaviors apart; co-articulation is not likely to be a problem as the transient between primitives is much shorter than a normal periodic behavior. Third, a complex movement may have discrete and rhythmic parts. Those should be separable based on frequency analysis and duration. Finally, there may be superimposed discrete and rhythmic movement primitives in the complex movement. When analyzing such data in the Fourier domain, a subtraction of the most basic Fourier terms from the trajectory should uncover the discrete movement. All the ideas yet await an algorithm and experimental realization, and there has been no previous work to examine such methods in the light of biological information processing.

Several other approaches to complex movement recognition (and movement generation) have been suggested, inspired more by statistical data analysis than by the quest for what would be an ideal motor primitive, as we tried to do. Pook and Ballard (1993) created a robotic system that initially discretized demonstrated movement and force trajectories by means of vector quantization, and then trained a hidden Markov model (HMM)[7] with these discrete states as observables to find the transition probabilities among a given set of movement primitives, i.e., to create a graph similar to that in Fig. 6.7. An experimental implementation could be realized for one complex task (egg

---

[7] HMMs are essentially state machines where it is not known in which state the system is. Instead, the hidden state information needs to be inferred from some other observable features that are provided to the learning system.

flipping), albeit generalization of the suggested methods to different tasks or changes in the environment was not further explored. Inamura *et al.* (2004) followed a similar approach using HMMs. A movement primitive was defined in terms of Gaussian trajectory clusters in joint position space, and the HMM methodology was employed to recognize sequences through these primitives. The transition probabilities between primitives could afterwards be used to generate movements by sampling trajectories from the stochastic HMM. As motor primitives in terms of Gaussian clusters in joint space can be created automatically, the authors suggested that this method might be some form of proto-symbol formation, which could be exploited for the purpose of bootstrapping communication in future work. One more HMM approach to movement sequence recognition and movement sequence generation was suggested by Amit and Mataric (2002). Movement primitives were assumed to be given a priori, and a two-stage learning system associated goals with each primitive and sequences through the set of primitives.

Inspired by theories of computational motor control, Miyamoto *et al.* (1996; Miyamoto and Kawato, 1998) employed a movement parameterization in terms of fifth-order time-dependent splines, and characterized movement primitives based on typical sequences of spline nodes. Such sequences can be detected in observed actions by first fitting them with splines, and then searching the spline node representation for known primitives (Kawato *et al.*, 1994; Wada and Kawato, 1994, 1995). The spline representation lends itself naturally to movement generation in an optimal control framework (Kawato, 1999). Robustness of this method towards spatial and temporal scaling of movements, however, was not further explored. Miyamoto *et al.*'s work was a precursor for a more refined approach of a reciprocally constrained system of movement recognition and movement generation, suggested by Wolpert and Kawato (1998) and adapted for sequence learning in Samejima *et al.* (2003). Reinforcement learning was employed to create complex movements from basic control primitives, and action recognition was possible with the help of predictive forward models, as sketched in Section 6.2.2. A limitation of this approach may lie in its slow learning performance due to the inherent limitation of the current state-of-the-art reinforcement learning algorithms.

Instead of HMMs, motor sequences can also be encoded in recurrent neural networks, i.e., neural networks with closed loopy connectivity. Paine and Tani (2004) trained a well-designed recurrent network system on observed movement data such that the same network could represent multiple primitives depending on the setting of certain parameters. The network topology also supported the development of specialized neurons, called mirror neurons, which classified which motor primitive was currently active. After training, when exposed to new movement observation, these mirror neurons could be used to automatically parse the observed action into motor primitives, i.e., periods of quasi-constant activation of these mirror neurons. Another recurrent network approach was proposed by Billard (Billard and Mataric, 2001), who employed integrated-and-fire neurons to code movement primitives in the spirit of associative memory neural networks. Teaching and recognition of complex movement on humanoid robots could be realized in this fashion (Billard and Schaal, 2002; Schaal *et al.*, 2003a).

6.3.4 Summary

In order to create complex actions out of motor primitives, some mechanisms are needed to organize sequencing and superposition of primitives. The framework of DMPs offers a particular path into this topic, but alternative approaches exist as well in the literature. Currently, research in this area is still in its infancy, with various feasibility studies, but no approach that does not immediately have a large number of computational and functional problems. The framework of MDP may offer the analytically most tractable approach to sequencing and superposition of motor primitives, although neural network methods may provide interesting alternatives.

6.4 The development of hierarchical behaviors

A hierarchical behavior refers to an abstraction of a sequence of motor primitives, which becomes a new entity that can be used in the same way as a basic motor primitive. Theoretically, such an abstraction process could be applied recursively, leading to a multi-level nested hierarchy of motor primitives and actions. Given that the topic of sequencing and superposition of motor primitives is rather immature, currently there does not exist a large amount of research projects in hierarchical action generation and recognition in the spirit of the MSH. Action generation has primarily been addressed by research in reinforcement learning (Barto and Mahadevan, 2003), but this approach remains computationally complex and requires a fair amount of prestructuring, as the grouping of primitives into abstract actions usually needs to be done manually. An interesting recent example of related research is in Osentoski *et al.* (2004) who used a hierarchical HMM to model activity in navigation tasks, with Gaussian position clusters as a simpler form of motor primitives. The authors found that the hierarchical model had superior performance in modeling and classifying action when the observed behavior was complex. In the same spirit, a recurrent neural network approach to learning hierarchical behavior was investigated by Tani and Nolfi (1999). The idea of this approach was that mixture-model networks at each level of the hierarchy learned to specialize the sequences of activation of the lower level of the hierarchy, where the lowest level is the level of motor primitives. The authors demonstrated in simulations of a navigation task that the system acquired increasingly more complex behavioral abstractions, starting from simple "move around the corner" behaviors, to "traverse the hallway of the north wing of the building." In recent work (Paine and Tani, 2004), the authors also combined this hierarchical behavior approach with a mirror-system-inspired sequential behavior learning system. The effort of structuring and training such neural networks, however, remains complicated. Future work will have to demonstrate how hierarchical behaviors can be brought together with well-developed theories of motor primitives, like the DMP approach.

6.5 Biological parallels

So far, the sections of this chapter have focused on providing a formal basis of the processes that seem to be important for developing a computational theory of MSH. Our dynamic systems approach to motor primitives, movement imitation, and movement recognition, and, in more hypothesized form, to complex and hierarchical behaviors makes explicit, and sometimes potentially to idealistic, suggestions for the necessary ingredients. In this section, we will thus add some discussions of how far away from biological realism we have strayed in our search for theoretically sound formalisms.

6.5.1 Dynamic systems approach

First, one may ask whether our choice of an approach in terms of dynamic systems theory was appropriate. Among the many computational approaches to motor control, there seem to be two major distinct classes. One seeks to model motor phenomena in terms of some underlying fundamental organizational principle, usually in terms of optimality criteria like minimum energy consumption, minimum jerk, minimum torque change, minimum variance, etc. (Hogan, 1984; Flash and Hogan, 1985; Hasan, 1986; Stein *et al.*, 1986; Uno *et al.*, 1989; Kawato, 1990; Hoff, 1994; Harris and Wolpert, 1998; Todorov and Jordan, 2002). In contrast, the other class of computational models, i.e., dynamic systems models, focus on motor coordination as the emergent property of well-tuned interacting dynamic systems (Kugler and Turvey, 1987; Turvey, 1990; Kelso, 1995).

Inherent to these two classes of computational modeling is also a preference on the kind of motor patterns that are examined. Dynamic systems approaches are more concerned with rhythmic movement generation, which, in vertebrates, is primarily associated with pattern generator circuits in the spinal cord and the brainstem. Yet, it has been difficult to locate elements of such circuits in higher vertebrates because of the complexity of these nervous structures and their additional modulation by higher brain centers in the mature system (Marder, 2000). In humans, studies of the development of infant locomotion and microstimulation in patients with spinal cord injury have provided some behavioral evidence for pattern generators, again assumed to be on the spinal or brainstem level (Dimitrijevic *et al.*, 1998; Pinter and Dimitrijevic, 1999; Lamb and Yang, 2000). Due to the ubiquitous presence of rhythmic movements, a significant part of behavioral research in motor control employs the rhythmic pattern generation metaphor as a guiding model (Kugler and Turvey, 1987; Turvey, 1990; Kelso, 1995).

On the other hand, optimization approaches have almost exclusively been conducted on discrete and visually guided reaching, and have strong connections to the neuroscience of primate arm movements (Andersen *et al.*, 1997; Kalaska *et al.*, 1997; Sabes, 2000; Flash and Sejnowski, 2001). Several prominent computational models of arm movement are based on discrete strokes between a start and an end point (Morasso, 1981; Abend *et al.*, 1982), or, to generate more complex trajectories, a start and an end point with intermediate via-points (Flash and Hogan, 1985; Uno *et al.*, 1989), using either extrinsic (e.g.,

Cartesian) or intrinsic (e.g., joint or muscle space) coordinates to plan and execute the trajectory between these landmarks (Kawato, 1999). From this viewpoint, rhythmic arm movement is a sequence of movements between recurrent via-points (Feldman, 1980; Soechting and Terzuolo, 1987; Latash, 1993; Flash and Sejnowski, 2001).

However, recent behavioral evidence has cast some doubt that this subsumption of rhythmic arm movement by discrete movement is appropriate. For instance, a series of studies (Sternad and Schaal, 1999; Schaal and Sternad, 2001; Richardson and Flash, 2002) demonstrated that certain kinematic features of the hand trajectory in rhythmic arm movement, previously interpreted as a sign of segmented movement generation, can actually be accounted for by oscillator-based and/or optimally smooth movement generation, i.e., without the need of any segmentation. Experiments that tested the possibility of superposition of rhythmic and discrete patterns in arm movements equally concluded that rhythmic and discrete movement might be two different movement regimes (Sternad *et al.*, 2000, 2002; Wei *et al.*, 2003). Similar conclusions were obtained in a Fitts movement paradigm, where it was found that in rhythmic performance significantly higher movement speed can be tolerated at the same level of goal accuracy as in discrete movement (Smits-Engelsman *et al.*, 2002). Some first computational models have been suggested to make the difference between rhythmic and discrete movement explicit and to explore the general value of such theories in computational neuroscience and behavioral experiments (Sternad *et al.*, 2000; de Rugy and Sternad, 2003; Ijspeert *et al.*, 2003; Sternad and Dean, 2003). Among the most compelling evidence against the idea that discrete movement subsumes rhythmic movement is a recent functional magnetic resonance imaging (fMRI) study that demonstrated that rhythmic and discrete movement activate different brain areas. Figure 6.9 illustrates the summary results from this experiment, where subjects performed either periodic wrist flexion–extension oscillations, or discrete flexion-to-extension or extension-to-flexion point-to-point movements with the same wrist. The major findings were that while rhythmic movement activated only a small number of unilateral primary motor areas (M1, S1, PMdc, SMA, pre-SMA, CCZ, RCZp, cerebellum), discrete movement activated a variety of additional contralateral non-primary motor areas (BA 7, BA 40, BA 44, BA 47, PMdr, RCZa) and, moreover, showed very strong bilateral activity in both the cerebrum and cerebellum (Schaal *et al.*, 2004).[8] Control experiments examined whether in discrete movement the much more frequent movement initiation and termination and the associated cognitive effort could account for the observed differences. Only BA 40, BA 44, RCZa, and the cerebellum were potentially involved in such issues, leaving BA 7, BA 47, and PMdr as well as a large amount of bilateral activation a unique feature in discrete movement. Since rhythmic movement

---

[8] Abbreviations are (Picard and Strick, 2001): CCZ, caudal cingulate zone; RCZ, rostral cingulate zone, divided in an anterior (RCZa) and posterior (RCZp) part; SMA, caudal portion of the supplementary motor area, corresponding to SMA proper; pre-SMA, rostral portion of the supplementary motor area; M1, primary motor cortex; S1, primary sensory cortex; PMdr, rostral part of the dorsal premotor cortex; PMdc, caudal part of the dorsal premotor cortex; BA, Brodman area; BA 7, precuneus in parietal cortex; BA 8, middle frontal gyrus; BA 9, middle frontal gyrus; BA 10, anterior frontal lobe; BA 40, inferior parietal cortex; BA 44, Broca's area; BA 47, inferior frontal gyrus.

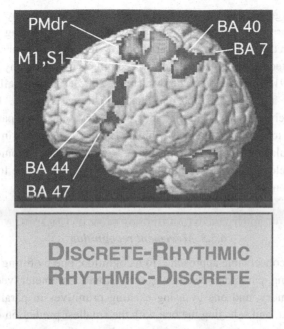

Figure 6.9 Differences in brain activation between discrete and rhythmic wrist movements.

activates significantly fewer brain areas than discrete movement, it was concluded that it does not seem to be warranted to claim that rhythmic movement is generated on top of a discrete movement system, i.e., rhythmic arm movement is *not* composed of discrete strokes.

As a summary of all these different experimental results, one can conclude that dynamic systems models for computational models of motor control in higher vertebrates remain a viable approach, and within this framework, a distinction between periodic and discrete movement arises naturally due to the core behaviors of dynamics system, i.e., limit cycles and point attractors. Our DMP model made these two movement regimes explicit.

6.5.2 Biological realism of dynamic motor primitives

Turning to the explicit realization of DMPs, one may ask whether the suggested computations can be realized in a nervous system. We are suggesting three ingredients. First, an output system, which, without the non-linear function term is just a simple linear damped-spring mass system which can be realized by a single neuron – for example, the well-known Hodgkin–Huxley equations (Hodgkin and Huxley, 1952) as a model of the membrane potential of a single neuron are far more complex than our model. The same statement holds for the canonical system, which is either another linear spring-damper or an oscillator – oscillator equations can be created by two neurons with mutual inhibitory connections, as suggested almost 100 years ago (Brown, 1914). Many parts of the brain can most likely learn non-linear transformations from one processing stage to the next one, as suggested in the non-linear function of our model. The basis function approach we use can be interpreted as a population code, as seemingly used in many parts of the brain (Georgopoulos, 1991; Shams and von der Malsburg, 2002). The cerebellum might also be a natural candidate for learning such non-linear functions (Kawato, 1999). Often, the cerebellum is conceived of a system that adds, through a secondary pathway, additional modulatory signals to a primary pathway (Shidara *et al.*, 1993), quite similar to our output system that is modulated through the non-linear function and the canonical system. Thus, the computational elements employed in our work do not seem to be too far-fetched for biological realization.

6.5.3 Movement recognition

We suggested two conceivable approaches to recognition: one by fitting primitive parameters and comparing parameter vectors against stored parameter vectors of existing movements in memory, and one by using existing primitives in parallel to predict the observed movement, and selecting the one with the smallest prediction error for recognition. The first of these approaches requires manipulation of parameters of the DMPs, and it is not clear how a nervous system would access such parameters, e.g., as stored in the synapses of the parallel fibers to Purkinje cell connection in the cerebellum. Recognition by prediction seems to be closer to biological possibilities, and such models have been hypothesized in some previous work (Gallese and Goldman, 1998; Wolpert and Kawato, 1998). Potentially, the nervous system may also learn other forms of classifiers, directly based on neural presentations of observed movement (Perrett *et al.*, 1989b), such that complex computations can largely be avoided during recognition – future work will have to shed more light on this issue.

6.5.4 The origin of movement primitives

Particularly in the light of MSH, an interesting question becomes how a motor system with motor primitives would bootstrap itself. In our DMP approach, we need rather few "innate" components: a discrete canonical system, a periodic canonical system, and an output system. All these systems have to be wired up such that the non-linear function can modulate the output system appropriately, and that the non-linear function itself can be divided into modules for different motor primitives. From this point onwards, new primitives can be created as needed, i.e., when previously learned primitives are insufficient to deal with a new task. It is easy to realize a rudimentary system of this form in technical applications, although there will be many open parameters to determine threshold when to add a new primitive, when to modify an existing one, etc. Direct biological connections to such processes are speculative, although some data exist about a modular specialization in the cerebellum (Imamizu *et al.*, 2000), which might be suitable for our framework.

6.6 Discussion

In Chapter 1 of this book, Arbib laid out seven steps in the evolution of language within the MSH framework: S1, grasping; S2, a mirror system for grasping; S3, a simple imitation system for grasping; S4, a complex imitation system for grasping; S5, protosign; S6, protospeech; and finally, S7, language. The developments presented in the current chapter tried to create a computational foundation for these steps. We selected a dynamic systems framework for modeling dynamic movement primitives (DMPs), the most basic element that is needed in S1 to generate action in a way that can account for a large number of phenomena observed in human movement, including generalization of movement to different situations, perception–action coupling, robustness towards perturbation, periodic and discrete movement, etc. Movement recognition could be addressed in this approach as a means of mapping observed movement to existing motor primitives, which is one of the core components of MSH as needed in S2. Bootstrapping a simple imitation system for S3 was in our formulation rather simple: all that was required was a means of supervised learning of motor primitives from demonstrated movements, a process that could theoretically be taken over by the cerebellum. A valid piece of criticism might be that we did not take into account object specificity, as observed in mirror neurons, but from a computational point of view, adding this component would be rather simple by means of associating certain primitives with certain objects. Thus, our DMP modeling approach can address S1 to S3 without major obstacles, i.e., creating movement (where grasping is just a special case), recognizing simple movement, and imitating simple movement. From stage S4 onwards, a new level of complexity is entered, that Arbib attributes only to the hominid line of animals. Interestingly, also from the view of computational theory, imitation of complex motor acts, i.e., parsing of sequential movement into its constituent motor primitives, and the generation of complex action from primitive components, is a much more complicated topic with rather little impressive or conclusive prior work in the literature. Some form of Markov chain approaches seem to be a useful start, as also explored in biological models (Fagg and Arbib, 1998), but flexible and theoretically sound learning methods for such behaviors are missing. Thus, stage S4 seems to be the next main target that computational modelers have to address, and there is a fair amount of ongoing efforts towards this goal in various communities. Moving on to stage S5, i.e., protosign, will be an interesting challenge. But maybe, the same way as it is possible to associate objects with a motor primitive or complex action, one could also associate expected behaviors of others with them, which could then lead to the recognition or interpretation of the act of others, such that protosign could develop. This kind of thinking is related to some of Meltzhoff's ideas (Meltzoff and Moore, 1995) of how infants use imitation for a simple form of communication, and first realizations of such ideas may be feasible in the coming years. Stages S6 and S7, i.e., protospeech and language, are yet not quite within the grasp of modelers.

Acknowledgments

This research was supported in part by National Science Foundation grants ECS-0325383, IIS-0312802, IIS-0082995, ECS-0326095, ANI-0224419, a NASA grant AC#98–516, an AFOSR grant on Intelligent Control, the ERATO Kawato Dynamic Brain Project funded by the Japanese Science and Technology Agency, and the ATR Computational Neuroscience Laboratories.

References

Abend, W., Bizzi, E, and Morasso, P., 1982. Human arm trajectory formation. *Brain* **105**: 331–348.

Adamovich, S. V., Levin, M. F., and Feldman, A. G., 1994. Merging different motor patterns: coordination between rhythmical and discrete single-joint. *Exp. Brain Res.* **99**: 325–337.

Aggarwal, J. K., and Cai, Q., 1999. Human motion analysis: a review. *Comput. Vision Image Understand.* **73**: 428–440.

Alissandrakis, A., Nehaniv, C. L., and Dautenhahn, K., 2002. Imitating with ALICE: learning to imitate corresponding actions across dissimilar embodiments. *IEEE Transactions on Systems, Man, and Cybernetics, Part A: Systems and Humans* **32**: 482–496.

Amit, R., and Mataric, M., 2002. Learning movement sequences from demonstration. *Proceedings International Conference on Development and Learning*, Cambridge, MA, June 12–15 2002, pp. 302–306.

Andersen, R. A., Snyder, L. H., Bradley, D. C., and Xing, J, 1997. Multimodal representation of space in the posterior parietal cortex and its use in planning movements. *Annu. Rev. Neurosci.* **20**: 303–330.

Arbib, M. A. 1981. Perceptual structures and distributed motor control. In V. B. Brooks (ed.) *Handbook of Physiology*, Section 2, *The Nervous System*, vol. 2, *Motor Control*, Part 1. Bethesda, MD: American Physiological Society, pp. 1449–1480.

Bagnell, J., Kadade, S., Ng, A., and Schneider, J. 2003. Policy search by dynamic programming. In S. Thrun, L. K. Saul and B. Schölkopf (eds.) *Advances in Neural Information Processing Systems*, vol. 16. Cambridge, MA: MIT Press, pp. 831–838.

Barto, A. G., and Mahadevan, S., 2003. Recent advances in hierarchical reinforcement learning. *Discrete Event Dynamic Systems* **13**: 341–379.

Bell, A. J., and Sejnowski, T. J., 1997. The "independent components" of natural scenes are edge filters. *Vision Res.* **37**: 3327–3338.

Bellman, R., 1957. *Dynamic Programming*. Princeton, NJ: Princeton University Press.

Bernstein, N. A., 1967. *The Control and Regulation of Movements*. London: Pergamon Press.

Bertsekas, D. P., and Tsitsiklis, J. N., 1996. *Neuro-Dynamic Programming*. Bellmont, MA: Athena Scientific.

Billard, A., and Mataric, M., 2001. Learning human arm movements by imitation: evaluation of a biologically inspired architecture. *Robot. Auton. Syst.* **941**: 1–16.

Billard, A., and Schaal, S., 2002. Computational elements of robot learning by imitation. *Proceedings Central Section Meeting, American Mathematical Society.* Providence, RI, October 12–13, 2002.

Bishop, C. M., 1995. *Neural Networks for Pattern Recognition.* New York: Oxford University Press.

Brooks, R. A., 1986. A robust layered control system for a mobile robot. *IEEE J. Robot. Autom.* **2**: 14–23.

Brown, T. G., 1914. On the nature of the fundamental activity of the nervous centres; together with an analysis of rhythmic activity in progression, and a theory of the evolution of function in the nervous system. *J. Physiol.* **48**: 18–46.

Bullock, D., and Grossberg, S., 1988. Neural dynamics of planned arm movements: emergent invariants and speed–accuracy properties during trajectory formation. *Psychol. Rev.* **95**: 49–90.

Bullock, D., Grossberg, S., and Guenther, F. H., 1993. A self-organizing neural model of motor equivalent reaching and tool use by a multijoint arm. *J. Cogn. Neurosci.* **5**: 408–435.

Burridge, R. R., Rizzi, A. A., and Koditschek, D. E., 1999. Sequential composition of dynamically dexterous robot behaviors. *Int. J. Robot. Res.* **18**: 534–555.

Byrne, R. W., 2003. Imitation as behaviour parsing. *Phil. Trans. Roy. Soc. London B* **358**: 529–536.

Craig, J. J., 1986. *Introduction to Robotics.* Reading, MA: Addison-Wesley.

Dautenhahn, K., and Nehaniv, C. L. (eds.), 2002. *Imitation in Animals and Artifacts.* Cambridge, MA: MIT Press.

Dayan, P., Hinton, G. E., Neal, R. M., and Zemel, R. S., 1995. The Helmholtz machine. *Neur. Comput.* **7**: 889–904.

de Rugy, A., and Sternad, D., 2003. Interaction between discrete and rhythmic movements: reaction time and phase of discrete movement initiation during oscillatory movements. *Brain Res.* **994**: 160–174.

Demiris, J., and Hayes, G., 1999. Active and passive routes to imitation. *Proceedings AISB 1999 Symposium of Imitation in Animals and Artifacts*, pp. 81–87.

Dimitrijevic, M. R., Gerasimenko, Y., and Pinter, M. M., 1998. Evidence for a spinal central pattern generator in humans. *Ann. N. Y. Acad. Sci.* **860**: 360–376.

Dyer, P., and McReynolds, S. R., 1970. *The Computation and Theory of Optimal Control.* New York: Academic Press.

Fagg, A. H., and Arbib, M. A., 1998. Modeling parietal–premotor interactions in primate control of grasping. *Neur. Networks* **11**: 1277–1303.

Feldman, A. G., 1980. Superposition of motor programs. I. Rhythmic forearm movements in man. *Neuroscience* **5**: 81–90.

Flash, T., and Henis, E., 1991. Arm trajectory modification during reaching towards visual targets. *J. Cogn. Neurosci.* **3**: 220–230.

Flash, T., and Hogan, N., 1985. The coordination of arm movements: an experimentally confirmed mathematical model. *J. Neurosci.* **5**: 1688–1703.

Flash, T., and Sejnowski, T., 2001. Computational approaches to motor control. *Curr. Opin. Neurobiol.* **11**: 655–662.

Gallese, V., and Goldman, A., 1998. Mirror neurons and the simulation theory of mind-reading. *Trends Cogn. Sci.* **2**: 493–501.

Gavrila, D. M., 1997. The visual analysis of humant movements: a survey. *Comput. Vision Image Understand.* **73**: 82–98.

Georgopoulos, A. P., 1991. Higher order motor control. *Annu. Rev. Neurosci.* **14**: 361–377.

Harris, C. M., and Wolpert, D. M., 1998. Signal-dependent noise determines motor planning. *Nature* **394**: 780–784.

Haruno, M., Wolpert, D. M., and Kawato, M., 2001. Mosaic model for sensorimotor learning and control. *Neur. Comput.* **13**: 2201–2220.

Hasan, Z., 1986. Optimized movement trajectories and joint stiffness in unperturbed, inertially loaded movements. *Biol. Cybernet.* **53**: 373–382.

Henis, E. A., and Flash, T., 1995. Mechanisms underlying the generation of averaged modified trajectories. *Biol. Cybernet.* **72**: 407–419.

Hinton, G. E., Dayan, P., Frey, B. J., and Neal, R. M., 1995. The wake–sleep algorithm for unsupervised neural networks. *Science* **268**: 1158–1161.

Hodgkin, A. L., and Huxley, A. F., 1952. A quantitative description of membrane current and its application to conduction and excitation in nerve. *J. Physiol.* **117**: 500–544.

Hoff, B., 1994. A model of duration in normal and perturbed reaching movement. *Biol. Cybernet.* **71**: 481–488.

Hoff, B., and Arbib, M. A. 1992. A model of the effects of speed, accuracy, and perturbation on visually guided reaching. In R. M. Caminiti, P. B. Johnson and Y. Burnod (eds.) *Experimental Brain Research Series*, vol. 22. Berlin: Springer-Verlag, pp. 285–306.

Hogan, N., 1984. An organizing principle for a class of voluntary movements. *J. Neurosci.* **4**: 2745–2754.

Hollerbach, J. M., 1984. Dynamic scaling of manipulator trajectories. *Trans. Am. Soc. Mech. Eng.* **106**: 139–156.

Ijspeert, A., Nakanishi, J., and Schaal, S., 2001. Trajectory formation for imitation with nonlinear dynamical systems. *Proceedings IEEE Int. Conference on Intelligent Robots and Systems*, Weilea, HI, pp. 729–757.

2002. Movement imitation with nonlinear dynamical systems in humanoid robots. *Proceedings Int. Conference on Robotics and Automation*, Washington, May 11–15, 2002.

2003. Learning attractor landscapes for learning motor primitives. In S. Becker, S. Thrun and K. Obermayer (eds.) *Advances in Neural Information Processing Systems*, vol. 15. Cambridge, MA: MIT Press, pp. 1547–1554.

Imamizu, H., Miyauchi, S., Tamada, T, *et al.*, 2000. Human cerebellar activity reflecting an acquired internal model of a new tool. *Nature* **403**: 192–195.

Inamura, T., Toshima, I., and Nakamura, Y., 2002. Acquisition and embodiment of motion elements in closed mimesis loop. *Proceedings Int. Conference on Robotics and Automation*, Washington, May 11–15, 2002, pp. 1539–1544.

Inamura, T., Iwaki, T., Tanie, H., and Nakamura, Y., 2004. Embodied symbol emergence based on mimesis theory. *Int. J. Robot. Res.* **23**: 363–377.

Ivry, R. B., Spencer, R. M., Zelaznik, H. N., and Diedrichsen, J., 2002. The cerebellum and event timing. *Ann. N. Y. Acad. Sci.* **978**: 302–317.

Jordan, I. M., and Rumelhart D., 1992. Supervised learning with a distal teacher. *Cogn. Sci.*, **16**: 307–354.

Kalaska, J. F., Scott, S. H., Cisek, P., and Sergio, L. E., 1997. Cortical control of reaching movements. *Curr. Opin. Neurobiol.* **7**: 849–859.

Kawamura, S., and Fukao, N., 1994. Interpolation for input torque patterns obtained through learning control. *Proceedings Int. Conference on Automation, Robotics and Computer Vision*, Singapore, pp. 183–191.

Kawato, M., 1990. Computational schemes and neural network models for formation and control of multijoint arm trajectory. In W. T. Miller, R. S. Sutton and P. J. Werbos (eds.) *Neural Networks for Control.* Cambridge, MA: MIT Press, pp. 197–228.

  1996. Bi-directional theory approach to integration. In J. Konczak and E. Thelen (eds.) *Attention and Performance*, vol. 16. Cambridge, MA: MIT Press, pp. 335–367.

  1999. Internal models for motor control and trajectory planning. *Curr. Opin. Neurobiol.* **9**: 718–727.

Kawato, M., and Wolpert, D., 1998. Internal models for motor control. *Novartis Found. Symp.* **218**: 291–304.

Kawato, M., Gandolfo, F., Gomi, H., and Wada, Y., 1994. Teaching by showing in kendama based on optimization principle. *Proceedings Int. Conference on Artificial Neural Networks*, vol. 1, pp. 601–606.

Keating, J. G., and Thach, W. T., 1997. No clock signal in the discharge of neurons in the deep cerebellar nuclei. *J. Neurophysiol.* **77**: 2232–2234.

Kelso, J. A. S., 1995. *Dynamic Patterns: The Self-Organization of Brain and Behavior.* Cambridge, MA: MIT Press.

Kugler, P. N., and Turvey, M. T., 1987. *Information, Natural Law, and the Self-Assembly of Rhythmic Movement.* Hillsdale, NJ: Lawrence Erlbaum.

Lamb, T., and Yang, J. F., 2000. Could different directions of infant stepping be controlled by the same locomotor central pattern generator? *J. Neurophysiol.* **83**: 2814–2824.

Latash, M. L., 1993. *Control of Human Movement.* Champaign, IL: Human Kinetics.

Liberman, A. M., 1996. *Speech: A Special Code.* Cambridge, MA: MIT Press.

Lohmiller, W., and Slotine, J. J. E., 1998. On contraction analysis for nonlinear systems. *Automatica* **6**: 683–696.

Marder, E., 2000. Motor pattern generation. *Curr. Opin. Neurobiol.* **10**: 691–698.

Mataric, M., 1998. Behavior-based robotics as a tool for synthesis of artificial behavior and analysis of natural behavior. *Trends Cogn. Sci.* **2**: 82–86.

Meltzoff, A. N., and Moore, M. K. 1995. Infant's understanding of people and things: from body imitation to folk psychology. In J. L. Bermúdez, A. Marcel and N. Eilan (eds.) *The Body and the Self.* Cambridge, MA: MIT Press, pp. 43–69.

Miyamoto, H., and Kawato, M., 1998. A tennis serve and upswing learning robot based on bi-directional theory. *Neur. Networks* **11**: 1331–1344.

Miyamoto, H., Schaal, S., Gandolfo, F., *et al.*, 1996. A Kendama learning robot based on bi-directional theory. *Neur. Networks* **9**: 1281–1302.

Mohajerian, P., Mistry, M, and Schaal, S., 2004. Neuronal or spinal level interaction between rhythmic and discrete motion during multi-joint arm movement. *Abstracts 34th Meeting Soci. Neuroscience*, San Diego, CA, October, 23–27.

Morasso, P., 1981. Spatial control of arm movements. *Exp. Brain Res.* **42**: 223–227.

Mussa-Ivaldi, F. A., 1988. Do neurons in the motor cortex encode movement direction? An alternative hypothesis. *Neurosci. Lett.* **91**: 106–111.

Nakanishi, J., Morimoto, J., Endo, G., *et al.*, 2004. Learning from demonstration and adaptation of biped locomotion. *Robot. Auton. Syst.* **47**: 79–91.

Nakanishi, J., Cory, R., Mistry, M., Peters, J., and Schaal, S., 2005. Comparative experiments on task space control with redundancy resolution. *Proceedings IEEE Int. Conference on Intelligent Robots and Systems*, pp. 1575–1582.

Olshausen, B. A., and Field, D. J., 1996. Emergence of simple-cell receptive
field properties by learning a sparse code for natural images. *Nature* **381**: 607–609.

Osentoski, S., Manfredi, V., and Mahadevan S., 2004. Learning hierarchical models of
activity. *Proceedings IEEE Int. Conference on Intelligent Robots and Systems*,
Sendai, Japan.

Oztop, E., Wolpert, D., and Kawato, M., 2005. Mental state inference using visual control
parameters. *Brain Res. Cogn. Brain Res.* **22**: 129–151.

Paine, R. W., and Tani, J., 2004. Motor primitive and sequence self-organization in a
hierarchical recurrent neural network. *Neur. Networks* **17**: 1291–1309.

Perrett, D., Harries M., Mistlin, A. J., and Chitty, A. J. 1989a. Three stages in the
classification of body movements by visual neurons. In H. B. Barlow (ed.) *Images
and Understanding*. Cambridge, UK: Cambridge University Press, 94–107.

Perrett, D. I., Harries, M. H., Bevan, R., *et al.*, 1989b. Frameworks of analysis for the
neural representation of animate objects and actions. *J. Exp. Biol.* **146**: 87–113.

Peters, J., Vijayakumar, S., and Schaal, S., 2003. Reinforcement learning for humanoid
robotics. *Proceedings, 3rd IEEE–RAS Int. Conference on Humanoid Robots*,
Karlsruhe, Germany, September 29–30, 2003.

Picard, N., and Strick, P. L., 2001. Imaging the premotor areas. *Curr. Opin. Neurobiol.* **11**:
663–672.

Pinter, M. M., and Dimitrijevic, M. R., 1999. Gait after spinal cord injury and the central
pattern generator for locomotion. *Spinal Cord* **37**: 531–537.

Pook, P. K., and Ballard, D. H., 1993. Recognizing teleoperated manipulations.
Proceedings IEEE Int. Conference on Robotics and Automation, Atlanta, GA,
pp. 913–918.

Richardson, M. J., and Flash, T., 2002. Comparing smooth arm movements with the
two-thirds power law and the related segmented-control hypothesis. *J. Neurosci.*
22: 8201–8211.

Rizzi, A. A., and Koditschek, D. E., 1994. Further progress in robot juggling: solvable
mirror laws. *Proceedings IEEE Int. Conference on Robotics and Automation*,
San Diego, CA, pp. 2935–2940.

Rizzolatti, G., Fadiga, L., Gallese, V., and Fogassi, L., 1996. Premotor cortex and the
recognition of motor actions. *Cogn. Brain Res.* **3**: 131–141.

Roberts, P. D., and Bell, C. C., 2000. Computational consequences of temporally
asymmetric learning rules. II. Sensory image cancellation. *J. Comput. Neurosci.*
9: 67–83.

Sabes, P. N., 2000. The planning and control of reaching movements. *Curr. Opin.
Neurobiol.* **10**: 740–746.

Samejima, K., Doya, K., and Kawato, M., 2003. Inter-module credit assignment in
modular reinforcement learning. *Neur. Networks* **16**: 985–994.

Schaal, S., 1997. Learning from demonstration. In M. C. Mozer, M. Jordan and T. Petsche
(eds.) *Advances in Neural Information Processing Systems*, vol. 9. Cambridge, MA:
MIT Press, pp. 1040–1046.

1999. Is imitation learning the route to humanoid robots? *Trends Cog. Sci.* **3**: 233–242.

2002a. Arm and hand movement control. In M. A. Arbib (ed.) *The Handbook of
Brain Theory and Neural Networks*, 2nd edn. Cambridge, MA: MIT Press,
pp. 110–113.

2002b. Learning robot control. In M. A. Arbib (ed.) *The Handbook of Brain Theory and
Neural Networks*, 2nd edn. Cambridge, MA: MIT Press, pp. 983–987.

Schaal, S., and Atkeson, C. G., 1993. Open loop stable control strategies for robot juggling. *Proceedings IEEE Int. Conference on Robotics and Automation*, Atlanta, GA, pp. 913–918.

Schaal, S., and Sternad, D., 1998. Programmable pattern generators. *Proceedings 3rd Int. Conference on Computational Intelligence in Neuroscience*, Research Triangle Park, NC, pp. 48–51.

2001. Origins and violations of the 2/3 power law in rhythmic 3D movements. *Exp. Brain Res.* **136**: 60–72.

Schaal, S., Atkeson, C. G., and Botros, S., 1992. What should be learned? *Proceedings 7th Yale Workshop on Adaptive and Learning Systems*, New Haven, CT, pp. 199–204.

Schaal, S., Ijspeert, A., and Billard, A., 2003a. Computational approaches to motor learning by imitation. *Phil. Trans. Roy. Soc. London B* **358**: 537–547.

Schaal. S., Peters, J., Nakanishi, J., and Ijspeert, A., 2003b. Control, planning, learning, and imitation with dynamic movement primitives. *Proceedings Workshop on Bilateral Paradigms on Humans and Humanoids, IEEE Int. Conference on Intelligent Robots and Systems*, Las Vegas, NV, October 27–31, 2003.

Schaal, S., Sternad, D., Osu, R., and Kawato, M., 2004. Rhythmic movement is not discrete. *Nature Neurosci.* **7**: 1137–1144.

Schweighofer, N., Arbib, M. A., Spoelstra, J., and Kawato M., 1998. Role of the cerebellum in reaching movements in humans. II. A neural model of the intermediate cerebellum. *Eur. J. Neurosci.* **10**: 95–105.

Shams, L., and von der Malsburg, C., 2002. The role of complex cells in object recognition. *Vision Res.* **42**: 2547–2554.

Shidara, M., Kawano, K., Gomi, H., and Kawato, M., 1993. Inverse-dynamics model eye movement control by Purkinje cells in the cerebellum. *Nature* **365**: 50–52.

Smits-Engelsman, B. C., Van Galen, G. P., and Duysens, J., 2002. The breakdown of Fitts' law in rapid, reciprocal aiming movements. *Exp. Brain Res.* **145**: 222–230.

Soechting, J. F., and Terzuolo, C. A., 1987. Organization of arm movements in three-dimensional space: Wrist motion is piecewise planar. *Neuroscience* **23**: 53–61.

Sosnik, R., Hauptmann, B., Karni, A., and Flash, T., 2004. When practice leads to co-articulation: the evolution of geometrically defined movement primitives. *Exp. Brain Res.* **156**: 422–438.

Staude, G, and Wolf, W., 1997. Quantitative assessment of phase entrainment between discrete and cyclic motor actions. *Biomed. Tech.* **42**: 478–481.

Stein, R. B., Ogusztöreli, M. N., and Capaday, C. 1986. What is optimized in muscular movements? In N. L. Jones, N. McCartney and A. J. McComas (eds.) *Human Muscle Power*. Champaign, Il: Human Kinetics, pp. 131–150.

Sternad, D., and Dean, W. J., 2003. Rhythmic and discrete elements in multi-joint coordination. *Brain Res.* **989**: 152–171.

Sternad, D., and Schaal, D., 1999. Segmentation of endpoint trajectories does not imply segmented control. *Exp. Brain Res.* **124**: 118–136.

Sternad, D., Dean, W. J., and Schaal, S. 1999. Interaction of discrete and rhythmic dynamics in single-joint movements. In M. Grealy and J. Thomson (eds.) *Studies in Perception and Action*, vol. 5, Hillsdale, NJ: Lawrence Erlbaum, pp. 282–287.

2000. Interaction of rhythmic and discrete pattern generators in single joint movements. *Hum. Mov. Sci.* **19**: 627–665.

Sternad, D., de Rugy, A., Pataky, T., and Dean, W. J., 2002. Interaction of discrete and rhythmic movements over a wide range of periods. *Exp. Brain Res.* **147**: 162–174.

Strogatz, S. H., 1994. *Nonlinear Dynamics and Chaos: With Applications to Physics, Biology, Chemistry, and Engineering.* Reading, MA: Addison-Wesley.

Sutton, R. S., and Barto, A. G., 1998. *Reinforcement Learning: An Introduction.* Cambridge, MA: MIT Press.

Sutton, R. S., Precup, D., and Singh, S., 1999. Between MDPs and semi-MDPs: a framework for temporal abstraction in reinforcement learning. *Artif. Intell.* **112**: 181–211.

Tani, J., and Nolfi, S., 1999. Learning to perceive the world as articulated: an approach for hierarchical learning in sensory-motor systems. *Neur. Networks* **12**: 1131–1141.

Tesauro, G. 1992. Temporal difference learning of backgammon strategy. In D. Sleeman and P. Edwards (eds.) *Proceedings 9th International Workshop Machine.* Morgan Kaufmann, San Mateo, CA, pp. 9–18.

Tevatia, G., and Schaal, S., 2000. Inverse kinematics for humanoid robots. *Proceedings Int. Conference on Robotics and Automation*, San Fransisco, CA, pp. 294–299.

Thach, W. T., 1998. A role for the cerebellum in learning movement coordination. *Neurobiol. Learn. Mem.* **70**: 177–188.

Todorov, E., and Jordan, M. I., 2002. Optimal feedback control as a theory of motor coordination. *Native Neurosci.* **5**: 1226–1235.

Turvey, M. T., 1990. Coordination. *Am. Psychol.* **45**: 938–953.

Uno, Y., Kawato, M., and Suzuki, R., 1989. Formation and control of optimal trajectory in human multijoint arm movement: minimum torque-change model. *Biol. Cybernet.* **61**: 89–101.

Vijayakumar, S., and Schaal, S., 2000. Locally weighted projection regression: an O(n) algorithm for incremental real time learning in high dimensional spaces. *Proceedings 17th Int. Conference on Machine Learning*, vol. 1, Stanford, CA, pp. 288–293.

Viviani, P., 1986. Do units of motor action really exist? In *Experimental Brain Research Series*, vol. 15. Berlin: Springer-Verlag, pp. 828–845.

Wada, Y., and Kawato, M., 1994. Trajectory formation of arm movement by a neural network with forward and inverse dynamics models. *Systems Comput. in Japan* **24**: 37–50.

  1995. A theory for cursive handwriting based on the minimization principle. *Biol. Cybernet.* **73**: 3–13.

Wei, K., Wertman, G., and Sternad, D., 2003. Interactions between rhythmic and discrete components in a bimanual task. *Motor Control* **7**: 134–155.

Williamson, M., 1998. Neural control of rhythmic arm movements. *Neur. Networks* **11**: 1379–1394.

Wolpert, D. M., 1997. Computational approaches to motor control. *Trends Cogn. Sci.* **1**: 209–216.

Wolpert, D. M., and Kawato, M., 1998. Multiple paired forward and inverse models for motor control. *Neur. Networks* **11**: 1317–1329.

Wolpert, D. M., Miall, R. C., and Kawato, M., 1998. Internal models in the cerebellum. *Trends Cogn. Sci.* **2**: 338–347.

7

The role of vocal tract gestural action units in understanding the evolution of phonology

Louis Goldstein, Dani Byrd, and Elliot Saltzman

7.1 Introduction: duality of patterning

Language can be viewed as a structuring of cognitive units that can be transmitted among individuals for the purpose of communicating information. Cognitive units stand in specific and systematic relationships with one another, and linguists are interested in the characterization of these units and the nature of these relationships. Both can be examined at various levels of granularity. It has long been observed that languages exhibit distinct patterning of units in syntax and in phonology. This distinction, a universal characteristic of language, is termed *duality of patterning* (Hockett, 1960). Syntax refers to the structuring of words in sequence via hierarchical organization, where words are meaningful units belonging to an infinitely expandable set. But words *also* are composed of structured cognitive units. Phonology structures a small, closed set of recombinable, non-meaningful units that compose words (or signs, in the case of signed languages). It is precisely the use of a set of non-meaningful arbitrary discrete units that allows word creation to be productive.[1]

In this chapter we outline a proposal that views the evolution of syntax and of phonology as arising from different sources and ultimately converging in a symbiotic relationship. Duality of patterning forms the intellectual basis for this proposal. Grasp and other manual gestures in early hominids are, as Arbib (Chapter 1, this volume) notes, well suited to provide a link from the iconic to the symbolic. Critically, the iconic aspects of manual gestures lend them a *meaningful* aspect that is critical to evolution of a system of symbolic units. However, we will argue that, given duality of patterning, *phonological* evolution crucially requires the emergence of effectively *non-meaningful* combinatorial units. We suggest that vocal tract action gestures are well suited to play a direct role in phonological evolution because, as argued by Studdert-Kennedy (2002a), they are

---

[1] Phonological units, in addition to being discrete and recombinable, must yield sufficient sensory perceptibility and distinctiveness to be useful for communication.

Action to Language via the Mirror Neuron System, ed. Michael A. Arbib. Published by Cambridge University Press. © Cambridge University Press 2006.

inherently non-iconic[2] and non-meaningful yet particulate (discrete and combinable). The vocal organs form the innate basis for this particulate nature (Studdert-Kennedy, 1998; Studdert-Kennedy and Goldstein, 2003). In our proposal the lack of iconicity, rather than being a weakness of vocal gestures for language evolution (cf. Arbib, 2005; Chapter 1, this volume), is *advantageous* specifically for phonological evolution in that little or no semantic content would have been needed to be "bleached out" of these mouth actions to allow them to serve as the necessarily non-meaningful phonological units – they are ideally suited to phonological function.

A reconsideration of spoken language evolution that attends to syntactic and phonological evolution as potentially distinct may cast new light on the issues. In one proposed evolutionary scenario (Arbib, Chapter 1, this volume), the original functional retrieval of symbolic meaning from manual action required, at one point, a way of distinguishing similar limb, hand, or face movements. The combination of intrinsically distinct (but meaningless) vocal actions with members of a set of iconic manual gestures could have provided the necessary means of disambiguating otherwise quite similar and continuously variable manual protowords (see also Corballis (2003) and Studdert-Kennedy and Lane (1980)). Ultimately, vocal gestures became the primary method of distinguishing words, and manual gestures were reduced in importance to the role we see them play in contemporary spoken communication (McNeill, 1992). Thus, the syntactic organization of meaningful actions and the phonological organization of particulate non-meaningful actions can be speculated to have evolved symbiotically.[3]

The existence of duality of patterning as a requisite characteristic of human languages indicates that both components of the evolution of language – the syntactic and the phonological – are intrinsic to the structure and function of language. Significantly, however, syntax and phonology may have originally evolved through different pathways. This would be consistent with the fact that when their fundamental organizing principles are described as abstract formal systems, there is little overlap between the two.[4]

7.2 Phonology and language evolution

A hallmark property of human language, signed or spoken, is that a limited set of meaningless discrete units can recombine to form the large number of configurations that are the possible word forms of a language. How did such a system evolve? This hallmark property requires particulate units (Studdert-Kennedy, 1998; see Abler, 1989) that can

---

[2] MacNeilage and Davis (2005) argue that there are non-arbitrary mappings in some cases between vocal gestures and meaning, such as the use of nasalization for words meaning "mother." However, their proposed account for the non-arbitrariness is contextual not iconic per se.

[3] Clearly, a radical change over evolutionary time to spoken language as the primary communication mode would have required enormous restructuring in the neural system(s) involved to move from a dependence on the manual–optic–visual chain to the vocal–acoustic–auditory chain of communication. However, we are suggesting that the systems evolved in parallel complementary ways, and the existence of signed languages with syntactic structures comparable to spoken language indicates that a manual–optic–visual chain remains accessible for communication, given a suitably structured linguistic system, namely one having duality of patterning.

[4] Except where they directly interact with one another, in the prosodic form of utterances.

function as atoms or integral building blocks; further, it requires a "glue" that can bond these atomic units together into larger combinatorial structures or molecules. From an evolutionary perspective, we wish to consider the possibility that the basis for these (units and glue) was already largely present in the vocal tract and its control before the evolution of phonology, thus explaining how such a system might have readily taken on (or over) the job of distinguishing words from one another.

Traditional phonological theories have analyzed word forms in terms of segments (or phonemes) and symbolic bi- or univalent features differentiating these segments in linguistically relevant dimensions. Both segments and features are assumed to be abstract cognitive units, the complete set of which is defined by universal properties of phonological contrast and with specific segments/features defining lexical contrast within a particular phonological system. However, speech scientists have long noticed an apparent mismatch between the proposed traditional sequence of concatenated discrete units and the physical, observable characteristics of speech, which lacks overt boundaries (e.g., silences) between segments, syllables, words, and often phrases (e.g., Harris, 1951; Hockett, 1955; Liberman *et al.*, 1959).

In response to this mismatch, the framework of articulatory phonology (Browman and Goldstein, 1992, 1995; Byrd, 1996a; Byrd and Saltzman, 2003) has been forwarded as an account of how spoken language is structured. This approach is termed *articulatory phonology* because it pursues the view that the units of contrast, i.e., the phonological units, are isomorphic with the units of language production, i.e., the phonetic units. This framework views the act of speaking as decomposable into atomic units of vocal tract action – which have been termed in the field of phonetics for the last two decades *articulatory gestures*. (It's important to note here that in presenting this framework, we will be using the term articulatory *gesture* to refer to goal-directed vocal tract actions, specifically the formation of auditorily important constrictions in the vocal tract; we will not generally be using the term to refer to anything manual, unlike many of the other papers in this volume.) Articulatory gestures are actions of distinct vocal *organs*, such as the lips, tongue tip, tongue dorsum, velum, glottis (Fig. 7.1). Articulatory gestures are simultaneously units of action and units of information (contrast and encoding). This approach, like some other current approaches to phonological evolution (see, e.g., MacNeilage and Davis, 2005), views phonological structure as arising from the structural and functional characteristics and constraints of body action in the environment.

Under this account, plausible ingredients for the particulate units and bonding ("glue") necessary as precursors to the evolution of phonology are hypothesized to be, respectively, *actions* of vocal tract organs existing outside the linguistic system (plausibly actions of oral dexterity such as involved in sucking and swallowing and/or actions producing affectual vocalizations) and the *dynamic* character of their interactions as they ultimately cohere into structured units.

As action units, gestures can be described further in a twofold manner. The first is in terms of the articulatory motions themselves. More specifically, a gesture can be defined as an equivalence class of goal-directed movements by a set of articulators in the vocal

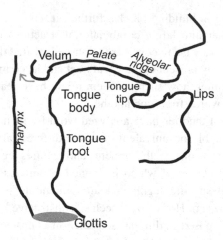

Figure 7.1 Constricting organs of the vocal tract (tongue tip, tongue body, tongue root, lips, velum, and glottis), and some of the potential places of constriction for these organs (palate, alveolar ridge, and pharynx).

tract (e.g., Saltzman and Munhall, 1989). For example, the bilabial gestures for /p/, /b/, /m/ are produced by a family of functionally equivalent movement patterns of the upper lip, lower lip, and jaw that are actively controlled to attain the speech-relevant goal of closing the lips. Here the upper lip, lower lip, and jaw comprise the *lips* organ system, or effector system, and the gap or aperture between the lips comprises the controlled variable of the organ/effector system. The second manner in which gestures serve as action units is that they embody a particular type of dynamical system, a point-attractor system that acts similarly to a damped mass-spring system, that creates and releases constrictions of the end-effectors that are being controlled. Point attractors have properties useful for characterizing articulatory gestures. For example, regardless of their initial position or unexpected perturbations, articulatory gestures can reach their target (i.e., equilibrium position) successfully. When activated, a given gesture's point-attractor dynamics creates a pattern of articulatory motion whose details reflect the ongoing context, yet whose overall effect is to attain the constriction goal in a flexible and adaptable way.

   Point-attractor models have been used to characterize many skilled human movements. Reviews of the use of point-attractor models for skilled movement can be found in Shadmehr (1995), Mussa-Ivaldi (1995), and Flash and Sejnowski (2001). In other words, we view the control of these units of action in speech to be no different from that involved in controlling skilled movements generally, e.g., reaching, grasping, kicking, pointing, etc. Further, just as a given pattern of articulatory gestures can be modulated by expressive or idiosyncratic influences, the pointing movements of an orchestral conductor can also be so modulated.

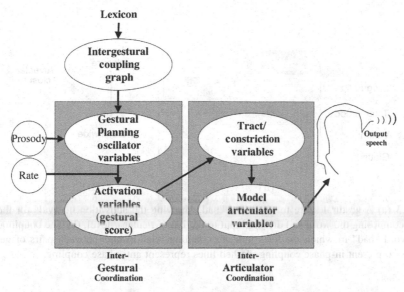

Figure 7.2 The organization of the task-dynamic model of speech production (Saltzman and Munhall, 1989; Browman and Goldstein, 1992; Nam and Saltzman, 2003).

7.2.1 Overview of a gestural, task-dynamic model of speech production

The organization of the gesture-based task-dynamic model of speech production that we have developed is shown in Fig. 7.2 (Saltzman, 1986, 1995; Saltzman and Kelso, 1987; Saltzman and Munhall, 1989; Browman and Goldstein, 1992; Nam and Saltzman, 2003).

The spatiotemporal patterns of articulatory motion emerge as behaviors implicit in a dynamical system with two functionally distinct but interacting levels (associated with the corresponding models shown as boxes in Fig. 7.2). The *interarticulator* coordination level is defined according to both *model articulator* (e.g., lips and jaw) variables and goal space or *tract-variables* (which are constriction based) (e.g., lip aperture (LA) and lip protrusion (LP)). The *intergestural* level is defined according to a set of *planning oscillator* variables and *activation* variables. The activation trajectories shaped by the intergestural level define a *gestural score* (an example of which is shown in Fig. 7.3a for the word "bad") that provides driving input to the interarticulator level.

The gestural score represents an utterance as a set of invariant gestures in the form of context-independent sets of dynamical parameters (e.g., target, stiffness, and damping coefficients) that characterize a gesture's point-attractor dynamics and are associated with corresponding subsets of model articulator, tract-variable, and activation variables. Each activation variable reflects the strength with which the associated gesture (e.g., lip closure) "attempts" to shape vocal tract movements at any given point in time. The tract-variable and model articulator variables associated with each gesture specify, respectively, the particular vocal-tract constriction (e.g., lips) and articulatory synergy

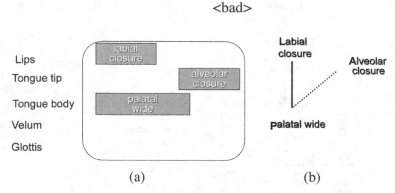

(a)   (b)

Figure 7.3 (a) A gestural score for the word "bad" showing the activation intervals for the three gestures composing the word and driving input to the interarticulator level. (b) The coupling graph for the word "bad" in which the lines indicate coupling relationships between pairs of gestures. Solid lines represent in-phase coupling, dashed lines represent anti-phase coupling.

(e.g., upper lip, lower lip, and jaw) whose behaviors are affected directly by the associated gesture's activation. The interarticulator level accounts for the observed spatiotemporal coordination among articulators in the currently active gesture set as a function of their dynamical parameter specifications. While each gesture is modeled with invariant point-attractor dynamics, the concurrent activation of multiple gestures will result in correspondingly context-dependent patterns of articulator motion. Thus, invariance in phonological specification lies not at the level of articulatory movements but in the speech tasks that those movements serve.

The intergestural level can be thought of as implementing a dynamics of *planning* – it determines the patterns of relative timing among the activation waves of gestures participating in an utterance as well as the shapes and durations of the individual gesture activation waves. Each gesture's activation wave acts to insert the gesture's parameter set into the interarticulator dynamical system defined by the set of tract-variable and model articulator coordinates (see Saltzman and Munhall (1989) for further details).

In the current model,[5] intergestural timing is determined by the *planning oscillators* associated with the set of gestures in a given utterance. The oscillators for the utterance are coupled in a pairwise, bidirectional manner specified in a *coupling graph* that is part of the lexical specification of a word. (There are also prosodic gestures (Byrd and Saltzman, 2003) that are not part of the lexical specification, but these will not be discussed further here.) For example, a coupling graph for the word "bad" is shown in Fig. 7.3b, where the

---

[5] In the original version of the model (e.g., Browman and Goldstein, 1990), the activation variables in gestural scores were determined by a set of rules that specified the relative phasing of the gestures and calculated activation trajectories based on those phases and the time constants associated with the individual gestures. The gestural score then unidirectionally drove articulatory motion at the interarticulator level. Thus, intergestural timing was not part of the dynamical system per se, and such a model was not capable of exhibiting dynamical coherence, such as can be seen, for example, in the temporal adjustment to external perturbation (Saltzman *et al.*, 2000).

lines indicate coupling relationships between pairs of gestures (to be discussed further in Section 7.2.4). A set of equations of motion for the coupled oscillator system is implemented using the task-dynamical coupled oscillator model of Saltzman and Byrd (2000), as extended by Nam and Saltzman (2003). The steady-state output of the coupled oscillator model is a set of limit-cycle oscillations with stabilized relative phases. From this output, the activation trajectories of the gestural score are derived as a function of the steady-state pattern of interoscillator phasings and a speech rate parameter. The settling of the coupled oscillator system from initial relative phases to the final steady-state phases (over several cycles) can be conceived as a real-time planning process. Nam (in press) has found that the model's settling time in fact correlates well with speakers' reaction time to begin to produce an utterance across variations in phonological structure.[6]

This method of controlling intergestural timing can be related to a class of generic recurrent connectionist network architectures (e.g., Jordan, 1986, 1990, 1992; see also Lathroum, 1989; Bailly *et al.*, 1991). As argued in Saltzman *et al.* (2006), an advantage of an architecture in which each gesture is associated with its own limit cycle oscillator (a "clock") and in which gestures are coordinated in time by coupling their clocks is that such networks will exhibit hallmark behaviors of coupled non-linear oscillators – entrainment, multiple stable modes, and phase transitions, all of which appear relevant to speech timing. As we attempt to show later, it is also possible to draw a connection between such behavioral phenomena and the qualitative properties of syllable structure patterning in languages. In addition, we will show that an explicit model based on these principles can also reproduce some subtle quantitative observations about intergestural timing and its variability as a function of syllable structure.

7.2.2 The basis of particulation in speech

Next, we must consider what makes articulatory gestures discrete (particulate) and what causes them to cohere structurally. Articulatory gestures control independent constricting devices or organs, such as the lips, tongue tip, tongue dorsum, tongue root, velum, and

---

[6] It is reasonable to ask why we would propose a model of speech production as apparently complex as this, with planning oscillators associated with each point-attractor vocal tract constriction gesture. Other models of speech production, at approximately this level, do not include such components. For example, in Guenther's (1995) model of speech production, timing is controlled by specifying speech as a chain-like *sequence* of phonemic targets in which articulatory movement onsets are triggered whenever an associated preceding movement either achieves near-zero velocity as it attains its target or passes through other kinematically defined critical points in its trajectory such as peak tangential velocity (e.g., Bullock *et al.*, 1993; Guenther, 1994, 1995). Several kinds of observations made over the last 15 years appear to present problems for simpler models of this kind when they attempt to account for the temporal structure of speech – regularities in relative timing between units, stochastic variability in that timing, and systematic variability in timing due to rate, speaking style, and prosodic context. For example, one observation that poses a challenge to such a serial model is the possibility of temporal sliding of some (but not all) production units with respect to one another (Suprenant and Goldstein, 1998). Another challenging observation is the existence of systematic differences in segment-internal gestural timing as a function of syllable position (Krakow, 1993) or as a function of prosodic boundaries (Sproat and Fujimura, 1993; Byrd and Saltzman, 1998). While it is, of course, possible that the simple serial model could be supplemented in some way to produce these phenomena, they emerge naturally in a model in which gesture-sized units are directly timed to one another in a pairwise fashion (Browman and Goldstein, 2000; Byrd and Saltzman, 2003; Nam and Saltzman, 2003).

glottis. Articulatory gestures of distinct organs have the capacity to function as discretely different. Even neonates (with presumably no phonological system) show sensitivity to the partitioning of the orofacial system into distinct organs (Meltzoff and Moore, 1997). Young infants will protrude their tongues or lips selectively in response to seeing an experimenter move those organs, and will, when initiating such an imitation, cease to move organs that are not participating in the imitative movement (Meltzoff and Moore, 1997). Imitation of the organ's action is not always accurate and the infant will show improvement across attempts, but the organ choice is correct. While there is yet little direct evidence that this kind of somatotopic organization of the orofacial system is deployed in speech behavior, the basis for it is strongly supported by the mimicry data.

Particulate units are discrete and can therefore be combined without loss of their distinguishing characteristics (Abler, 1989). Studdert-Kennedy (1998) hypothesizes that the distinct organs of the vocal system are the basis for particulation in language. Indeed, it can be shown that contrasts between articulatory gestures of distinct organs are the primary contrasts used by the phonologies of human languages to differentiate word forms (Goldstein and Fowler, 2003), and that the ability to perceive as distinct the actions of distinct vocal organs may not decline in adulthood (Best and McRoberts, 2003), in contrast to the often noted loss of ability to discriminate contrasts not present in one's native language. Also, there is a recent report (Polka *et al.*, 2001) of a consonant contrast that is not discriminated in an adult-like fashion at birth (unlike almost all others that have been tested) and that requires ambient language experience to develop full discriminability, and this is a within-organ contrast ("d" vs. "th").

Within one of these organs, articulatory gestures can be differentiated by the degree and location of a vocal tract constriction goal. For example, *tick*, *sick*, and *thick* all begin with a constriction made by the tongue tip organ but their constrictions can be differentiated in terms of constriction degree (*tick* versus *sick*) or in terms of constriction location (*sick* versus *thick*). Example gestural scores, schematically capturing the action representation of these words, are shown in Fig. 7.4. Each rectangle represents a gestural activation interval, which denotes the time interval during which a given gesture actively shapes movements of its designated organ/effector system (shown at the left of each row). Within the activation interval for each gesture is indicated the location within the vocal tract at which the organ (shown at left) is required to create a constriction if there is more than a single possibility (e.g., at the alveolar ridge or at the teeth) and the narrowness of that constriction (closure, narrow, or wide).

While these constriction parameters are, in principle, numerical continua, it is possible to model the emergence of discrete regions of these continua through self-organizing systems of agents that attune to one another (e.g., Browman and Goldstein, 2000; de Boer, 2000a, 2000b; Goldstein and Fowler, 2003; Oudeyer, 2003, 2005). These models divide a continuous constriction dimension into a large number of units that are evenly distributed across the continuum at the start of the simulation. The location of the units shifts during the simulation as a function of an agent's experience of its own productions and those of

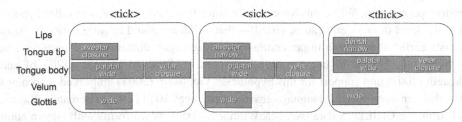

Figure 7.4 Gestural scores for "tick," "sick," and "thick."

its partner(s) via positive feedback – experience of a particular continuum value makes the agent more likely to produce that value again. Over time, the values of the units clump into a small number of modes. Critical to the clumping is a kind of categorical perception – an exact match is not required to inform the agent that some particular value has occurred in its experience. If exact matches are required (a behavior not found in real language perception), no clumping occurs.

When the agents interact through a constriction-acoustics map that harbors non-linearities (like those uncovered by Stevens, 1989), the nature of the clumping will be constrained by those non-linearities, and repeated simulations will produce modes in similar locations (Goldstein, 2003; Oudeyer, 2003). Languages would be expected to divide continua in these cases into similar regions, and this seems to be the case (e.g., constriction degree is associated with such non-linearities and is divided into comparable stop–fricative–glide regions in most languages). Other cases do not involve such non-linearities, and there may be different modal structures in different languages. For example, Hindi has a bimodal distribution of tongue tip constriction location values (dental vs. retroflex), where English has a single mode. It will be difficult, therefore, for English speakers to perceive the contrasting Hindi values as they are part of a single mode in English – they will be referred to the same mode. This can be the basis for Kuhl's *perceptual magnet effect* (Kuhl *et al.*, 1992) and the *single category* case of Best's (1995) *perceptual assimilation model* (which attempts to account for which non-native contrasts will be difficult to discriminate and which easy). It may also be difficult to acquire enough experience to bifurcate a well-learned mode into two. While speakers may learn to produce the contrasting second-language values by splitting a native language mode on the basis of explicit instruction and orosensory feedback, speakers' perceptions (even of their own productions) may still be influenced by their native language's modal structure.

Evolutionarily, this view predicts that systematic differentiation of an organ's constriction goals evolved *later* than systematic use of distinct organs. Distinct organs and the sounds produced when they move during vocal expiration existed independently of and in advance of any phonology (though neural machinery for linking action and perception of distinct organs may have evolved at some point), while the process of differentiation appears to require some interaction among a set of individuals already engaged in coupling their vocal actions to one another (and therefore possibly already using a primitive phonology). While this hypothesis cannot be tested directly, a parallel hypothesis at the level of ontogeny can be tested[7] – that children should acquire between-organ contrasts earlier than within-organ contrasts because organ differentiation requires that the infant must attune to her language environment. Studdert-Kennedy (2002b) and Goldstein (2003) find support for this hypothesis. Goldstein (2003) employed recordings of six children in an English language environment, ages $10:11:98^{8}$) (Bernstein-Ratner, 1984, from the CHILDES database, MacWhinney, 2000). Word forms with known adult targets were played to judges who classified initial consonants as English consonants. The results indicate that for all six children, the oral constricting organ in the child's production (lips, tongue tip, tongue body) matched the adult target with greater than chance frequency, even when the segment as a whole was not perceived by judges as correct. That is, the errors shared the correct organ with the adult form, and differed in some other properties, usually in constriction degree or location. Some children also showed significant matching of glottis and velum organs with adult targets. However, no child showed matching of within-organ differentiation of constriction degree (i.e., stop, fricative, glide) with greater than chance frequency. These results support the organ hypothesis. The child is matching her own organs to those perceived and, in doing so, is using organs to differentiate lexical items.

In sum, organs provide a basis for discreteness or particularity of articulatory gestures in space. Within-organ goals can particulate through self-organization in a population of agents, though children join a community in which the "choices" have already been made.

7.2.3 The coherence of gestures

Next, we must turn to a consideration of how articulatory gestures cohere into words. Word forms are organized "molecules" composed of multiple articulatory gestures (the "atomic" units). Gestural molecules are systematically patterned and harbor substructures that are cohesive (i.e., resistant to perturbation) (see, e.g., Saltzman *et al.*, 1995, 1998, 2000) and recurrent (appear repeatedly within a language's lexicon). The gestures composing a word can be organized in configurations that can differ somewhat from language to language (but as we argue below, these patterns are guided by a basic set of universal principles). Contrasting lexical items can be formed by replacing one atom with a different one, for example "bad" and "dad" in Fig. 7.5.

Note that the discreteness properties of gestures imply that these two words are phonologically and articulatorily equivalent except within the window of time in which

---

[7] See MacNeilage and Davis (2005) for an extended discussion of the validity of a recapitulationist position: "In both ontogeny and phylogeny, sound patterns are characterized by . . . [a stage in which they] are highly subject to basic constraints of biomechanical inertia and . . . [a stage of] partially overcoming these constraints in the course of developing lexical openness." They take the view that an advantage of vocal-origins theories of phylogeny is that they "can begin with the known outcome of language evolution – *current speech* – and use its *linguistic structure*, as seen in the course of ontogeny and in the structure of current languages, as a basis for inferring its phylogeny."

[8] $x:y$ means x years and y months of age.

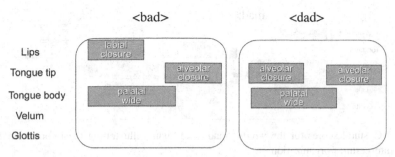

Figure 7.5 Gestural scores for "bad" and "dad," illustrating the informational significance of gestural selection.

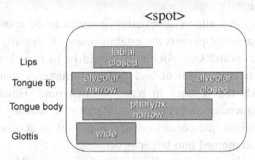

Figure 7.6 A gestural score for "spot" indicating the structure of the word initial consonants – two oral gestures and a single laryngeal abduction gesture.

the lips are being controlled in one and the tongue tip in the other. In this way, lexical differentiation is assured by gestural selection.

The organization of gestures into molecules appears to be an independent layer of structure that speakers of a language learn – it cannot be reduced to a concatenation of traditional segmental units. While traditional segments can be viewed as sets of gestures, it is not the case that the gestural molecule for a given word corresponds to the concatenation of the gesture sets for a sequence of segments. For example, in Fig. 7.6, we see the gestural score for "spot" in which a single gesture for voicelessness (a wide glottis) is coordinated with the oral gestures required to produce the two initial consonants – they do not each have a separate glottal gesture (as they do at the beginning of words like "saw" and "pa") and thus cannot be understood as the simple concatenation of an /s/ segment and a /p/ segment. Rather, the [sp] is an organized collection of two oral gestures and one glottal gesture centered between them. (This organization, incidentally, is the source of the lack of aspiration of voiceless stops seen in [sp], [st], and [sk] clusters, where segmental concatenation would otherwise lead one to expect aspiration.)

The pattern of intergestural relative timing for a given molecule is informationally significant and is represented by the pattern of activation intervals in a gestural score. For

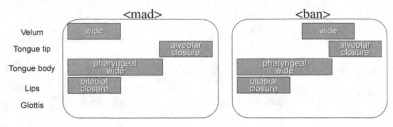

Figure 7.7 Gestural scores for the words "mad" and "ban," illustrating the informational significance of intergestural organization.

example, Fig. 7.7 shows that "mad" and "ban" are composed of the same gestures, but in a
different organization.

What is the "glue" that allows articulatory gestures to be coordinated appropriately
within a word form and that permits the establishment of lexically distinct coordination
patterns across word forms? One possibility would be to hypothesize that gestures are
organized into hierarchical segment and syllable structures that could serve as the
scaffolding that holds the gestures in place through time. However, these relatively
complex linguistic structures could only exist as part of an already developed phonology
and could not be available pre-phonologically as part of an account of how articulatory
gestures begin to be combined into larger structures.

Is there an alternative hypothesis in which the glue could exist in advance of a
phonology? One such proposal is the syllable "frame" hypothesis, offered by MacNeilage
and Davis (e.g., MacNeilage and Davis, 1990; MacNeilage, 1998) in which proto-
syllables are defined by jaw oscillation alone. However, as will be discussed below,
important aspects of syllable-internal structure are not addressed by such a frame-based
account. In contrast to hierarchical structure or jaw oscillation hypotheses, our proposal
is that relatively complex molecular organizations in adult phonologies are controlled
dynamically via coupling relations among individual gestures (Saltzman and Munhall,
1989). This is shown schematically in the coupling graph for "bad" in Fig. 7.3b, where the
lines indicate coupling relationships between pairs of gestures (solid and dashed are
different modes of coupling as discussed below, in-phase and anti-phase, respectively).

Thus, learning to pronounce a particular language includes not only learning the atomic
gestural components (the relevant tract variables/organs and the gestures' dynamic
parameter specifications) of the language but also tuning the overall dynamical system
to coordinate these atomic units into larger molecular structures. Coupled, non-linear
(limit-cycle) oscillators are used to control (or plan) the timing of gestural activations
within the molecules. The timing of the activation variables for a molecule's component
gestures (i.e., the molecule's gestural score) is specified or planned according to the
steady-state pattern of relative timing among the planning oscillators associated with
the individual gestures. Gestural molecules are represented using coupling graphs in
which nodes represent gestures and internode links represent the intergestural coupling

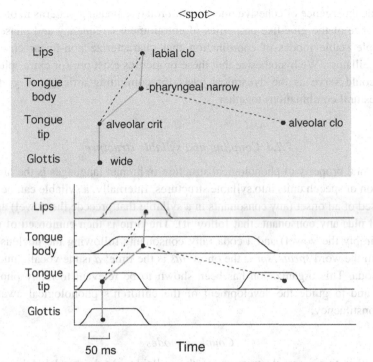

Figure 7.8 The coupling graph for "spot" (top) in which the tongue tip (fricative) gesture and the lip closure gesture are coupled (in-phase) to the tongue body (vowel) gesture, while they are also coupled to one another in the anti-phase mode. The pattern of gestural activations that results from the planning model is also shown (bottom). Lines indicate coupling relationships between pairs of gestures – solid and dashed are different in-phase and anti-phase coupling modes, respectively.

functions. In turn, these coupling graphs are used to parameterize a planning network for controlling gestural activations.

In the planning network, we adopt coupling functions originally developed by Saltzman and Byrd (2000) to control the relative phasings between pairs of gestures, and have generalized their model to molecular ensembles composed of multiple gestural oscillators (Nam and Saltzman, 2003), which can, in principle, exhibit competing phase specifications. For example, in the coupling graph for "spot" in Fig. 7.8 (top), both the tongue tip (fricative) gesture and the lip closure gesture are coupled (in-phase) to the tongue body (vowel) gesture, while they are also coupled to one another in the anti-phase mode. The basis for these couplings and evidence for them are discussed in the following sections. In Fig. 7.8 (bottom) the pattern of gestural activations that results from the planning model is added.

This generalized, competitive model has provided a promising account of intergestural phasing patterns within and between syllables, capturing both the mean relative phase values and the variability of these phase values observed in actual speech data (e.g., Byrd,

1996b). The emergence of cohesive intergestural relative phasing patterns in our model is a consequence of the glue-like properties of entrainment (frequency and phase locking) and multiple stable modes of coordination that characterize non-linear ensembles of coupled oscillators. We hypothesize that these properties exist pre-(or extra-)phonologically and could serve as the dynamical basis for combining articulatory gestures and holding gestural combinations together.[9]

7.2.4 Coupling and syllable structure

A fundamental property of phonological structure in human languages is the hierarchical organization of speech units into syllable structures. Internally, a syllable can be analyzed as composed of an onset (any consonants in a syllable that precede the vowel) and a rime (the vowel plus any consonants that follow it). The rime is then composed of a nucleus (usually simply the vowel) and a coda (any consonants following the nucleus). So, for example, in the word *sprats*, *spr* is the onset; *ats* is the rime; *a* is the vocalic nucleus, and *ts* is the coda. This organization has been shown to be relevant to many phonological processes and to guide the development of the children's phonological awareness of syllable constituency.

Coupling modes

We argue that this internal structure of the syllable can be modeled by a coupled dynamical system in which action units (gestures) are coordinated into larger (syllable-sized) molecules. In fact, we will show how syllable structure is implicit in coupling graphs such as the one shown in Fig. 7.8. The key idea is that there are intrinsically stable ways to coordinate, or phase, multiple actions in time, and in a model in which timing is controlled by coupling oscillators corresponding to the individual action units, the stable coordination possibilities can be related to stable modes of the coupled oscillators.

Ensembles of coupled non-linear oscillators are known to harbor multiple stable modes (Pikovsky *et al.*, 2003). Christiaan Huygens, a seventeenth-century Dutch physicist, noticed that pendulum clocks on a common wall tended to synchronize with each other. They come to exhibit the same frequency (1 : 1 frequency-locking) and a constant relative

---

[9] As stated earlier, gestural molecules corresponding to words can be characterized as being composed of organized gestural substructures (e.g., generally corresponding to segments and syllables) that are both recurrent in a language's lexicon and internally cohesive. We hypothesize that the internal cohesion of these substructures may be attributed to their underlying dynamics – the dynamics of (sub)systems of coupled nonlinear oscillators. We further hypothesize that their recurrence in a lexicon can be attributed to their functional utility. When viewed from this perspective, these substructures – more accurately, their corresponding coupling graphs – appear to play the role of network motifs (e.g., Milo *et al.*, 2002, 2004). Network motifs have been defined as recurring subpatterns of "interconnections occurring in complex networks at numbers that are significantly higher than those in randomized networks" (Milo *et al.*, 2002, p.824), that perform specific functions in the networks and that have been identified in networks of systems as seemingly diverse as ecological food webs, biological neural networks, genetic transcription networks, the World Wide Web, and written texts' word-adjacency networks. Understanding the combinatorial properties of these molecules then becomes the challenging problem of understanding the dynamical properties of their corresponding, underlying subgraphs, i.e., of understanding their graph-dynamics (see Farmer (1990) and Saltzman and Munhall (1992) for a discussion of graph-dynamics and its relation to the dynamics of a system's state-variables and parameters).

phase (phase-locking). In human bimanual coordination, limbs that start out oscillating at slightly different frequencies will similarly entrain in frequency and phase (e.g., Turvey, 1990). Certain modes are spontaneously available for phase-locking in interlimb coordination; these are 0° (in-phase) and 180° (anti-phase). (These two modes can be found in many types of simple systems.) Further, as frequency increases, abrupt transitions are observed from the less stable of these two spontaneous modes – 180° – to the more stable, in-phase mode – 0° (Haken *et al.*, 1985). Other phase-locks can be learned only with difficulty. In principle, arbitrary coupling relations can be learned, but we hypothesize that phonological systems make use of intrinsically stable modes where possible, and that the early evolution of phonological systems took advantage of these modes to begin to coordinate multiple speech actions.

Central to understanding syllable structure using a coupling model is the hypothesis that there are two basic types of gesture (in terms of their intrinsic properties, as discussed in the next section) – consonant and vowel – and that the internal structure of syllables results from different ways of coordinating gestures of these basic types. Consonant and vowel gestures can be hypothesized to be coordinated in either of the intrinsically stable modes: in-phase (the most stable) and anti-phase. We hypothesize that syllable-initial consonants and their following vowels are coordinated in phase with one another, and we can call this the *onset relation*.

Since the planning oscillators in our model determine the relative timing of the onsets of two coordinated gestures, in-phase coupling implies that the onsets of the two gestures should be synchronous. In the case of syllables that begin with a single consonant, we can see direct evidence for this in the speech kinematics. Figure 7.9 shows time functions of vocal tract variables, as measured using X-ray microbeam data, for the phrase "pea pots." Boxes delimit the times of presumed active control for the oral constriction gestures for /p/, /a/, and /t/, which are determined algorithmically from the velocities of the observed tract variables. As the figure shows, the onset of the lip gesture is approximately synchronous with the onset of the vowel gesture (within 25 ms).[10]

We hypothesize that this synchronous coordination emerges spontaneously in development because it is the most stable mode. Support for this is the early preference for consonant–vowel (CV) syllables (e.g., Stoel-Gammon, 1985). In contrast, we hypothesize that the coordination of a vowel with its following consonants – the coda relation – is an anti-phase coordination. As shown in Fig. 7.8, the onset of the /t/ in "pot" occurs at a point late in the control for the vowel gestures.

When *multiple* consonants occur in an onset, such as in the consonant cluster at the beginning of the word "spot," we assume that *each* of the consonants is coupled in-phase with the vowel (the syllable nucleus) – this is what makes them part of the onset. However, the consonant gestures must be at least partially sequential in order for the resulting form to be perceptually recoverable. Therefore, we have hypothesized

---

[10] In data from another subject with a receiver further back on the tongue, closer synchrony is observed. However, for this subject we do not have the parallel "spot" utterances for comparison below.

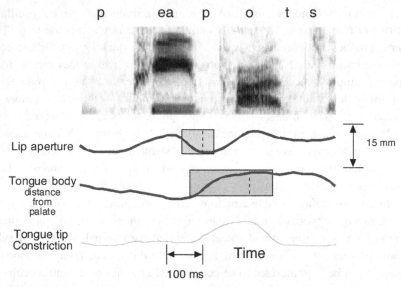

Figure 7.9 Time functions of vocal tract variables, as measured using X-ray microbeam data, for the phrase "pea pots" showing the in-phase (synchronous within 25 ms) coordination of the lip gesture for the /p/ in "pots" and the /a/ gesture for the vowel in "pots." Tract variables shown are *lip aperture* (distance between upper and lower lips), which is controlled for lip closure gestures (/p/ in this example) and *tongue tip constriction degree* (distance of the tongue tip from the palate), which is controlled in tongue tip gestures (/t/ and /s/ in this example). Also shown is the time function for the distance of the tongue body from the palate, which is small for /i/ and large for the vowel /a/, when the tongue is lowered and back into the pharynx. (The actual controlled tract variable for the vowel /a/ is the degree of constriction of the tongue root in pharynx, which cannot be directly measured using a technique that employs transducers on the front of the tongue only. So distance of the tongue body from the palate is used here as a rough index of tongue root constriction degree.) Boxes delimit the times of presumed active control for the oral constriction gestures for /p/ and /a/. These are determined algorithmically from the velocities of the observed tract variables. The left edge of the box represents gesture *onset*, the point in time at which the tract variable velocity towards constriction exceeds some threshold value. The right edge of the box represents the gesture *release*, the point in time at which velocity away from the constricted position exceeds some threshold. The line within the box represents the time at which the constriction *target* is effectively achieved, defined as the point in time at which the velocity towards constriction drops below threshold.

(Browman and Goldstein, 2000), that *multiple, competing* coupling relations can be specified in the network of oscillators in the coupling graph. For example, in the case of "spot," the oral constriction gestures of /s/ and /p/ are coupled in-phase to the vowel gesture and simultaneously anti-phase to one another, as shown in the coupling graph in Fig. 7.8. The coupled oscillator planning model (Nam and Saltzman, 2003) predicts that the onset of the vowel gesture should occur midway between the onset of the tongue tip gesture for /s/ and the lip gesture for /p/. As shown in Fig. 7.10 for the phrase "pea spots," kinematic data (for the same speaker as in Fig. 7.9) supports this prediction. Nam and

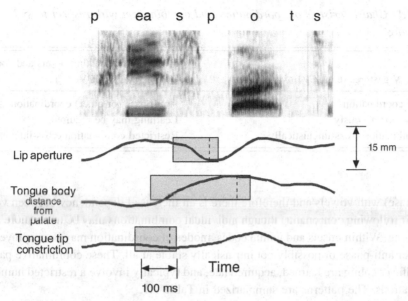

Figure 7.10 Kinematic data (for the same speaker as in Fig. 7.9) for the phrase "pea spots" showing that the onset of the vowel gesture occurs midway between the onset of the tongue tip gesture for /s/ and of the lip gesture for /p/. Boxes delimit the times of presumed active control for the oral constriction gestures for /s/, /p/, and /a/. (See Fig 7.9 caption for further details.)

Saltzman (2003) show that not only can the relative timing of onsets and codas be predicted by coupling networks with this kind of topology, but also their differential timing variability (Byrd, 1996b).

Coupling, combinatoriality, and valence

Traditional hierarchical syllable structure captures (but does not explain) systematic differences in combinatorial freedom of different syllable constituents. Syllable onsets and rimes combine relatively freely in most languages, as is seen in the English examples: *sight, blight, light, right, . . .; sip, blip, lip, rip.* Combinations of vowel and coda consonants are however somewhat more restricted than onset plus vowel combinations. For example, English allows a wider array of consonant sequences after a lax vowel: *ram, rap, ramp*, than after a tense vowel: *tame, tape, *taimp* (*indicates a sequence that is not permitted in the language). And finally, combinations of consonants within syllable onsets (or within syllable codas) are also more restricted (and disallowed in many languages) than onset-vowel combinations: *sl sn pl pr* are allowed in English onsets, but **pn *sr * tl* are not.

We hypothesize that coupling and phonological combinatoriality are related – free combination occurs just where articulatory gestures are coordinated in the most stable, in-phase mode. Consequently, onset gestures combine freely with vowel gestures because of the stability and availability of the in-phase mode. Coda gestures are in a less stable mode

Table 7.1 *Characteristics of coordination and combination with respect to syllable positioning*

Initial C, V gestures (e.g. [*CV*])	Final V, C and within onsets and codas (e.g., [*VC*], [*CCVCC*])
In-phase coordination	Anti-phase (or other) coordination
Emerges spontaneously	Learning may be required
Free combination cross-linguistically	Restricted combination cross-linguistically

(anti-phase) with vowels and therefore there is an increased dependency between vowels and their following consonants; though individual combinations may be made more stable by learning. Within onsets and within codas, modes of coordination may be employed that are either anti-phase or possibly not intrinsically stable at all. These coordinative patterns of specific coupling are learned, acquired late, and typically involve a restricted number of combinations. The patterns are summarized in Table 7.1.

What is it about consonant and vowel gestures that allows them to be initiated synchronously (in-phase) while multiple consonant gestures (for example, the /s/ and /p/ in the "spots") are produced sequentially (anti-phase)? A minimal requirement for a gestural molecule to emerge as a stable, shared structure in a speech community is that the gestures produced are generally recoverable by other members of the community. If the tongue tip and lip gestures for /s/ and /p/ were to be produced synchronously, then the /s/ would be largely "hidden" by the /p/ – no fricative turbulence would be produced. The resulting structure would sound much the same as a lip gesture alone and almost identical to a synchronously produced /t/ and /p/ gesture. Similar considerations hold for other pairs of consonant gestures with constriction degrees narrow enough to produce closures or fricative noise. It is hard to imagine how structures with these properties could survive as part of the shared activity of community members – what would guarantee that everyone is producing the same set of gestures, rather than one of the several others that produce nearly identical sounds (and even nearly identical facial movements). However, vowel and consonant gestures can be produced in-phase and still both be recovered because of two key differences between them. (1) Vowel gestures are less constricted than (stop or fricative) consonant gestures, so the vowel can be produced during consonant production without interfering with the acoustic properties of the consonant that signal its presence (because the more narrow a constriction, the more it dominates the source and resonance properties of an acoustic tube). (2) Vowel gestures are formed more slowly and are active longer than consonant gestures, so they dominate the acoustics of the tube during a time with no overlapping or only weakly competing consonant gestures. Mattingly (1981) makes a very similar point, arguing that CV structures are efficient in that they allow parallel transmission of information, while remaining recoverable.

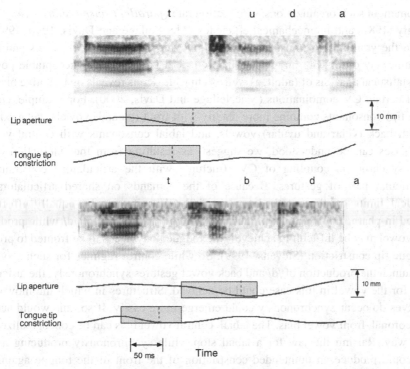

Figure 7.11 Example words /tuda/ "suffice" (top) and /tbda/ "begin" (bottom) from Tashlhiyt Berber, spoken in Morocco. The tongue tip and lip gestures are produced in-phase (synchronous at onset) in /tuda/, but not in /tbda/, where the production of the lip closure is initiated at the release of the [t]'s tongue tip gesture–anti-phase coordination. (See Fig. 7.9 caption for further details.)

To illustrate this point, consider an example from Tashlhiyt Berber (spoken in Morocco), a language that allows long strings of consonant gestures; words can even consist entirely of consonants. The Berber words /tuda/ "suffice" and /tbda/ "begin" are quite similar, differing primarily in the constriction degree of the lip gesture that follows the initial tongue tip closure and its duration – the lip gesture for /u/ is much less narrow than that for /b/ and is longer. As the kinematic data in Fig. 7.11 show,[11] the speaker produces the tongue tip and lip gestures in-phase (synchronous at onset) in /tuda/ (top part of figure), but not in /tbda/ (bottom part), where the production of the lip closure is initiated at the release of the [t]'s tongue tip gesture – anti-phase coordination.

Vowel gestures and consonant gestures are typically produced in-phase, as in this example and as in Fig. 7.8, but for multiple consonant gestures, this kind of coordination does not readily occur. Thus, the distinguishing properties of vowel and consonant gestures, together with the stability of in-phase coupling gives rise to their *valence* – they combine freely with each other in onset relations. Other reasons also support the

[11] These data were collected in collaboration with Catherine Browman, Lisa Selkirk, and Abdelkrim Jabbour.

development of such organizations: information rate (*parallel transmission*), as argued by Mattingly (1981) and biomechanical efficiency (MacNeilage and Davis, 1990, 1993).

While the grammars of most languages allow free combination of onsets and rimes (combining any onset with any rime will produce a grammatically acceptable possible word), statistical analysis of (adult, as well as child) lexicons reveals quantitative biases in favor of certain CV combinations (MacNeilage and Davis, 2000). For example, coronal (tongue tip) consonants combine more readily with front (palatal) vowels, velar consonants with back (velar and uvular) vowels, and labial consonants with central vowels. These biases can be understood, we suggest, as resulting from the interaction of the desired synchronous coupling of CV structures with the articulatory constraints of the particular pair of gestures. Because of the demands on shared articulators and anatomical limits on those articulators, not all CV pairs can be equally effectively produced in-phase. For example, producing a tongue tip gesture for /d/ while producing a back vowel may be difficult to achieve – the tongue body needs to be fronted to produce the tongue tip constriction (Stevens, 1999). So while control regimes for such a syllable might launch the production of /d/ and back vowel gestures synchronously, the articulator motion for the vowel in this case could be delayed. Structures in which the movements themselves do occur synchronously could emerge more easily. If so, this would account for the coronal–front vowel bias. The labial–central vowel bias can be conceptualized in a similar way. Raising the jaw for a labial stop while synchronously producing a front vowel could produce an unintended constriction of the front of the tongue against the palate, and synchronous production with a back vowel could produce an unintended velar constriction. Careful modulation of the amount of jaw raising in these contexts or reposturing of the tongue would be required to prevent this. Movements for labial consonants and central vowels, however, can be synchronized without these problems.

Evidence that these biases are indeed due to difficulties in synchronization can be found in the fact that the pattern of biases observed in adult languages in CV structures are absent in VC structures (MacNeilage *et al.*, 2000). VC structures are hypothesized to be coupled in an anti-phase mode (above, p. 233), rather than an in-phase mode, so synchronization difficulties would not be predicted to be relevant. While these bias patterns are exhibited in VC structures in children's early words (Davis *et al.*, 2002), this could be due to the overwhelming preference for reduplicative C sequences at this stage, so the CV biases are effectively transferred to the VC. This interpretation of the CV bias is compared to that of MacNeilage and Davis (below, pp. 237–239), along with a general comparison of views on the origins of the syllable and syllable structure.

Segment-level valences

While closure gestures of the lips, tongue tip, and tongue body cannot be produced synchronously with one another without compromising recoverability, there are other gestures (in addition to vowel gestures) that can successfully be produced synchronously with these constrictions. These are gestures of the velum (lowering) and glottis (abduction or closure). The nature of velic and glottal gestures is such that their acoustic consequences are fairly independent of a concurrently produced lip, tongue tip, or tongue body constriction, so both gestures are readily recoverable even if they are synchronous at both onset and offset. And of course such combinations of gestures occur frequently among languages and are usually described as constituting single *segments*. The basis for the single-segment transcription may lie in the fact that these combinations are typically produced in-phase and are roughly equivalent in duration, so they occupy the same interval of time.

When a multi-gestural segment, such as a nasal stop (e.g., [m]), is produced in a syllable onset, all three necessary gestures can be (and are) produced in-phase – the oral closure, the velum lowering, and the vowel gesture – and this is clearly the most stable structure for that combination of gestures. Direct articulatory evidence for synchronous production of these events in onsets has been presented by Krakow (1993). If the nasal consonant is in coda, however, the oral constriction gesture for the consonant must be produced in an anti-phase relation to the vowel, as argued above, so any of the remaining possibilities of coordinating the three gestures is less stable than the onset pattern, and it is not clear which of these suboptimal patterns would be the most stable. If the velum lowering gesture is synchronized with the oral closure, then it necessarily is anti-phase with the vowel. Or if it is synchronized with the vowel, then it is anti-phase with respect to the consonantal oral gesture. This leads to the predictions that (a) the segment-internal coordination of gestures may be different in coda than in onset, and that (b) more than one stable pattern may be evidenced in coda cross-linguistically.

Evidence is available in support of both predictions. With respect to (a), a number of recent articulatory studies have demonstrated systematically different segment-internal gestural timing relations depending on whether a segment appears in the onset or coda of the syllable (see Krakow (1999) for a review). For example, velum and oral constriction gestures for nasals in English are coordinated sequentially in coda, rather than the synchronous coordination observed in onset. The multiple gestures of the English liquid consonant [l] (tongue tip raising and tongue rear backing) show a very similar difference in coordination pattern (Delattre, 1971; Sproat and Fujimura, 1993; Browman and Goldstein, 1995; Krakow, 1999) as a function of syllable position. These differences in coordination result in what have traditionally been described as examples of allophonic variation – nasalization of vowels before coda nasals and velarization of /l/ in coda. While in traditional formal analyses the allophonic variants for nasals and /l/s are unrelated to each other, they are in fact both superficial consequences of the deeper generalization that onset and coda relations involve distinct coupling modes (Browman and Goldstein, 1995).

Evidence for (b), the availability of multiple stable structures in coda, is found in the fact that languages may differ in which of the predicted stable patterns they employ. Direct articulatory evidence for anti-phase coordination between the velum and oral closure gestures has been found in English, and similar indirect evidence is available from many languages in the nasalization of vowels in the context of a coda nasal (Schourup, 1973; Krakow, 1993). However, such nasalization of vowels in the context of coda nasals is not present in aerodynamic data from French (Cohn, 1993) and various Australian languages (Butcher, 1999). For the latter languages, the velum lowering and oral closure gestures are apparently coordinated synchronously in coda.

In the within-segment case, we can also see a correlation between combinatoriality and the stability of coordination mode, like that observed in the case of CV versus VC coordination. In onset, oral closures can combine relatively freely with glottal gestures or with velic gestures. Such combinatoriality is part of the basis for the traditional decomposition of segments into *features*. In coda, however, such combinatoriality may be lacking in some languages. Many languages, for example, allow only nasal consonants in coda, requiring that an oral closure and a velic lowering gesture occur together, or do not occur at all (for example Mandarin Chinese). Interestingly, the coda consonants in Australian languages that show relatively synchronous coordination of oral constrictions and velic lowering in coda (Butcher, 1999) are highly combinatorial. There are four to six different places of articulation (labial, dorsal, and as many as four distinct coronals), and these can all occur with or without nasalization in coda. Butcher attributes the synchronous coordination to the functional requirement for keeping the vowel in pre-nasal position completely non-nasalized, so that the formant transitions effectively distinguish the many places of articulation (Tabain *et al.*, 2004). However, in a broader perspective, this observation can be seen as combinatoriality co-occurring with synchronous coordination.

Language differentiation

As is clear from several examples presented in the last two sections, patterns of coordination of speech gestures may differ from language to language. CV syllables have been argued to be the most stable coordination pattern, as the component actions are coordinated in-phase. Indeed such patterns occur in every language, and there are languages in which CV is the only syllable available (e.g., Hawai'ian). However, there are languages that also employ coda consonants (anti-phase coordination) and sequences of consonants, including languages like Georgian where sequences of three or four consonants can begin a syllable (Chitoran, 2000) and like Tashlhiyt Berber where syllables can be composed exclusively of consonants (Dell and Elmedloui, 1985, 2002). Indeed Tashlhiyt Berber can be analyzed as having no restrictions on segment sequencing (no phonotactic constraints). How do such hypothetically suboptimal structures (from the point of view of coordination stability) arise, and how are they sustained?

It is a reasonable speculation that CV syllables were the first to arise in the evolution of phonology (MacNeilage and Davis, 2005). Gestures such as lip smacks and tongue smacks, observed in non-human primates, could have spontaneously synchronized with phonated vowel-like constrictions to form early CV or CVCV words. Once a lexicon evolved, however, other forces could come into play. First, there is an inverse power law relation between word-length (measured in segments) and word frequency (Zipf, 1949).

Frequently used words can become shorter by loss of one or more of their gestures, for example, the final V in a CVCV structure. The result would be CVC structure. Loss of the first of the vowels would result in a CCV structure. Examples of vowel loss of these types do occur in historical sound change in languages (apocope and syncope, respectively); less stable coordination patterns could arise this way. While they are suboptimal with respect to stability, they satisfy a competing constraint on word form. In this way languages with different patterns of coordination and combination could arise. Second, the existence of a well-learned lexicon of molecules imparts its own stability to the shared forms. So even if a coda consonant is not the most intrinsically stable organization, repeated instances of this in the words of a language make it a sufficiently stable mode. In fact it is possible to view the occurrence of speech errors as the result of competition between the intrinsic modes and the learned patterns associated with particular lexical items (Goldstein *et al.*, in press).

The regularities underlying the coordination patterns (syllable structures) of a particular language form part of its phonological grammar, an account of speakers' knowledge of the possible word forms of the language and of the regular correspondences among words or morphemes in different contexts. One model of phonological grammar, *optimality theory*, calls on a set of rank-ordered constraints (Prince and Smolensky, 2004). Lower-ranked constraints may be violated to satisfy higher-ranked constraints. The hierarchy of constraint ordering can differ from language to language. Constraints include preferences for certain types of structures ("markedness" constraints) and for keeping the various phonological forms of a particular lexical item as similar as possible ("faithfulness" constraints). Cross-linguistic differences in constraint ranking result in different segmental inventories, different syllable structures, or different alternations in a word's form as a function of context. In the domain of syllables and syllabification, constraints include injunctions against coda consonants and against onset clusters. In recent work, some markedness constraints have been explicitly cast as constraints on gestural coordination (Gafos, 2002), and the preference for synchronous coordination has been argued to play a role in phonological grammar (Pouplier, 2003; Nam, in press) and could be the basis for the "Align" family of constraints that require edges of certain structures to be coincident. One phonological process that can be related to a preference for the most stable coordination pattern is *resyllabification*. A word-final consonant before a word beginning with a vowel can be resyllabified with the following vowel (so that *keep eels* is pronounced like *key peels*), and increases in speaking rate can lead to a greater tendency for such resyllabifications (Stetson, 1951). This phenomenon can be analyzed as an abrupt transition to a more stable coordination mode (Tuller and Kelso, 1991; de Jong, 2001).

The emergence of the syllable: mandible-based frame or synchronous coupling?

Like the present view, the frame-content theory (e.g., MacNeilage, 1998; MacNeilage and Davis, 1999, 2000, 2005) attempts to explain the evolution of phonological structure and its emergence in infants on the basis of general anatomical, physiological, and evolutionary principles, rather than on domain-specific innate endowments. In their theory, the syllable develops out of (biomechanical) mandibular oscillation, an organized behavior already present in non-human primates subserving the function of mastication. The infant's babbling and early words are hypothesized to result almost exclusively from jaw oscillation with no independent control of the tongue and lips at the timescale of consonants and vowels. This oscillation constitutes the syllable frame (one cycle of oscillation is a syllable), while control over individual consonants and vowels is the content, which is hypothesized to develop later. The theory predicts that the CV pattern-ing in early syllables should not be random, but rather there should be systematic associations as described above: coronals with front vowels, dorsals with back vowels, labials with central vowels. The basis of the prediction is that if the tongue happens to be in a relatively advanced static posture when the jaw begins to oscillate, the front part of the tongue will make contact with the palate when the jaw rises, thus producing (pas-sively) a coronal constriction. When the jaw lowers, the fronted tongue will produce a shape appropriate for a front vowel. When there is no tongue advancement, the lips (rather than the tongue tip), will form a passive constriction, and the lowered jaw will produce a central vowel shape. A retracted tongue will produce a dorsal constriction and back vowel when the jaw oscillates. These predictions are borne out in data of CV combinations in babbling (Davis and MacNeilage, 1995) and early words (Davis *et al.*, 2002); the predicted combinations consistently show ratios of observed to expected frequencies of greater than 1, while most other combinations show ratios less than 1.

As attractive and elegant as this theory is, it seems to us to suffer from some empirical problems connecting data and theory. While it is true that the preferred CV syllables *could* be produced by moving only the jaw,[12] there is no direct evidence that the infants are in fact producing them in that way. And there is some indirect evidence that they are not. First, there are many syllables that the infants produce that are not of one of the preferred types – the preferences are reliable but are relatively small in magnitude. Some independent control of tongue and/or lips is required to produce these other patterns. Second, there appears to be no developmental progression from exclusively preferred CVs early on to a more varied set later. Perhaps this is just a question of the appropriate analyses having not yet been performed. A preliminary test for age grading (comparing 6, 9, and 12 months) using the Haskins babbling database does not reveal such a developmental trend, but more data needs to be examined. Finally, MacNeilage *et al.* (2000) show that similar observed-to-expected ratios are observed in the lexicons of ten (*adult*) languages. But we know that adults do *not* produce CV syllables by moving only the jaw (even when they produce the preferred syllable types). They have independent control over vowel and consonant constrictions. While it is possible that mandibular oscillation is part of the explanation for this CV preference in adult languages (more on this below) and that the preferences are inherited from childhood, the empirical

---

[12] In MacNeilage and Davis (2005), the claim that certain syllables are produced exclusively by jaw movement is made in explaining why labials should be a preferred consonant – they can be produced with only jaw activity.

point here is that the existence of these CV preferences cannot constitute *evidence* for a jaw-only motor control strategy, since the preferences exist in adult languages but the jaw-only strategy does not.

Another weakness of the frame-content theory compared to the current view is that it provides no account of differences between onset and rime consonants – in their timing, their variability, and their combinatoriality. As we have argued above, these can be accounted for in a principled way in the oscillator coupling model. Relatedly, the coupling model predicts the lack of VC associations in adult languages (as discussed above), while this does not follow from the frame-content theory.

We outlined above (p. 234) an alternative account of the CV preferences in adult languages, based on the hypothesis that synchronous production of consonant and vowel constrictions can be successfully achieved more easily in the preferred CV patterns than in other combinations. This could account for the preferences in infants as well, if we assume, contra MacNeilage and Davis, that infants are synchronizing multiple actions. This account would not suffer the same problem as that of MacNeilage and Davis, in that no assumption of jaw-only control is made. Nonetheless, there is at least one potentially explanatory aspect of MacNeilage and Davis' jaw-based account that the current account lacks – an explanation for the overall duration of a syllable. In their account, syllable production rate should correspond to an oscillatory mode (natural frequency) of the jaw. Thus, it would be desirable to integrate the accounts in some way. One possibility is to hypothesize that infants are synchronously coupling constriction-directed activity of the tongue and lips along with jaw oscillation. Given the more massive jaw, its natural frequency would be expected to dominate in the coupling.

It is also interesting to consider how the jaw could play a role in determining the preferred CV patterns in this hybrid account. The preferred CV combinations not only afford in-phase productions of consonant and vowel, but they also have the characteristic that the compatibility of the vowel and consonant constrictions is such that when they are produced synchronously, jaw raising and lowering can assist in the production of both constrictions. In contrast, in a synchronous combination of a non-preferred pattern (for example coronal consonant and back vowel), competing demands on the jaw from the two synchronous constrictions would cancel each other out, and less jaw movement would be expected. In general, since the jaw is massive, system efficiency would be enhanced by being able to take advantage of jaw movement in production of consonant and vowel constrictions.

7.3 Phonology as gestural structure: implications for language evolution

The hypothesis that stored lexical forms are molecules built through coupling of dynamical vocal tract constriction gestures can provide an approach for understanding: how a combinatorial phonological system might have evolved, why phonologies tend to have the kinds of units that they have, and why phonologies tend to have the types of combinatorial structure that they have.

If we assume action recognition to be important in language evolution (and acquisition), an understanding of speech production as relying on dynamically defined and organized vocal tract actions has a direct bearing. We agree with Davis *et al.* (2002) that "mental representation cannot be fully understood without consideration of activities available to the body for building such representations . . . [including the] dynamic characteristics of the production mechanism." Phonological structures can emerge through self-organization as individuals interact in a community, though certain biological preconditions are necessary (Studdert-Kennedy and Goldstein, 2003). Emergent representations rely on a universal (i.e., shared) set of organs moving over time. This initial organ-based representational space constrains possible future phonologies. Later differentiation of this representational space allows for phonological grammar to emerge (Studdert-Kennedy and Goldstein, 2003).

A necessary precondition for the evolution of language is that there exist a compatibility and complementarity of perception and action (Turvey, 1990; Liberman, 1996). Such complementarity can be demonstrated in a much wider range of contexts than just language. Perception guides action in many obvious ways (e.g., obstacle avoidance in locomotion and in reaching), but action systems are also involved in perception (at least in humans), particularly when the perceived act is self-produced or is a type of action that could be self-produced, for example, an act produced by a conspecific. Galuntucci *et al.* (in press) have summarized evidence for such effects in a variety of domains.

7.3.1 Mirror neurons and the complementarity of perception and action

Given the functionally integrated nature of perception and action (see, e.g., Barsalou, 1999; Hommel *et al.*, 2001, cited in Dale *et al.*, 2003), it is not surprising to find common neural units active during both the perception and performance of the same or related acts (cf. Rizzolatti and Arbib, 1998). But there is no reason to think that complementarity is limited to those tasks for which mirror neurons have (so far) been uncovered or that the ones uncovered so far are in any way primary (see the discussion in Dale *et al.*, 2003; Oztop *et al.*, this volume). Also, while there has been a tendency to focus on the role of *visual* and particularly manual information as providing the sensory input to such mirror neurons, recent evidence has shown that auditory information associated with the performance of some acts may also elicit responses in mirror neurons active during the performance of those acts (Kohler *et al.*, 2002; see also Romanski and Goldman-Rakic, 2002), that mouth movements and tactile stimulation of the mouth yield a response by specific inferior F5 neurons (Rizzolatti *et al.*, 1981), and that there are mirror neurons relating ingestive behavioral and orofacial communicative acts (Ferrari *et al.*, 2003). This is consistent with behavioral evidence of the functional equivalence in perception of the multiple lawful sensory consequences of an act (for example the "McGurk" effect in speech perception: McGurk and MacDonald, 1976; Massaro *et al.*, 1996; Fowler and Dekle, 1991; Rosenblum and Saldaña, 1996).

Types of action and action recognition

Two kinds of action/action recognition scenarios might be relevant in considering the path followed in the evolution of language. First, an act *on* the environment, for example a grasp, might be executed in the presence of a perceiver. Such an act is generally visible, often limb-related, and may result in the movement of an object by an agent. The recognition of this act might be considered to proceed from the visual to the conceptual, and its semantic aspects are relatively transparently recoverable from the sensory information available to the perceiver. A different kind of action to consider might be an act *in*, rather than *on*, the environment – a body-internal act like vocalization would be an example. Such an act would likely (though not necessarily) involve body-internal, largely non-visible actions and result in a perceivable signal with at least the potential of arbitrary semantic content. Its recognition by a perceiver would, in the case of vocalization, proceed from the auditory to the conceptual, and if semantically arbitrary, would be *mediated* by a lexicon of sorts. The first type of action – action *on* the environment – may well be the evolutionary source of syntax, that is, the linguistic characterization of how meaningful words are patterned to convey information. The second type of action – action *in* the environment – is a possible evolutionary source of phonology, the linguistic characterization of how vocal tract actions are patterned to create words. It is this latter *phonological* type of action and action recognition that we focus on – in particular, how human articulatory gestural actions can encode words and the implications of these considerations for understanding the evolution of phonology. But we suggest that both these evolutionary processes (syntax and phonology) could have gone on in parallel and that vocal gestures and manual gestures could be produced concurrently. Thus, we do not view the process of language evolution as having necessarily undergone a massive shift from manual to vocal (a shift that MacNeilage and Davis (2005) also find implausible). Rather the modalities have been complementary.

Action tasks and perceptual objects

The compatibility and complementarity of perception and action suggests that it is profitable to decompose an animal's behavior into functionally defined *tasks* that can be given a formal description integrating both their motor and multi-modal sensory consequences. For example, in the case of speech, we hypothesize that gestural task-space, i.e., the space of articulatory constriction variables, is the domain of confluence for perception-oriented and action-oriented information during the perception and production of speech. Such a task-space would provide an informational medium in which gestures become the objects shared by speech production and perception, atomic or molecular (Liberman, 1996; Goldstein and Fowler, 2003; Galantucci *et al.*, in press). In this regard, speech's task-space defines a domain for coupling acting and perceiving in a manner similar to that proposed for the locomotory navigation task-space of Fajen and Warren (2003) and to the common coding principle for perception and action of

Prinz (1997). It is possible that a mirror neuron system could be the neurophysiological instantiation of this cognitive coupling, though the identification of such a specific system in humans remains an outstanding challenge.

From this point of view, evolution of a new function (such as language) can be seen as a process in which previously existing tasks (or coherent parts thereof) are recruited and combined into a new pattern of organization (see Farmer, 1990; Saltzman and Munhall, 1992; and also footnote 9). This is the familiar self-organization approach to the ontogenesis of locomotion skills (Thelen, 1989) applied at a longer timescale: "development proceeds . . . as the opportunistic marshalling of the available components that best befit the task at hand. Development is function-driven to the extent that anatomical structure and neurological mechanisms exist only as components until they are expressed in a context. Once assembled in context, behavior is, in turn, molded and modulated by its functional consequences" (Thelen, 1989, p.947). In the case of language, we offer the possibility that elements of manual tasks (e.g., from grasping) and orofacial tasks (e.g., from food ingestion or emoting) are recombined in the evolution of language. Each of the tasks provides an important component – syntactic and phonological, respectively – of evolving language use.

Grasping tasks can be employed (in the absence of an actual object actually being present) to represent actions symbolically, thus providing a basis for reference and semantics. Pantomime can convey a great deal of information prior to the development of arbitrary conventions (Arbib, 2005; Chapter 1, this volume). Such information would have been crucial in developing the use of symbols, indexicality ("mine," "yours," "theirs"), and possibly concepts of events and complex cause-and-effect. This type of action was likely necessary for the evolution of syntax, i.e., the structured patterning of words to convey meaning. But pantomimic actions formed with the hands (and attached limbs) without more sophisticated elaboration (such as that found in signed *languages*, in which the actions are elaborated in signing space and are no longer pantomimic) do not provide a ready source of discreteness to differentiate similar actions with different symbolic meaning (e.g., similar-looking objects; actions that took place in the past or might occur in the future). The fingers are, of course, intrinsically discrete units, but it is hard to use them independently while still performing a grasping-related action with the hand as a whole. Further, there are several ways in which the grasping task does not provide an ideal scaffolding, at least alone, for the evolution of language (see, e.g., Corballis, 2002, 2003): a reliance on (proto)sign is problematic when there is no direct line of sight between individuals, a situation which must be assumed to have existed for our ancestors, in the trees, in the dark, or over moderate distance. In fact, it is in just these circumstances that communicating information about potential danger or food is likely to be most vital. Falk (2004; cited in MacNeilage and Davis, 2005) suggests that mothers' need to be able to undertake parental care at a distance fostered dyadic vocal communication. Manual communication is also problematic with occupied hands, such as during foraging, grooming, tool use, or child-care – all, one would think, frequent activities of (somewhat) smart early primates. (See, however, Emmorey, this volume, for an alternative view.)

Tasks engaging the organs of the face and mouth might have been recruited to supplement manual tasks in the evolution of spoken language. The partitioning of the face and mouth into distinct organs and the association of their actions with reliable structuring of optic and acoustic media (lip-smack vs. tongue-smack; lip protrusion vs. no lip protrusion) afford the ability to distinguish ambiguous meanings by using discrete non-iconic actions. The relative independence of the organs when producing constrictions of the vocal tube, the existence of intrinsically stable modes of interorgan action coupling, and the physics of sound generation allows articulatory gestures to combine readily into larger combinations with distinctive structuring of the acoustic and visual media. The combinatorial possibilities would presumably increase with anatomical changes in hom inid vocal tract, such as we see in ontogeny, with the tongue body and tongue tip becoming independently useful in constriction formation (Vihman, 1996). Further, organ-specific mirror neurons for orofacial actions in the F5 of macaque monkeys have been discovered with selective tuning to particular orofacial organs or their combination (Ferrari *et al*., 2003); different neurons appear to be sensitive to different orofacial tasks: grasping, ingesting of food, communicative acts.

Thus, we see a possible direct source of phonological evolution, i.e., the patterning of (non-meaningful) particulate action units to form (meaningful) words, *as distinct from* the development of symbolic thinking and/or syntax. This development could be expected to enhance syntactic evolution by (a) allowing a larger and more discrete set of word forms and grammatical markers and (b) freeing the specifically syntactic and semantic processes from the parallel responsibility of generating multiple word forms. Eventually, the rich phonological differentiation of words could have made accompanying manual iconic acts redundant (see also Corballis, 2003).

7.4 Summary

Whether or not our particular speculations deserve further serious consideration, we think it is important to draw attention to the more general consideration of the potential independence of phonological evolution in theorizing as to how language evolved. Arbib (2005; Chapter 1, this volume) addresses the question of language evolution by treating it as a progression from protosign and protospeech to languages with full-blown syntax and compositional semantics, but this view says little about the phonology of protosign and critically neglects to consider the emergence of duality of patterning as a hallmark characteristic of language.

In this chapter we have proposed that the evolution of syntax and of phonology arose from different sources and ultimately converged in a symbiotic relationship. We argue that phonological evolution crucially requires the emergence of particulate combinatorial units. We suggest that articulatory *gestures* are well-suited to play a direct role in phonological evolution because, as argued by Studdert-Kennedy (2002a), they are typically non-iconic and non-meaningful, yet discrete. The lack of iconicity of vocal gestures, rather than being a weakness for language evolution, is *advantageous specifically for phonological evolution*. The combination of syntactically meaningful actions with phonologically particulate non-meaningful actions can be speculated to have symbiotically generated the evolution of language. The existence of duality of patterning as a hallmark characteristic of human languages indicates that both components of the evolution of language – the syntactic and the phonological – are robustly present and necessary. Significantly, however, syntax and phonology may have originally evolved along different pathways.

Acknowledgments

The authors gratefully acknowledge the support of the National Institutes of Health and thank Michael Arbib, Barbara Davis, and Karen Emmorey for their helpful comments. This work was supported by NIH grants DC-03172, DC-00403, and DC-03663.

References

Abler, W. L., 1989. On the particulate principle of self-diversifying systems. *J. Soc. Biol. Struct.* **12**: 1–13.

Arbib, M. A., 2005. From monkey-like action recognition to human language: an evolutionary framework for neurolinguistics. *Behav. Brain Sci* **28**: 105–124.

Bailly, G., Laboissière, R., and Schwartz, J. L., 1991. Formant trajectories as audible gestures: an alternative for speech synthesis. *J. Phonet.* **19**: 9–23.

Barsalou, L., 1999. Perceptual symbol systems. *Behav. Brain Sci.* **22**: 577–660.

Bernstein-Ratner, N., 1984. Phonological rule usage in mother–child speech. *J. Phonet.* **12**: 245–254.

Best, C. T., 1995. A direct realist perspective on cross-language speech perception. In W. Strange and J. J. Jenkins (eds). *Cross-Language Speech Perception*. Timonium, MD: York Press, pp. 171–204.

Best, C. T., and McRoberts, G. W., 2003. Infant perception of nonnative contrasts that adults assimilate in different ways. *Lang. Speech* **46**: 183–216.

Browman, C. P., and Goldstein, L., 1990. Tiers in articulatory phonology, with some inplications for casual speech. In J. Kingston and M. E. Beckman (eds.) *Papers in Laboratory Phonology*, vol. 1, *Between the Grammar and Physics of Speech*. Cambridge, UK: Cambridge University Press, pp. 341–376.

1992. Articulatory phonology: an overview. *Phonetica* **49**: 155–180.

1995. Dynamics and articulatory phonology. In T. van Gelder (ed.) *Mind as Motion Explorations in the Dynamics of Cognition*. Cambridge, MA: MIT Press, pp. 175–194.

2000. Competing constraints on intergestural coordination and self-organization of phonological structures. *Bull. Commun. Parlée* **5**: 25–34.

Bullock, D., Grossberg, S., and Mannes, C., 1993. A neural network model for cursive script production. *Biol. Cybernet.* **70**: 15–28.

Butcher, A. R., 1999. What speakers of Australian aboriginal languages do with their velums and why: the phonetics of the nasal/oral contrast. *Proceedings 16th International Congress of Phonetic Sciences*, Berkeley, CA, pp. 479–482.

Byrd, D., 1996a. A phase window framework for articulatory timing. *Phonology* **13**: 139–169.

1996b. Influences on articulatory timing in consonant sequences. *J. Phonet.* **24**: 209–244.

Byrd, D., and Saltzman, E., 1998. Intragestural dynamics of multiple phrasal boundaries. *J. Phonet.* **26**: 173–199.

2003. The elastic phrase: dynamics of boundary-adjacent lengthening. *J. Phoneti.* **31**: 149–180.

Chitoran, I., 2000. Some evidence for feature specification contraints on Georgian consonant sequencing. In O. Fujimura, B. Joseph and B. Palek (eds.) *Proceedings of LP 98*, pp. 185–204.

Cohn, A. C., 1993. The status of nasalized continuants. In M. Huffman and R. Krakow (eds.) *Nasal, Nasalization, and the Velum.* San Diego, CA: Academic Press, pp. 329–367.

Corballis, M. C., 2002. *From Hand to Mouth: The Origins of Language.* Princeton, NJ: Princeton University Press.

2003. From mouth to hand: gesture, speech, and the evolution of handedness. *Behavi. Brain Sci.* **26**: 199–260.

Dale, R., Richardson, D. C., and Owen, M. J., 2003. Pumping for gestural origins: the well may be rather dry. *Behav. Brain Sci.* **26**: 218–219.

Davis, B. L., and MacNeilage, P. F., 1995. The articulatory basis of babbling. *J. Speech Hear. Res.* **38**: 1199–1211.

Davis, B. L. and MacNeilage, P. F., and Matyear, C. L., 2002. Acquisition of serial complexity in speech production: A comparison of phonetic and phonological approaches to first word production. *Phonetica* **59**: 75–107.

de Boer, B., 2000a. Self-organization in vowel systems. *J. Phonet.* **28**: 441–465.

2000b. Emergence of vowel systems through self-organisation. *A.I. Commun.* **13**: 27–39.

de Jong, K., 2001. Rate induced re-syllabification revisited. *Lang. Speech* **44**: 229–259.

Delattre, P., 1971. Consonant gemination in four languages: an acoustic, perceptual, and radiographic study, Part I. *Int. Rev. Appl. Linguist.* **9**: 31–52.

Dell, F., and Elmedlaoui, M., 1985. Syllabic consonants and syllabification in Imdlawn Tashlhiyt Berber. *J. Afri. Lang. Linguist.* **7**: 105–130.

2002. *Syllables in Tashlhiyt Berber and in Moroccan Arabic.* Dordrecht, Nethelands: Kluwer.

Fajen, B. R., and Warren, W. H., 2003. Behavioral dynamics of steering, obstacle avoidance, and route selection. *J. Exp. Psychol. Hum. Percep. Perform.* **29**: 343–362.

Falk, D., 2004. Prelinguistic evolution in early hominids: whence motherese. *Behav. Brain Sci.* **27**: 491–503.

Farmer, J. D., 1990. A Rosetta Stone for connectionism. *Physica D* **42**: 153–187.

Ferrari, P. F., Gallese, V., Rizzolatti, G., and Fogassi, L., 2003. Mirror neurons responding to the observation of ingestive and communicative mouth actions in the monkey ventral premotor cortex. *Eur. J. Neurosci.* **17**: 1703–1714.

Flash, T., and Sejnowski, T., 2001. Computational approaches to motor control. *Curr. Opin. Neurobiol.* **11**: 655–662.

Fowler, C. A., and Dekle, D. J., 1991. Listening with eye and hand: cross-modal contributions to speech perception. *J. Exp. Psychol. Hum. Percept. Perform.* **17**: 816–828.

Fowler, C. A., Galantucci, B., and Saltzman, E., 2003. Motor theories of perception. In M. Arbib (ed.) *The Handbook of Brain Theory and Neural Networks*, 2 edn. Cambridge, MA: MIT Press, pp. 705–707.

Gafos, A., 2002. A grammar of gestural coordination. *Nat. Lang. Linguist. Theory* **20**: 269–337.

Galuntucci, B., Fowler, C. A., and Turvey, M., in press. The motor theory of speech perception reviewed. *Psychonom. Bull. Rev.*

Goldstein, L., 2003. Emergence of discrete gestures. *Proceedings 15th International Congress of Phonetic Sciences*, pp. 85–88.

Goldstein, L., and Fowler, C. A., 2003. Articulatory phonology: a phonology for public language use. In N. Schiller and A. Meyer (eds.) *Phonetics and Phonology in Language Comprehension and Production*. Berlin: Mouton de Gruyter. pp. 159–208.

Goldstein, L., Pouplier, M., Chen, L., Saltzman, E., and Byrd, D., in press. Dynamic action units slip in speech production errors. *Cognition.*

Guenther, F. H., 1994. A neural network model of speech acquisition and motor equivalent speech production. *Biol. Cybernet.* **72**: 43–53.

1995. Speech sound acquisition, coarticulation, and rate effects in a neural network model of speech production. *Psychol. Rev.* **102**: 594–621.

Haken, H., Kelso, J. A. S., and Bunz, H., 1985. A theoretical model of phase transitions in human hand movements. *Biol. Cybernet.* **51**: 347–356.

Harris, Z. S., 1951. *Methods in Structural Linguistics*. Chicago, IL: University of Chicago Press.

Hockett, C., 1955. *A Manual of Phonology*. Bloomington, IN: Indiana University Press.

1960, The origin of speech. *Sci. American* **203**: 88–111.

Hommel, B., Musseler, J., Aschersleben, G., and Prinz, W., 2001. The theory of event coding (TEC): a framework for perception and action planning. *Behavi. Brain Sci.* **24**: 849–937.

Jordan, M. I., 1986. *Serial Order in Behavior: A Parallel Distributed Processing Approach*, Technical Repant No. 8604. San Diego, CA: University of California, Institute for Cognitive Science.

1990. Motor learning and the degrees of freedom problem. In M. Jeannerod (ed.) *Attention and Performance*. vol. 13 Hillsdale, NJ: Lawrence Erlbaum, pp. 796–836.

1992. Constrained supervised learning. *J. Math. Psychol.* **36**: 396–425.

Kohler, E., Keysers, C., Umiltà, M. A., *et al.*, 2002. Hearing sounds, understanding actions: action representation in mirror neurons. *Science* **297**: 846–848.

Krakow, R. A., 1993. Nonsegmental influences on velum movement patterns: syllables, sentences, stress, and speaking rate. In M. A. Huffman and R. A. Krakow (eds.) *Nasals, Nasalization, and the Velum*. New York: Academic Press, pp. 87–116.

1999. Physiological organization of syllables: a review. *J. Phonet.* **27**: 23–54.

Kuhl, P. K., Williams, K. A., Lacerda, F., Stevens, K. N., and Lindblom, B., 1992. Linguistic experience alters phonetic perception in infants by 6 months of age. *Science* **255**: 606–608.

Lathroum, A., 1989. Feature encoding by neural nets. *Phonology* **6**: 305–316.

Liberman, A. M., 1996. *Speech: A Special Code*. Cambridge, MA: MIT Press.

Liberman, A. M., Ingemann, F., Lisker, L., Delattre, P. C., and Cooper, F. S., 1959. Minimal rules for synthesizing speech. *J. Acoust. Soc. America* **31**: 1490–1499.

MacNeilage, P. F., 1998. The frame/content theory of evolution of speech production. *Behav. Brain Sci.* **21**: 499–546.

MacNeilage, P. F., and Davis, B. L., 1990. Acquisition of speech production: achievement of segmental independence; In N. Hardcastle and A. Marchal (eds.) *Speech Production and Speech Modeling*. Dordrecht, Netherlands: Kluwer, pp. 55–68.

1993. A motor learning perspective on speech and babbling. In B. de Boysson-Bardies, S. Schoen, P. Jusczyk, P. MacNeilage, and J. Morton (eds.) *Changes in Speech and Face Processing in Infancy: A Glimpse at Developmental Mechanisms of Cognition*. Dordrecht, Netherlands: Kluwer pp. 341–352.

1999. Euolution of the form of spoken words. *Evol. Commun.* **3**: 3–20.

2000. Origin of the internal structure of word forms. *Science* **288**: 527–531.

2005. The frame/content theory of evolution of speech: a comparison with a gestural origins alternative. *Interaction Studies: Interaction Stud.* **6**: 173–199.

MacNeilage, P. F., Davis, B. L., Kinney, A., and Matyear, C. L., 2000. The comparison of serial organization patterns in infants and languages. Issue, *Infant Devel.* **71**: 153–163.

MacWhinney, B., 2000. *The CHILDES Project: Tools for Analyzing Talk*, 3rd edn. Mahwah, NJ: Lawrence Erlbaum.

Massaro, D. W., Cohen, M. M., and Smeele, P. M., 1996. Perception of asynchronous and conflicting visual and auditory speech. *J. Acoust. Soc. America* **100**: 1777–1786.

Mattingly, I. G., 1981. Phonetic representation and speech synthesis by rule. In. T. Myers, J. Laver and J. Anderson (eds.) *The Cognitive Representation of Speech*. Amsterdam: North Holland, pp. 415–420.

McGurk, H., and MacDonald, J., 1976. Hearing lips and seeing voices. *Nature* **264**: 746–747.

McNeill, D., 1992. *Hand and Mind: What Gestures Reveal about Thought*. Chicago, IL: University of Chicago Press.

Meltzoff, M., and Moore, K., 1997. Explaining facial imitation: a theoretical model. *Early Devel. Parent.* **6**: 179–192.

Milo, R., Shen-Orr, S., Itzkovitz, S., *et al.*, 2002. Network motifs: Simple building blocks of complex networks. *Science* **298**: 824–827.

Milo, R., Itzkovitz, S., Kashtan, N., *et al.*, 2004. Superfamilies of evolved and designed networks. *Science*, **303**: 1538–1542.

Mussa-Ivaldi, F. A., 1995. Geometrical principles in motor control. In M. Arbib (ed.) *The Handbook of Brain Theory and Neural Networks*. Cambridge, MA: MIT Press, pp. 434–438.

Nam, H., in press. A competitive, coupled oscillator model of moraic structure: split-gesture dynamics focusing on positional asymmetry. In J. Cole and J. Hualde (eds). *Papers in Laboratory Phonology*, vol. 9.

Nam, H., and Saltzman, E., 2003. A competitive, coupled oscillator of syllable structure. *Proceedings 12th International Congress of Phonetic Sciences*, Barcelona, pp. 2253–2256.

Oudeyer, P.-Y., 2003. L'auto-organisation de la parole. Ph.D. dissertation, University of Paris VI.

2005. The self-organization of speech sounds. *J. Theoret. Biol.* **233**: 435–449.

Pikovsky, A., Rosenblum, M., and Kurths, J., 2003. *Synchronization*. Cambridge, UK: Cambridge University Press.

Polka, L., Colantonio, C., and Sundara, M., 2001. A cross-language comparison of /d/-/D/ perception: evidence for a new developmental pattern. *J. Acoust. Soc. America* **109**: 2190–2201.

Prince, A., and Smolensky, P., 2004. *Optimality Theory: Constraint Interaction in Generative Grammar*. Oxford, UK: Blackwell.

Prinz, W., 1997. Perception and action planning. *Eur. J. Cogn. Psych.* **9**: 129–154.

Pouplier, M., 2003. The dynamics of error. *Proceedings 15th International Congress of the Phonetic Sciences*, pp. 2245–2248.

Rizzolatti, G., and Arbib, M. A., 1998. Language within our grasp. *Trends Neurosci.* **21**: 188–194.

Rizzolatti, G., Scandolara, C., Gentilucci, M., and Camarda, R., 1981. Response properties and behavioral modulation of "mouth" neurons of the postarcuate cortex (area 6) in macaque monkeys. *Brain Res.* **255**: 421–424.

Romanski, L. M., and Goldman-Rakic, P. S., 2002. An auditory domain in primate prefrontal cortex. *Nature Neurosci.* **5**: 15–16.

Rosenblum, L. D., and Saldaña, H. M., 1996. An audiovisual test of kinematic primitives for visual speech perception. *J. Exp. Psychol. Hum. Percept. Perform.* **22**: 318–331.

Saltzman, E. L., 1986. Task dynamic coordination of the speech articulators: a preliminary model – Generation and modulation of action patterns. In H. Heuer and C. Fromm (eds.) *Experimental Brain Research*, New York: Springer-Verlag, pp. 129–144.

1995. Dynamics and coordinate systems in skilled sensorimotor activity. In R. Port and T. van Gelder (eds.) *Mind as Motion*. Cambridge, MA: MIT Press, pp. 150–173.

Saltzman, E., and Byrd, D., 2000. Task-dynamics of gestural timing: phase windows and multifrequency rhythms. *Hum. Mov. Sci.* **19**: 499–526.

Saltzman, E. L., and Kelso, J. A. S., 1987. Skilled actions: a task dynamic approach. *Psychol. Rev.* **94**: 84–106.

Saltzman, E. L., and Munhall, K. G., 1989. A dynamical approach to gestural patterning in speech production. *Ecol. Psychol.* **1**: 333–382.

1992. Skill acquisition and development: the roles of state-, parameter-, and graph-dynamics. *J. Motor Behav.* **24**: 49–57.

Saltzman, E., Löfqvist, A., Kinsella-Shaw, J., Kay, B., and Rubin, P., 1995. On the dynamics of temporal patterning in speech. In F. Bell-Berti and L. Raphael (eds.) *Producing Speech: Contemporary Issues for Katherine Safford Harris*. Woodbury, NY: American Institute of Physics, pp. 469–487.

Saltzman, E., Löfqvist, A., Kay, B., Kinsella-Shaw, J., and Rubin, P., 1998. Dynamics of intergestural timing: a perturbation study of lip–larynx coordination. *Exp. Brain Res.* **123**: 412–424.

Saltzman, E., Löfqvist, A., and Mitra, S., 2000. "Glue" and "clocks": intergestural cohesion and global timing. In M. Broe and J. Pierrehumbert (eds.) *Papers in Laboratory Phonology*, vol. 5 Cambridge, UK: Cambridge University Press, pp. 88–101.

Saltzman, E., Nam, H., Goldstein, L., and Byrd, D., 2006. The distinctions between state, parameter and graph dynamics in sensorimotor control and coordination. In M. L. Latash and F. Lestienne (eds.) *Motor Control and Learning*. New York: Springer Publishing, pp. 63–73.

Schourup, A., 1973. A cross-language study of vowel nasalization. *Ohio State Univ. Working Papers Linguist.* **15**: 190–221.

Shadmehr, R., 1995. Equilibrium point hypothesis. In M. Arbib (ed.) *The Handbook of Brain Theory and Neural Networks*. Cambridge, MA: MIT Press, pp. 370–372.

Sproat, R., and Fujimura, O., 1993. Allophonic variation in English /l/ and its implications for phonetic implementation. *J. Phonet.* **21**: 291–311.

Stetson, R. H., 1951. *Motor Phonetics*. Boston, MA: College-Hill Press.

Stevens, K. N., 1989. On the quantal nature of speech. *J. Phonet.* **17**: 3–45.

1999. *Acoustic Phonetics*. Cambridge, MA: MIT Press.

Stoel-Gammon, C., 1985. Phonetic inventories, 15–24 months: a longitndinal study. *J. Speech Hearing Res.* **18**: 505–512.

Studdert-Kennedy, M., 1998. The particulate origins of language generativity. In J. Hurford, M. Studdert-Kennedy, and C. Knight, (eds.) *Approaches to the Evolution of Language*. Cambridge, UK: Cambridge University Press, pp. 202–221.

2002a. Mirror neurons, vocal imitation, and the evolution of particulate speech. In V. Gallese and M. Stamerov (eds.) *Mirror Neurons and the Evolution of Brain and Language*. Amsterdam: Benjamins, pp. 207–227.

2002b. Evolutionary implications of the particulate principle: imitation and the dissociation of phonetic form from semantic function. In C. Knight, M. Studdert-Kennedy and J. R. Hurford (eds.) *The Evolutionary Emergence of Language: Social Function and the Origin of Linguistic Form*. Cambridge, UK: Cambridge University Press, pp. 161–176.

Studdert-Kennedy, M., and Goldstein, L., 2003. Launching language: the gestural origin of discrete infinity. In M. H. Christiansen and S. Kirby (eds.) *Language Evolution: The States of the Art*. Oxford, UK: Oxford University Press, pp. 235–254.

Studdert-Kennedy, M., and Lane, H., 1980. Clues from the difference between signed and spoken languages. In U. Bellugi and M. Studdert-Kennedy (eds.) *Biological Constraints on Linguistic Form*. Berlin: Verlag Chemie, pp. 29–40.

Suprenant, A., and Goldstein, L., 1998. The perception of speech gestures. *J. Acoust. Soc. America* **104**: 518–529.

Tabain, M., Breen, J. G., and Butcher, A. R., 2004. VC vs. CV syllables: a comparison of Aboriginal languages with English. *J. Int. Phonet. Ass.* **34**: 175–200.

Thelen, E., 1989. The (re)discovery of motor development: learning new things from an old field. *Devel. Psychol.* **25**: 946–949.

Tuller, B., and Kelso, J. A. S., 1991. The production and perception of syllable structure. *J. Speech Hear. Res.* **34**: 501–508.

Turvey, M., 1990. Coordination. *Am. Psychologist* **45**: 938–953.

Vihman, M., 1996. *Phonological Development: The Origins of Language in the Child*. Cambridge, MA: Blackwell.

Zipf, G. K., 1949. *Human Behavior and the Principle of Least Effort*. New York: Addison-Wesley.

8

Lending a helping hand to hearing: another motor theory of speech perception

Jeremy I. Skipper, Howard C. Nusbaum, and Steven L. Small

> ... any comprehensive account of how speech is perceived should encompass audiovisual speech perception. The ability to see as well as hear has to be integral to the design, not merely a retro-fitted after-thought.
> *Summerfield (1987)*

8.1 The "lack of invariance problem" and multisensory speech perception

In speech there is a many-to-many mapping between acoustic patterns and phonetic categories. That is, similar acoustic properties can be assigned to different phonetic categories or quite distinct acoustic properties can be assigned to the same linguistic category. Attempting to solve this "lack of invariance problem" has framed much of the theoretical debate in speech research over the years. Indeed, most theories may be characterized as to how they deal with this "problem." Nonetheless, there is little evidence for even a single invariant acoustic property that uniquely identifies phonetic features and that is used by listeners (though see Blumstein and Stevens, 1981; Stevens and Blumstein, 1981).

Phonetic constancy can be achieved in spite of this lack of invariance by viewing speech perception as an active process (Nusbaum and Magnuson, 1997). Active processing models like the one to be described here derive from Helmholtz who described visual perception as a process of "unconscious inference" (see Hatfield, 2002). That is, visual perception is the result of forming and testing hypotheses about the inherently ambiguous information available to the retina. When applied to speech, "unconscious inference" may account for the observation that there is an increase in recognition time as the variability or ambiguity of the speech signal increases (Nusbaum and Schwab, 1986; Nusbaum and Magnuson, 1997). That is, this increase in recognition time may be due to an increase in cognitive load as listeners test more hypotheses about alternative phonetic interpretations of the acoustic signal. In the model presented here, hypothesis testing is carried out when attention encompasses certain acoustic properties or other sources of sensory information

Action to Language via the Mirror Neuron System, ed. Michael A. Arbib. Published by Cambridge University Press. © Cambridge University Press 2006.

(e.g., visual cues) or knowledge (e.g., lexical knowledge or context) that can be used to discriminate among alternative linguistic interpretations of the acoustic signal. For example, when there is a change in talker, there is a momentary increase in cognitive load and attention to acoustic properties such as talker pitch and higher formant frequencies (Nusbaum and Morin, 1992). Similarly, attention can encompass lexical knowledge to constrain phonetic interpretation. For example, Marslen-Wilson and Welsh (1978) have shown that when participants shadow words that contain mispronunciations, they are less likely to correct errors when they are the first or second as opposed to the third syllable within the word.

By this active process, the "lack of invariance problem" becomes tractable when the wealth of contextual information that naturally accompanies speech is taken into consideration. One rich source of contextual information is the observable gestures that accompany speech. These visible gestures include, for example, movement of a talker's arms, eyebrows, face, fingers, hands, head, jaw, lips, mouth, tongue, and/or torso. These gestures represent a significant source of visual contextual information that can actively be used by the listener during speech perception to help interpret linguistic categories. That is, listeners can test hypotheses about linguistic categories when attention encompasses observable movements, constraining the number of possible interpretations.

Indeed, a large body of research suggests that visual contextual information is readily used during speech perception. The McGurk–MacDonald effect is perhaps the most striking demonstration of this (McGurk and MacDonald, 1976). During the McGurk–MacDonald effect, for example, an "illusory" /ta/ is heard when the auditory track of the syllable /pa/ is dubbed onto the video of a face or mouth producing the syllable /ka/. This is an example of a "fusion" of the auditory and visual modalities. Another effect, "visual capture," occurs when listeners *hear* the visually presented syllable (/ka/ in this example). Both "fusion" and "visual capture" are robust and relatively impervious to listener's knowledge that they are experiencing the effect. Almost as striking as the McGurk–MacDonald effect, however, are studies that demonstrate the extent to which normal visual cues affect speech perception. Adding visible facial movements to speech enhances speech recognition as much as removing up to 20 dB of noise from the auditory signal (Sumby and Pollack, 1954). Multisensory enhancement in comprehension with degraded auditory speech is anywhere from two to six times greater than would be expected for comprehension of words or sentences in the auditory or visual modalities when presented alone (Risberg and Lubker, 1978; Grant and Greenberg, 2001).

It is not simply the case that adding any visual information results in the improvement of speech perception when the visual modality is present. Understanding is impaired when visual cues are phonetically incongruent with heard speech (Dodd, 1977). Also, it is not simply the case that these kinds of effects are limited to unnatural stimulus conditions. Visual cues contribute to understanding clear but hard-to-comprehend speech or speech spoken with an accent (Reisberg *et al.*, 1987). Finally, it is not the case that visual cues simply provide complementary information temporally correlated with the acoustic signal. Information from the auditory and visual modalities is not synchronous, and auditory and visual information unfold at different rates. Visual cues from articulation can precede acoustic information by more than 100 ms. The acoustic signal can be delayed by up to 180 ms from typical audiovisual timing without causing a decrease in the occurrence of the McGurk–MacDonald effect (Munhall *et al.*, 1996).

Nor are the visual contextual cues that accompany and aid speech perception limited to lip and mouth movements. Head movements that accompany speech improve the identification of syllables as compared to absent or distorted head movements (Munhall *et al.*, 2004). Furthermore, listeners use head and eyebrow movements to discriminate statements from questions (Bernstein *et al.*, 1998; Nicholson *et al.*, 2002) and to determine which words receive stress in sentences (Risberg and Lubker, 1978; Bernstein *et al.*, 1998).

Observable manual gestures also participate in speech perception and language comprehension. Manual gestures are coordinated movements of the torso, arm, and hand that naturally and spontaneously accompany speech. These manual gestures, referred to as "gesticulations," are to be distinguished from deliberate manual movements like emblems, pantomime, and sign language, which will not be discussed here. Rather, the present model is primarily concerned with manual gesticulations, which are grossly categorized as imagistic or non-imagistic. Imagistic manual gesticulations (e.g., iconic and metaphoric gestures as described by McNeill (1992)) describe features of actions and objects (e.g., making one's hand into the shape of a mango to describe its size). Non-imagistic manual gesticulations (e.g., deictic and beat gestures as described by McNeill (1992)) by contrast carry little or no propositional meaning and emphasize aspects of the discourse like, for example, syllable stress.

Manual gesticulations are intimately time-locked with and accompany nearly three-quarters of all speech productions (McNeill, 1992). Therefore, it is not surprising to find that manual gesticulations can be utilized to aid speech perception. Indeed, when speech is ambiguous people gesticulate more and rely more on manual gesticulations for understanding (Rogers, 1978; Records, 1994). In instructional settings people perform better on tasks when the instructor is gesticulating compared to the absence of manual gesticulations (Kendon, 1987). Furthermore, instructors' observable manual gesticulations promote learning in students and this can occur by providing information that is redundant with accompanying speech or by providing additional non-redundant information (Singer and Goldin-Meadow, 2005).

Collectively, these studies suggest that speech perception is intrinsically multisensory even though the auditory signal is usually sufficient to understand speech. This should make sense because evolution of the areas of the brain involved in language comprehension probably did not occur over the telephone or radio but in multisensory contexts. Similarly, development occurs in multisensory contexts and infants are sensitive to multisensory aspects of speech stimuli from a very young age (Kuhl and Meltzoff, 1982). By the proposed active process, attention encompasses observable body movements in these multisensory contexts to test hypotheses about the interpretation of a particular stretch of utterance. This requires that perceivers bring to bear upon these observable movements their own knowledge of the meaning of those movements.

The physiological properties of mirror neurons suggest one manner in which this might occur. That is, mirror neurons have the physiological property that they are active during both the execution of a movement and the observation of similar goal-directed movements (Rizzolatti *et al.*, 2002). This suggests that action understanding is mediated by relating observed movements to one's own motor plans used to elicit those movements (without the actual movement occurring). Though these physiological properties have only been directly recorded from neurons in the macaque, brain-imaging studies suggest that a "mirror system" exists in the human (for a review see Rizzolatti and Craighero, 2004). The mirror system is defined here as a distributed set of regions in the human brain used to relate sensed movements to one's own motor plans for those movements.

It is here proposed that the mirror system can be viewed as instantiating inverse and forward models of observed intended actions with mirror neurons being the interface between these two types of models (Arbib and Rizzolatti, 1997; Miall, 2003; Iacoboni, 2005; Oztop *et al.*, this volume). An inverse model is an internal representation of the (inverse) relationship between an intended action or goal and the motor commands needed to reach those goals (Wolpert and Kawato, 1998). A forward model predicts the effects of specific movements of the motor system (Jordan and Rumelhart, 1992). With respect to language, forward models are thought to map between overt articulation and predicted acoustic and somatosensory consequences of those movements (Perkell *et al.*, 1995). In this capacity, forward models have been shown to have explanatory value with respect to the development of phonology (Plaut and Kello, 1999) and speech production (Guenther and Ghosh, 2003; Guenther and Perkell, 2004) and adult motor control during speech production (Guenther and Ghosh, 2003; Guenther and Perkell, 2004).

By the present model, graphically depicted in Fig. 8.1, heard and observed communicative actions in multisensory environments initiate inverse models that transform the goal of the heard and observed actions into motor commands to produce those actions. These inverse models are paired with forward models (see Wolpert and Kawato, 1998; Iacoboni, 2005), which are motor predictions (i.e., hypotheses), in which motor commands are executed in simulation, that is, without overt movement, but which nonetheless have sensory consequences. It is proposed that these sensory consequences are compared with activated alternative linguistic interpretations of the speech signal and help mediate selection of a linguistic category. It is argued that these linguistic categories are varied and dependent on the type of observed movement encompassed by attention. Observed mouth movements provide information about segmental phonetic categories whereas eyebrow movements and non-imagistic manual gesticulations provide cues about both segmental and suprasegmental (i.e., prosodic) phonetic categories. Observed imagistic manual gesticulations additionally provide cues to semantic content which can sometimes provide further constraint on interpretation of which lexical item (i.e., word) was spoken.

The pairing of inverse and forward models is denoted with the phrase "inverse-forward model pairs" (IFMPs). IFMPs are viewed as a basic building-block of the mirror system as it has been defined here. In the absence of visible gestures (e.g., when on the telephone), it is proposed that the auditory signal alone generates IFMPs that can be used to

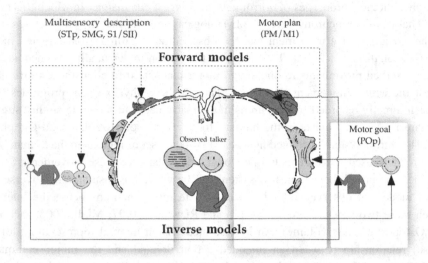

Figure 8.1 Diagram of inverse (solid lines) and forward (dashed lines) model pairs associated with the facial gestures (light gray) and manual gesticulations (dark gray) of a heard and observed talker (center). A multisensory description of the observed gestures (in posterior superior temporal (STp) areas) results in inverse models that specify the motor goals of those movements (in the pars opercularis (POp) the human homologue of macaque area F5 where mirror neurons have been found). These motor goals are mapped to motor plans that can be used to reach those goals (in premotor (PM) and primary motor cortices (M1)). Forward models generate predictions of the sensory states associated with executing these motor commands. Sensory (in STp areas) and somatosensory (in parietal cortices including the supramarginal gyrus (SMG) and primary and secondary somatosensory cortices (SI/SII)) predictions are compared (white circles) with the current description of the sensory state. The result is an improvement in the ability to perceive speech due to a reduction in ambiguity of the intended message of the observed talker.

disambiguate auditory speech. Thus, IFMPs in both unisensory and multisensory contexts are instances in which perception is mediated by gestural knowledge. When the auditory signal is presented alone, however, there are fewer cues from which to derive IFMPs. That is, visual cues are a significant source of added information and it is expected that more IFMPs are involved in multisensory communicative contexts. Because of this, in the absence of visual cues, other cues, like knowledge of other linguistic categories, may be more effective in solving the lack of invariance problem. That is, it is argued that there are multiple routes through which speech perception might be more or less mediated. One such route is more closely associated with *acoustic* analysis of the auditory signal and interpretation of this analysis in terms of other linguistic categories, for example, words and sentences, and their associated meaning. It is argued that speech perception in the absence of visual cues may place more emphasis or weight on this route in instances when the acoustic signal is relatively unambiguous. The addition of visual cues may shift this weighting to the gestural route because visual cues provide an added source of information. In addition, in both unisensory and multisensory contexts it is proposed that relative weight shifts to the gestural route as ambiguity of the speech signal increases.

This model is distinct from the motor theory of speech perception (Liberman and Mattingly, 1985; see also Goldstein, Byrd, and Saltzman, this volume) that claims to solve the lack of invariance problem by positing that speech perception is directly mediated by a gestural code. That is, speech perception occurs by references to invariant motor programs for speech production. Thus, *all* speech is directly transduced into a gestural code and it was suggested by Liberman and Mattingly (1985) that there is no auditory processing of speech. The present model makes a different claim that speech is not solely mediated by gestural codes but, rather, speech can be mediated by both acoustic and gestural codes. In the present model mediation by gestural codes is not restricted to articulatory commands. Mediation by the mirror system is expected to be most prominent during multisensory speech when the anatomical concomitant of speech sounds, facial movements and non-imagistic manual gesticulations provide cues that can be used to aid interpretation of the acoustic signal. Gestural codes are also thought to become more prominent and mediate perception during periods of variability or ambiguity of the speech signal; that is, when hypothesis testing regarding phonetic categories is necessary for disambiguation. In this sense, this model is related to the analysis-by-synthesis model of speech perception (Stevens and Halle, 1967). In Stevens and Halle's model, analysis-by-synthesis, and, thus, presumably the activation of the mirror system, occurs to aid interpretation of acoustic patterns, for example, when there is strong lack of invariance. By contrast, the motor theory of speech perception claims that speech is always mediated by a gestural code and, thus, the mirror system would presumably always mediate perception.

Indeed, speech perception *may* occur without invoking gestural codes. That is, the available evidence suggests that perception is not necessarily determined by activity in motor cortices or the mirror system. One type of evidence derives from studies in which behavioral attributes of speech perception in humans, like categorical perception, are demonstrated in other animals. For example, Kluender *et al.* (1987) have shown that trained Japanese quail can categorize place of articulation in stop consonants. As these birds have no ability to produce such sounds it is unlikely that they are transducing heard sounds into gestural codes. It is theoretically possible, therefore, that humans can also make such distinctions based on acoustic properties alone (see Miller (1977) for a more detailed argument). Similarly, infants categorically perceive speech without being able to produce speech (Jusczyk, 1981) though it is not possible to rule out a nascent influence of the motor system on perception before speech production is possible. Such behavioral findings are at odds with the claims of the motor theory. (See Diehl *et al.* (2004) for a review of gestural and general auditory accounts of speech perception and challenges to each.)

Neurobiological evidence also does not support the claim that speech perception is always mediated by a gestural code. Corresponding to the classic view based on the analysis of brain lesions (see Geschwind, 1965), speech perception and language comprehension are not significantly impaired by destruction of motor cortices though some deficits can be demonstrated. Supporting this, neuroimaging studies find inconsistent evidence that the motor system is active when speech stimuli are presented in the auditory modality alone and listeners' tasks involve only passive listening (Zatorre *et al.*, 1996; Belin *et al.*, 2000, 2002; Burton *et al.*, 2000; Zatorre and Belin, 2001).

8.2 Explication of the model

The model that has now been briefly introduced builds on and is indebted to previous accounts for motor control (e.g., Wolpert and Kawato, 1998), speech production (Guenther and Ghosh, 2003; Guenther and Perkell, 2004), auditory speech perception (Callan *et al.*, 2004; Hickok and Poeppel, 2004), and audiovisual speech perception (Skipper *et al.*, 2005a). In this section specific aspects are expanded upon and some of the assumptions of the model are exposed.

An important source of constraint on the model comes from the functional anatomy of the perceptual systems, specifically the existence of two global anatomical streams (or pathways) in both the visual (Ungerleider and Mishkin, 1982; Goodale and Milner, 1992; Jeannerod, 1997) and auditory (Kaas and Hackett, 2000; Rauschecker and Tian, 2000) systems (see Arbib and Bota, this volume, for further discussion). This anatomical constraint has proven useful with respect to the ventral and dorsal auditory streams for developing theoretical notions about the neurobiology of auditory speech perception (for an extensive treatment see Hickok and Poeppel, 2004). The present model presents a unified interpretation of the physiological properties of both the auditory and visual dorsal and ventral streams as they relate to speech perception and language comprehension in both unisensory and multisensory communication contexts. These overall anatomical and neurophysiological features of the model are visually depicted in Fig. 8.2.

Specifically, it is hypothesized that both the auditory and visual ventral systems and the frontal regions to which they connect are involved in bidirectional mappings between perceived sensations, interpretation of those sensations as auditory and visual categories, and the associated semantic import of those categories. The functional properties of the latter streams are referred to with the designation "sensory–semantic." By contrast the dorsal streams are involved in bidirectional transformations between perceived sensations related to heard and observed movements and the motor codes specifying those movements. These functional properties are referred to with the designation "sensory–motor."

In the following sections the ventral and dorsal streams are discussed independently. Then the dorsal streams are discussed in more depth as these streams' functional properties are such that they comprise what has here been defined as the mirror system, which implements IFMPs. Though the streams are discussed separately, they are not thought to be functionally or anatomically modular. Rather, the streams represent cooperating and competing (Arbib *et al.*, 1998, ch. 3) routes in the brain through which speech perception might be mediated. That is, perception is mediated by both streams or is more or less mediated by one or the other of the streams. Therefore, discussion of the streams

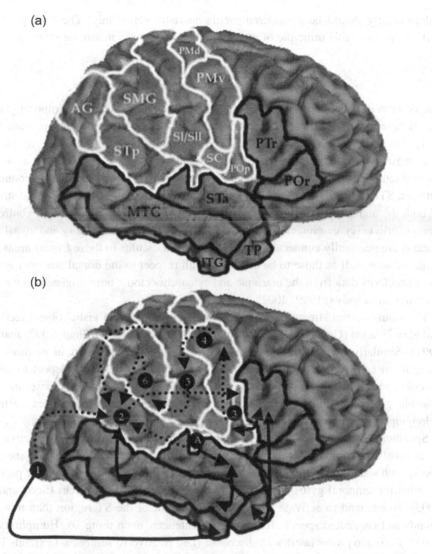

Figure 8.2 (a) Brain regions defining the model presented in the text. Regions outlined in black are key structures comprising the ventral auditory and visual streams involved in "sensory–semantic" processing. These are visual areas (not outlined), inferior temporal gyrus and sulcus (ITG), middle temporal gyrus and sulcus (MTG), anterior superior temporal structures (STa), temporal pole (TP), pars orbitalis (POr), and the pars triangularis (PTr). Regions outlined in white are key structures comprising the dorsal auditory and visual streams involved in "sensory–motor" processing. These are visual areas (not outlined), posterior superior temporal (STp) areas, supramarginal gyrus (SMG), somatosensory cortices (SI/SII), dorsal (PMd) and ventral (PMv) premotor cortex, and the pars opercularis (POp). Also shown is the angular gyrus (AG). (b) Schematic of connectivity in the ventral and dorsal streams and an example "inverse-forward model pair" (IFMP) as it relates to structures in the dorsal streams. Solid and dotted black lines represent proposed functional connectivity between the ventral and dorsal streams respectively. Actual anatomic connections are presumed to be bidirectional. IFMPs are thought to be implemented by the dorsal streams. Numbers

independently should be considered for its heuristic value only. The final part of this section discusses this principle of cooperation and competition among streams.

8.2.1 Ventral "sensory–semantic" streams

The cortices of the auditory ventral stream are defined as the superior temporal gyrus and sulcus anterior to the transverse temporal gyrus (STa), including the planum polare, middle and inferior temporal gyri, and the temporal poles. The cortices of the visual ventral stream are defined as V1 (primary visual cortex), V2, V4, and the areas of the inferior temporal cortex. The ventral auditory and visual streams interact by connectivity between STa and inferotemporal areas (Seltzer and Pandya, 1978). The frontal structures of both the auditory and visual streams are defined as orbital, medial, and ventrolateral frontal cortices (cytoarchitectonic areas 10, 45, 46, 47/12). The auditory and visual ventral streams are minimally connected via the uncinate fasciculus to these frontal areas. These definitions (as well as those to be discussed with respect to the dorsal streams) are based on connectivity data from the macaque and cytoarchitectonic homologies with the human (Petrides and Pandya, 1999, 2002).

The visual ventral stream has been implicated in non-verbal visual object recognition and identification (Ungerleider and Mishkin, 1982; Goodale and Milner, 1992; Jeannerod, 1997). Similarly, the auditory ventral stream has been implicated in auditory object recognition and identification (Rauschecker and Tian, 2000). With respect to language function, research indicates that a similar, that is, "sensory–semantic," interpretation is possible given functional homologies between temporal and frontal areas defined as belonging to the ventral streams.

Specifically, in the temporal ventral streams, bilateral STa cortices are active during tasks involving complex acoustic spectrotemporal structure, including both speech and non-speech sounds. Intelligible speech seems to be confined to more anterior portions of the superior temporal gyrus and especially the superior temporal sulcus (Scott and Wise, 2003). Words tend to activate a more anterior extent of the ST region than non-speech sounds and connected speech, for example, sentences, even more so (Humphries *et al.*, 2001). Activation associated with discourse (i.e., relative to sentences) extends into the

Caption for Figure 8.2 (cont.)
correspond to processing steps associated with IFMPs associated with observable mouth movements. These are visual processing of observable mouth movements (1) in terms of biological motion (2), which generates an inverse model (2–3) that specifies the observed movement in terms of the goal of that movement by mirror neurons (3). The motor goal of the movement is mapped to the parametric motor commands that could generate the observed movement in a somatotopically organized manner, in this case the mouth area of premotor cortex (3–4). These motor commands yield forward models that are predictions of both the auditory (4–2) and somatosensory (4–5–6) consequences of those commands had they been produced. These predictions can be used to constrain auditory processing (A–2) by supporting an interpretation of the acoustic signal.

temporal poles (Tzourio *et al.*, 1998). Collectively, it is these STa structures along with more inferior temporal lobe structures that are sensitive to semantic manipulations and grammatical structure (see Bookheimer (2002) for a review of neuroimaging studies of semantic processing).

In the frontal lobe, the anterior aspects of the inferior frontal gyrus, specifically the pars triangularis of Broca's area (cytoarchitectonic area 45) and the pars orbitalis (cytoarch-itectonic area 47/12) are activated by tasks intended to assess higher-level linguistic processing related to semantic manipulations and grammatical structure. Specifically, these regions are involved in strategic, controlled, or executive aspects of processing semantic information in words or sentences (Gold and Buckner, 2002; Devlin *et al.*, 2003) and show reduction of activation during semantic priming (Wagner *et al.*, 2000). These regions may also be involved in strategic, controlled, or executive aspects of processing syntactic information (Stromswold *et al.*, 1996; Embick *et al.*, 2000).

Thus, based on these shared functional homologies, it is argued that regions that have been defined as encompassing the ventral auditory and visual streams function together to interpret perceived sensations as auditory and visual objects or categories along with the associated semantic import of those categories. Auditory objects include linguistic cat-egories such as phonetic, word, phrase level, and discourse representations and their associated meanings derived from the acoustic signal. It is thus the ventral pathways that are most closely associated with the compositional and meaningful aspects of language function.

The present refinement, however, removes this strong sensory modality dependence associated with each stream independently. This allows for auditory objects like words to have multisensory properties of visual objects. That is, due to intrinsic connectivity and learned associations, interaction of the ventral auditory and visual streams allows auditory objects like words to become associated with corresponding visual objects. Thus, repre-sentations of words can be associated with visual features of objects with the result that words can activate those features and vice versa. Indeed, both behavioral and electro-physiology studies support this contention (Federmeier and Kutas, 2001).

The association of words with their features is extended here to accommodate imagistic manual gesticulations. That is, it is proposed that *both* non-verbal semantic knowledge of, for example, an observed mango and imagistic manual gesticulations representing a mango in a communication setting can activate words and lexical neighbors correspond-ing to mangos (e.g., durians, rambutans, lychees, etc.). Some evidence supports this claim. When people are in a tip-of-the-tongue state they produce imagistic manual gesticulations and observers can reliably determine what word the observed talker was attempting to communicate with these gesticulations (Beattie and Coughlan, 1999). These activated lexical items, associated with the observed mangos or manually gesticulated "virtual" mangos, detract attention away from acoustic/phonetic analysis (i.e., STa cortices) or attract attention to a specific interpretation of acoustic segments. This type of cortical interaction is proposed to underlie results like those of Marslen-Wilson and Welsh (1978) reviewed above.

8.2.2 Dorsal "sensory–motor" streams

The cortices of the auditory dorsal stream are defined as the superior temporal gyrus and sulcus posterior to the transverse temporal gyrus, including the planum temporale, and extending posterior to the angular gyrus. These regions will be referred to as the posterior superior temporal (STp) areas. Also included in the dorsal stream are inferior parietal areas including somatosensory cortices and the supramarginal gyrus. Collectively these areas have variously been referred to as Wernicke's area. The cortices of the visual dorsal stream are defined as V1, V2, V3, STp areas, and inferior and superior parietal areas. Visual motion areas (e.g., middle temporal and medial superior temporal cortices) are here included within STp areas. Note that STp areas and inferior parietal cortices are defined as a common locus of both the auditory and visual dorsal streams. Frontal structures of both streams are defined to be more posterior and ventral to those frontal areas of the ventral streams. These frontal structures include the pars opercularis (cytoarchitectonic area 44) of the inferior frontal gyrus, premotor cortex (cytoarchitectonic area 6), and primary motor cortex (cytoarchitectonic area 4). Posterior temporal–parietal structures of the dorsal streams are minimally reciprocally connected to these frontal areas via the arcuate fasciculus.

The visual dorsal stream has been implicated in visual guidance of action in contrast to object-centered processing occurring in the ventral visual stream (Goodale and Milner, 1992; Jeannerod, 1997). With respect to language function, research indicates that a similar, that is, "sensory–motor," interpretation is possible given functional homologies between cortices that have been defined as belonging to both dorsal streams.

Specifically, STp and inferior parietal areas are activated during acoustic and phonological analyses and the storage and manipulation of speech stimuli (Zatorre *et al.*, 1996; Jonides *et al.*, 1998; Binder *et al.*, 2000; Burton *et al.*, 2000; Wise *et al.*, 2001; Jancke *et al.*, 2002). However, unlike STa cortices, which are also involved in acoustic and phonetic analysis of stimuli, the posterior ST and inferior parietal cortices are involved in both overt and covert articulation, reading, and naming (Bookheimer *et al.*, 1995; Paus *et al.*, 1996; Wise *et al.*, 2001). The perception of speech sounds overlaps the production of the same sounds in these regions (Buchsbaum *et al.*, 2001).

Furthermore, the STp areas are activated by the visual modality in a manner that ventral stream areas are not. Specifically, they are activated by the observation of non-linguistic but biologically relevant movements and by implied movements of the eyes, mouth, hands, and limbs (see Allison *et al.* (2000) for a review). In addition to non-linguistic movements, the STp areas are activated during comprehension and production of sign language (Corina *et al.*, 1999; Braun *et al.*, 2001; MacSweeney *et al.*, 2002) and during lip-reading more so than non-linguistic lip movements (Campbell *et al.*, 2001). With both the auditory and visual modalities present, the STp areas becomes increasingly more active as the level of visual information increases from auditory to auditory and facial gestures (Mottonen *et al.*, 2002; Olson *et al.*, 2002; Jones and Callan, 2003; Wright *et al.*,

2003; Skipper *et al.*, 2005a) to auditory and facial gestures and manual gesticulations (Josse *et al.*, 2005).

The posterior frontal regions to which these temporal/parietal areas are connected also play a specific role in acoustic and phonological analyses and the storage and manipulation of speech stimuli (E. Paulesu *et al.*, 1993; P. Paulesu *et al.*, 1993; Awh *et al.*, 1995; Schumacher *et al.*, 1996; Zatorre *et al.*, 1996; Poldrack *et al.*, 1999; Burton *et al.*, 2000; Heim *et al.*, 2003). Posterior frontal regions are also well known to be involved in the preparation for and the production of speech (among other movements). Broca's area of the inferior frontal gyrus, comprising the pars opercularis and the pars triangularis, along with premotor cortex, has long been viewed as supporting the mechanism by which auditory forms are coded into articulatory forms in service of speech production (Geschwind, 1965; Wise *et al.*, 1999; Riecker *et al.*, 2000; Wildgruber *et al.*, 2001). It is argued, however, that it is the more posterior aspects of these areas that are more involved in speech production and phonology whereas, as reviewed in the previous section, the more anterior aspects (e.g., the pars triangularis or cytoarchitectonic area 45) are more involved in "sensory–semantic" transformations (Devlin *et al.*, 2003). Finally, the frontal areas of the dorsal streams also become increasingly more active as the level of visual information increases from auditory to audiovisual sensory modalities (Skipper *et al.*, 2005a).

Thus, it is argued, based on the shared audiovisual–motor functional properties of the regions defined as comprising the dorsal auditory and visual streams, that these regions work together to transform perceived sensations into motor plans that can be used to guide action. This should be contrasted with the ventral streams that are more involved in auditory and visual object processing associated with larger linguistic categories (e.g., sentences) that does not take place in the dorsal streams.

It is the dorsal streams that contain mirror neurons and comprise the mirror system as it relates to speech. Indeed, mirror neurons are thought to reside in the human pars opercularis, the proposed homologue of the macaque premotor area F5 where mirror neurons were discovered (Rizzolatti *et al.*, 2002). In addition mirror neurons have been discovered in macaque area PF, corresponding to human inferior parietal cortex (Fogassi *et al.*, 1998). The evidence suggests that the pars opercularis in the human has similar mirror neuron properties with regard to execution, imitation, and observation of hand (Iacoboni *et al.*, 1999; Binkofski *et al.*, 2000; Koski *et al.*, 2002) and mouth (Buccino *et al.*, 2004; Skipper *et al.*, 2004, 2005a, 2005b, 2005c) movements. A similar argument has been made for the existence of inferior parietal mirror neurons in the human (see Iacoboni, 2005). More generally, many studies have now demonstrated the existence of a mirror system for relating action perception to execution in the areas of the dorsal streams that have been described here (for a review see Rizzolatti and Craighero, 2004). During action observation there is usually a strong activation of STp, inferior parietal, pars opercularis, premotor, and motor areas. Furthermore, this mirror system is somatotopically organized in premotor and parietal cortices, with distinct loci for observation of mouth, hand, and foot actions corresponding to loci associated with execution of these movements respectively (Buccino *et al.*, 2001).

8.2.3 Dorsal "sensory–motor" streams, the mirror system, and inverse-forward model pairs

A more detailed presentation of the dorsal "sensory–motor" streams with respect to the implementation of IFMPs is now undertaken. As defined, a common locus of both auditory and visual dorsal streams is the STp areas. To review, STp areas receive communicatively relevant auditory and visual information about gestures from both the auditory and visual streams. It is proposed that STp areas are an early site of audiovisual integration as suggested by functional neuroimaging of multisensory language (Sams *et al.*, 1991; Calvert *et al.*, 2000; Surguladze *et al.*, 2001; Mottonen *et al.*, 2002; Olson *et al.*, 2002; Skipper *et al.*, 2005a). Early integration provides a sensory description of heard and observed gestures which initiate inverse models that are the beginning of a late stage of "sensory–motor" integration that concludes with the comparison of the forward modal's prediction with the nascent representations being processed in sensory cortices.

Like Iacoboni (2005), we argue here that connections of STp areas with inferior parietal and frontal areas form the physiological basis of inverse models. Inverse models map from the desired sensory consequences of an action to the motor commands for that action. In the present model, the desired sensory consequences are those specified by the observed communicative gestures. Though the observed gestures are specified in terms of their motor commands, actual movement does not occur above the level of awareness though measurable electrical activity in specific muscles may change. Some behavioral evidence is consistent with the idea that sensation elicits inverse models related to speech. Listening to phonemes can change productions of those phonemes (Cooper, 1979). Similarly, when feedback from a talker's voice is artificially delayed there are corresponding disruptions in speech production (Houde and Jordan, 1998). Finally, the reviewed neurobiological evidence regarding the existence of mirror neurons in the macaque and a mirror system in humans supports the existence of inverse models during both observation and overt production of movements.

The connections from frontal areas back to STp and inferior parietal areas (i.e., somatosensory and supramarginal areas) form the basis of forward models. Forward models map the current state of the motor system to the predicted state of the motor system through reafferent sensory inflow. This occurs through reciprocal connections between motor and temporal and parietal cortices. This mapping operates on motor codes to produce internal representations of their sensory effects, including, at least, auditory and somatosensory effects.

Neurobiological evidence supports the existence of forward models and is consistent with the idea that forward models have sensory effects. Speech production has been demonstrated to alter activity in auditory cortices as a function of the expected acoustic feedback (Numminen *et al.*, 1999; Houde *et al.*, 2002). Delayed auditory feedback during speech production results in greater activation of bilateral STp and inferior parietal areas (Hashimoto and Sakai, 2003). Evidence from non-human primates suggests that this might consist of inhibitory responses preceding and excitatory responses following vocalization (Eliades and Wang, 2003). Eliades and Wang (2003) suggested that feedback is not simply due to hearing one's own vocalization but also involves feedback connections from the production system. Consistent with this, it has also been found that human STp and inferior parietal areas are active during speech production in the *absence* of feedback from the talker's voice (Paus *et al.*, 1996; Hickok *et al.*, 2000; Shergill *et al.*, 2002). Paus and collegues (1996) maintain that this is evidence for reafference during speech production. Finally, reafference is not limited to speech production but occurs for other movements as demonstrated for finger movements by Iacoboni and colleagues (2001).

Such studies are critical with respect to the present model because the claim is that language *perception* activates inverse and corresponding forward models and, in the absence of overt production, would require functional pathways (i.e., the anatomical pathways are known to exist) in which forward models have sensory effects. These studies demonstrate that reafference occurs in STp and inferior parietal areas. It is proposed that the former is related to auditory while the latter is related to somatosensory feedback.

Also similar to the work of Iacoboni (2005) and Miall (2003), it is proposed that mirror neurons provide a crucial interface between inverse and forward models. With respect to the dorsal streams, the output of activated inverse models occurs at the level of the pars opercularis of the inferior frontal gyrus where mirror neurons putatively exist (Rizzolatti *et al.*, 2002). This activation initiates the appropriate forward models. The present model differs from that of Iacoboni (2005) in that forward models are proposed to be the result of interaction between the pars opercularis, premotor, and motor cortex and sensory cortices. That is, the level of representation of mirror neurons is such that they do not encode the actual dynamics of the movement or the effector required to perform a specific action (Rizzolatti *et al.*, 2002). Inverse models activate the abstract goal of observed actions, which are encoded into the actual movement dynamics through interaction with premotor and primary motor cortices. This occurs in a somatotopically organized manner (Buccino *et al.*, 2001).

There are multiple effectors and movements observable in multisensory communication environments. These determine the specific inverse models that determine which somatotopically paired forward models are elicited and therefore which sensory consequences result. Inverse models corresponding to heard and/or observed facial movements would be expected to activate forward models associated with the sensory consequences of activating premotor and primary motor cortices associated with articulation. More specifically, the distribution of premotor and primary motor cortices are proposed to be specific to the heard and observed syllable for the same reason that producing different syllables would require coordination of different muscles and, therefore, mediation by non-identical neuronal assemblies. For example, hearing and observing /ta/ and /ka/ would be expected to activate neuronal assemblies involved in controlling muscles associated with lowering the mandible and lip position, elevating the tongue, and closing the velopharyngeal port. The two syllables, however, differ on how parted the lips are (i.e., more so for /ka/) and which part of the tongue is elevated and where it makes contact (i.e., the tip and alveolar ridge for /ta/ and the middle and soft palate and molars for /ka/). Given their similar profiles, these syllables would be expected to activate very similar yet somewhat somatotopically distinct areas, especially in motor areas more central to controlling the tongue. This can be contrasted with /pa/ which has a very different articulatory profile in which the lip movements (i.e., bilabial closure) are more prominent, the tongue is mostly irrelevant, and the mandible is lowered at the end of production. Thus, the distribution of activity for hearing and observing /pa/ would be expected to be maximally different from /ka/ and /ta/ in motor areas that are more associated with the lips and tongue.

Sensory consequences of the activated areas associated with these articulatory patterns would occur in auditory and somatosensory cortices related to the sound and propriocep-tive feedback from making these articulatory movements respectively (though no sound or movement is actually produced). The specificity of the auditory and somatosensory cortices activated by feedback would be a function of the observed movement. For example, forward models corresponding to mouth and upper face movements would likely have sensory consequences in areas more associated with phonological and pros-odic processing of auditory stimuli respectively. Similarly, the activated somatosensory cortices would be those associated with sensory feedback from making lower and upper face movements respectively. Inverse models corresponding to observed non-imagistic manual gesticulations would be expected to activate forward models associated with the sensory consequences of activating the more dorsal hand premotor and primary motor cortex. These would have somatosensory consequences dependent on the specific movement.

It is proposed that, other than somatosensory consequences, observation of non-imagistic manual gesticulations can also have auditory consequences. That is, observed actions can elicit "inappropriate" somatotopically organized IFMPs. For instance, some non-imagistic manual gesticulations are intimately time-locked with speech and share features with the co-occurring speech. For example, the most effortful manual gesticula-tions tend to co-occur with the most prominent syllables occurring in the accompanying speech (Kendon, 1994). Certain non-imagistic gesticulations almost always occur with pitch-accented syllables (Renwick *et al.*, 2001). These non-imagistic manual gesticula-tions activate mirror neurons that code for only the goal of the action. Thus, activity in mirror neurons can be mapped to motor areas that code for manual productions *or* speech productions because either set of areas can achieve the desired goal. When mapped to speech production areas, resulting forward models will have acoustic sensory conse-quences. Behavioral findings support this property of the model in that both the observa-tion and execution of manual movements has been shown to affect the acoustic properties of speech production (Gentilucci *et al.*, 2004a, 2004b; these papers are discussed in detail in Arbib, chapter 1, this volume). Furthermore, this aspect of the model conforms to the finding that mirror neurons encode the goals of actions and not necessarily the specific effector used to achieve that goal. That is, some mirror neurons in the macaque have been shown to encode an action performed by either the hand or mouth given that the goal of the action is similar (Gallese *et al.*, 1996). Thus, in the human, it is argued that when the goal of observed manual gesticulations and the auditory modality specify correlated features, mirror system activity can result in "inappropriate" somatotopically organized activity which can, in turn, have cross-effector and, thus, cross-sensory modality effects. This aspect of the model can in principle apply to other observed movements.

Sensory feedback signals associated with somatotopically organized forward models are compared with the current context being processed in sensory cortices, leading to support for a particular interpretation. If correct (e.g., as determined by efficient speech perception and language comprehension), that particular pairing of inverse and forward model is weighted such that future encounters with the same cues will elicit activation of the correct IFMPs. In the short term, the result is an improvement in speech perception due to a reduction in phonetic and/or semantic uncertainty. For example (Fig. 8.2b), movement of the lips and mouth begin before acoustic cues are heard. Therefore, IFMPs can predict the particular acoustic context and lend support to the interpretation of the acoustic pattern much earlier than when visual cues are absent. This may allow cognitive resources to be shifted to other aspects of the acoustic signal (e.g., meaning). Head, eyebrow, and non-imagistic manual gesticulations may function to constrain interpretation by providing cues to both sentential and suprasegmental aspects of the acoustic signal.

8.2.4 Cooperation and competition between streams

Though processing stream models like the present model have heuristic value, to conceptualize the two streams as completely separate or as not interacting at multiple levels would be wrong. Rather, an orchestral metaphor of the streams may be more suited for thinking about processing streams in a manner that is less dichotomous. That is, our appreciation of an orchestra is determined by the functioning of the whole orchestra (i.e., cooperation among sections). At times, however, this appreciation may be more or less determined by, for example, the string or woodwind sections (i.e., competition between sections). Similarly, the dorsal and ventral streams are probably cooperative and competitive in mediating perception. This section serves not to discuss all of the ways in which the streams are cooperative but, rather, serves to illustrate the cooperative principle by discussing functional interactions that bridge the two streams. Interactions between the dorsal streams and lexical knowledge (the ventral streams) are discussed. Imagistic manual gesticulations will also be discussed as they embody both "sensory–semantic" and "sensory–motor" properties. Finally, this section also serves to illustrate the competitive principle by discussing how both the dorsal and ventral streams may more or less mediate perception depending on the auditory and visual contexts.

Broca's area may be one important locus with respect to the interaction among the dorsal and ventral streams. Earlier it was argued that the pars triangularis and pars opercularis of Broca's area have relatively dissociable functions. The pars triangularis functions more in the "sensory–semantic" domain and pars opercularis functions more in "sensory–motor" domain. These two regions, however, are clearly connected though they have different cytoarchitectonic properties (Petrides and Pandya, 2002).

It is claimed that reciprocal interactions at the level of Broca's area allows for activity in the ventral streams associated with "sensory–semantic" representations to increase (or perhaps decrease) the level of activity in dorsal streams areas associated with "sensory–motor" representations and vice versa. For example, in a multisensory communication setting, an observed talker may be saying, "I have a mango" while displaying a mango. The observed mango may activate visual features of the mango and, in turn, the word and lexical neighbors of the word mango, a process that is mediated by temporal areas and maintained by the pars triangularis of the ventral streams. The heard and observed lip movements for the /mæ/ in mango may activate IFMPs associated with the syllable in the pars opercularis. This activity is expected to further raise the level of activity for the word mango through interaction with the pars triangularis, which would result in an improvement in performance associated with lexical access. The converse is also true in that the threshold for the /mæ/ is already raised as the word mango is active in the pars triangularis enforcing the activated IFMPs associated with hearing and observing /mæ/ resulting in a performance increase with respect to speech perception.

It was earlier argued that, like displaying the mango in the above example, imagistic manual gesticulations can activate visual features of the mango (i.e., a "virtual" mango) and, in turn, the word and lexical neighbors of the word mango. Mangoes are, however, also associated with actions that might be utilized to manipulate them. For example, mangoes can be picked up, peeled, nibbled, etc. It is proposed that both the observation of a mango being, for example, moved to the mouth and the observed imagistic manual gesticulation of the hand being moved to the mouth in the absence of the mango, can initiate motor plans for eating that can activate the appropriate corresponding words (e.g., "eating") and vice versa.

Though clearly more study is necessary, there is some evidence that imagistic manual gesticulations could activate the words that represent those acts and vice versa (Beattie and Coughlan, 1999). There is more evidence available for the corresponding claim that words encode motor features when those words represent or afford motor acts. Specifically, the available data suggest that deficits in naming caused by damage to temporal lobe areas follow an anterior–posterior distinction that may mirror the association of words with their features. That is, more anterior temporal lobe damage results in deficits in naming non-manipulable objects whereas more posterior damage results in naming deficits associated with manipulable objects (Damasio *et al.*, 1996; Tranel, 2001). Thus, the more posterior organization for naming manipulable objects may occur because manipulable objects are associated with motor plans for their manipulation. Indeed, premotor cortex is more active during naming of graspable objects (e.g., tools, fruits, or clothing) relative to non-manipulable objects (Martin *et al.*, 1996; Grafton *et al.*, 1997; Chao and Martin, 2000; Gerlach *et al.*, 2002). Some evidence even suggests that verbs (e.g., run vs. type vs. smile) activate motor cortices in a somatotopic manner (Pulvermüller *et al.*, 2000, 2001).

This principle that words and possibly imagistic manual gesticulations can encode motor aspects of their representation in terms of actual movement dynamics is an important bridge between the dorsal and ventral streams. We maintain that areas of the ventral streams mediate representations of linguistic categories (e.g., words). Associated semantic features, however, can activate both the dorsal or ventral streams. When associated semantic features involve action performance in some manner the dorsal streams are more active. Similarly, there is more processing in ventral visual areas, for example, for animals relative to tools (Martin *et al.*, 1996; Mummery *et al.*, 1998; Moore and Price, 1999; Perani *et al.*, 1999), perhaps because we encode the latter more in terms of their visual properties and the former more in terms of their "sensory–motor" properties. Indeed, it has been shown, for example, that color words activate visual areas involved in color processing (Martin *et al.*, 1995).

Thus, during discourse it would be expected that both the dorsal and ventral streams cooperate to yield perception. That is, words variously represent objects and actions and these are associated with cortical areas that encode the referents to those actions and objects. Therefore, both streams must cooperate to yield comprehension. Streams, however, may also compete with each other.

That is, by the present model, it is also possible for one or the other of the streams to become more strongly active and thus play a larger role in mediating perception. The McGurk MacDonald illusion itself will serve to illustrate how this might occur. A McGurk–MacDonald stimulus results in more "combination" responses when the visual modality is relatively more visually salient (i.e., is more likely to be classified correctly without the auditory modality). For example, approximately 98% of participants classify an audio ba-ba paired with a visual ga-ga as da-da. This is the fusion response. By contrast, approximately 55% of participants classify an auditory ga-ga and a visual ba-ba as either ba-ga or ga-ba (McGurk and MacDonald, 1976). This is the "combination" response. The same approximate percentages hold for audio pa-pa and visual ka-ka and audio ka-ka and visual pa-pa pairings respectively. Similarly, visual capture occurs with McGurk–MacDonald stimuli in which the visual stimulus is very salient and the auditory stimulus is relatively ambiguous. Thus, when compared to the stimuli that induce the "fusion" response, stimuli inducing "combination" and "visual capture" responses indicate that as the visual stimulus becomes more salient participants are more likely to classify syllables based on their visual content.

It is proposed that the fused perception is mediated by similar weighting of activity in both the dorsal and ventral processing streams. When, however, the visual stimulus becomes more salient, as in the combination or visual capture perceptions, activity levels increase in the dorsal streams with the result that perception is more mediated by these streams.

This proposal follows from the proposal that speech perception can be conceptualized as both an auditory and a gestural process. When the acoustic track is relatively unambiguous, acoustic processing of the stimulus may be sufficient for perception. The ventral streams mediate acoustic processing of stimuli. In such instances activity in the dorsal streams corresponding to IFMPs may serve only to bias perception. As the ambiguity of the auditory stimulus increases or the ambiguity of the visual stimulus decreases, however, the number of IFMPs should increase as described above. In these instances, the dorsal streams will play a stronger role in mediating perception.

This proposal is supported by selective adaptation studies that show that selective adaptation occurs for the auditory component of McGurk–MacDonald stimulus for *unambiguous* auditory stimuli (Roberts and Summerfield, 1981; Saldaña and Rosenblum, 1994). For ambiguous auditory stimuli (e.g., a sound intermediate between /aba/ and /ada/), however, it has been shown that visual speech can recalibrate auditory speech identification in the direction of the visual stimulus (Bertelson *et al.*, 2003; Vroomen *et al.*, 2004). The authors argue that these adaptation phenomena occur through two different brain mechanisms (Bertelson *et al.*, 2003; Vroomen *et al.*, 2004). By the present model, in the situation in which the auditory stimulus is ambiguous, perception is more mediated by processing occurring in the dorsal streams.

A similar proposal applies for non-imagistic and imagistic manual gesticulations. As the ambiguity of the auditory stimulus increases there is a corresponding increase in the reliance on manual gesticulations to aid in perception (Rogers, 1978; Records, 1994). Non-imagistic manual gesticulations are expected to aid perception by increasing activity in the dorsal streams and thus causing these streams to increase their role in mediating perception. Imagistic manual gesticulations, however, would be expected to increase activity in both the dorsal and ventral streams because these gestures have the additional quality of having semantic content. Furthermore, imagistic manual gesticulations encoding objects may yield a greater relative influence over activity in the ventral streams whereas imagistic manual gesticulations encoding actions may yield a greater relative influence over the dorsal streams. Most imagistic manual gesticulations, however, can be about both objects and their actions. Thus, gesture is an exceptionally powerful component of language comprehension.

In the following sections neurobiological evidence that provides further support for aspects of this model are discussed. Neuroimaging studies of speech perception and language comprehension in multisensory contexts are reviewed. These studies are then contrasted with neuroimaging studies in which only the auditory signal is presented. This provides a more detailed neurobiological argument that complements the argument given in the introduction that speech perception is mediated by both auditory and gestural codes but may be more mediated by acoustic analysis when the auditory signal is unambiguous. When the auditory signal becomes more ambiguous, however, perception becomes more mediated by gestural codes.

8.3 Neurobiological investigations of speech perception

8.3.1 Inverse-forward model pairs associated with observable gestures during multisensory speech perception

During multisensory speech perception, IFMPs begin with inverse models corresponding to heard and observed gestures. These inverse models necessitate evidence for activity in the motor system not related to task demands other than those of listening and watching (e.g., button-pressing). Indeed, when participant's only task is to watch and listen to discourse (Skipper *et al.*, 2005a) or syllables (Callan *et al.*, 2001; Skipper *et al.*, 2004, 2005b, 2005c) spoken by a talker videotaped from the neck up, there is robust activity in the motor system. While watching and listening to interesting audiovisually presented stories containing naturally accompanying head, eyebrow, and mouth movements the dorsal streams are active to a much greater extent than listening to the same stories presented in the auditory modality alone (Skipper *et al.*, 2005a). These regions of the dorsal streams were STp and inferior parietal areas, pars opercularis, premotor cortex, adjacent primary motor cortex, somatosensory cortex, and the cerebellum. Similarly, watching and listening to audiovisually presented syllables containing mouth movements activates the same areas of the dorsal streams (Skipper *et al.*, 2004, 2005c).

Another aspect of the present model is that areas of the dorsal streams active during audiovisual speech perception of syllables in which facial gestures are present are not simply any frontal areas but, rather, those involved in actually producing those syllables. That is, IFMPs occurring during audiovisual speech perception should activate roughly the same circuitry involved in speech production. In one study, after participants listened to and watched syllables, they were asked to produce the same syllables (Skipper *et al.*, 2004, 2005b). In a separate study, participants watched and listened to syllables or watched and listened to the same syllables and imitated them (Skipper *et al.*, 2005c). In both studies activation associated with audiovisual speech perception and speech production of the same sounds overlapped in the dorsal streams. That is, overlap was in the STp and inferior parietal areas, pars opercularis, premotor cortices, primary motor cortex, subcentral gyrus and sulcus, insula, and the cerebellum (Fig. 8.3a).

The model also claims that, when available, the visual aspects of the facial gestures are being used through mediation by the dorsal streams to aid in or improve speech perception. This is supported by the pattern of activity resulting from the overlap between perception of visual syllables alone and speech production which is very similar to the overlap associated with audiovisual speech perception and speech production (Fig. 8.3b). This claim, however, requires that the motor system is not coincidentally active. Rather, the motor system should show sensitivity to the visual content of the observed speech. Indeed, activity in motor system is modulated by the visual saliency of audiovisual speech. It was shown that as the amount of visually distinguishable phonemes in audiovisual stories increases, there is a concomitant increase in STp areas and premotor cortical

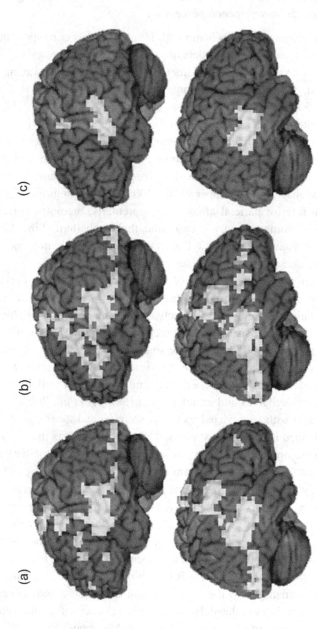

Figure 8.3 Overlap between (a) audiovisual, (b) visual alone, and (c) auditory alone speech perception and speech production and speech imitation. In three separate conditions, participants watched and listened to audiovisually presented syllables, watched the syllables without the accompanying auditory track, or heard the syllables without the accompanying video track. The talker presented in the audiovisual and visual alone conditions was filmed from the neck up. In two further conditions participants produced syllables or imitated audiovisually presented syllables. Overlap was determined by the logical conjunction of images each thresholded at $p < 0.05$ corrected for multiple comparisons. Overlap associated with audiovisual speech perception was in the pars opercularis of the inferior frontal gyrus, dorsal and ventral premotor cortices, primary motor cortex, subcentral gyrus and sulcus, insula, supramarginal gyrus, and posterior aspects of the superior temporal gyrus and sulcus. Overlap associated with visual speech perception alone was in the same areas. By contrast, overlap associated with auditory speech perception alone was only in posterior aspects of the superior temporal gyrus and sulcus bilaterally and the left ventral premotor cortex.

activity (Skipper *et al.*, 2005a). Furthermore, there was no such increase during auditory story-listening alone. This supports the inference that it is the visual content of the stories per se and not the lexical or acoustic properties (or corresponding gestural code for those acoustic properties) that induced this activity. Similarly, highly visually salient syllables produce more intense motor activation when compared with less visible syllables during audiovisually presented speech. Furthermore, the same relationship exists during imitation of the same audiovisual syllables (Skipper *et al.*, 2005c). This indicates a tight coupling between the perception and production systems with respect to observable facial gestures.

Other aspects of the presented model were more explicitly tested using the McGurk–MacDonald fusion effect (Skipper *et al.*, 2004, 2005b, unpublished data). Specifically, it was hypothesized that if activity in motor cortices corresponds to forward models or hypotheses then that activity can be theoretically uncoupled from what is actually presented to the senses. That is, because inverse and corresponding forward models are "models" or "hypotheses" they do not necessarily need to be actual replications or re-representations of the heard and observed world. Cortical activity associated with the congruent audiovisual syllables /pa/, /ka/, and /ta/ was compared to that associated with the McGurk–MacDonald fusion effect, /ta/, elicited when an auditory /pa/ is dubbed onto a visual /ka/. Participants only task was to watch and listen to these syllables and no motor response was required. Activity in the motor system for those participants who perceived the McGurk–MacDonald fusion effect (i.e., /ta/) was most correlated with activity corresponding to the congruent audiovisual stimulus that corresponded to their perception (i.e., /ta/) rather than the stimuli that they were actually presented (i.e., auditory /pa/ or visual /ka/). There was a similar trend for those participants who experienced visual capture (i.e., they heard the visual stimulus /ka/). That is, motor system activity was more correlated with activity corresponding to the audiovisual stimulus /ka/ than either /ta/ or the actually presented auditory track (i.e., /pa/).

The notion that hypotheses or predictions of forward models affect processing in auditory and somatosensory cortices was also tested. It was found that at stimulus onset, activity in auditory and somatosensory cortices when participants perceived the McGurk–MacDonald fusion effect was most highly correlated with the stimulus that corresponded to the input sensory modality. That is, activity was most correlated with /pa/. However, subsequent to motor system activity these areas become more like /ta/. This indicates that the forward models biased activity in the STp and inferior parietal areas.

Also tested was the prediction that different IFMPs in the dorsal streams would activate different motor plans in the same way that producing different syllables requires coordination of different muscles and is, therefore, mediated by non-identical neuronal assemblies. By the proposed model, parametric specification of the motor plan occurs in the interaction between pars opercularis and premotor and primary motor cortices. Thus, it is expected that evidence for this contention will be found in the latter areas. In the same study in which participants passively listened to the McGurk–MacDonald stimulus, they later classified this syllable as /pa/, /ka/, or /ta/ (participants were not aware that they would be doing this after passive watching and listening). It was found that when participants classified the McGurk–MacDonald stimulus as /ta/ or /ka/ spatially adjacent but distinct areas in right inferior and superior parietal lobules, left somatosensory cortices, left ventral premotor, and left primary motor cortex were active. This result, however, could have been confounded with the task or the unnatural McGurk stimulus. To show that this is not the case, discriminant analysis (Haxby *et al.*, 2001) was performed on the congruent /pa/, /ka/, or /ta/ syllables from the passive condition in the same regions. It was shown that these syllables are distinguishable from one another lending support to the claim that different syllables activate different hypotheses with correspondingly different distributions in motor areas.

Finally, the present model is not limited to facial gestures. Josse *et al.* (2005) looked for evidence for the relative mediation of the perception of the manual gesticulations that naturally accompany multisensory communication by both the dorsal and ventral streams as proposed by the present model. Specifically, it was predicted that the addition of manual gesticulations to facial gestures in audiovisual language comprehension should result in relatively more activation of the dorsal streams as more potential IFMPs are being initiated. That is, IFMPs can be derived for observable facial and manual gesticulations as opposed to facial gestures alone. Furthermore, there should be more activation in more dorsal aspects of motor cortex as arm and hand somatotopy is probabilistically more dorsal than somatotopy associated with the face. It was also predicted that manual gesticulations would result in more activation of the ventral streams due to the addition of an added source of redundant semantic content associated with imagistic manual gesticulations.

Indeed, activation of STp areas and premotor cortex was greater while listening to stories with manual gesticulations than to audiovisual stories that were not accompanied by manual gesticulations. Furthermore, the premotor cortex was not only more active but extended more dorsally. Critically, left STa cortex extending to the temporal pole and pars triangularis was more active when speech was accompanied by manual gesticulations than not. Thus, areas affected by manual gesticulations were both in the dorsal and ventral streams.

8.3.2 Inverse-forward model pairs associated with auditory alone and multisensory speech perception

The research reviewed thus far indicates that when facial gestures are present, audiovisual speech perception results in activation of cortical areas involved in speech production and that this is related to IFMPs. That is, the use of facial gestures and manual gesticulations in the process of speech perception is associated with mediation by the dorsal streams. This section argues that in the absence of these gestures, when presented with a clear auditory speech signal, speech perception can be thought of as an auditory process mostly mediated by the ventral streams. It is also argued that as the auditory stimulus becomes more variable or ambiguous, activity in the dorsal streams becomes more pertinent in mediating perception due to an increased use of IFMPs to disambiguate speech. This is true whether during auditory or audiovisual speech perception.

Activation associated with the overlap of auditory speech perception alone and speech production (Fig. 8.3c) supports the contention that auditory speech perception alone can be thought of as an auditory process in which perception is more mediated by the ventral streams (Skipper *et al.*, 2005a). As can be seen, the relative activation of motor cortices during speech perception when the visual modality is not present is less than would be expected simply based on overall levels of stimulation. If amount of stimulation were the underlying factor it could be reasonably argued that motor areas would remain fairly consistent but overall amount of activation would decrease. This is not the case. Auditory speech perception alone leads to restricted activation mostly in the superior aspect of ventral premotor cortex in studies in which participants' only task is to listen to speech sounds (Wilson *et al.*, 2004; Skipper *et al.*, 2005a) (see Fig. 8.3c). This does not substantially change when statistical thresholds are manipulated to be less conservative (Skipper *et al.*, 2005a). Furthermore, this is consistent with most imaging studies which show that passive listening to sounds is relatively inconsistent in activating the pars opercularis or premotor cortices (Zatorre *et al.*, 1996; Belin *et al.*, 2000, 2002; Burton *et al.*, 2000; Zatorre and Belin, 2001).

It seems that the only good way to activate the motor system when the auditory modality is presented alone is by repetitive bombardment with a circumscribed set of syllables (Skipper *et al.*, 2004, 2005b, 2005c; Wilson *et al.*, 2004). When, however, participants are presented with discourse, frontal activity tends to be in prefrontal cortices associated with the ventral streams rather than the premotor cortices associated with the dorsal streams (Skipper *et al.*, 2005a). That is, it seems that the number of active motor cortices during auditory perception alone decreases as the level of discourse increases. The only other proven method of activating the motor system during auditory speech perception alone is to force participants to perform a meta-linguistic or memory task about the acoustic signal (Burton *et al.*, 2000).

These findings are not simply thought to be a methodological issue. Rather, by the present model, existing neuroimaging data with respect to auditory speech perception alone makes sense. Activation of motor cortices is inconsistent during auditory speech perception alone because activity in the ventral streams is more involved in mediating acoustic analysis of speech and language comprehension. What activation of motor cortices can be found decreases as a function of the discourse level because during discourse there are many sources of contextual information other than the acoustic speech signal that can be used to achieve the goal of language comprehension. That is, fewer IFMPs are necessary to disambiguate the acoustic input given the availability of other sources of information in discourse. Furthermore, active tasks activate the dorsal streams because the controlled application of linguistic knowledge is required to complete the meta-linguistic tasks frequently used in neuroimaging studies (e.g., "are these the same or different phonemes"). To make a difficult perceptual judgement about linguistic categories requires IFMPs relative to passive perception of auditory speech sounds in which acoustic processing of the speech signal is generally (but not always) sufficient.

Transcranial magnetic stimulation (TMS) studies, a methodology that allows non-invasive stimulation of a small area of cortex, may be interpreted as further evidence for the position that when visible gestures are present, the dorsal streams are more involved in mediating perception whereas mediation during auditory speech alone is less strong. Specifically, Sundara *et al.* (2001) found that observation of speech movements enhances motor-evoked potential (MEP) amplitude in muscles involved in production of the observed speech but not while listening to the sound alone. However, Watkins *et al.* (2003) found that both hearing and, separately, viewing speech enhances the size of MEPs from the lips. Similarly, Fadiga *et al.* (2002) found enhanced MEPs from the tongue for words that involved tongue movements over those that do not. The difference between the three studies is that Sundara *et al.* (2001) used sub-threshold whereas the latter two used supra-threshold stimulation. The latter two may have, therefore, engaged or bolstered the effect of IFMPs through supra-threshold stimulation that would not normally be prominent in mediating perception during auditory speech perception alone. Nonetheless, these studies at least suggest that there is differential (i.e., less) involvement of the motor system during perception of unambiguous auditory speech.

A similar conclusion might be reached by extrapolating from the macaque to the human brain with respect to mirror neurons. It seems that the number of auditory mirror neurons is quite small and the number of auditory neurons that do not have visual properties is even smaller (Kohler *et al.*, 2002). Auditory mirror neurons may amount to as few as 10% of all mirror neurons. There are more "mouth-mirror neurons" (Ferrari *et al.*, 2003) and mirror neurons coding visible actions than auditory-mirror neurons alone. It may be that more anterior prefrontal regions belonging to the ventral streams are more involved in acoustic analysis of sounds (Romanski *et al.*, 2005). Caution is, of course, warranted because the human brain and body have changed markedly to make speech possible.

Neurobiological evidence also supports the contention that as the auditory stimulus becomes more ambiguous, activity in the dorsal streams becomes more pertinent in mediating perception due to the increased reliance on IFMPs to disambiguate speech. Indeed, when testing a similar model, Callan and colleagues (2004) showed that second-language learners produce more activity in "brain regions implicated with instantiating forward and inverse" models. That is, the authors found greater activity in regions that they define as (and that have been defined here as) composing the dorsal streams (i.e., STp and inferior parietal areas and Broca's area). Native speakers, however, produced greater activation in STa cortices "consistent with the hypothesis that native-language speakers use auditory phonetic representations more extensively than second-language speakers." Furthermore, Callan and colleagues (2003) show that difficult distinctions for non-native speakers (e.g., the r–l distinction for Japanese speakers) activate motor cortices to a much larger extent than an easy distinction. Similar results have been shown when participants listen to speech in a second language, in contrast to one's first language (Dehaene *et al.*, 1997; Hasegawa *et al.*, 2002).

Another study was conducted to test the role of IFMPs in knowledge-based application of expectations during speech perception using another type of ambiguous stimulus, sine wave speech (SWS) sentences (Rémez *et al.*, 1981; Wymbs *et al.*, 2004). SWS use sine waves to model the center frequencies of the first three formants of speech and lack any typical speech cues and are not typically heard as speech but rather as bird chirps, robot sounds, or alien noises (Rémez *et al.*, 1981). If expecting language, however, listeners understand these noises correctly as sentences. By the present model, IFMPs (i.e., knowledge of possible interpretations used to specify expectations) must be implemented when SWS is heard as speech. Indeed, when participants begin hearing SWS as speech activity increases in STp, inferior parietal, and motor areas of the dorsal streams.

8.4 Discussion: Towards another motor theory of speech perception

An active model of speech perception and its relationship to cortical structures and functions has been proposed. In this model listeners test hypotheses about alternative interpretations of the acoustic signal by having attention encompass certain contextual information that can constrain linguistic interpretation. It was proposed that mirror neurons and the mirror system of the dorsal auditory and visual pathways form the basis for inverse and forward models used to test these hypotheses. Inverse models map heard and observed speech productions and manual gesticulations to abstract representations of speaking actions that would have been activated had the observer been the one producing the actions. These representations, a physiological property of mirror neurons in the pars opercularis, are then mapped in a somatotopically appropriate manner to premotor and primary motor cortices. Forward models then map the resulting motor commands to a prediction of the sensory consequences of those commands through reafference to sensory areas. These predictions are compared to processing in the various sensory modalities. This aids in the recognition by constraining alternative linguistic interpretations. According to this model, both facial gestures and manual gesticulations that naturally accompany multisensory language play an important role in language understanding. It was argued that perception in multisensory contexts is more mediated by IFMPs than auditory speech alone. It was argued that in both auditory speech alone and multisensory listening conditions, however, when speech variability or ambiguity exists, IFMPs are utilized to aid in recognition.

The focus of the presented model was the cortical structures associated with mirror neurons, the mirror system, and inverse and forward models in speech perception. Future research will need to further explore the functional properties of each area comprising dorsal streams. More importantly, however, research will need to more explicitly focus on the relative role played by these structures (and other structures not discussed, e.g., the insula and subcentral gyrus) through network-based analysis of neurobiological data. Future instantiations of the model will need to work out a more detailed account of the mechanisms and cortical structures involved in attentional networks related to active processing in speech (Nusbaum and Schwab, 1986). Tentatively, it is thought that frontal cortices more rostral and medial to the frontal cortices discussed with respect to the dorsal and ventral streams in concert with superior parietal areas are important in mediating this process (see Posner and DiGirolamo (2000) for a review). Future work will also need to establish a clear role in this model for cortical–subcortical interactions and the cerebellum. The cerebellum is believed to instantiate inverse and forward models (Miall, 2003). With respect to auditory speech perception, inverse and forward models deriving from the cerebellum may be used as an update signal to cortical regions (Callan *et al.*, 2004).

More generally, this model accounts for a fundamental problem in speech research: understanding how listeners recover the intended linguistic message from the acoustic input. In part, it is suggested that this problem has been partially misconstrued by the emphasis placed on research focusing on the acoustic signals alone. Spoken language evolved as a communication system within the context of face-to-face interaction. It seems quite reasonable that, in trying to understand the basic mechanisms of spoken language processing, the system be investigated in an ecologically reasonable setting.

This view, though in part a motor theory itself, is somewhat different from the perspective taken by the motor theory of speech perception (Liberman and Mattingly, 1985) in spite of surprisingly similar conclusions regarding why visual input is important in speech perception. From the perspective of motor theory, the information presented by the mouth movements and acoustic signals resulting from speech perception are equivalent – both specify underlying articulatory gestures that are themselves the targets of speech recognition. While it is agreed that the visual information about mouth movements does indeed specify to the perceiver intended articulatory gestures, the present view diverges from motor theory in the assumption that this information is not the same as that derived from the acoustic signal. Instead the proposed view is that these sources of information are convergent on a single linguistic interpretation of utterance. Thus, mouth movements and acoustics may be thought of as complementary aspects of speech rather than two pictures of the same thing.

Other theories have proposed that auditory and visual information are integrated in the process of determining linguistic interpretations of an utterance (e.g., Massaro, 1998). However, such theories do not distinguish the qualities and processing attributed to these sources of information. The proposed view is that acoustic and visual–gestural information play different roles in the perceptual process and these differences are reflected in the anatomical and functional cortical networks that mediate their use. As outlined in the present model, visual information about gestures serves to provide hypotheses about the interpretation of the acoustic signal and can form the foundation for hypothesis testing to reject alternative, incorrect interpretations.

Acknowledgments

Many thanks to Michael Arbib for making this a more informed, readable, and thoughtful article. JIS greatly appreciated support from Bernadette Brogan, E. Chen, Shahrina Chowdhury, Uri Hasson, Goulven Josse, Philippa Lauben and Leo Stengel, Nameeta Lobo, Matthew Longo, Robert Lyons, Xander Meadow, Bryon Nastasi, Ana Solodkin, Ryan Walsh, and Nicholas "The Wheem" Wymbs.

References

Allison, T., Puce, A., and McCarthy, G., 2000. Social perception from visual cues: role of the STS region. *Trends Cogn. Sci.* **4**: 267–278.

Arbib, M., and Rizzolatti, G., 1997. Neural expectations: a possible evolutionary path from manual skills to language. *Commun. Cogn.* **29**: 393–424.

Arbib, M. A., Érdi, P., and Szentágothai, J., 1998. *Neural Organization: Structure, Function and Dynamics*. Cambridge, MA: MIT Press

Awh, E., Smith, E. E., and Jonides, J., 1995. Human rehearsal processes and the frontal lobes: PET evidence. *Ann. N.Y. Acad. Sci.*, **769**: 97–117.

Beattie, G., and Coughlan, J., 1999. An experimental investigation of the role of iconic gestures in lexical access using the tip-of-the-tongue phenomenon. *Br. J. Psychol.* **90**: 35–56.

Belin, P., Zatorre, R. J., Lafaille, P., Ahad, P., and Pike, B., 2000. Voice-selective areas in human auditory cortex. *Nature* **403**. 309–312.

Belin, P., Zatorre, R. J., and Ahad, P., 2002. Human temporal-lobe response to vocal sounds. *Brain Res. Cogn. Brain Res.* **13**: 17–26.

Bernstein, L. E., Eberhardt, S. P., and Demorest, M. E., 1998. Single-channel vibrotactile supplements to visual perception of intonation and stress. *J. Acoust. Soc. America* **85**: 397–405.

Bertelson, P., Vroomen, J., and de Gelder, B., 2003. Visual recalibration of auditory speech identification: a McGurk aftereffect. *Psychol. Sci.* **14**: 592–597.

Binder, J. R., Frost, J. A., Hammeke, T. A., *et al.*, 2000. Human temporal lobe activation by speech and nonspeech sounds. *Cereb. Cortex* **10**: 512–528.

Binkofski, F., Amunts, K., Stephan, K. M., *et al.*, 2000. Broca's region subserves imagery of motion: a combined cytoarchitectonic and fMRI study. *Hum. Brain Map.* **11**: 273–285.

Blumstein, S. E., and Stevens, K. N., 1981. Phonetic features and acoustic invariance in speech. *Cognition* **10**: 25–32.

Bookheimer, S., 2002. Functional MRI of language: new approaches to understanding the cortical organization of semantic processing. *Annu. Rev. Neurosci.* **25**: 151–188.

Bookheimer, S. Y., Zeffiro, T. A., Blaxton, T., Gaillard, W., and Theodore, W., 1995. Regional cerebral blood flow during object naming and word reading. *Hum. Brain Map.* **3**: 93–106.

Braun, A. R., Guillemin, A., Hosey, L., and Varga, M., 2001. The neural organization of discourse: an H_2 ^{15}O-PET study of narrative production in English and American sign language. *Brain* **124**: 2028–2044.

Buccino, G., Binkofski, F., Fink, G. R., *et al.*, 2001. Action observation activates premotor and parietal areas in a somatotopic manner: an fMRI study. *Eur. J. Neurosci.* **13**: 400–404.

Buccino, G., Lui, F., Canessa, N., *et al.*, 2004. Neural circuits involved in the recognition of actions performed by nonconspecifics: an fMRI study. *J. Cogn. Neurosci.* **16**: 114–126.

Buchsbaum, B., Hickok, G., and Humphries, C., 2001. Role of left posterior superior temporal gyrus in phonological processing for speech perception and production. *Cogn. Sci.* **25**: 663–678.

Burton, M. W., Small, S. L., and Blumstein, S. E., 2000. The role of segmentation in phonological processing: an fMRI investigation. *J. Cogn. Neurosci.* **12**: 679–690.

Callan, D. E., Callan, A. M., Kroos, C., and Vatikiotis-Bateson, E., 2001. Multimodal contribution to speech perception revealed by independent component analysis: a single-sweep EEG case study. *Brain Res. Cogn. Brain Res.* **10**: 349–353.

Callan, D. E., Tajima, K., Callan, A. M., *et al.*, 2003. Learning-induced neural plasticity associated with improved identification performance after training of a difficult second-language phonetic contrast. *Neuroimage* **19**: 113–124.

Callan, D. E., Jones, J. A., Callan, A. M., and Akahane-Yamada, R., 2004. Phonetic perceptual identification by native- and second-language speakers differentially activates brain regions involved with acoustic phonetic processing and those involved with articulatory-auditory/orosensory internal models. *Neuroimage* **22**: 1182–1194.

Calvert, G. A., Campbell, R., and Brammer, M. J., 2000. Evidence from functional magnetic resonance imaging of crossmodal binding in the human heteromodal cortex. *Curr. Biol.* **10**: 649–657.

Campbell, R., MacSweeney, M., Surguladze, S., *et al.*, 2001. Cortical substrates for the perception of face actions: an fMRI study of the specificity of activation for seen speech and for meaningless lower-face acts (gurning). *Brain Res. Cogn. Brain Res.* **12**: 233–243.

Chao, L. L., and Martin, A., 2000. Representation of manipulable man-made objects in the dorsal stream. *Neuroimage* **12**: 478–484.

Church, R. B., and Goldin-Meadow, S., 1986. The mismatch between gesture and speech as an index of transitional knowledge. *Cognition* **23**: 43–71.

Cooper, W. E., 1979. *Speech Perception and Production: Studies in Selective Adaptation.* Norwood, NJ: Ablex.

Corina, D. P., McBurney, S. L., Dodrill, C., *et al.*, 1999. Functional roles of Broca's area and SMG: evidence from cortical stimulation mapping in a deaf signer. *Neuroimage* **10**: 570–581.

Damasio, H., Grabowski, T. J., Tranel, D., Hichwa, R. D., and Damasio, A. R., 1996. A neural basis for lexical retrieval. *Nature* **380**: 499–505.

Dehaene, S., Dupoux, E., Mehler, J., *et al.*, 1997. Anatomical variability in the cortical representation of first and second language. *Neuroreport* **8**: 3809–3815.

Devlin, J. T., Matthews, P. M., and Rushworth, M. F., 2003. Semantic processing in the left inferior prefrontal cortex: a combined functional magnetic resonance imaging and transcranial magnetic stimulation study. *J. Cogn. Neurosci.* **15**: 71–84.

Diehl, R. L., Lotto, A. J., and Holt, L. L., 2004. Speech perception. *Annu. Rev. Psychol.* **55**: 149–179.

Dodd, B., 1977. The role of vision in the perception of speech. *Perception* **6**: 31–40.

Eliades, S. J., and Wang, X., 2003. Sensory-motor interaction in the primate auditory cortex during self-initiated vocalizations. *J. Neurophysiol.* **89**: 2194–2207.

Embick, D., Marantz, A., Miyashita, Y., O'Neil, W., and Sakai, K. L., 2000. A syntactic specialization for Broca's area. *Proc. Natl. Acad. Sci. USA* **97**: 6150–6154.

Fadiga, L., Craighero, L., Buccino, G., and Rizzolatti, G., 2002. Speech listening specifically modulates the excitability of tongue muscles: a tms study. *Eur. J. Neurosci.* **15**: 399–402.

Federmeier, K. D., and Kutas, M., 2001. Meaning and modality: influences of context, semantic memory organization, and perceptual predictability on picture processing. *J. Exp. Psychol. Learn. Mem. Cogn.* **27**: 202–224.

Ferrari, P. F., Gallese, V., Rizzolatti, G., and Fogassi, L., 2003. Mirror neurons responding to the observation of ingestive and communicative mouth actions in the monkey ventral premotor cortex. *Eur. J. Neurosci.* **17**: 1703–1714.

Fogassi, L., Gallese, V., Fadiga, L., and Rizzolatti, G., 1998. Neurons responding to the sight of goal directed hand/arm actions in the parietal area PF (7b) of the macaque monkey. *Soc. Neurosci. Abstr.* **24**, 257.

Gallese, V., Fadiga, L., Fogassi, L., and Rizzolatti, G., 1996. Action recognition in the premotor cortex. *Brain* **119**: 593–609.

Gentilucci, M., Santunione, P., Roy, A. C., and Stefanini, S., 2004a. Execution and observation of bringing a fruit to the mouth affect syllable pronunciation. *Eur. J. Neurosci.* **19**: 190–202.

Gentilucci, M., Stefanini, S., Roy, A. C., and Santunione, P., 2004b. Action observation and speech production: study on children and adults. *Neuropsychologia* **42**: 1554–1567.

Gerlach, C., Law, I., and Paulson, O. B., 2002. When action turns into words: activation of motor-based knowledge during categorization of manipulable objects. *J. Cogn. Neurosci.* **14**: 1230–1239.

Geschwind, N., 1965. The organization of language and the brain. *Science* **170**: 940–944.

Gold, B. T., and Buckner, R. L., 2002. Common prefrontal regions coactivate with dissociable posterior regions during controlled semantic and phonological tasks. *Neuron* **35**: 803–812.

Goodale, M. A., and Milner, A. D., 1992. Separate visual pathways for perception and action. *Trends Neurosci.* **15**: 20–25.

Grafton, S. T., Fadiga, L., Arbib, M. A., and Rizzolatti, G., 1997. Premotor cortex activation during observation and naming of familiar tools. *Neuroimage* **6**: 231–236.

Grant, K. W., and Greenberg, S., 2001. Speech intelligibility derived from asynchronous processing of auditory-visual information. *Proceedings of the Workshop on Audio-Visual Speech Processing*, Scheelsminde, Denmark, pp. 132–137.

Guenther, F. H., and Ghosh, S. S., 2003. A model of cortical and cerebellar function in speechz. *Proceedings of the 15th International Congress of Phonetic Sciences*, pp. 169–173.

Guenther, F. H., and Perkell, J. S., 2004. A neural model of speech production and its application to studies of the role of auditory feedback in speech. In B. Maassen, R. Kent, H. Peters, P. Van Lieshout, and W. Hulstijn (eds.) *Speech Motor Control In Normal and Disordered Speech*. Oxford, UK: Oxford University Press, pp. 29–49.

Hasegawa, M., Carpenter, P. A., and Just, M. A., 2002. An fMRI study of bilingual sentence comprehension and workload. *Neuroimage* **15**: 647–660.

Hashimoto, Y., and Sakai, K. L., 2003. Brain activations during conscious self-monitoring of speech production with delayed auditory feedback: an fMRI study. *Hum. Brain Map.* **20**: 22–28.

Hatfield, G., 2002. Perception as unconscious inference. In D. Heyer and R. Mausfield (eds.) *Perception and the Physical World: Psychological and Philosophical Issues in Perception*. New York: John Wiley, pp. 115–143.

Haxby, J. V., Gobbini, M. I., Furey, M. L., *et al.*, 2001. Distributed and overlapping representations of faces and objects in ventral temporal cortex. *Science* **293**: 2425–2430.

Heim, S., Opitz, B., Muller, K., and Friederici, A. D., 2003. Phonological processing during language production: fMRI evidence for a shared production-comprehension network. *Brain Res. Cogn. Brain Res.* **16**: 285–296.

Hickok, G., and Poeppel, D., 2004. Dorsal and ventral streams: a framework for understanding aspects of the functional anatomy of language. *Cognition* **92**: 67–99.

Hickok, G., Erhard, P., Kassubek, J., *et al.*, 2000. A functional magnetic resonance imaging study of the role of left posterior superior temporal gyrus in speech production: implications for the explanation of conduction aphasia. *Neurosci. Lett.* **287**: 156–160.

Houde, J. F., and Jordan, M. I., 1998. Sensorimotor adaptation in speech production. *Science* **279**: 1213–1216.

Houde, J. F., Nagarajan, S. S., Sekihara, K., and Merzenich, M. M., 2002. Modulation of the auditory cortex during speech: an meg study. *J. Cogn. Neurosci.* **14**: 1125–1138.

Humphries, C., Willard, K., Buchsbaum, B., and Hickok, G., 2001. Role of anterior temporal cortex in auditory sentence comprehension: an fMRI study. *Neuroreport* **12**: 1749–1752.

Iacoboni, M., 2005. Understanding others: imitation, language, empathy. In S. Hurley and N. Chater (eds.) *Perspectives on Imitation: From Cognitive Neuroscience to Social Science, vol. 1, Mechanisms of Imitation and Imitation in Animals.* Cambridge, MA: MIT Press, pp. 77–99.

Iacoboni, M., Woods, R. P., Brass, M., *et al.*, 1999. Cortical mechanisms of human imitation. *Science* **286**: 2526–2528.

Iacoboni, M., Koski, L. M., Brass, M., *et al.*, 2001. Reafferent copies of imitated actions in the right superior temporal cortex. *Proc. Natl Acad. Sci. USA* **98**: 13995–13999.

Jancke, L., Wustenberg, T., Scheich, H., and Heinze, H. J., 2002. Phonetic perception and the temporal cortex. *Neuroimage* **15**: 733–746.

Jeannerod, M., 1997. *The Cognitive Neuroscience of Action.* Oxford, UK: Blackwell.

Jones, J. A., and Callan, D. E., 2003. Brain activity during audiovisual speech perception: an fMRI study of the McGurk effect. *Neuroreport* **14**: 1129–1133.

Jonides, J., Schumacher, E. H., Smith, E. E., *et al.*, 1998. The role of parietal cortex in verbal working memory. *J. Neurosci.* **18**: 5026–5034.

Jordan, M., and Rumelhart, D., 1992. Forward models: supervised learning with a distal teacher. *Cogn. Sci.* **16**: 307–354.

Josse, G., Suriyakham, L. W., Skipper, J. I., *et al.*, 2005. Language-associated gesture modulates areas of the brain involved in action observation and language comprehension. Poster presented at *The Organization for Human Brain Mapping*, Toronto, Canada.

Jusczyk, P. W., 1981. Infant speech perception: a critical appraisal. In P. D. Eimas and J. L. Miller (eds.) *Perspectives on the Study of Speech.* Hillsdale, NJ: Lawrence Erlbaum, pp. 113–164.

Kaas, J. H., and Hackett, T. A., 2000. Subdivisions of auditory cortex and processing streams in primates. *Proc. Natl Acad. Sci. USA* **97**: 11793–11799.

Kahl, R. (ed.), 1971. *Selected Writings of Hermann von Helmholtz.* Middletown, CT: Wesleyan University Press.

Kendon, A., 1987. On gesture: its complementary relationship with speech. In A. Siegman and S. Feldstein (eds.) *Nonverbal Communication.* Hillsdale, NJ: Lawrence Erlbaum, pp. 65–97.

1994. Do gestures communicate? A review. *Res. Lang. Soc. Interact.* **27**: 175–200.

Kluender, K. R., Diehl, R. L., and Killeen, P. R., 1987. Japanese quail can learn phonetic categories. *Science* **237**: 1195–1197.

Kohler, E., Keysers, C., Umiltá, M. A., *et al.*, 2002. Hearing sounds, understanding actions: action representation in mirror neurons. *Science* **297**: 846–848.

Koski, L., Wohlschlager, A., Bekkering, H., *et al.*, 2002. Modulation of motor and premotor activity during imitation of target-directed actions. *Cereb. Cortex* **12**: 847–855.

Kuhl, P. K., and Meltzoff, A. N., 1982. The bimodal perception of speech in infancy. *Science* **218**: 1138–1141.

Liberman, A. M., and Mattingly, I. G., 1985. The motor theory of speech perception revised. *Cognition* **21**: 1–36.

MacSweeney, M., Woll, B., Campbell, R., *et al.*, 2002. Neural correlates of British sign language comprehension: spatial processing demands of topographic language. *J. Cogn. Neurosci.* **14**: 1064–1075.

Marslen-Wilson, W., and Welsh, A., 1978. Processing interactions and lexical access during word recognition in continuous speech. *Cogn. Psychol.* **10**: 29–63.

Martin, A., Haxby, J. V., Lalonde, F. M., Wiggs, C. L., and Ungerleider, L. G., 1995. Discrete cortical regions associated with knowledge of color and knowledge of action. *Science* **270**, 102–105.

Martin, A., Wiggs, C. L., Ungerleider, L. G., and Haxby, J. V., 1996. Neural correlates of category-specific knowledge. *Nature* **379**: 649–652.

Massaro, D. W., 1998. *Perceiving Talking Faces: From Speech Perception to a Behavioral Principle.* Cambridge, MA: MIT Press.

McGurk, H., and MacDonald, J., 1976. Hearing lips and seeing voices. *Nature* **264**: 746–748.

McNeill, D., 1992. *Hand and Mind: What Gestures Reveal about Thought.* Chicago, IL: University of Chicago Press.

Miall, R. C., 2003. Connecting mirror neurons and forward models. *Neuroreport* **14**: 2135–2137.

Miller, J. D., 1977. Perception of speech sounds by animals: evidence for speech processing by mammalian auditory mechanisms. In T. H. Bullock (ed.) *Recognition of Complex Acoustic Signals.* Berlin: Dahlem Konferenzen, pp. 49–58.

Moore, C. J., and Price, C. J., 1999. A functional neuroimaging study of the variables that generate category-specific object processing differences. *Brain* **122**: 943–962.

Mottonen, R., Krause, C. M., Tiippana, K., and Sams, M., 2002. Processing of changes in visual speech in the human auditory cortex. *Brain Res. Cogn. Brain Res.* **13**: 417–425.

Mummery, C. J., Patterson, K., Hodges, J. R., and Price, C. J., 1998. Functional neuroanatomy of the semantic system: divisible by what? *J. Cogn. Neurosci.* **10**: 766–777.

Munhall, K. G., Gribble, P., Sacco, L., and Ward, M., 1996. Temporal constraints on the McGurk effect. *Percept. Psychophys.* **58**: 351–362.

Munhall, K. G., Jones, J. A., Callan, D. E., Kuratate, T., and Vatikiotis-Bateson, E., 2004. Visual prosody and speech intelligibility: head movement improves auditory speech perception. *Psychol. Sci.* **15**: 133–137.

Nicholson, K. G., Baum, S., Cuddy, L. L., and Munhall, K. G., 2002. A case of impaired auditory and visual speech prosody perception after right hemisphere damage. *Neurocase* **8**: 314–322.

Numminen, J., Salmelin, R., and Hari, R., 1999. Subject's own speech reduces reactivity of the human auditory cortex. *Neurosci. Lett.* **265**: 119–122.

Nusbaum, H. C., and Magnuson, J., 1997. Talker normalization: phonetic constancy as a cognitive process. In K. Johnson and J. W. Mullennix (eds.) *Talker Variability in Speech Processing.* San Diego, CA: Academic Press, pp. 109–132.

Nusbaum, H. C., and Morin, T. M., 1992. Paying attention to differences among talkers. In Y. Tohkura, Y. Sagisaka, and E. Vatikiotis-Bateson (eds.) *Speech Perception, Production, and Linguistic Structure.* Tokyo: Ohmasha Publishing, pp. 113–134.

Nusbaum, H. C., and Schwab, E. C., 1986. The role of attention and active processing in speech perception. In E. C. Schwab and H. C. Nusbaum (eds.) *Pattern Recognition by Humans and Machines, vol. 1, Speech Perception.* New York: Academic Press, pp. 113–157.

Olson, I. R., Gatenby, J. C., and Gore, J. C., 2002. A comparison of bound and unbound audio-visual information processing in the human cerebral cortex. *Brain Res. Cogn. Brain Res.* **14**: 129–138.

Paulesu, E., Frith, C. D., and Frackowiak, R. J., 1993a. The neural correlates of the verbal component of working memory. *Nature* **362**: 342–345.

Paulesu, P., Frith, C. D., Bench, C. J., *et al.*, 1993b. Functional anatomy of working memory: The articulatory loop. *J. Cereb. Blood Flow Metab.* **13**: 551.

Paus, T., Perry, D. W., Zatorre, R. J., Worsley, K. J., and Evans, A. C., 1996. Modulation of cerebral blood flow in the human auditory cortex during speech: role of motor-to-sensory discharges. *Eur. J. Neurosci.* **8**: 2236–2246.

Perani, D., Schnur, T., Tettamanti, M., *et al.*, 1999. Word and picture matching: a PET study of semantic category effects. *Neuropsychologia* **37**: 293–306.

Perkell, J. S., Matthies, M. L., Svirsky, M. A., and Jordan, M. I., 1995. Goal-based speech motor control: a theoretical framework and some preliminary data. *J. Phonet.* **23**: 23–35.

Petrides, M., and Pandya, D. N., 1999. Dorsolateral prefrontal cortex: comparative cytoarchitectonic analysis in the human and the macaque brain and corticocortical connection patterns. *Eur. J. Neurosci.* **11**: 1011–1036.

2002. Comparative cytoarchitectonic analysis of the human and the macaque ventrolateral prefrontal cortex and corticocortical connection patterns in the monkey. *Eur. J. Neurosci.* **16**: 291–310.

Plaut, D. C., and Kello, C. T., 1999. The emergence of phonology from the interplay of speech comprehension and production: a distributed connectionist approach. In B. MacWhinney (ed.) *The Emergence of Language.* Mahwah, NJ: Lawrence Erlbaum, pp. 381–415.

Poldrack, R. A., Wagner, A. D., Prull, M. W., *et al.*, 1999. Functional specialization for semantic and phonological processing in the left inferior prefrontal cortex. *Neuroimage* **10**: 15–35.

Posner, M. I., and DiGirolamo, G. J., 2000. Attention in cognitive neuroscience: an overview. In M. S. Gazzaniga (ed.) *The New Cognitive Neuroscience*, 2nd edn. Cambridge, MA: MIT Press, pp. 621–632.

Pulvermüller, F., Harle, M., and Hummel, F., 2000. Neurophysiological distinction of verb categories. *Neuroreport* **11**: 2789–2793.

2001. Walking or talking? Behavioral and neurophysiological correlates of action verb processing. *Brain Lang.* **78**: 43–168.

Rauschecker, J. P., and Tian, B., 2000. Mechanisms and streams for processing of "what" and "where" in auditory cortex. *Proc. Natl Acad. Sci. USA* **97**: 11800–11806.

Records, N. L., 1994. A measure of the contribution of a gesture to the perception of speech in listeners with aphasia. *J. Speech Hear. Res.* **37**: 1086–1099.

Reisberg, D., McLean, J., and Goldfield, A., 1987. Easy to hear but hard to understand: a lipreading advantage with intact auditory stimuli. In B. Dodd and R. Campbell (eds.)

Hearing by Eye: The Psychology of Lipreading. Hillsdale, NJ: Lawrence Erlbaum, pp. 97–114.

Rémez, R. E., Rubin, P. E., Pisoni, D. B., and Carrell, T. I., 1981. Speech perception without traditional speech cues. *Science* 212: 947–950.

Renwick, M., Shattuck-Hufnagel, S., and Yasinnik, Y., 2001. The timing of speech-accompanying gestures with respect to prosody. *J. Acoust. Soc. America* 115: 2397–2397.

Riecker, A., Ackermann, H., Wildgruber, D., *et al.*, 2000. Articulatory/phonetic sequencing at the level of the anterior perisylvian cortex: a functional magnetic resonance imaging (fMRI) study. *Brain Lang.* 75: 259–276.

Risberg, A., and Lubker, J., 1978. Prosody and speechreading. *Speech Transmission Lab. Q. Progr. Rep. Status Report* 4: 1–16.

Rizzolatti, G., and Craighero, L., 2004. The mirror-neuron system. *Annu. Rev. Neurosci.* 27: 169–192.

Rizzolatti, G., Fogassi, L., and Gallese, V., 2002. Motor and cognitive functions of the ventral premotor cortex. *Curr. Opin. Neurobiol.* 12: 149–154.

Roberts, M., and Summerfield, Q., 1981. Audiovisual presentation demonstrates that selective adaptation in speech perception is purely auditory. *Percept. Psychophys.* 30: 309–314.

Rogers, W. T., 1978. The contribution of kinesic illustrators towards the comprehension of verbal behavior within utterances. *Hum. Commun. Res.* 5: 54–62.

Romanski, L. M., Averbeck, B. B., and Diltz, M., 2005. Neural representation of vocalizations in the primate ventrolateral prefrontal cortex. *J. Neurophysiol.* 93: 734–747.

Saldaña, H. M., and Rosenblum, L. D., 1994. Selective adaptation in speech perception using a compelling audiovisual adaptor. *J. Acoust. Soc. America* 95: 3658–3661.

Sams, M., Aulanko, R., Hamalainen, M., *et al.*, 1991. Seeing speech: visual information from lip movements modifies activity in the human auditory cortex. *Neurosci. Lett.* 127: 141–145.

Schumacher, E. H., Lauber, E., Awh, E., *et al.*, 1996. PET evidence for an amodal verbal working memory system. *Neuroimage* 3: 79–88.

Scott, S. K., and Wise, R. J. S., 2003. PET and fMRI studies of the neural basis of speech perception. *Speech Commun.* 41: 23–34.

Seltzer, B., and Pandya, D. N., 1978. Afferent cortical connections and architectonics of the superior temporal sulcus and surrounding cortex in the rhesus monkey. *Brain Res.* 149: 1–24.

Shergill, S. S., Brammer, M. J., Fukuda, R., *et al.*, 2002. Modulation of activity in temporal cortex during generation of inner speech. *Hum. Brain Map.* 16: 219–227.

Singer, M. A., and Goldin-Meadow, S., 2005. Children learn when their teacher's gestures and speech differ. *Psychol. Sci.* 16: 85–89.

Skipper, J. I., van Wassenhove, V., Nusbaum, H. C., and Small, S. L., 2004. Hearing lips and seeing voices in the brain: motor mechanisms of speech perception. Poster presented at *11th Annual Meeting of the Cognitive Neuroscience Society*, San Francisco, CA.

Skipper, J. I., Nusbaum, H. C., and Small, S. L., 2005a. Listening to talking faces: motor cortical activation during speech perception. *Neuroimage* 25: 76–89.

Skipper, J. I., Nusbaum, H. C., van Wassenhove, V., *et al.*, 2005b. The role of ventral premotor and primary motor cortex in audiovisual speech perception. Poster presented at *The Organization for Human Brain Mapping*, Toronto, Canada.

Skipper, J. I., Wymbs, N. F., Lobo, N., Cherney, L. R., and Small, S. L., 2005c. Common motor circuitry for audiovisual speech imitation and audiovisual speech perception. Poster presented at *The Organization for Human Brain Mapping*, Toronto, Canada.

Stevens, K. N., and Blumstein, S. E., 1981. The search for invariant acoustic correlates of phonetic features. In P. D. Eimas and J. L. Miller (eds.) *Perspectives on the Study of Speech*. Hillsdale, NJ: Lawrence Erlbaum, pp. 1–39.

Stevens, K. N., and Halle, M., 1967. Remarks on analysis by synthesis and distinctive features. In W. Walthen-Dunn (ed.) *Models for the Perception of Speech and Visual Form*. Cambridge, MA: MIT Press, pp. 88– 102.

Stromswold, K., Caplan, D., Alpert, N., and Rauch, S., 1996. Localization of syntactic comprehension by positron emission tomography. *Brain Lang.* **52**: 452–473.

Sumby, W. H., and Pollack, I., 1954. Visual contribution of speech intelligibility in noise. *J. Acoust. Soc. America* **26**: 212–215.

Summerfield, A. Q., 1987. Some preliminaries to a comprehensive account of audio-visual speech perception. In B. Dodd and R. Campbell (eds.) *Hearing by Eye: The Psychology of Lip Reading*. Hillsdale, NJ: Lawrence Erlbaum, pp. 3–51.

Sundara, M., Namasivayam, A. K., and Chen, R., 2001. Observation-execution matching system for speech: a magnetic stimulation study. *Neuroreport* **12**: 1341–1344.

Surguladze, S. A., Calvert, G. A., Brammer, M. J., *et al.*, 2001. Audio-visual speech perception in schizophrenia: an fMRI study. *Psychiatry Res.* **106**: 1–14.

Tranel, D., 2001. Combs, ducks, and the brain. *Lancet* **357**: 1818–1819.

Tzourio, N., Nkanga-Ngila, B., and Mazoyer, B., 1998. Left planum temporale surface correlates with functional dominance during story listening. *Neuroreport* **9**: 829–833.

Ungerleider, L. G., and Mishkin, M., 1982. Two cortical visual systems. In D. J. Ingle, M. A. Goodale, and R. J. W. Mansfield (eds.) *Analysis of Visual Behavior*. Cambridge, MA: MIT Press, pp. 549–586.

Vroomen, J., van Linden, S., Keetels, M., de Gelder, B., and Bertelson, P., 2004. Selective adaptation and recalibration of auditory speech by lipread information: dissipation. *Speech Commun.* **44**: 55–61.

Wagner, A. D., Koutstaal, W., Maril, A., Schacter, D. L., and Buckner, R. L., 2000. Task-specific repetition priming in left inferior prefrontal cortex. *Cereb. Cortex* **10**: 1176–1184.

Watkins, K. E., Strafella, A. P., and Paus, T., 2003. Seeing and hearing speech excites the motor system involved in speech production. *Neuropsychologia* **41**: 989–994.

Wildgruber, D., Ackermann, H., and Grodd, W., 2001. Differential contributions of motor cortex, basal ganglia, and cerebellum to speech motor control: effects of syllable repetition rate evaluated by fMRI. *Neuroimage* **13**: 101–109.

Wilson, S. M., Saygin, A. P., Sereno, M. I., and Iacoboni, M., 2004. Listening to speech activates motor areas involved in speech production. *Nature Neurosci.* **7**: 701–702.

Wise, R. J., Greene, J., Buchel, C., and Scott, S. K., 1999. Brain regions involved in articulation. *Lancet* **353**: 1057–1061.

Wise, R. J., Scott, S. K., Blank, S. C., *et al.*, 2001. Separate neural subsystems within 'Wernicke's area'. *Brain* **124**: 83–95.

Wolpert, D. M., and Kawato, M., 1998. Multiple paired forward and inverse models for motor control. *Neur. Networks* **11**: 1317–1329.

Wright, T. M., Pelphrey, K. A., Allison, T., McKeown, M. J., and McCarthy, G., 2003. Polysensory interactions along lateral temporal regions evoked by audiovisual speech. *Cereb. Cortex* **13**: 1034–1043.

Wymbs, N. F., Nusbaum, H. C., and Small, S. L., 2004. The informed perceiver: neural correlates of linguistic expectation and speech perception. Poster presented at *11th Annual Meeting of the Cognitive Neuroscience Society*, San Francisco, CA.

Zatorre, R. J., and Belin, P., 2001. Spectral and temporal processing in human auditory cortex. *Cereb. Cortex* **11**: 946–953.

Zatorre, R. J., Meyer, E., Gjedde, A., and Evans, A. C., 1996. PET studies of phonetic processing of speech: review, replication, and reanalysis. *Cereb. Cortex* **6**: 21–30.

Part IV

From mirror system to syntax and Theory of Mind

9

Attention and the minimal subscene

Laurent Itti and Michael A. Arbib

9.1 Introduction

The Mirror System Hypothesis (MSH), described in Chapter 1, asserts that recognition of manual actions may ground the evolution of the language-ready brain. More specifically, the hypothesis suggests that manual praxic actions provide the basis for the successive evolution of pantomime, then protosign and protospeech, and finally the articulatory actions (of hands, face and – most importantly for speech – voice) that define the phonology of language. But whereas a praxic action just *is* a praxic action, a communicative action (which is usually a compound of meaningless articulatory actions; see Goldstein, Byrd, and Saltzman, this volume, on duality of patterning) is *about something else*. We want to give an account of that relationship between the sign and the signified (Arbib, this volume, Section 1.4.3).

Words and sentences can be about many things and abstractions, or can have social import within a variety of speech acts. However, here we choose to focus our discussion by looking at two specific tasks of language in relation to a visually perceptible scene: (1) generating a description of the scene, and (2) answering a question about the scene. At one level, vision appears to be highly parallel, whereas producing or understanding a sentence appears to be essentially serial. However, in each case there is both low-level parallel processing (across the spatial dimension in vision, across the frequency spectrum in audition) and high-level seriality in time (a sequence of visual fixations or foci of attention in vision, a sequence of words in language). Our attempt to integrate these diverse processes requires us to weave together a large set of empirical data and computational models. The rest of this introduction is designed to provide a road map for the rest of the chapter to help the reader integrate the varied strands of our argument.

Section 9.2: We introduce the notion of minimal and anchored subscenes as providing the link between the structure of a visual scene and the structure of a sentence which describes or queries it.

Section 9.3: We review the general idea of "cooperative computation" as well as of perceptual and motor schemas, and illustrate it with two classical models: the VISIONS

Action to Language via the Mirror Neuron System, ed. Michael A. Arbib. Published by Cambridge University Press.
© Cambridge University Press 2006.

model of recognizing a visual scene, and the HEARSAY model of understanding a spoken sentence.

Section 9.4: We also note two background studies of visual attention to which we have contributed, in both of which the notion of "winner take all" plays a central role. The first, by Arbib and Didday (1975), is more comprehensive in its conceptual structure; the second, by Itti and Koch (2000, 2001), is restricted to low-level salience but has been worked out in detail and has yielded interesting results through detailed simulation.

Section 9.5: The unfolding of a description and the answering of a question require that attention be driven not solely by low-level salience but also by the search for objects (and actions) that are deemed relevant to the unfolding task. We thus present a conceptual model, based on a model implemented by Navalpakkam and Itti (2005), which builds on the concepts of Section 9.4 but with the addition of a task relevance map, short-term memory, top–down task-dependent biasing, and other subsystems. We make explicit the role of minimal and anchored subscenes in this processing, in particular by showing how continuous scene analysis yields sequences of temporally less volatile short-term memory representations that are amenable to verbalization.

Section 9.6: But of course we are not the first to investigate the relevance of visual attention to language. Henderson and Ferreira (2004) offer a rich collection of relevant review articles. We briefly note the contributions of Tanenhaus *et al.* (2004) and Griffin and Bock (2000).

Section 9.7: The theoretical framework that comes closest to what we seek is that offered by Knott (2003), which builds on the Itti–Koch model and links this to a system for execution and observation of actions. Knott translates the sensorimotor sequence of attention to the scene into the operations involved in constructing the syntactic tree for its description. We argue that the theory is more pertinent if we build the scene description on symbolic short-term memory, with items tagged for relevance, rather than on the eye movements that went into the building of that representation. Competitive queuing yields the sequence of "virtual attention shifts" for this internal representation.

Section 9.8: How can some of the new notions introduced in this chapter be explored experimentally? One difficulty is that the framework is oriented more towards dynamic scenes, in which change constantly occurs, than towards static images which may more easily be studied experimentally. To illustrate how our framework may prompt for new experiments on human comprehension of dynamic scenes, we conducted a pilot study of a person describing a range of videoclips and analyzed the relation between the subject's eye movements and the descriptions he generated.

In the remainder of the chapter, we review other research efforts relevant to developing a comprehensive computational model adequate to meeting the challenges of the data and modeling introduced above.

Section 9.9: We note interesting efforts (VITRA: Nevatia *et al.*, 2003) within AI on the recognition of events in dynamically changing scenes.

Section 9.10: We link Knott's brief description of motor control to the more general notion of forward and inverse models, then briefly summarize the discussion given by

Oztop, Bradley, and Arbib (this volume) relating the FARS and MNS models of manual action and action recognition, respectively, to this general framework. We return to the distinction between the sign and the signified to distinguish between producing or perceiving a word and perceiving the concept that it signifies.

Section 9.11: We also note the preliminary work by Vergnaud and Arbib giving a bipartite analysis of the verb in terms of mirror neurons and canonical neurons.

Section 9.12: Both Knott and Vergnaud operate within the framework of generative grammar. We review the attractions of an alternative framework, construction grammar (employed by Kemmerer, this volume), for our work. We show how "vision constructions" may synergize with "grammar constructions" in structuring the analysis of a scene in relation to the demands of scene description and question answering in a way which ties naturally into our concern with minimal and anchored subscenes.

Section 9.13: We revisit the Salience, Vision, and Symbolic Schemas (SVSS) model to discuss its extension to extract episodes from dynamically changing visual input. While the detailed discussion of language processing is outside the scope of this chapter, we do suggest how the integration of ideas from Levelt's (2001) model of language production and the HEARSAY model of speech understanding may set the stage for a cooperative computation model of language perception and production. We then outline how this might be integrated with our general analysis of attention-based visual perception, and of the generation and recognition of actions, in defining a truly general model for linking eye movements to the processes of scene description and question answering.

Section 9.14: We recall a highly schematic diagram developed by Arbib and Bota (2003) which sketches the relations between action generation, action recognition and language within the anatomical framework offered by the MSH. Where this diagram focuses on single actions, we review the insights offered by the functional analysis in this chapter towards extending the MSH to complex scenes.

9.2 Minimal and anchored subscenes

In this chapter, we outline a computational framework in which to probe the visual mechanisms required to recognize salient aspects of the environment, whether as a basis for praxic action (e.g., manipulation of objects or locomotion), or as part of the linguistic acts of scene description or question answering. The key concept introduced for this purpose is that of a *minimal subscene* as the basic unit of action recognition, in which an agent interacts with objects or others. We now define minimal subscenes and their expansion to anchored subscenes, and then briefly review processes of active, goal-oriented scene perception to be treated at greater length in the rest of the chapter.

9.2.1 Minimal subscenes defined

While the definition of what constitutes a minimal subscene is not fully rigorous, the notion is that, for whatever reason, an agent, action, or object may attract the viewer's attention more strongly than other elements that may be present in the scene. If that agent, action, or object is of interest by some measure, then attention will be directed "top–down" to seek to place this "anchor" in context within the scene:

- Given an *agent* as "anchor," complete the minimal subscene by including the actions in which the agent is engaged and the objects and (where appropriate) other agents engaged in these actions
- Given an *action* as "anchor," complete the minimal subscene by including the agents and objects involved in that action
- Given an *object* as "anchor," complete the minimal subscene by including the actions performed on that object and the agents and (where appropriate) other objects engaged in that action.

Importantly, which anchor is of interest and which minimal subscene is incrementally constructed during observation is dependent upon tasks and goals, so that the same visual input may yield different minimal subscenes depending on these key top–down factors. The minimal subscene thus is fundamentally an observer-dependent notion. Data observations only enrich it to the extent that they are relevant to the tasks and goals of the observer. Also noteworthy is that our definition does not require that all the elements of the minimal subscene be present in the visual world at every instant; thus, it essentially is a *short-term memory* construct, which evolves at a somewhat slower pace than the constantly streaming visual inputs that nourish it.

A minimal subscene as defined above may be extended by adding details to elements already in the subscene, or by adding new elements that bear some strong relationship to elements in the subscene. We refer to the result as an *anchored subscene* since it is defined by expanding the relationships which link elements of the original subscene to its anchor. However, attention may be caught by a salient object or action that is not part of the original subscene and that is not related or relevant to the current subscene, yet triggers the interest of the observer. This may anchor a new subscene to be added to the observer's inherently dynamic short-term memory (STM). Thus, by definition, minimal or anchored subscenes must at least include some neural representations for or links to the collection of featural representations (appearance, identity, properties), spatial representations (location, size, relationships), and dynamic representations (actions linking actors, objects, and actions) that have been observed and attended to in the recent past, and that have further been retained as potentially relevant to the current tasks and goals of the observer. This new subscene may either coexist with the previous one, for example if the actors of the previous subscene remain present, or it may take precedence and become the only current minimal subscene, for example after the elements of the previous subscene have drifted out of the field of view. Internal representations of the scene thus develop in STM and are refined as new data are accumulated by the sensors and evaluated against the internal beliefs and goals of the observer. Conversely, aspects of STM may be discarded either due to some process of forgetting, or because they are no longer relevant to the goals of the system or its changing relationship with the environment.

A significant contribution of the present work is to examine how these notions, which we have thus far introduced in relation to perception, link to sentences which describe a subscene or raise questions about the scene. We view the minimal subscene as providing a middle-ground representation between patterns of neural activation in response to the observation of a dynamic visual scene involving some actors and actions, and more abstract symbolic descriptions of the scene in some human language. We study how such an *action–object frame* relates to the *verb–argument structure* for "Who is doing what and to whom." For example, if Harry's forearm moves up and down as he holds a hammer, then we may have an action Hit(Harry, Nail, Hammer) which falls under the general action–object frame of Action(Actor, Patient, Instrument) and is expressible in English as "Harry is hitting the nail with a hammer." Of course, there are other ways to express this same idea in English, such as "Harry hammered the nail." As made clear by our tripartite account of "minimal subscene," the viewer's attention might first be focused on Harry, say, and extend to recognition of his action and the objects involved, or start with the hammering movement and extend this to complete the minimal subscene. The anchor of the scene will in general determine the focus of the sentence.

9.2.2 Processes of active, goal-oriented scene perception

Our understanding of the brain mechanisms at play during active, goal-oriented scene perception has significantly progressed over recent years, through psychophysical, electrophysiological, imaging, and modeling studies. The current state of understanding is sketched as in Fig. 9.1. Humans rapidly extract a wealth of information, often referred to as the "gist of the scene," from the first glance at a new scene. We use the term gist in the sense of some overall classification – such as "a beach scene," "a landscape," "a suburban street scene," or "a battle zone" – which can provide spatial priors as to the most likely regions of interest. Gist thus refers to scene representations that are acquired over very short time-frames, for example as observers view briefly flashed photographs and are asked to describe what they see (Potter, 1975; Biederman *et al.*, 1983; Oliva and Schyns, 1997).[1] With very brief exposures (100 ms or below), gist representations are typically limited to a few general semantic attributes (e.g., indoors, outdoors, office, kitchen) and a coarse evaluation of distributions of visual features (e.g., highly colorful, grayscale, several large masses, many small objects) (Sanocki and Epstein, 1997; Rensink, 2000). Gist may be computed in brain areas which have been shown to preferentially respond to types of "places," that is, visual scene types with a restricted spatial layout (Epstein *et al.*, 2000). Spectral contents and color diagnosticity have been shown to influence gist perception (Oliva and Schyns, 1997, 2000), leading to the development of computational models that approximate the computation of gist by classifying an image along a number of semantic and featural dimensions, based on a one-shot global analysis of the spectral composition of a scene (Torralba, 2003).

---

[1] The first-impression notion of gist is to be contrasted with the sense of gist as summarizing the key aspects of a scene that can only be extracted after careful analysis, as in "John is finally proposing to Mary but she doesn't seem happy about it."

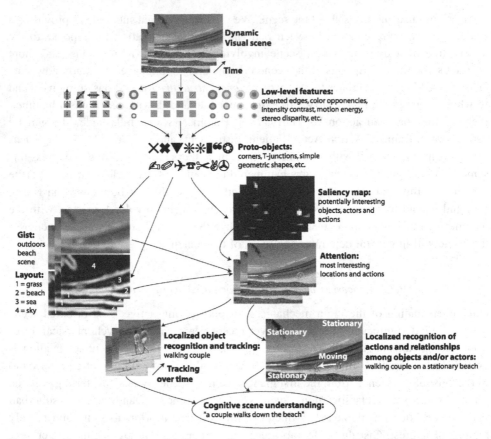

Figure 9.1 Overview of how task influences visual attention: it primes the desired features that are in turn used to compute the gist, layout, and the bottom–up salience of scene locations. Further, task influences top–down processes that predict the relevance of scene locations based on some prior knowledge. Finally, the gist, layout, bottom–up salience, and top–down relevance of scene locations are somehow combined to decide the focus of attention.

In our conceptual framework, the gist representation cooperates with visual attention to focus on cognitively salient visual targets. The result will be updating of the minimal or anchored subscene representation after each shift of attention to a new scene element that is related to elements already represented within that subscene, or the initiation or updating of another subscene in case no such relation exists. The focusing of attention onto specific scene elements has been shown to be mediated by the interplay between bottom–up (dependent upon uninterpreted properties of the input images) and top–down factors (which bring volition, expectations, and internal beliefs of the observers into play). Attention is thus attracted bottom–up towards conspicuous or salient scene elements (e.g., a bright flickering light in an otherwise dark environment) and top–down towards locations which the observer believes are or soon will be important (e.g., the expected end point of a ball's trajectory). We posit that this interplay between reactive and proactive modes of attention plays a critical role in the elaboration of the minimal subscene. Indeed, studies of change blindness (Rensink *et al.*, 1997) and inattentional blindness (Simons, 2000) suggest that human observers are unlikely to notice or report scene elements which an outsider might expect to gain their attention, either because they were distracted by salient bottom–up stimuli, or because they were strongly focused top–down onto a visual inspection task. Thus, attention may gate the addition of new constituents to subscene representations.

Here we must add that, while much of present visual scene analysis focuses on the recognition of elements in a single static view, our real concern (as becomes clear in Sections 9.8 and 9.9) is with dynamically changing scenes. In such a framework, the gist will play a broader role being updated as the dynamics of the scene and its interpretation unfold.

Recognizing objects, actors, and actions is another obvious necessary step in the construction of subscenes. Unless an attended entity is recognized, it will be difficult to evaluate how relevant it is to current behavioral goals, and whether it should be immediately discarded or integrated into the current subscene representation. Beyond the instantaneous recognition of the scene element that is the current focus of attention, some memory trace of that element, if deemed relevant and recognized with high confidence, must be stored if the subscene is to be built incrementally by integrating across attention and gaze shifts, and possibly later reported on. Many studies have explored the mechanisms and limitations of the STM processes involved in transiently holding some representation of the constituents of the minimal subscene. These studies have yielded a wide range of putative representations and levels of detail, ranging from the "world as an outside memory" hypothesis (O'Regan, 1992) where virtually no visual details are retained internally in memory because they could be easily fetched back from the outside world itself if desired, to collections of four to six "object files" (Treisman and Gelade, 1980; Irwin and Andrews, 1996; Irwin and Zelinsky, 2002) which crudely abstract the visual details of retained scene elements, to more complete representations with rich detail (Hollingworth and Henderson, 2002; Hollingworth, 2004).

Finally, cognition and reasoning must influence the construction of the minimal subscene in at least two ways. First, they determine in real-time the evolution of top–down influences on attention, by evaluating previous objects of attention and recognition against goals, and inferring a course of action. For example, if your attention was just caught by a man's face, but you want to know about his shoes, just look down. Second, in many situations, it seems intuitively reasonable to assume that not all attended and recognized scene elements will automatically become part of the minimal subscene. Instead, some cognitive evaluation of the behavioral relevance of an object of attention is likely to play a role in deciding whether it will be retained or immediately forgotten. For example, as you scan a crowd for a loved one, you will probably forget about most of the persons who you attended to but who were false positives during your search. Many factors, possibly including familiarity (e.g., what is my cousin doing in that crowd?) or an element of surprise (who is that man with a green face?) may augment evaluation of attended scene elements when determining whether or not they will be integrated into the minimal subscene representation or barred from it and forgotten.

This brief overview suggests a highly active and dynamic view of goal-oriented scene perception. Instead of a snapshot of the scene such as gist may be, the minimal subscene – and the anchored subscene which extends it – is an evolving representation, shaped by a combination of attending, recognizing, and reasoning over time. Before diving into the details of how the orchestration of all the factors just mentioned may contribute to the minimal subscene, one challenge is to define a suitable neurocognitive framework to support its implementation.

9.3 Cooperative computation and schema theory

Visual organization in the primate brain relies heavily on topographic neural maps representing various transformed versions of the visual input, distributed population coding for visual features, patterns, or objects present in the scene, and highly dynamic representations shaped both bottom–up (by visual inputs) and top–down (by internal cognitive representations). To relate these highly redundant, distributed and dynamic neural representations to corresponding linguistic representations, we appeal to the general idea of *cooperative computation* as well as the perceptual and motor schemas of *schema theory*, and illustrate them with two classical models: the VISIONS model of recognizing a visual scene, and the HEARSAY model of understanding a spoken sentence.

9.3.1 Schemas for perceptual structures and distributed motor control

Arbib (1981; Arbib et al., 1998, ch. 3) offers a version of *schema theory* to complement neuroscience's terminology for levels of structural analysis with a framework for analysis of function in terms of *schemas* (units of functional analysis). Central to our approach is *action-oriented perception*, as the active "organism" (which may be an animal or an embodied computational system) seeks from the world the information it needs to pursue its chosen course of action. A *perceptual schema* not only determines whether a given "domain of interaction" (an action-oriented generalization of the notion of object) is present in the environment but can also provide parameters concerning the current relationship of the organism with that domain. *Motor schemas* provide the control systems which can exploit such parameters and can be coordinated to effect the wide variety of action.

A schema is, in basic cases, what is learned about some aspect of the world, combining knowledge with the processes for applying it; a brain may deploy more than one *instance* of the processes that define a given schema. In particular, schema instances may be combined (possibly including those of more abstract schemas, including coordinating schemas) to form *schema assemblages*. For example, an assemblage of perceptual schema instances may combine an estimate of environmental state with a representation of goals and needs. A *coordinated control program* is a schema assemblage which processes input via perceptual schemas and delivers its output via motor schemas, interweaving the activations of these schemas in accordance with the current task and sensory environment to mediate more complex behaviors.

Figure 9.2a shows the original coordinated control program (Arbib (1981), analyzing data of Jeannerod and Biguer (1982)). As the hand moves to grasp an object, it is *preshaped* so that when it has almost reached the object, it is of the right shape and orientation to enclose some part of the object prior to gripping it firmly. Moreover (to a first approximation; see, e.g., Hoff and Arbib (1993) for a revised analysis), the movement can be broken into a fast phase and a slow phase. The output of three perceptual schemas is available for the control of the hand movement by concurrent activation of two motor schemas: *Reaching* controls the arm to transport the hand towards the object and *Grasping* first preshapes the hand. Once the hand is preshaped, it is (according to Fig. 9.2) only the completion of the fast phase of hand transport that "wakes up" the final stage of *Grasping* to shape the fingers under control of tactile feedback. (This model anticipates the much later discovery of perceptual schemas for grasping in a localized area (AIP) of parietal cortex and motor schemas for grasping in a localized area (F5) of premotor cortex; see Chapter 1 and below.) Jeannerod (1997) surveys the role of schemas and other constructs in the cognitive neuroscience of action; schemas have also played an important role in the development of behavior-based robots (Arkin, 1998).

Figure 9.2a clearly separates perceptual and motor schemas. But this raises the question as to why we do not combine perceptual and motor schemas into a single notion of schema that integrates sensory analysis with motor control. Indeed, there are cases where such a combination makes sense. However, recognizing an object (an apple, say) may be linked to many different courses of action (to place it in one's shopping basket; to place it in a bowl; to pick it up; to peel it; to cook with it; to eat it; to discard a rotten apple, etc.). Of course, once one has decided on a particular course of action then specific perceptual and motor subschemas must be invoked. But note that, in the list just given, some items are apple-specific whereas others invoke generic schemas for reaching and grasping. It was considerations like this that led me (Arbib, 1981) to separate perceptual and motor schemas – a given action may be invoked in a wide variety of circumstances; a given perception may precede many courses of action. There is no one grand "apple schema" which links all "apple perception strategies" to "every action that involves an apple." Moreover, in the schema-theoretic approach, "apple perception" is not mere categorization – "this is an apple" – but may provide access to a range of parameters relevant to interaction with the apple at hand. Thus this approach views the brain as encoding a varied network of perceptual and motor schemas and coordinated control programs built upon them, perhaps with the mediation of coordinating schemas. Only rarely (as in the case of certain basic actions) will the perceptual and motor schemas be integrated into a "mirror schema" (Fig. 9.2b).

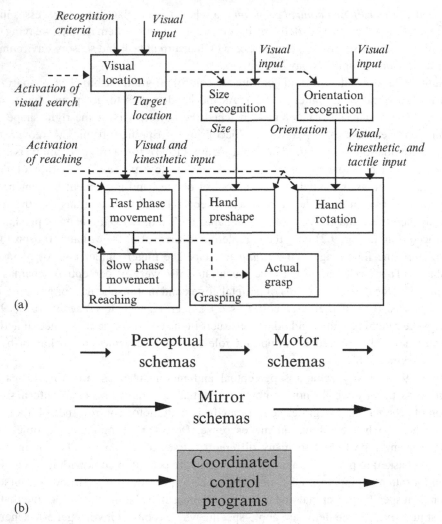

Figure 9.2 (a) Hypothetical coordinated control program for reaching and grasping. Different perceptual schemas (top half of figure) provide input for the motor schemas (bottom half of figure) for the control of "reaching" (arm transport ≈ reaching) and "grasping" (controlling the hand to conform to the object). Note too the timing relations posited here between subschemas within the "Reaching" motor schema and those within the motor schema for "Grasping." Dashed lines, activation signals; solid lines, transfer of data. (Adapted from Arbib, 1981.) (b) A general diagram emphasizing that the linkage between perception and action may involve perceptual schemas talking to motor schemas, "mirror" schemas which can both recognize and generate a class of actions (integrating a perceptual and motor schema in one), or coordinated control programs which are assemblages of perceptual, motor, and coordinating schemas.

9.3.2 VISIONS: schemas for visual scene understanding

An early example of schema-based interpretation for visual scene analysis is the VISIONS system (Arbib, 1989; Draper *et al.*, 1989). While this is an "old" system, it allows us to describe how distributed computation may be used in visual scene perception in a way that we can build upon in presenting our current approach to minimal subscenes and language. In VISIONS, there is no extraction of gist – rather, the gist is prespecified so that only those schemas are deployed relevant to recognizing a certain kind of scene (e.g., an outdoor scene with houses, trees, lawn, etc.). Low-level processes take an image of such an outdoor visual scene and extract and build a representation in the *intermediate database* – including contours and surfaces tagged with features such as color, texture, shape, size, and location. An important point is that the segmentation of the scene in the intermediate database is based not only on bottom–up input (data-driven) but also on top–down hypotheses (e.g., that a large region may correspond to two objects, and thus should be resegmented; or that two continuous regions may correspond to parts of the same object and should be merged). These are the features on which bottom–up attention (see Fig. 9.5 below) can operate, but VISIONS has a limited stock of schemas and so applies perceptual schemas across the whole intermediate representation to form confidence values for the presence of objects like houses, walls and trees. The knowledge required for interpretation is stored in long-term memory (LTM) as a network of schemas, while the state of interpretation of the particular scene unfolds in STM as a network of schema instances (Fig. 9.3a). Note that this STM is not defined in terms of recency but rather in terms of continuing relevance. Our interest in VISIONS is that it provides a good conceptual basis for the elaboration of the concept of minimal subscene, although it lacks crucial components that include attention, representations of dynamic events and relationships, and rapidly changing top–down modulation as an increasing number of attended scene elements are evaluated against goals and tasks.

In the VISIONS system, interpretation of a novel scene starts with the bottom–up instantiation of several schemas (e.g., a certain range of color and texture might cue an instance of the foliage schema for a certain region of the image). When a schema instance is activated, VISIONS links it with an associated area of the image and an associated set of local variables. Each schema instance in STM has an associated confidence level which changes on the basis of interactions with other units in STM. The STM network makes context explicit: each object represents a context for further processing. Thus, once several schema instances are active, they may instantiate others in a "top–down" way (e.g., recognizing what appears to be a roof will activate an instance of the house schema which will in turn activate an instance of the wall schema to seek confirming evidence in the region below that of the putative roof). Ensuing computation is based on the competition and cooperation of concurrently active schema instances. Once a number of schema instances have been activated, the schema network is invoked to formulate hypotheses, set goals, and then iterate the process of adjusting the activity level of schemas linked to the image until a coherent scene interpretation of (part of) the scene is obtained. Cooperation

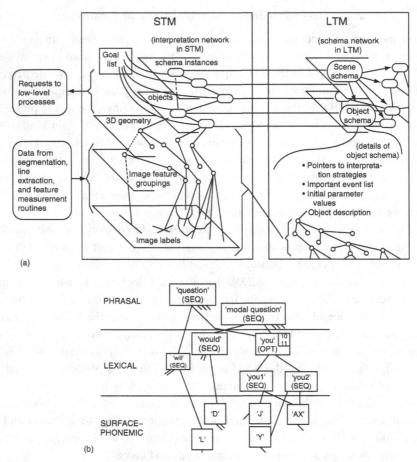

Figure 9.3　(a) The VISIONS paradigm for cooperative computation in visual scene analysis. Interpretation strategies are stored in schemas which are linked in a schema network in long-term memory (LTM). Under the guidance of these schemas, the intermediate representation (data concerning edges, region boundaries, color, texture, etc.) is modified and interpreted by a network of schema instances which label regions of the image and link them to a three-dimensional geometry in short-term memory (STM). (From Arbib, 1989, after Weymouth, 1986.) (b) The HEARSAY paradigm for cooperative computation in speech understanding.

yields a pattern of "strengthened alliances" between mutually consistent schema instances that allows them to achieve high activity levels to constitute the overall solution of a problem. As a result of competition, instances which do not meet the evolving consensus lose activity, and thus are not part of this solution (though their continuing sub-threshold activity may well affect later behavior). Successful instances of perceptual schemas become part of the current short-term model of the environment.

The classic VISIONS system had only a small number of schemas at its disposal, and so could afford to be lax about scheduling their application. However, for visual systems operating in a complex world, many schemas are potentially applicable, and many features of the environment are interpretable. In this case, "attention" – the scheduling of resources to process specific parts of the image in particular ways – becomes crucial. We emphasize that attention includes not only *where* to look but also *how* to look. We shall return to the theme of attention in Section 9.5, but first let us place this notion of cooperative computation in a broader perspective.

Returning to Fig. 9.3a, we make five points:

a. We regard gist as priming the appropriate "top-level" schema – e.g., suburban scene, city scene, beach scene, etc.
b. We note that the lower levels ("image feature groupings" – basically, the "scene layout") are very close to the "intermediate database" – with local features replaced by interpretable regions which can be linked (with more or less confidence) to the (parameterized) schema instances that constitute their interpretation. Three-dimensional geometry may emerge from this, whether linked to a schema instance (the shape of a house) or not (as in the unfolding of landscape).
c. We note the need to interpose a level below object schemas for "prototype objects" – feature patterns that are relatively easy to detect bottom up, yet which greatly focus the search for schemas compatible with a given region (cf. Rensink, 2000).
d. We stress that STM does not hold a unique schema instance for each region. Rather, schema instances may compete and cooperate with shifting confidence values till finally a group of them passes some threshold level of confidence to constitute the interpretation of the scene.
e. Finally, we will below distinguish between two aspects of STM: the first is very close to the intermediate database used by VISIONS, which evolves rather rapidly, fairly closely following any changes in the incoming visual inputs; the other is a more symbolic component, which is the basis for minimal and anchored subscene representations, evolving at a slower pace as either new scene elements both confidently identified and deemed relevant to the current task enrich the subscene, or elements previously in the subscene but no longer relevant fade away.

9.3.3 From vision to action

We now "reflect" VISIONS, analyzing the scene in terms of opportunities for action – motor schemas which then compete for realization (Arbib and Liaw, 1995). In addition, motor schemas are affected top–down by goals and drive states, and middle–out by the priming effect of other motor schemas. While only a few perceptual schemas may be active for the current focus of attention, STM will be updated as new results come in from this focal processing. STM is now more dynamic and task-oriented and must include a representation of goals and needs, linking instances of perceptual schemas to motor schemas, providing parameters and changing confidence levels, so as to provide suitable input to STM. As their activity levels reach threshold, certain motor schemas create patterns of overt behavior. To see this, consider a driver instructed to "Turn right at the red barn." At first the person drives along looking for something large and red, after which the perceptual schema for barns is brought to bear. Once a barn is identified, the emphasis shifts to recognition of spatial relations appropriate to executing a right turn "at" the barn, but constrained by the placement of the roadway, etc. The latter are an example of *affordances* in the sense of Gibson (1979), i.e., information extracted from sensory systems concerning the possibility of interaction with the world, as distinct from recognition of the type of objects being observed. For example, optic flow may alert one to the possibility of a collision without any analysis of what it is that is on a collision course.

In the VISIONS system, schemas in LTM are the passive codes for processes (the programs for deciding if a region is a roof, for example), while the schema instances in STM are active copies of these processes (the execution of that program to test a particular region for "roofness"). By contrast, it may be that in analyzing the brain, we should reverse the view of activity/passivity of schemas and instances: the active circuitry is the *schema*, so that only one or a few instances can apply data-driven updating at a time, while the *schema instance* is a parameterized working memory of the linkage of a schema to a region of the scene, rather than an active process.

Such considerations offer a different perspective on the neuropsychological view of working memory offered by Baddeley (2003). The initial three-component model of working memory proposed by Baddeley and Hitch (1974) posits a *central executive* (an attentional controller) coordinating two subsidiary systems, the *phonological loop*, capable of holding speech-based information, and the *visuospatial sketchpad*. The latter is viewed as passive, since the emphasis of Baddeley's work has been on the role of working memory in sentence processing. Baddeley (2003) added an episodic LTM to the Baddeley–Hitch model, with the ability to hold language information complementing the phonological loop and (the idea is less well developed) an LTM for visual semantics complementing the visuospatial sketchpad. He further adds an *episodic buffer*, controlled by the central executive, which is assumed to provide a temporary interface between the phonological loop and the visuospatial sketchpad and LTM. The Arbib–Liaw scheme seems far more general, because it integrates dynamic visual analysis with the ongoing control of action. As such, it seems better suited to encompass Emmorey's notion (see the section "Broca's area and working memory for sign language," in Emmorey, 2004) that sign language employs a visuospatial phonological short-term store. With this, let us see how the above ideas play out in the domain of speech understanding.

9.3.4 HEARSAY: schemas for speech understanding

Jackendoff (2002) makes much use of the artificial intelligence (AI) notion of blackboard in presenting his architecture for language. HEARSAY-II (Lesser *et al.*, 1975) was perhaps the first AI system to develop a blackboard architecture, and the architecture of the VISIONS computer vision system was based on the HEARSAY architecture as well as on neurally inspired schema theory. While obviously not the state of the art in computer-based speech understanding, it is of interest here because it foreshadows features of Jackendoff's architecture. Digitized speech data provide input at the *param-*

eter level; the output at the *phrasal level* interprets the speech signal as a sequence of words with associated syntactic and semantic structure. Because of ambiguities in the spoken input, a variety of hypotheses must be considered. To keep track of all these hypotheses, HEARSAY uses a dynamic global data structure, called the *blackboard*, partitioned into various levels; processes called *knowledge sources* act upon hypotheses at one level to generate hypotheses at another (Fig. 9.3b). First, a knowledge source takes data from the *parameter level* to hypothesize a phoneme at the *surface-phonemic level*. Many different phonemes may be posted as possible interpretations of the same speech segment. A lexical knowledge source takes phoneme hypotheses and finds words in its dictionary that are consistent with the phoneme data – thus posting hypotheses at the *lexical level* and allowing certain phoneme hypotheses to be discarded. These hypotheses are akin to the schema instances of the VISIONS system (Fig. 9.3a).

To obtain hypotheses at the *phrasal level*, knowledge sources embodying syntax and semantics are brought to bear. Each hypothesis is annotated with a number expressing the current confidence level assigned to it. Each hypothesis is explicitly linked to those it supports at another level. Knowledge sources cooperate and compete to limit ambiguities. In addition to data-driven processing which works upward, HEARSAY also uses hypothesis-driven processing so that when a hypothesis is formed on the basis of partial data, a search may be initiated to find supporting data at lower levels. For example, finding a verb that is marked for plural, a knowledge process might check for a hitherto unremarked "s" at the end of a preceding noun. A hypothesis activated with sufficient confidence will provide context for determination of other hypotheses. However, such an *island of reliability* need not survive into the final interpretation of the sentence. All we can ask is that it forwards the process which eventually yields this interpretation.

Arbib and Caplan (1979) discussed how the knowledge sources of HEARSAY, which were scheduled serially, might be replaced by schemas distributed across the brain to capture the spirit of "distributed localization" of Luria (e.g., 1973). Today, advances in the understanding of distributed computation and the flood of brain imaging data make the time ripe for a new push at a neurolinguistics informed by the understanding of cooperative computation. It is also worth relating the discussion of VISIONS and HEARSAY to the evolutionary theme, "from action to language," of the present volume. While non-humans have well-developed mechanisms for scene analysis and for integrating that analysis to their ongoing behavior, they lack the ability to link that capability for "praxic behavior" to "communicative behavior" that can convert the perception of agents, actions, and objects and their relationships with each other and ongoing behavior into a structured set of symbols which can be used to express certain details of these relationships to others. Thus while Fig. 9.3 illustrates the parallels between the blackboard architectures for visual (VISIONS) and speech (HEARSAY) perception, the evolution of human brain mechanisms that support language perception and production by exploiting variants of the mechanisms of perception and production relevant to praxic action is in no sense direct.

We close this section by summarizing its main contributions. Section 9.3.1 introduced the general framework of schema theory (perceptual and motor schemas; schema assemblages; coordinated control programs). Section 9.3.2 showed how the VISIONS system could represent a static visual scene by a hierarchically structured network of schema instances linked to regions of the image. In Section 9.3.3, we suggested how the framework offered by VISIONS might be extended to include motor schemas and the planning of action, as well as perceptual schemas for scene recognition. Complementing VISIONS, the HEARSAY system (Section 9.3.4) demonstrated how to represent a speech stream by a hierarchically structured network of schema instances linked to time intervals of the spoken input. The role of attention in scheduling resources for scene recognition was implicit in our description of how in VISIONS activation of a roof schema instance might lead to activation of an instance of the wall schema to check the region below that linked to the roof schema. In the next section, we present two "classic" models which make explicit the role of attention in vision. We then build on this to present a framework for modeling the goal-directed and action-oriented guidance of attention. Here the emphasis is on recognition of a single static scene. Our challenge in later sections will be to combine the insights of these two classical systems to describe how to represent a dynamic visual scene (or multi-modal sensory data more generally) by a hierarchically structured network of schema instances each linked to a space–time region of the image. For example, a person may only be tracked over a certain time interval as they move from place to place; an action will extend across a certain region of time but may be quite localized.

9.4 Integrating attention and dynamic scene analysis

We now turn to two background studies of visual attention to which we have contributed, in both of which the notion of "winner take all" (WTA) plays a central role. The first, by Arbib and Didday (1975), is more comprehensive in its conceptual structure; the second, by Itti and Koch (2000, 2001), is restricted to low-level salience but has been worked out in detail and has yielded interesting results through detailed simulation.

9.4.1 The Arbib–Didday model

Our ancient "two visual system" model of visual perception (Arbib and Didday, 1975) has the following list of components (and see also Fig. 9.4):

1. The superior colliculus (h) responds to visual input to direct the eyes (a) to turn to foveate a new target. The target is chosen by a winner-take-all mechanism (inspired by frog prey selection) in response to bottom–up salience cues (as developed in more detail in the Itti–Koch model of the next section) and top–down attention cues (as developed in the SVSS model of Section 9.5).
2. Arbib and Didday use the terms "slide" and "slide-box" (d) where we would now speak of "schemas" and "STM" (a "linked collection of relevant schemas," as in VISIONS). In fact, the

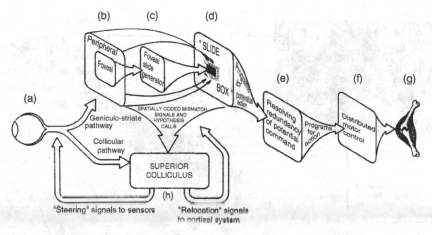

Figure 9.4   A "two visual system" model of visual perception (Arbib and Didday, 1975).

slide-box here may be seen as combining the functions of the intermediate database and STM in VISIONS. Foveal attention (b) provides the detailed input to appropriately add slides to, or adjust slides (c) in, the slide-box.

3. As the eyes move, the foveal input must be redirected appropriately – (c) → (d) – to address the appropriate spatially tagged region of the slide-box (cf. the notion of "dynamic remapping"; Dominey and Arbib, 1992; Medendorp et al., 2003). Moreover, the model of Fig. 9.4 is designed to process dynamic scenes, so that the state of the slide-box will depend as much on the retention of earlier hypotheses as on the analysis of the current visual input.

4. A crucial point (anticipating Arbib and Liaw, 1995) is that perceptual schemas are not ends in themselves but are linked to motor schemas for potential action (the other half of (d)) with respect to entities in the observed scene.

5. Since more motor schemas may initially be activated than can be executed, a process of "resolving redundancy of potential command" (McCulloch, 1965) is required (e) to determine which actions are indeed executed (f, g) – just as the superior colliculus (h), the "motor side" of vision in the present model, must execute a winner-take-all computation to select the next focus of attention.

9.4.2   The Itti–Koch model

In our neuromorphic model of the bottom–up guidance of attention in primates (Fig. 9.5) (Itti and Koch, 2000, 2001), the input video stream is decomposed into eight feature channels at six spatial scales. After surround suppression, only a sparse number of locations remain active in each map, and all maps are combined into the unique *saliency map*. This map is scanned by the focus of attention in order of decreasing saliency through the interaction between a winner-take-all mechanism (which selects the most salient location) and an inhibition-of-return mechanism (which transiently suppresses recently attended locations from the saliency map). Because it includes a detailed low-level vision front-end, the model has been applied not only to laboratory stimuli, but also to a wide variety of natural scenes (e.g., Itti *et al.*, 2001), predicting a wealth of data from psychophysical experiments.

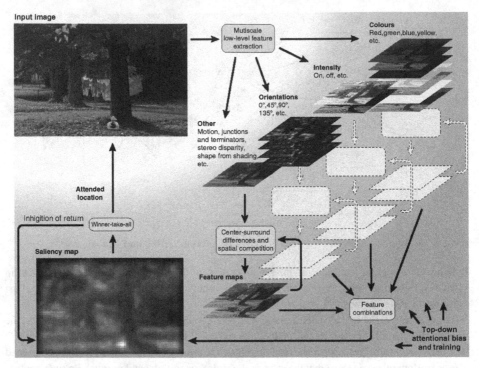

Figure 9.5   Overview of Itti and Koch's (2001, 2001) model of visual attention guided by bottom–up salience.

In this model, inhibition of return is largely automatic and by default disengages covert attention from each target shortly after the target has been acquired and attended to. An extension of the model (Itti *et al.*, 2003) adds a simple mechanism for the control of overt attention (eye movements) in rapidly changing environments. In this context, automatic inhibition of return is not desirable, as the slower oculomotor system cannot keep up with the ensuing rapid shifts of covert attention (up to 20 items per second). Thus, for tasks that require eye movement control rather than rapid covert search, we disable inhibition of return, so that covert attention locks onto the most salient location in the dynamic display. As this winning location may drift or even jump from one object to another according to the dynamic video inputs, covert and resulting overt shifts of attention are still experienced, but typically at a slower pace which the oculomotor system can follow. In certain environments containing a distinctive actor or object moving against a fixed background, the model's behavior becomes dominated by tracking this actor or object. As suggested by our pilot human data in Section 9.8, it is probable that implementing some trade-off mechanism between tracking and rapid attention shifts employing inhibition of return will be necessary to adequately model the perception of changing scenes in a videoclip. Bottom–up saliency is only one of the many factors which contribute to the guidance of attention and eye movements onto a visual scene. Attention indeed is complemented by rapid analysis of the scene's gist and layout to provide priors on where objects of current interest may be located, and facilitate their recognition (Oliva and Schyns, 1997; Hollingworth and Henderson, 1998; Henderson and Hollingworth, 1999; Rensink, 2000; Itti and Koch, 2001; Torralba, 2003). As in the VISIONS example, when specific objects are searched for, low-level visual processing can be biased not only by the gist (e.g., "outdoor suburban scene") but also for the features of that object (Moran and Desimone, 1985; Ito and Gilbert, 1999). This top–down modulation of bottom–up processing results in an ability to guide search towards targets of interest (Wolfe, 1994). Task affects eye movements (Yarbus, 1967), as do training and general expertise (Nodine and Krupinski, 1998; Moreno *et al.*, 2002; Savelsbergh et al., 2002). Finally, eye movements from different observers exhibit different idiosyncrasies, which may result from possibly different internal world representations, different search strategies, and other factors (Hollingworth, 2004; Hollingworth and Henderson, 2004).

This model has been restricted to the bottom–up guidance of attention towards candidate locations of interest. Thus, it does not address the issue of what might hold attention. A static fixation may last long enough for us to categorize an object or decide it is not of interest. Return may then be inhibited until either fading memory or relations with another object yield further attention to it. Movement itself is a strong salience cue. For a moving object of interest, "what will it do next?" may be enough to hold attention, but once attention shifts, inhibition of return will then apply to the region of its imminent trajectory. Clearly, these general observations pose interesting challenges for both psychophysical and neurophysiological experiments and future modeling.

In most biological models of attention and visual search, such as the "guided search" model (Wolfe, 1994), the interaction between top–down commands derived from symbolic reasoning and low-level vision have been restricted to two simple effects: feature-based biasing (e.g., boost neurons tuned to a feature of interest like vertical motion, throughout the visual field) and spatial biasing (e.g., boost neurons responding to a given spatial region) of the kind supported by monkey physiology (Treue and Martinez-Trujillo, 1999). However, several computer models have attacked the problem of more complex top–down influences. We have already seen the hypothesis-driven strategies employed by the VISIONS system, based on knowledge stored in schemas about their relationship (spatial or hierarchical) with other schemas. The model of Rybak *et al.* (1998) stores and recognizes scenes using scanpaths (i.e., sequences of vectorial eye movements) learned for each scene or object to be recognized. When presented with a new image, the model attempts to replay one of its known scanpaths, and matches stored local features to those found in the image at each fixation (cf. Noton and Stark, 1971). However, Arbib and Didday (1975) argued that scanpaths are generated on-line from visual memory stored as retinotopic arrays, an approach consistent with the rapid generation of successive fixations performed by the model of Fig. 9.5.

Schill et al. (2001) employ an optimal strategy for deciding where next to shift attention (based on maximizing information gain). But such computer vision models (a) do not address the problem of handling scenes that are both dynamic and novel and (b) lack biological correlates. On the other hand, Dominey *et al.* (1995) extended a model of the cooperation of frontal eye fields and superior colliculus in the production of saccadic eye movements to show how top–down influences could be learned, via reinforcement learning, which can bias the effect of salience. In each case, cerebral cortex projected to the basal ganglia which in turn modulated activity in superior colliculus. In one case, the projection from inferotemporal cortex to striatum was adapted to associate each specific visual pattern with a leftward or rightward bias on the winner-take-all selection of dynamic targets. In a second study, the order of presentation of targets yielded, via patterns of connections from prefrontal cortex to basal ganglia, to a bias on the order in which the targets would be attended to if they were later presented simultaneously. Turning from sequences of saccades to sequences of hand actions, Arbib (Chapter 1, this volume, Section 1.2.1) and Arbib and Bota (this volume, Section 5.4) discuss the transition from single actions to overlearned sequences, and the transition from overlearned sequences to sequences as the context-dependent expression of hierarchical structures, respectively.

9.5 Goal-directed and action-oriented guidance of attention

To start building a scalable model, we must address the key question "What draws the viewer's attention to a specific object or action, and then expands that attention to determine the minimal subscene containing it?" The unfolding of a description and the answering of a question require that attention be driven not solely by low-level salience but also by the search for objects (and actions) that are deemed relevant to the unfolding task. We are motivated here by Yarbus's (1967) classic study, which revealed subjects' dramatically different patterns of eye movements when inspecting a given static scene under different task guidelines. In this spirit, we extend the Itti–Koch model (Fig. 9.5) to a model of goal-directed and action-oriented guidance of attention in real-world static and dynamic scenes. The idea is to only memorize objects/events determined to be task-relevant, developing the current minimal subscene in symbolic STM. We first present a general conceptual architecture to achieve this, then briefly examine experimental and simulation results obtained with a partial implementation (Navalpakkam and Itti, 2005).

9.5.1 The Salience, Vision, and Symbolic Schemas (SVSS) model

Our present conceptual model (Fig. 9.6), the Salience, Vision, and Symbolic Schemas (SVSS) model, uses a Task Relevance Map (TRM) to hold the locations currently deemed relevant to the task requirements. The TRM is a topographic map which acts as a top–down mask or filter applied to bottom–up activation. Thus, the attractiveness of a bottom–up salient region may be reduced if that region occurs at a location expected top–down to be of little behavioral relevance. Conversely, sensitivity to bottom–up salience is increased in visual regions expected to contain relevant targets. An initial implementation of some of the concepts outlined here will be described in Section 9.5.2, which focuses mostly on describing a new scene and less on answering complex questions about a scene.

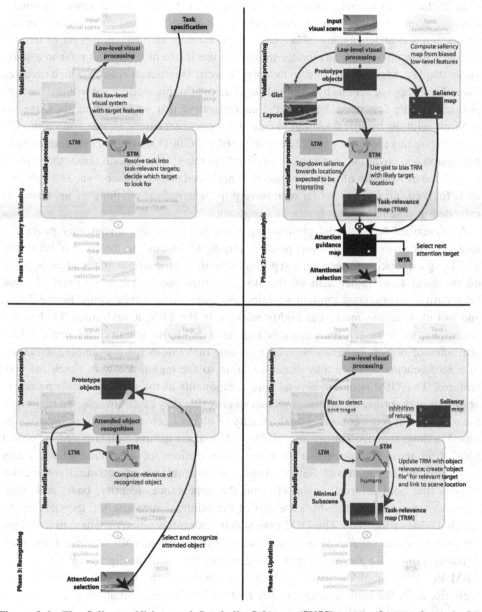

Figure 9.6 The Salience, Vision, and Symbolic Schemas (SVSS) system for top–down (task-dependent) attention to objects in a static visual scene. Each cycle of activity in this model passes through four phases: *Preparatory task biasing* (preparatory top–down biasing of low-level vision once the task is known even before visual input is received); *Feature analysis* (extraction of low-level visual features that will serve in computing gist, salience maps, and task-relevance maps); *Recognizing* (selecting the most salient and relevant location in the scene, attempting to recognize it, and updating the short-term memory (STM) based on how it relates to currently relevant entities); and *Updating* (by which recognized objects, actors or actions on a given attention shift are used to update both STM and task-relevance map (TRM), so that preparation for the next shift of attention

In Section 9.13 we will consider more fully the extension of this model to handle dynamic scenes (changing over time) and how it may be linked to processes for the production and understanding of language. Here we note an interesting point of terminology. Within the vision community, we tend to use the term "scene" to refer to a static visual display, whereas when we go to a play, a scene is rather an episode which involves a set of overlapping actions by a limited set of actors. In this section, we will use "scene" primarily in the first sense, but with the understanding that we are setting the stage (to continue with the theatrical setting) to consider scenes that extend in space and time. We are here entering the territory of Zacks and Tversky's (2001) analysis of event structure in perception and cognition. A scene/event/episode in this spatiotemporal sense, too, will be considered as made up of a set of minimal or anchored subscenes/subevents/subepisodes each focused on a single action or relationship, where now we think of an action as extended in time rather than frozen in a snapshot.

As described below, rapid preliminary analysis of gist and layout may provide a spatial prior as to the location of possible targets of interest, in the form of an initial TRM (e.g., if looking for people, and presented with a scene whose gist is a beach, focus on the sand areas rather than on the sky). As time passes by, more detailed focal information accumulated through attention and gaze shifts serves to update the TRM, and to link locations marked as highly relevant in the TRM to additional STM representations, such as schema instances that may hold the identity and some visual attributes of objects currently deemed relevant. In VISIONS, for example, activation of a roof schema instance may direct attention to the region below to check for wall features. The TRM is one of several core components of the minimal subscene representation: it describes the spatial emphasis an observer may cast onto selected regions of the scene, but it is not concerned with identity of objects, actors, and actions that may be found in these regions. This identity and visual attribute information is stored in "object files" (the term for minimalist short-term descriptions of a specific visual entity employed by Treisman and Gelade, 1980), or assemblages of activated schema instances that hold information in STM on the appearance, identity, pose, and other aspects of scene elements which are part of the minimal subscene (cf. the state of the Arbib–Didday slide-box). The TRM may also invoke information on where to look next for objects whose salience is defined in relation to the objects already found and their dynamic properties; for example, when attending to a ball in flight, highlight in the TRM the location of the expected future end point of the ball's trajectory. In this scenario, as in the static example of the roof and walls, perceptual schemas triggered by visual inputs (the ball or the roof) may prime additional schemas (trajectory extrapo-

Caption for Figure 9.6 (cont.)

can begin and be dependent on the current partial evaluation of the minimal subscene). Volatile processing refers to low-level representations which are updated in real-time from visual inputs, while non-volatile processing operates at a slower timescale, with one update occurring only with each attention shift. (Modified from Navalpakkam and Itti, 2005.)

lation schema to estimate the ball's end point, or house schema prompting where to look for walls), resulting both in a spatial priming of the related scene locations in the TRM and possibly in priming of related low-level visual features.

Where VISIONS offers no principled model of scheduling of schema instances, SVSS links this scheduling to an account of top–down mechanisms of attention which interact with salience, developing a theme of the Arbib–Didday model. Where VISIONS links schema instances to retinotopically defined regions of the intermediate representation in STM, SVSS extends STM by including a *symbolic* component which links certain "symbolic schemas" (which may or may not correspond to items in the lexicon) to those more "visual" schemas in the task representation and the currently active set of minimal and anchored subscenes. There are two notions at work here:

a. The first is to emphasize that our perception of a scene may depend both on specifically visual relations and on more conceptual knowledge about objects and actions. In this way, LTM encodes a number of symbolic concepts and their possible interrelationships that seem to operate at the posited level of symbolic STM – for example, that a man has two hands. This complements the more specifically VISIONS-style use of LTM which enables recognition of a hand (activation of a perceptual schema for hand) to prime a schema for recognizing a human, whereas activation of a schema for a human could activate two instances of the hand schema with appropriate spatial constraints for possible incorporation into the unfolding schema assemblage.

b. The second is to consider STM in this extended sense as the bridge between the richness of the visual representation it contains and two different entities: (1) the verbal description of the scene, which is what we emphasize here; and (2) STM which is believed to result from hippocampal "tagging" of certain episodes for consolidation for subsequent recall. This link to STM provides some of the motivation for our theory, but cannot be treated at any length in this chapter. In the terminology of Arbib (Chapter 1, this volume, Section 1.1.3), we may relate STM to the cognitive structures (Cognitive Form; schema assemblages) from which some aspects are selected for possible expression, while the symbolic component of STM underwrites the semantic structures (hierarchical constituents expressing objects, actions and relationships) which constitute a Semantic Form. A selection of the ideas in the Semantic Form must be expressed in words whose markings and ordering provide a "phonological" structure, the Phonological Form.

We shall later turn to results from a partial implementation (Navalpakkam and Itti, 2005) of the SVSS architecture, but first we outline the mode of operation of the general scheme. Here we focus on analysis of a single static visual input. Computation is initiated by Phase 1 below, then cycles through Phases 2, 3, and 4 until Phase 1 is reactivated to update the task.

9.5.1.1 Phase 1: Preparatory task biasing

In situations where a task is defined in advance, e.g., "Describe who is doing what and to whom in the following scene" or "What is Harry hammering in the following scene?", the task definition is encoded as entities in symbolic STM, using prior knowledge stored in LTM. For instance, "Who is doing what and to whom?" would prime the STM for humans and actions involving humans; "What is Harry hammering?" would prime the STM for Harry, hammers, hammering actions, and objects that can be hammered on, like nails. One possible strategy for priming STM would be to provide an assemblage of primed schema instances for those entities deemed relevant to the task, either because they have been explicitly mentioned in the task definition (e.g., Harry, hammering), or because they are known from the LTM to be associated with the explicitly mentioned entities (e.g., a hammer, a nail). The issue then is to determine whether these primed instances can be linked to specific regions of the intermediate database (the preprocessed visual input) and given confidence levels which pass some threshold for accepting that the schema instance provides an acceptable interpretation of the region. To prioritize the search for these entities, one may compute a relevance score for each; for example, looking first for a hammer, then checking that Harry is using it, finally determining the object of the hammering may be a reasonable prioritization to answer "What is Harry hammering?". Such relevance-based scoring may be implemented through stronger output to the attentional mechanism of the schema instances which represent the more relevant objects, actors, or actions. As such, it is to be distinguished from the confidence level. The former reflects the priority for checking out a hypothesis; the latter reflects the extent to which the hypothesis has been confirmed. In general, before the scene is shown, little or no prior knowledge of spatial localization is available as to where the entities that are currently relevant and held in STM will appear once the scene is shown; thus the TRM at this initial stage is generally uniform. In specific situations, some initial spatial bias in the TRM may, however, already be possible; for example if the visual stimulus is presented on a relatively small computer monitor, the TRM may already assign low relevance to regions outside the monitor's display area.

As we have seen, the model can, in preparation for the analysis of the visual inputs, already prime STM to bias its saliency-based visual attention system for the learned low-level visual features of the most relevant entity, as stored in visual LTM. (We use the usual shorthand here, omitting the phrase "the computer representation of the neural code for" when we speak of keywords and entities, etc.) For example, if a hammer is currently the most relevant entity, knowing that hammers typically feature a thin elongated handle may facilitate focusing attention towards a hammer, simply by enhancing the salience of oriented line segments of various orientations in bottom–up processing, while toning down non-diagnostic features, for example color which may vary from hammer to hammer.

In summary, we here propose a mechanism where prior knowledge stored in LTM may combine with task definition so as to populate the STM with a prioritized collection of task-relevant targets and possibly how they are related to each other. Next, the STM determines the current most task-relevant target as the desired target. To detect the desired target in the scene, the learned visual representation of the target is recalled from LTM and biases the low-level visual system with the target's features.

9.5.1.2 Phase 2: Feature analysis

As the visual scene is presented and its gist is rapidly analyzed, the STM may impose additional biases onto the TRM, highlighting likely locations of the desired entity given the gist of the scene; for example, Harry and his hammer are more likely to be found near the ground plane than floating in the sky. The low-level visual system that is biased by the target's features computes the biased salience map. Cues from the TRM, together with the saliency map computed from biased low-level features (Phase 1), combine to yield the attention guidance map, which highlights spatial locations which are both salient and/or relevant. To select the focus of attention, we deploy a winner-take-all competition that chooses the most active location in the attention-guidance map. It is important to note that there is no intelligence in this selection and all the intelligence of the model lies in STM and its deployment of schema instances on the basis of information stored in LTM. Given a current bottom–up input in the form of a saliency map biased to give higher weight to the features of desired objects, and a current top–down mask or filter in the form of a TRM, attention simply goes to the location where the combination of bottom–up and top–down inputs is maximized.

Regarding how bottom–up salience and top–down relevance may combine, it is interesting to note that non-salient locations may be relevant based on knowledge of spatial relationships between current objects of attention and desired relevant targets; for example, "Where is John headed?", "What is Mary looking at?", "What is the hammer going to hit?", or "Where is that thrown ball going to land?". We may term these special locations, which are of potential interest not because of any visual input but purely based on knowledge of spatial relationships, as "top–down salient." Top–down salience complements bottom–up salience: a location is bottom–up salient if it has distinctive or conspicuous visual appearance, based on low-level visual analysis; conversely, a location is top–down salient if it is expected to be of interest, no matter what its visual appearance. Integrating top–down salience to the present framework may be achieved by allowing top–down salience to provide additive inputs to the TRM and resulting attention-guidance map, in addition to the bottom–up salience gated by relevance inputs described above. Top–down salience may result from the activation of expectation schemas; for example, a moving hammer may activate an expectation schema for what the object of the hammering may be, resulting in a spatial highlight in the form of a cone of top–down salience below the hammer. Similarly, a flying ball may activate an expectation schema which will result in highlighting the expected endpoint of the ball.

9.5.1.3 Phase 3: Recognizing

Once a target is acquired, further processing is required to verify that the attended scene element meets the task requirements. This requires matching against stored LTM representations to recognize the entity at the focus of attention. Here, the low-level features or intermediary visual representations in the form of "prototype objects" (Rensink, 2000), which are intermediary between low-level features like edges or corners and full objects like a face, are bound together at the attended location to yield transiently coherent object representations (Treisman and Gelade, 1980; Rensink, 2000). The bound object representations are then analyzed for recognition, employing LTM to match the visual attributes of the bound representations to LTM traces of objects. If above-threshold confidence is reached that a given object or actor is present at the currently attended location, then the STM is called upon (in the next phase) to estimate the task-relevance and confidence level of the recognized entities, and to decide whether they are worth retaining or should be discarded; if no reliable recognition is achieved, one may either just ignore that location, mark it as being puzzling and as deserving further analysis later, or one may trigger other behaviors, such as slight head motion to change viewpoint, or an exploration of neighboring locations, in an attempt to achieve recognition.[2]

9.5.1.4 Phase 4: Updating

Recognition of an entity (object, actor, or action) as its confidence level passes some threshold serves to update the STM and TRM in two manners:

1. If the recognized entity is found to be relevant (e.g., because it was one of the relevant entities already in working memory), its location is marked as relevant in the TRM, the entity is marked as found in the STM, and symbolic schemas may be activated to index key attributes of the entity in a form that may (but need not) link to language mechanisms. Which entity is next most relevant to look for is determined from the updated contents of STM.
2. If the recognized entity was irrelevant (e.g., either because it was unrelated to any of the entities in STM – as in building up a minimal or anchored subscene – or lacks independent interest to serve as a possible anchor for a new subscene), its location is inhibited in the TRM and its lowered activity in STM will reduce its likely impact on further processing.

In this phase, the STM updates its state (e.g., if looking for a man's face, and a hand has been attended to, the STM may add a symbolic schema attesting that it has found a hand, which may help deciding where to look next to find the face by activating an expectation schema and associated top–down salience). In addition, the computed relevance of the currently attended entity may influence the system's behavior in several ways. For instance, it may affect the duration of fixation. In a last step, the STM inhibits the current focus of attention from continuously demanding attention (inhibition of return in the saliency map). Then, in preparation for a subsequent iteration of Phase 2, the visual features for the new most-relevant entity in STM are retrieved and used to bias the low-level visual system.

This completes one iteration, and each iteration involves one shift of attention. Subsequent shifts of attention will replay Phases 2–4 and incrementally build the TRM and populate the symbolic STM until the task is complete. Upon completion, the TRM shows all task-relevant locations and the STM contains all task-relevant targets as high-confidence instances of perceptual and symbolic schemas, structured as one (or a concurrent set of) anchored subscene(s) representations, in a form appropriate to ground a variety of cognitive processes, including language production as in describing the scene or answering questions.

The STM and TRM are where SVSS, which is concerned with information processing from the retina to the minimal or anchored subscene, interfaces with models concerned with producing verbal descriptions of the scene from the subscene representations, for example using computational architectures which do for production what HEARSAY

---

[2] We trust that the reader has understood we are using the words "ignore" and "puzzling" and other such terms in this section as shorthand for detailed processes of computational analysis which in humans would usually be accomplished by the brain's neural networks without correlates in consciousness.

does for speech understanding. Although we have emphasized static scenes and object recognition in the above outline of SVSS, it is clear that it extends to recognition and linkage of agents, actions, and objects that motivates this chapter. In this spatiotemporal extension, our subscene representation links a highly processed but still spatially structured visual representation of the visual world with a more symbolic aggregate of the recent past and expected future which provides one "now" that can be translated into language.

9.5.2 *Implementing a prototype*

Our prototype software implementation of SVSS (Navalpakkam and Itti, 2005) emphasizes three aspects of biological vision: biasing attention for low-level visual features of desired targets, recognizing these targets using the same low-level features, and incrementally building a visual map of the task-relevance of every scene location. Task definitions are given to the system as unstructured lists of keywords; thus, there currently is no attempt in the prototype at parsing and exploiting the linguistic structure of more complex task definitions. LTM is represented as an ontology in the sense used by the database research community, i.e., a collection of symbols forming a graph where links represent several possible types of relationships between symbols. To each symbol is also attached visual description information, which allows both recognition of that entity in an image, and biasing of attention for the low-level features of the entity. Relationships currently implemented are: *is-a, includes, part-of, contains, similar-to*, and *related-to*. In addition, a link weight expresses the learned frequency of occurrence of a given relationship in the world; for example, a "pen" *is-a* "holdable-object" with a strength of 0.99 because nearly all pens are holdable, while a "car" *is-a* "holdable-object" with a strength of only 0.05 because most cars (except small toy cars) cannot be held. Thus, if looking for holdable objects, the system is more likely to be biased towards pens than towards cars. In the prototype, the ontology is small and hand-coded, with arbitrary weight values assigned to the various links in the graph. In future implementations, these weights would have to be learned, and the ontology possibly expanded, based on experience and training.

The STM of the prototype is initialized as a sub-ontology which is a copied portion of the long-term ontology, to include the concepts explicitly specified in the task definition as well as all related concepts down to a given weight threshold. Given this initial STM, a computation of relevance of the various entities to the task at hand gives rise to a single most-desired entity. Using visual attributes of entities stored in LTM, the visual attributes of the most-desired entity are retrieved. In the prototype, the visual properties of entities in LTM are encoded as a hierarchical collection of feature vectors, with one feature vector describing each learned view of an object (e.g., different photographs of a specific hammer), then combined to yield one feature vector for each object instance (a specific hammer), and finally for each object (a hammer). Feature vectors currently are very impoverished descriptions of each view, instance, or object, based on simple properties such as the amount (mean plus standard deviation) of various colors, various orientations, etc. typically observed for given objects. The prototype currently has no means of computing gist or top–down salience. Initially, hence, the TRM is initialized to a uniform unity value, and the low-level visual processing is biased for the low-level features of the most-desired target. This allows computation of the saliency map and guidance of attention to a given location. Recognition is based on matching the feature vector extracted at the attended location to its closest vector in the LTM. The recognized entity (if any passes the recognition confidence threshold) is then evaluated for relevance, using the LTM to evaluate how it may relate to the entities in STM and what the cumulative graph path weight is; for example, if the STM was interested in holdable-objects and a car is found, the relevance of the car to the current STM can be computed by finding a path from car to holdable-object in the LTM graph, and computing how strong that path is. The TRM is then updated using the computed relevance value, which may either enhance a relevant location or suppress an irrelevant one. Finally, if the attended object has a relevance score above threshold, its identity is remembered and attached to the corresponding location in the TRM. Thus, the output of the prototype is a TRM with a number of attached symbols at its most relevant locations.

This prototype presents obvious limitations, with its hand-coded ontology, weak object representation, and absence of gist and top–down salience. But it is a starting point in implementing the broader framework detailed above. The prototype was tested on three types of tasks: single-target detection in 343 natural and synthetic images, where biasing for the target accelerated target detection over twofold on average compared to a naïve bottom–up attention model which simply selects objects in a scene in order of decreasing bottom–up salience; sequential multiple-target detection in 28 natural images, where biasing, recognition, working memory, and LTM contributed to rapidly finding all targets; and learning a map of likely locations of cars from a videoclip filmed while driving on a highway. The model's performance on search for single features and feature conjunctions was shown to be consistent with existing psychophysical data. These results of our prototype suggest that it may provide a reasonable approximation to many brain processes involved in complex task-driven visual behaviors, and are described in details in Navalpakkam and Itti (2005).

9.6 Eye movements and language: a brief review

Before discussing explicit models which relate models like those of the previous two sections to language, we must note that, of course, we are not the first to investigate the relevance of visual attention to language. Henderson and Ferreira (2004) offer a collection of review articles that cover far more material than we can survey here. Instead we briefly note the contributions of Tanenhaus *et al.* (2004) and Griffin and Bock (2000). We note, without giving details, the related work of Tversky and Lee (1998) and Zacks and Tversky (2001) who explore "event structure" and the way events are structured spatially as a basis for relating that structure to language.

Griffin and Bock (2000) monitored the eye movements of speakers as they described black-and-white line drawings of simple transitive events with single sentences. However, the task was simpler than that we have set ourselves in that each drawing represented a single minimal subscene in which an agent acted in some way upon another agent or object. Eye movements indicated the temporal relationships among event apprehension (extracting a coarse understanding of the event as a whole), sentence formulation (the cognitive preparation of linguistic elements, including retrieving and arranging words), and speech execution (overt production). Their findings support the view that apprehension precedes formulation to provide the holistic representation that supports the creation of a sentence.

The experimental pictures depicted two types of events. Active events elicited predominantly active sentences in the experiment, regardless of which element was the agent. Passive–active events were those involving both a human and a non-human which were typically described with active sentences if the human is the agent and with passive sentences if the human is the patient ("The mailman is being chased by the dog" and "The mailman is chasing the dog"). Their analysis included a variety of conditions. Here, we simply note that Griffin and Bock (2000) found an orderly linkage between successive fixations in viewing and word order in speech. Significant interactions between event roles and time period indicated that speakers spent significantly more time fixating agents before subject onset than during speech but spent more time on patients during speech than before subject onset. We would suggest that this means that the subjects create their STM of the scene, linked to the image, before they start to speak, and that in general – whatever the original sequence of fixations – the agent (or the human in the active–passive sentences) serves as anchor for the minimal subscene that will be described in the spoken sentence. Indeed, Griffin and Bock (2000) assert that analysis of passive–active pictures implies that speakers did not simply follow the causal structure of the events by fixating agents early and patients later. Rather, when patients were encoded as sentential subjects, they were fixated longer before subject onset than after whereas agents were fixated less before subject onset than during speech. Thus, the distribution of fixation times anticipated the order of mention regardless of sentence structure.

Griffin and Bock (2000) conclude that their evidence that apprehension preceded formulation supports the view that a holistic process of conceptualization sets the stage for the creation of a to-be-spoken sentence. Their data reflect only the most basic kind of sentence formulation, in English, involving minimal scenes and the production of single clauses. Nonetheless, the results point to a language production process that begins with apprehension of the structure of the scene and proceeds through incremental formulation of the sentence which expresses it.

Tanenhaus *et al.* (2004) survey a rich set of experiments using eye movements to chart the role of referential domains in spoken language comprehension. The basic notion here is that a *referential domain* for a sentence provides the means to specify all the potential referents that satisfy the linguistic description. A definite noun phrase can then be used with multiple potential referents so long as the relevant domain defines a unique interpretation. For example, they observe that at a banquet you could ask the person sitting next to you to "Please pass the red wine" even if there were six bottles of the same red wine on the table, but only one was clearly within reach of the addressee. Tanenhaus *et al.* demonstrate that referential domains take into account behavioral goals expressed in a sentence. They do this by using the latency of eye movements to probe the effects of having ambiguity or not in the possible referents. They focus particularly on structural ambiguities which involve a choice between a syntactic structure in which the ambiguous phrase modifies a definite noun phrase and one in which it is a syntactic complement (argument) of a verb phrase. Just one example of these studies will have to suffice here. First note that "Pour the egg in the bowl over the flour" is temporarily ambiguous – it could be temporarily interpreted as "Pour the egg in the bowl" before it is discovered to be equivalent to the unambiguous instructions "Pour the egg that's in the bowl over the flour." Chambers *et al.* (2004) observed their subject's eye movements in response to both these forms when subjects observed a variety of displays, each of which showed an egg in a cup, a broken egg (no shell) in a bowl, an empty bowl, and a pile of flour. In the "compatible competitor" case, the egg in the cup was also broken; in the "incompatible competitor" case, the egg in the cup was in its shell, unbroken. "Pour the egg" can only apply once the egg has been broken – thus "in the bowl" serves to disambiguate "the egg" in the compatible competitor case (liquid egg in glass) but not in the incompatible competitor case (solid egg in glass). In short, the subject entertains the idea that "in the bowl" in "Pour the egg in the bowl" is the goal of the action if it seems unlikely that "in the bowl" could be an adjunct modifying the noun phrase, "the egg" because there is only one egg that is pourable.

When both potential referents matched the verb (e.g., the condition with two liquid eggs), there were few looks to the false goal (e.g., the bowl) and no differences between the ambiguous and unambiguous instructions. Thus, the prepositional phrase was correctly interpreted as a modifier. However, when the properties of only one of the potential referents matched the verb, participants were more likely to look to the competitor goal (the bowl) with the ambiguous instruction than with the unambiguous instruction. Thus, listeners misinterpreted the ambiguous prepositional phrase as introducing a goal only when a single potential referent (the liquid egg) was compatible with a pouring action. In short, Chambers *et al.* (2004) showed that the relevant referential domain defined can be dynamically updated based on the *action-based affordances* of objects.

By reviewing a number of other studies, Tanenhaus *et al.* (2004) show that not only actions but also intentions, real-world knowledge, and mutual knowledge can circumscribe referential domains, and that these context-specific domains affect syntactic ambiguity resolution (among other processes). In many of the studies, the subject has had time to inspect the visual display before responding to a sentence or instruction. In these circumstances, the recording of the subject's eye movements supports the view that they map linguistic input onto action-based representations from the earliest moments of processing that input. Moreover, they note evidence that behavioral context, including attention and intention, affects even basic perceptual processes (e.g., Gandhi *et al.*, 1998;

Colby and Goldberg, 1999) and that brain systems involved in perception and action are implicated in the earliest moments of language processing (e.g., Pulvermüller *et al.*, 2001). Given our concern in the MSH on the roots of language in manual actions and recognition of actions of the other, it is noteworthy that the final words of the Tanenhaus *et al.* (2004) review are: "there is now substantial evidence that social pragmatic cues such as joint attention and intentionality are critical in early language development (e.g., Bloom, 1997; Sabbagh and Baldwin, 2001), as well as evidence showing that non-linguistic gestures contribute to the understanding of speech (e.g., Goldin-Meadow, 1999; McNeill, 2000)."

9.7 Linking sensorimotor sequences to sentences

The theoretical framework which comes closest to what we seek in linking the SVSS model of Section 9.5.1 is that offered by Knott (2003), which we briefly review. It builds on the Itti–Koch model (Section 9.4.2) and links this to a system for execution and observation of actions. Knott translates the sensorimotor sequence of attention to the scene into the operations involved in constructing the syntactic tree for its description. Box 1 of Fig. 9.7 combines the Itti–Koch saliency model which selects an object for attention based on its saliency in the image with circuitry for classifying the selected object. Box 2 (Fig. 9.7) addresses how, once an object becomes the center of attention, the agent's direction of gaze establishes a frame for further processing. However, Knott cites Tipper *et al.* (1992, 1998) concerning "action-centered"' representations in which objects compete in virtue of their closeness to the starting position of the hand which will reach for them. Thus, somewhat contrary to the actual labeling of the diagram, the key point is that determination of which is the most salient retinal stimulus may be gated by the position of the target object relative to the hand. Another possibility is that the observer can modulate the saliency map centered not on his own hand, but on the position of an *observed* agent or object (cf. Perrett *et al.*, 1989; Jellema *et al.*, 2000). We see that these generalizations are amenable to the processes offered by SVSS for building minimal and anchored subscenes.

Box 3 of Fig. 9.7 is a model of the motor controller which takes two inputs (a goal motor state and a current motor state) and generates a goal-directed motor signal. Knott (2003) emphasizes the MOSAIC model (Haruno *et al.*, 2001) for this subsystem, but this is not essential to the argument. For Knott, the goal input is the agent-centered position of the target and the current motor state is the agent-centered position of the agent's arm. Different controllers are required to execute different motor programs (e.g., *touch* vs. *push*) in relation to a target. This description assumes the agent is the observer but Knott notes the evidence for a mirror system – the mechanism by which we recognize actions in others using the same representations that are used to control our own actions – suggests the possibility of augmenting Box 3 by a network for "biological motion recognition."

We will say more below about the relation of Fig. 9.7 to our own modeling, but here we note the key innovation of Knott's (2003) work, relating the sensorimotor model to

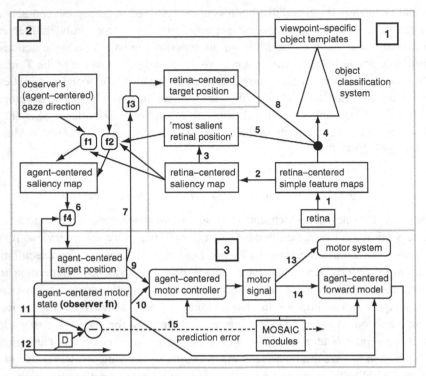

Figure 9.7 Sensorimotor circuit diagram (Knott, 2003) representing cognitive processes involved in executing a simple transitive action, such as reaching for a target object with the hand (functions in round boxes; data in square boxes).

syntactic representations. The idea is outlined in Fig. 9.8. Knott assumes that the logical form of a sentence is a cognitive process rather than a representation of the world, and proposes that a syntactic constituent denotes an episode of activity within a sensorimotor system of the kind that appears in Fig. 9.7. The bottom right-hand part of the syntactic tree of Fig. 9.8a will be familiar to most readers: the idea that a structure (here called VP) links the subject subj to another structure (V′) which combines the verb V and object obj. As we see in Fig. 9.8b, Knott specifies linkages of these syntactic elements to processes shown in Fig. 9.6.

Knott's analysis of the rest of Fig. 9.7 incorporates ideas of Koopman and Sportiche (1991) and Pollock (1989) for generative syntax, but the syntactic complexities are outside the scope of this chapter. Suffice to say that these relate to the idea that the formation of a syntactic tree moves elements from one position to another but preserves links (traces) between the final position of an element and its initial position. This is exemplified by the notation "Who$_i$ did John love [t$_i$]?" to indicate that the "Who" should be considered as having moved from the object position of "John loves X". The fact that the change from "loves" to "did . . . love" must also be explained gives a feel for why the

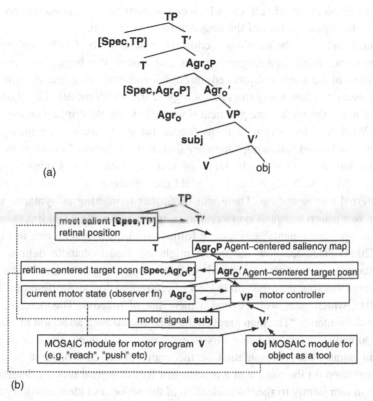

Figure 9.8 (a) Syntactic structure of a transitive clause. (b) The same structure with syntactic constituents associated with sensorimotor representations or sensorimotor functions. (Adapted from Knott, 2003.)

complexity of the syntactic tree may be necessary even if at first sight it seem unduly complicated. In any case, Fig. 9.8b embeds the basic VP tree in a larger tree which indicates (with dashed lines) the various movements that go into forming the final tree. Knott's aim is to characterize the notion of hierarchical position in a syntax tree in terms of the order in which representations become active during a sensorimotor action, with hierarchically high constituents becoming active before hierarchically lower ones.

The general idea, then, is the perception of a "sentence-sized event" involves a sequence of transitions between different attentional states, each of which generates a distinctive side effect in a medium for assembling linguistic representations. An agent is given a set of situations to observe, each accompanied by a sentence describing it, and must learn to generate appropriate sentences for similar situations. The agent processes each situation using a sensorimotor mechanism consisting of several different interacting components. Each of these components generates a side effect of its operation at a linguistic level of representation. The architecture of the perceptual mechanism imposes a (partial) order on these side effects, which can be construed as encoding certain aspects of the syntactic representation of the sentence to be expressed.

However, recall from the previous section the conclusion by Griffin and Bock (2000) that their evidence points to a language production process that begins with apprehension of the structure of the scene and proceeds through incremental formulation of the sentence which expresses it. Thus, using the terminology of the SVSS model (Fig. 9.6) we would argue that Knott's theory is more pertinent if we build scene description on the state of the Symbolic Working Memory, with items tagged for relevance or "communicative salience," that is achieved following "apprehension of the subscene" rather than on the eye movements that went into the building of that representation. Competitive queuing (Bullock and Rhodes, 2003) would then yield the sequence of "virtual attention shifts" for this internal representation. There will be further reordering as syntactic constraints reorder the words into accepted syntactic forms, much as Knott (2003) links sensorimotor processes to the "movements" engaged in formation of the syntactic tree in Fig. 9.8.

Knott (2004) discusses how the model might be used to handle definite reference. During discourse, one maintains a record of all tokens that are understood to be accessible to both discoursers (but note the subtle view of referential domains offered by Tanenhaus et al. (2004), which goes well beyond a mere list of tokens). The perceptual process underlying the sentence "The man grabbed the cup" begins with an action of reattention to an object already encountered and this is encoded as *the man*, not *a man*. This then triggers (in parallel) a representation of the man's local environment in a frame of reference centered on the man, and a mechanism for biological motion detection. These two events in turn jointly trigger identification of the action and identification of the target object. This corresponds to anchoring perception of a minimal subscene in an agent. We would also anchor it in a target object, but motion might be the salient cue that draws our attention to Harry's hand and thence to Harry and then to the theme of the motion – and the hand might not even be mentioned in the sentence "Harry is throwing a ball." But note that Knott is generating the sentence as attention shifts (like a scanpath) rather than after the scene-to-be-described has been chosen. Of course, one may also generate part of a sentence then update it as attention notes further aspects, or realizes they are needed for communicative clarity. In any case, our object is not to quibble with details of Knott's analysis, but rather to welcome it as a welcome signpost pointing us in the right direction for work integrating attention, scene description and action (Fig. 9.7) with language (Fig. 9.8).

9.8 Description of dynamic scenes: a pilot study

The previous sections offer a conceptual framework within which to model how processes responsible for object recognition, action recognition, and rapid evaluation of the setting or gist of a scene integrate with top–down goals, task demands, and working memory in expanding a symbolic representation of a scene. However, almost all the efforts reviewed in the preceding two sections focus on the processing of a single, static scene whereas our eventual goal is to extend the study to dynamic scenes, in which change constantly occurs. To this end, we conducted a pilot study of eye movements in generating sentence after sentence in describing a rapidly changing scene. We had a subject (MAA) view 12 videoclips, and narrate a description as he observed each videoclip (each about 30 s, divided into from two to six discrete scenes). Eye tracking apparatus allowed us to record his eye movements and superimpose them on a copy of the videoclip along with an audio recording of his narrative. Methods were as previously described (Itti, 2005). Briefly, the observer sat at a viewing distance of 80 cm from a 22" computer monitor and eye movements were recorded using a 240 Hz infrared-video-based system (ISCAN RK-464) following a 9-point calibration procedure.

We analyzed these data to extract a range of hypotheses about the interaction between attention, object and action recognition, gist extraction, interplay between tracking and inhibition of return, and what we call "narrative momentum," in which not only does the speaker try to complete the description of the present subscene but may also direct attention to a subsequent minimal subscene whose description will extend the story in the description so far. In addition, below we show how such data may be more quantitatively analyzed so as to answer specific questions, for example, were objects and actors reported by the observer more bottom–up salient when the observer looked at them than objects and actors not reported?

The aim of our first broad and rather qualitative analysis of the data is to set targets for future modeling of a variety of constituent processes and to suggest strategies for the design of more structured tests of these models. Here we present some example sentences (in boldface) used to describe very different scenes and add observations made in relating the verbal description to the movement of the visual target on the dynamic scene.

Videoclip 1

There's a document: Target traverses the field of view but this scene is too short to do any real reading. The subject notices someone moving across the scene, but despite this low-level salience, the sign holds his attention. . . **it looks as if there is a riot . . . they've got a refrigerator or something:** people at left are the first target and anchor recognition of the gist of the scene as a riot. The refrigerator is then the most salient object – the eyes track the refrigerator in an attempt to make sense of what is being done with it. **Now a jeep or something comes in. A guy with a gun is moving across:** the eyes follow the jeep moving away but there is no time to describe this before a salient motion captures attention and anchors a new minimal subscene.

Successive subscenes may build on schemas and/or expectations of a previous subscene, or involve novel elements, when an unexpected object or action captures the attention. In addition, there is a real competition at the start of each scene between completing the ongoing description of the previous scene and starting the description of the new scene. There is a working memory issue here since the subject is processing a representation of the prior scene while building the representation of the new scene.

Episodic memory is also at work since the subject can recognize a videoclip seen days or weeks previously.

Videoclip 2

Now we're in front of the embassy with a couple of soldiers guarding it: The embassy sign is targeted then two policemen. The scene shifts to show soldiers. We hypothesize that the later sight of soldiers pre-empted the completion of the initial description, which thus conflated the policemen and the soldiers from two minimal subscenes.

Videoclip 3

Looks like a snow scene with various rocks, a few trees. As we pan down it looks like it must be a hot pool because there's steam rising up . . . This is one of the few cases where the motion of the camera is reflected in the verbal description. Interestingly, the steam is seen before the hot pool, but the description mentions the *inferred* pool before the steam that "implies" it. **And my goodness there's a monkey sitting in the pool so maybe it's in Sapporo or something – up in the island.** Bottom–up salience of a rather slight movement compared to the rest of the scene drew the subject's attention to the monkey. Note the long inference from "monkey in the snow" to the likely location. However, the subject mistakenly recalled the name of the town Sapporo where he should have named the island, Hokkaido, and attempts repair with the last phrase.

In a second analysis, we provide for the first time a quantitative answer to the simple question of whether objects and actors which are retained into the minimal subscene representation and which are reported by the observer tended to be more bottom–up salient than objects and actors not reported. To this end, we parsed the verbal reports in conjunction with the videoclips onto which the scanpaths of the observer had been superimposed. This allowed us to mark those saccadic eye movements of the observer that were directed towards some objects, actors, or actions in the scene which were part of the verbal report. For example, as the observer described that "there is a monkey," we assigned a label of "monkey" to saccades directed towards the monkey in the corresponding video clip, and occurring within up to 5 s before or after the verbal report. To evaluate the bottom–up salience of saccade targets, we subsequently processed the video clips through the Itti–Koch model of bottom–up visual attention (Itti and Koch, 2000). At the onset of every saccade, we recorded the model-predicted salience at the future end point of the saccade, as well as at a random end point, for comparison (this random end point selection was repeated 100 times in our analysis, so that it provides a good on-line estimate of the overall sparseness of the dynamic saliency map at a given instant). For all saccades directed towards a reported scene element, and the first saccade directed towards each reported element, we compared the distribution of salience values at the end points of those saccades to what would have been observed by chance. We derived a score, which is zero if the targets of human saccades are not more salient than expected by chance, and greater than zero otherwise.

Figure 9.9a shows an example frame (left) and associated saliency map (right; brighter regions are more salient) of one clip, for the first saccade (arrow) directed towards "a man" also described in the narrative. The man was the most salient entity in the scene at that instant. Figure 9.9b shows another example from another videoclip, for the first saccade directed towards "a monkey," also reported in the narrative. In this example, the monkey is not the most salient item in the display, although it has non-zero salience.

Figure 9.9c shows the number of saccades histogrammed by normalized saliency at saccade targets, for human and random saccades. Comparing the saliency at human saccade targets to that for random targets shows whether humans tended to orient towards regions with different saliency than expected by chance. Overall, saliency maps were quite sparse (many random saccades landing onto locations with near-zero saliency), but humans avoided non-salient regions (though fewer human saccades landing on these regions as compared to highly salient regions), while humans tended to select the few locations of near-unity saliency more often than expected by chance. The reported KL score measures the difference between the human and random histograms: a KL score of 0.00 would indicate that humans were not attracted to salient image locations more than expected by chance. The KL score obtained here for all saccades is significantly higher than 0.00 (t-test, $p < 10^{-27}$) as humans were more attracted to locations with above-average saliency than to those with below-average saliency.

Figure 9.9d shows a similar analysis to that presented in Fig. 9.9c, but only for those saccades that were directed towards some object, actor, or action reported in the narrative. The resulting KL score is higher than in Fig. 9.9c, indicating that items reported in the narrative had on average higher saliency than non-reported items (t-test, $p < 10^{-13}$). Finally, Fig. 9.9e shows a similar analysis, but considering only the first saccade towards reported objects, actors, and actions, that is, the initial noticing of these items. The KL score is even higher, indicating that when they were first noticed, items that would eventually be reported were more salient than other items (t-test, $p < 0.0008$).

Thus, these pilot data (to be taken carefully as there is only one observer, although all results are statistically significant for that observer due to the reasonably large number of saccades) indicate that bottom–up salience may act as a facilitator, whereby more salient items actually not only attract gaze towards them, but are also more likely to be retained in the minimal subscene and verbally reported. Overall, these data demonstrated a rich variety of interactions between the factors driving attention (as reviewed above) and verbal reports. The merit of this first pilot study was thus to start cataloguing the interactions between visual scene understanding and language production when observers attempt to describe the minimal subscenes when they are exposed to complex scenes.

We next conducted experiments with 32 videoclips filmed at a busy outdoors mall and a group of nine observers, each asked *post hoc* what they thought the minimal subscene was for each videoclip. A working definition of the minimal subscene was provided and discussed with the subjects beforehand; subjects were instructed that the minimal subscene was the smallest set of objects, actors, and actions that are important in the scene,

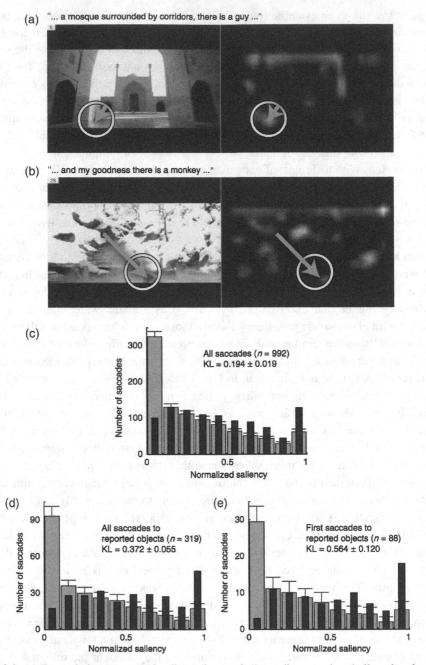

Figure 9.9 (a) Example human saccade, directed towards "a guy" reported verbally, who also was the most salient entity in the video clip at that instant. (b) Another example human saccade, directed towards "a monkey" also reported verbally, and who was not the most salient, although it was somewhat salient. (c)–(e) Histograms counting the number of human (thin dark bars) and random (wider light bars) saccades directed towards various salience values (± SD from 100-times repeated

and those that they would want to mention when requested to give a short, one-sentence summary of what happened in each videoclip.

Post hoc rather than simultaneous reporting eliminated some of the factors of the previous experiments (e.g., narrative momentum, switching between subscenes, lacking verbal bandwidth to report some fixated events) – clearly both a plus and a minus, but indicative of how one may manipulate the videoclips to focus on specific phenomena in each study. The study suggested the following broad categorization of factors influencing the selection of minimal subscenes:

(1) Factors derived from the gist, layout, and overall setting of the scene (e.g., objects belonging to the minimal subscene tended to be in the foreground, often occluding other objects; they usually occupied central positions within the scene's layout, and were present for extended time periods), including a number of extrinsic factors (the camera tended to follow the objects/ actors/actions of the minimal subscene, often zoomed towards them).
(2) Bottom–up salience (e.g., isolated people, actors moving against a general flow, or brightly colored objects) with the noted caveat that, often, salient objects attracted attention although they were quickly discarded as being irrelevant to the current minimal subscene.
(3) Cultural and learned factors (e.g., finger-pointing movements, facial expressions, postures), which made some events more relevant than others.
(4) Personal preferences could guide different observers towards different minimal subscenes (e.g., a man playing with some electronic gadget attracted technology-oriented observers while a dog attracted animal-lover observers in the same clip). This second pilot study suggested stimulus manipulations by which some of these factors may be selectively emphasized or suppressed.

9.9 Recognition of events in dynamically changing scenes

In this section, we briefly note interesting efforts within AI on the recognition of events in dynamically changing scenes. A classic example of a scene analysis system, VITRA (the Visual Translator) (Herzog and Waziuski., 1994), was able to generate real-time verbal commentaries while watching a televised soccer game. The low-level visual system recognizes and tracks all visible objects from video streams captured by a fixed overhead camera, and creates a geometric scene representation (the 22 players, the field, and the goal locations). This representation is analyzed by series of Bayesian belief networks that incrementally recognize plans and intentions. The model includes a non-visual notion of

Caption for Figure 9.9 (cont.)
random sampling). Salience scores are derived from comparing the human and random histograms (see text for details), with a score of 0 indicating that humans did not orient towards model-predicted salient locations more than expected by chance, and scores above 0 indicating that bottom–up salience attracted humans. Scores are reported for (c), all saccades (d), all saccades to reported objects and, (e) the first saccades to reported objects, and indicate that out of all the locations gazed to by the observers, those that were reported verbally were significantly more salient than those that were not reported. KL, Kullback–Liebler divergence between distributions of salience at random saccade end points and at human saccade end points.

salience which characterizes each recognized event on the basis of recency, frequency, complexity, importance for the game, and so on. The system finally generates a verbal commentary, which typically starts as soon as the beginning of an event has been recognized, but may be interrupted by new comments if highly salient events occur before the current sentence has been completed. However, VITRA is restricted to one highly structured environment and one specific task. Further, it is not a biologically realistic model, and cannot scale to unconstrained environments as it constantly tracks all objects and attempts to recognize all known actions.

Another computer system for the cooperation between low-level perceptual analysis and symbolic representations and reasoning is provided by the work of Zhao and Nevatia (2004) on tracking multiple humans in complex situations. A human's motion is decomposed into its global motion and limb motion. Multiple human objects in a scene are first segmented and their global motions tracked in three dimensions using ellipsoid human shape models. This approach is successful with a small number of people even when occlusion, shadows, or reflections are present. The activity of the humans (e.g., walking, running, standing) and three dimensional body postures are inferred using a prior locomotion model. Such analyses provide the basis for the hierarchical representation of events in video streams. To this end, Nevatia *et al.* (2003) developed an event ontology that represents complex spatiotemporal events common in the physical world by a composition of simpler events. The events are abstracted into three hierarchies. Primitive events are defined directly from the mobile object properties. Single-thread composite events are a number of primitive events with temporal sequencing. Multi-thread composite events are a number of single-thread events with temporal/spatial/logical relationships. This hierarchical event representation is the basis for their Event Recognition Language (ERL), which allows the users to define the events of interest conveniently without interacting with the low-level processing in the program. For example, complex event "Contact1" is a linear sequence of three simple events: "approaching a person," "stopping at the person," and "turning around and leaving." Similarly, complex event "Passing_by" is a linear occurrence of "approaching a person," and "leaving" without stopping in between.

Such artificial intelligence systems pose the challenge of trying to better understand how the functional processes they embody may map onto brain areas and processes.

9.10 Inverse models, forward models, and the mirror system

We here link the role of motor control in Fig. 9.7 (Knott, 2003) to the more general notion of forward and inverse models, then point the reader to the discussion given by Oztop, Bradley, and Arbib (this volume) relating the FARS and Mirror Neuron System (MNS) models of manual action and action recognition, respectively, to this general framework. We return to the distinction between the sign and the signified to distinguish between producing or perceiving a word and perceiving the concept that it signifies.

A *direct* or *forward model* of the effect of commands is a neural network that predicts the (neural code for) sensory effects in response to a wide set of (neurally encoded) commands. Conversely, an *inverse model* converts (the neural code for) a desired sensory situation into the code for a motor program that will generate the desired response. As described by Oztop, Bradley, and Arbib (this volume), Arbib and Rizzolatti (1997) used the notions of forward and inverse models to analyze the control of grasping:

(1) The *execution system* leads from "view of object" via parietal area AIP (visual recognition of affordances – possible ways to grasp an object) and F5 (motor schemas) to the motor cortex which commands the grasp to an observed object. This pathway (and the way in which prefrontal cortex may modulate it) implements an inverse model (mapping the desired sensory situation of having the object in one's grasp to the motor activity that will bring this about). This has been implemented in the FARS model (Fagg and Arbib, 1998; Arbib, Chapter 1, this volume) (Fig. 9.4).

(2) The *observation matching system* leads from "view of gesture" via gesture description (posited to be in superior temporal sulcus) and gesture recognition (mirror neurons in F5 or area 7b) to a representation of the "command" for such a gesture (canonical neurons in F5) – another access to an inverse model. This has been implemented in the MNS model (Oztop and Arbib, 2002; Oztop, Bradley, and Arbib this volume) (Fig. 9.6). The *expectation system* is a *direct* model, transforming an F5 canonical command into the expected outcome of generating a given gesture. The latter path may provide visual feedback comparing "expected gesture" and "observed gesture" for monkey's self-generated movements, and also create expectations which enable the visual feedback loop to serve for learning an action through imitation of the actions of others.

Oztop, Bradley, and Arbib (this volume) stress that these inverse and forward models may in reality be seen as encompassing a whole family of inverse and forward models. Thus, in recognizing an action (as in the MNS model), we are not so much employing "the" inverse model, one of multiple inverse models best matching the observed interaction of hand and object. Rather, the system may recognize that the current action can better be viewed as a combination of actions already within the repertoire. Similar ideas have been applied in the "sensory predictor" hypothesis of mirror neurons which views *mental simulation* as the substrate for inferring others' intentions (Oztop et al., 2005).

The reader interested in this discussion of forward and inverse models should also consult the chapter by Skipper, Nusbaum, and Small (this volume) which describes an active model of speech perception that involves mirror neurons as the basis for inverse and forward models used in the recognition of speech. In this view, both the facial and manual gestures that naturally accompany language play a role in language understanding.

9.11 Mirror neurons and the structure of the verb

The MNS model (briefly recalled in the previous section) provides the mechanisms whereby data on hand motion and object affordances are combined to identify the action. The claim is that, in concert with activity in region PF of the parietal lobe, the mirror neurons in F5 will encode the current action, whether executed or observed. However, F5 alone cannot provide the full neural representation of $Grasp_A$ (Agent, Object), where $Grasp_A$ is the current type of grasp. The full representation requires that the F5 mirror activity be bound to the inferotemporal activity encoding the identity of the object and activity in the superior temporal sulcus (or elsewhere) encoding the identity of the agent, with "neural binding" linking these encodings to the appropriate roles in the action–object frame. This clearly involves a working memory (WM) that maintains and updates the relationships of agent, action, and object (cf. our earlier critique of the approach of Baddeley (2003); the discussion of relevant neurophysiology by Arbib and Bota in Chapter 5 (this volume); and the review by Scalaidhe *et al.* (1999) of the variety of working memories implemented in macaque prefrontal cortex and with links to specific parietal circuitry). This takes us from the hand–action focus of FARS–MNS to the study of actions more generally. We will build on the above to speculate about brain mechanisms involved in recognition of more general action–object frames ("The car is turning the corner") of the form Action(Agent, Object) and Action(Agent, Instrument, Object). We expect these to involve mechanisms far removed from the F5–Broca's mirror system for grasping, and must thus seek to understand how they nonetheless access Broca's area (and other language mechanisms) when such minimal subscenes are linked to language.

A note of caution: from a rigorous philosophical point of view, it is inappropriate to view a car as an agent and its turning as an action. Nonetheless, it can be useful to regard such verbs as "generalized actions." Indeed, linguists generally agree that most verbs should be given a syntactic analysis that includes thematic roles generalized from the (Agent, Instrument, Object) discussed above (Williams, 1995). Since our concern will be with nouns and verbs used in describing visual scenes, we have taken the pragmatic stance of deciding to use "action" for "that which a verb denotes" and "object" for "that which a noun denotes." Of course, this can only go so far, and the verb *to be* does not denote an action. Such subtleties are secondary to the present account of brain mechanisms linking attention, action, and language.

With this, we devote the rest of the section to briefly noting the preliminary work by J.-R. Vergnaud (personal communication) and Arbib seeking to gain deeper insights into the structure of the verb by giving a bipartite analysis of the verb in terms of mirror neurons and canonical neurons. Much work has been done by others on the lexicon, showing how a "lemma" – the "idea" of a word plus syntactic information (gender, number, . . .) – may be selected and then transformed into an articulatory code (see, e.g., Levelt (2001) for an overview and Indefrey and Levelt (2000) for a meta-analysis of relevant brain imaging). We will instead focus on how the linkage of a verb to its arguments may be illuminated by reference to the MSH. Briefly, we start from the observation that the two immediate arguments of the verb have different *linguistic* properties. The verb and its "object" can form a tight semantic and structural unit, unlike the verb and its "subject." There are also more abstract properties that distinguish "subjects" from "objects," relating to quantificational scope, anaphora, coreference, control, weak cross-over, and incorporation (Kayne, 1994). Such asymmetry is explicit in the verbs of languages such as Eastern Armenian (Megerdoomian, 2001) where the structure of an elementary clause is similar to that of, e.g., "Mary made honey drip." where the verb now has two parts linked to the agent and object respectively (Mary made) and (drip honey). The Vergnaud–Arbib hypothesis is that verbs should in fact receive this bipartite analysis even when it is not evident (as in most English sentences) on the surface. The resultant phrase structure [$_S$NP$_S$[vPv[$_{vP}$V NPo]]] with v–V being the analysis of the verb in the sentence, is – we argue – the expression of a neural system whereby the human homologue of the F5 canonical system binding action to object underlies the merging of the V-component with the object NP$_o$; while the human homologue of the F5 mirror system binding agent to action underlies the merging of the subject NP$_S$ with the complex v–VP.

We close this section by reiterating crucial distinctions emphasized in analysis (Arbib, Chapter 1, this volume) of the MSH. We distinguish a mirror system for hand and face movements from mirror systems (posited but little studied to date) for other actions in the observer's own repertoire. Both of these are to be distinguished from mechanisms for the recognition of actions which are outside the observer's possible repertoire. MSH then hypothesizes that the production and recognition of words (phonological form) is based on an evolutionary refinement of the original mirror system for grasping, which has through evolution become multi-modal, integrating manual, facial, and vocal gestures for communication. Thus it is only verbs related to manual action which are intimately linked to the networks that support the "semantic form" for the related action. For most nouns and other kinds of verbs, the recognition of the corresponding objects and actions must be conducted elsewhere and linked to the phonological form through processes which embody syntax to express perceived relations between actions and objects in a hierarchical structure which is expressed as a linear sequence of words.

9.12 Construction grammar

Both Knott and Vergnaud operate within the framework of generative grammar. Kemmerer (this volume) employs an alternative framework, construction grammar, in his work. We briefly recall some key ideas about construction grammar then suggest how "vision constructions" may synergize with "grammar constructions" in structuring the analysis of a scene in relation to the demands of scene description and question answering in a way which ties naturally into our concern with minimal and anchored subscenes.

In the next few paragraphs, we provide a brief exposition of construction grammar based on that provided by Croft and Cruse (2004, ch. 9): their starting point is to contrast the approach of generative grammar with the use of *constructions* in the sense of traditional grammar. A basic example of such a construction is the *passive construction*:

(a) The ball was kicked by Harry.
(b) [Subject *be* Verb-PastParticiple *by* Oblique]

which combines the syntactic elements given in (b), including the subject noun phrase, the passive auxiliary verb *be* in some form, a past participle of a verb, and (optionally) a prepositional phrase with the preposition *by* and an oblique noun phrase. (A noun is said to be in oblique case when it is the predicate of a sentence or preposition but is not the subject of the sentence.) Semantically, the agent of the action in the passive construction is expressed by the object of the prepositional phrase, and the undergoer is expressed by the subject. Such a specific construction is contrasted with a generative description which would seek to explain properties of a given construction in terms of general rules of the various components, with any idiosyncratic properties derived from the lexicon.

Generative grammar distinguishes the lexicon from the grammar, and this is seen as having three components – phonological, syntactic, and semantic – with linking rules to map information from one component onto another. The "rule-breaking" within any particular language is restricted to idiosyncrasies captured within the lexicon. Moreover, the rules inside each component are considered to be so highly intertwined and self-contained that each represents a separate structure that can be considered as being relatively autonomous. As Croft and Cruse (2004) note, Minimalist theory (Chomsky, 1995) recasts the phonological component as an "articulatory–perceptual interface" which links the language faculty to the perceptual–motor system and recasts the semantic component as a "conceptual–intentional interface" which links the language faculty to other human conceptual activity. The lexicon remains as the repository of idiosyncratic information, and as such provides information linking the three components together (Chomsky, 1995, pp.235–236). At first glance, the emphasis of Chomsky's Minimalism on an "articulatory–perceptual interface" and a "conceptual–intentional interface" seems compatible with our view of language within a broader framework action and perception. However, closer inspection shows that the Minimalist Program is far removed from a model of the speaker or hearer using language. The Minimalist Program characterizes which strings of lexical items are "grammatically correct" as follows (Fig. 9.10): a set of lexical items is taken at random, the computational system then sees whether legal derivations can be built each of which combines all and only these elements. Spell-Out occurs when one of the legal derivations, if any, is chosen on the basis of some optimality criteria. The Computational System then transforms the result into two different forms, the Phonological Form, the actual sequence of sounds that constitutes the utterance, and the Logical Form, which provides an abstract semantics of the sentence. There is no attempt here to model actual sentence production or perception – the process starts with words chosen at random and only at the end do we see whether or not they can be arranged in some way that yields a semantic structure. Of course, we have seen that Knott (see above, Section 9.7) has offered a novel way of linking the formalism of generative grammar to sensorimotor structures. Thus while Fig. 9.9 suggests that the Minimalist Program is at best indirectly related to perception and production, the possibility of using insights from generative grammar to develop an action-oriented approach to language has not been ruled out.

However, construction grammar grew not out of a concern to model perception and production, but rather out of the need to find a place for idiomatic expressions like *kick the bucket, shoot the breeze, take the bull by the horns,* or *climb the wall* in the speaker's knowledge of his language. Nunberg *et al.* (1994, pp.492–493) identified *conventionality* as a necessary feature of an idiom – its meaning or use requires an agreed-upon convention and cannot be (entirely) predicted on the basis of its parts. But this suggests

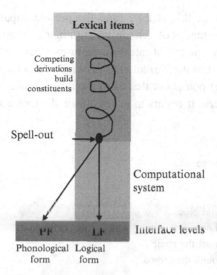

Figure 9.10 Derivations and the computational system: the Minimalist Program (whose descriptive adequacy may be compared to Kepler's "conic sections" *description* of planetary motion). Contrast Figure 1.2 of Arbib (this volume) which places language within a broader framework action and perception (and whose adequacy may be compared to Newton's dynamical *explanation* of planetary motion).

that the meaning of each idiom must thus be stored in each speaker's mind. Should we consider these meanings, then, as a supplement to the general rules of the syntactic and semantic components and their linking rules? In proposing the original construction grammar, Fillmore *et al.* (1988) took a more radical step. Instead of adding idioms to the componential model, they suggested that the tools they used in analyzing idioms could form the basis for *construction grammar* as a new model of grammatical organization.

Croft and Cruse (2004) give a careful analysis of how a wide range of idioms can be captured by constructions to draw two general observations: (1) a given construction will often turn out to be just one of a family of related constructions; (2) the number and variety of constructions uncovered in studies of idioms imply that speakers possess a huge range of specialized knowledge that augments general rules of syntax and semantic interpretation on the one hand. They further note that many linguists working outside construction grammar have also examined schematic idioms and constructions and teased out the rule-governed and productive linguistic behaviors specific to each family of constructions. However, the key point is that constructions, like the lexical items in the lexicon, cut across the separate components of generative grammar to combine syntactic, semantic, and even in some cases phonological information. The idea of construction grammar (Fillmore et al., 1988; Goldberg, 1995, 2003; Croft, 2001) is thus to abandon the search for separate rule systems within three separate components – syntactic, semantic, and phonological – and instead base the whole of grammar on the "cross-cutting" properties of constructions.

It is beyond the scope of this chapter (and the present capability of the authors) to adjudicate on the relative merits of generative grammar or construction grammar, or to suggest whether and how one might integrate features of each, but we do want to offer certain comments relevant to the current enterprise. To this end, we note that Kemmerer (Chapter 10, this volume) points out that even though *kick* is usually considered to be a prototypical transitive verb, it occurs in at least nine distinct active-voice constructions (Goldberg, 1995):

(1) Bill kicked the ball.
(2) Bill kicked the ball into the lake.
(3) Bill kicked at the ball.
(4) Bill kicked Bob the ball.
(5) Bill kicked Bob black and blue.
(6) Bill kicked Bob in the knee.
(7) Bill kicked his foot against the chair.
(8) Bill kicked his way through the crowd.
(9) Horses kick.

These sentences describe very different kinds of events, and each argument structure construction here provides clausal patterns that are directly associated with specific patterns of meanings. We refer the reader to Kemmerer's chapter for his analysis of "action verbs, argument structure constructions, and the mirror neuron system." Here, we want to briefly argue that the approach to language via a large but finite inventory of constructions motivates the return to visual scene interpretation armed with the notion that a large but finite inventory of "scene schemas" (from the visual end) may provide the linkage with constructions (from the language end) rich enough to encompass an exemplary set of questions we will ask and sentences subjects will generate. Each constituent which expands a "slot" within a scene schema or verbal construction may be seen as a hierarchical structure in which extended attention to a given component of the scene extends the complexity of the constituents in the parse tree of the sentence. This enforces the view that visual scene analysis must encompass a wide variety of basic "schema networks" in the system of high-level vision, akin to those relating *sky* and *roof*, or *roof*, *house*, and *wall* in the VISIONS system (Section 9.3.2 above). Of course, we do not claim that all sentences are limited to descriptions of, or questions about, visual scenes, but we do suggest that understanding such descriptions and questions can ground an understanding of a wide range of language phenomena – see Arbib (Chapter 1, this volume, Section 1.5.2) on "concrete sentences" versus "abstract sentences."

9.13 Towards integration

The detailed discussion of language processing is outside the scope of this chapter, but let us consider briefly the possible extension of the SVSS model of Section 9.5.1 above into a complete model in which the more symbolic schemas of STM are linked to a system for language production, as schematized in Fig. 9.11. As argued throughout this chapter, one of our main claims here is that the minimal or anchored subscene representation is the ideal interface between vision and language. In the case of meta-VISIONS, our top level below the gist will be a spatial array of schemas. We suggest that the representation will not be a set of schemas so much as a graph of schemas. Thus there will be a schema for each recognized object, whereas an action will be a schema that links the schemas for agent and patient, etc., as constrained by the scene schemas ("vision constructions") of the previous section. This provides the connection to construction grammar, in that a scene schema may be instantiated with more or less links, acting in the top–down way in which the VISIONS house-schema sets up hypotheses on the presence of walls and windows – but here with the linkage being defined more by its extent in time than by its extent in space. Other links may correspond to spatial relations, giving the concrete grounding for prepositions, etc.

The challenge for future research is to turn from the role of F5 and its homologues in recognition of manual actions to the study of the neural underpinnings of "general" action recognition (e.g., cars crashing). With this in place it should become possible to extend the Task Relevance Map (Section 9.5.1) to include actions and add them to the attentional specification. This takes us from keywords to relational structures. Thus Hit(o,o,o) instructs the system to find any instance of hitting in the scene, whereas Hit(o,ball,o) instructs the system to find only those hitting actions in which the instrument of the hitting is a ball. Similarly, the Task Relevance Map will not only link objects to locations, but will further provide links between objects representing any interactions between them. (Clearly, the same representation can accommodate spatial relations, such as a ball being on a table.)

Where the static version of the Itti–Koch model inhibits return of attention to an object for a short period after it is fixated, it is common in viewing a videoclip (as in our pilot studies) to track an object until its interest declines or its role in a minimal scene is determined. Also, it is no longer enough to say where an object is in a scene; one must also give its trajectory – i.e., where it is at each time it is in the scene. However, the ensuing minimal scene will in general incorporate only general characteristics of the trajectory rather than its temporal details. Similarly, each action will have a temporal duration, where action recognition may now be based on the *time course* of the relationship between two objects.

In the case of HEARSAY, we have at the top level a temporal sequence of sentences covering the single speech stream, and these in turn are linked to the sequence of words which best conform with that level. Sentences in turn link to semantic representations of whatever the sentence may mean, providing a basis for further behavior in relation to the scene. Such a representation must include not just the relevant objects (nodes), and actions and relations (links) but a set of dynamic via points which segment the continuous visual input (as filtered through selective attention) into episodes which can be integrated into narrative memory. Our prototype (Section 9.5.2) uses classical Artificial Intelligence tools to implement the long-term and working memories of the model. These should be

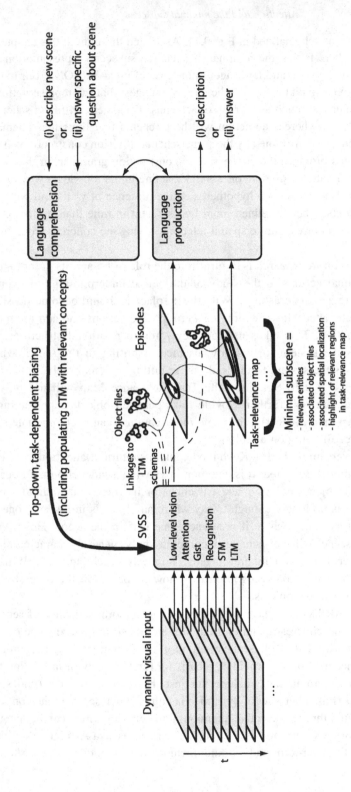

Figure 9.11 Minimal subscenes and symbolic schemas for episodes as the interface between vision and language.

translated into schema-based representations, with possible neural correlates (Rolls and Arbib, 2003).

Such efforts must also factor into the extension of our Task Relevance Map model to build a graph (in the sense of nodes with some connecting edges) in which each minimal subscene becomes encapsulated as an "episode-node," and these nodes can be connected by links expressing salient spatial, temporal and causal relations. Such a model would enrich the very active study of episodic memory. We note studies which relate hippocampal activity to ongoing action (Lacquaniti et al., 1997; Elsner et al., 2002). The extension of the Task Relevance Map will address the diversity of hypothetical internal scene representations in the literature, such as the "world as an outside memory" hypothesis (O'Regan, 1992), the "coherence theory" according to which only one spatiotemporal structure can be represented at a time (Rensink, 2000), a limited representation of five to six "objects files" in visual short term memory (Irwin and Andrews, 1996; Irwin and Zelinsky, 2002), and finally representations for many more previously attended objects in STM and LTM (Hollingworth and Henderson, 2002).

Figure 9.12 gives an overview of speech production and perception. Production of an utterance extends Levelt's scheme (Levelt, 1989, 2001; Levelt et al., 1999) for going from concept to phonological word; the complementary speech understanding system, for which HEARSAY provides our cooperative computation At the bottom level (not shown here), HEARSAY accesses the time-varying spectrogram of the speech signal. This is interesting because (a) the base level varies with time but, further (b) the representation at any time is a function of the recent past, not just the instantaneous present since a frequency estimate rests on a recent time window. Similarly, then, in developing our analysis of vision, we must note that the base level will include, e.g., activity in mediotemporal cortex that encodes motion, so we will have a spatial map of spatiotemporal jets (i.e., a collection of features centered on a particular point in space–time but extending in both space and time) as our base level. Note that this quickly gives way to quasi-symbolic representations. As we move up the hierarchy, we get symbolic representations (with confidence weights – a link to current treatments in terms of Bayesian models and Hidden Markov Models) that cover a greater and greater time-span of lower-level representations raising the issue of the span of working memory at each of these levels.

In considering the implications for vision, note that object recognition can be thought of as based on a single time – but, of course, reflecting movement up the pyramid of spatial extent. However, when we look at action recognition, we carry this spatial processing into the temporal realm. In the MNS model, we think of the firing of a mirror neuron as in some sense a confidence level for the action it represents (leaving aside population coding for now), and that this level (akin to the lexical level in HEARSAY, for those words that are verbs) has an anticipatory component in the sense that the labeling of an action may change as it proceeds but will often be correct prior to completion of the action. Moreover, just as recognizing one word creates expectations about words that follow, so may perceiving one action create expectations of what

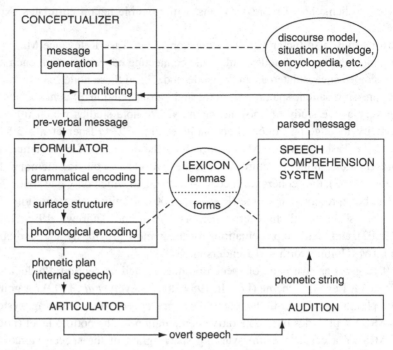

Figure 9.12 The left-hand side of the figure shows Levelt's scheme for going from concept to phonological word in a scheme for the overall production of whole messages or sentences; at right is the complementary speech understanding system, for which HEARSAY provides our cooperative computation place-holder. (Adapted from W. J. M. Levelt, *Speaking*, MIT Press 1989, p.9.)

further actions will unfold and towards what goal. Indeed, recent work in Parma reveals neurons that correlate with the next action of the sequence more than with the current action (Fogassi et al., 2005).

9.14 Action, perception, and the minimal subscene

Arbib and Bota (this volume, Section 5.6) built upon a schematic diagram (Fig. 9.13) developed by Arbib and Bota (2003) to sketch the relations between action generation, action recognition, and language within the framework of homologies between the brains of macaques and humans. They summarize the anatomical context for the generation and recognition of *single* actions and how these "lift" to communicative actions. Complementing this, this chapter has developed an overall framework for the functional analysis of how attention guides vision in the analysis of a dynamic visual scene, and how the structure of minimal subscenes or episodes (integrating the recognition of agents, actions, and objects) may be linked with language production both in describing a scene or in answering questions about the scene.

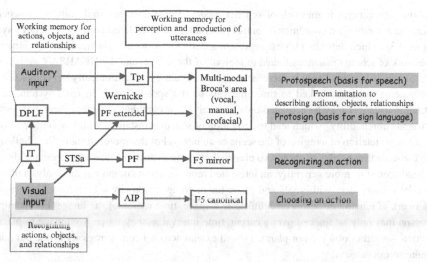

Figure 9.13 A high-level view of the cumulative emergence of three fronto-parietal systems: choosing a hand-related action, recognizing a hand-related action, and describing an action (in multiple modalities). This schematic builds on and modifies the schematic presented in Arbib (2001) to set goals for the modeling proposed here. See Arbib and Bota (this volume, Section 5.2.4) for further details.

Since the chapter has taken a long path through at times seemingly disparate elements, we review this framework by summarizing the key points of previous sections. This summary complements the section by section overview of the chapter given in Section 9.1.

(1) A minimal subscene may be anchored by an agent, action, or object and may then be extended to form an anchored subscene. A new focus of attention may (or may not) lead to the recognition of a new minimal/anchored subscene. In addition to overt shifts of attention to incorporate new objects and actions into a minimal subscene, there will be "internal" shifts which attend to new features of an agent, action, or object in fleshing out its description – as in going from "the man" to "the handsome old man on the left."

(2) Gist and layout provide a framework for more detailed scene analysis, whether of a static or dynamic scene.

(3) In linking vision and action to language, we will be particularly concerned with the relation between (i) building up a representation of minimal/anchored subscenes and scene description; and (ii) the role of top–down cues in visual attention involved in finding the appropriate subscene(s) on which to base the answer to a question.

(4) We stress that the Mirror System Hypothesis (MSH) does not ask us to conflate the sign with the signified. Recognizing an action is part of recognizing a scene; uttering the verb to describe that action is itself a different action. MSH asks us to consider the parallels between the mechanisms for producing and recognizing manual actions and the mechanisms for producing and recognizing linguistic actions.

(5) Within the general framework of schema theory (Fig. 9.2: perceptual and motor schemas; schema assemblages; coordinated control programs) we considered both the VISIONS system (Fig. 9.3a) which described how to represent a static visual scene by a hierarchically structured network of schema instances linked to regions of the image, and the HEARSAY system (Fig. 9.3b) which described how to represent a speech stream by a hierarchically structured network of schema instances linked to time intervals of the spoken input. In each system, certain schema instances may be activated only to receive low confidence value; others may serve as "islands of reliability" which lead to activation of schemas instances which become part of the final interpretation of a region of the scene or an interval of the speech stream. Our challenge is to combine the insights of these two classical systems to describe how to represent a dynamic visual scene (or, more generally, an integrated representation combining the analysis of multi-modal sensory data with goals and plans for ongoing action) by a hierarchically structured network of schema instances each linked to a space–time region of the image. For example, a person may only be tracked over a certain time interval as they move from place to place; an action – whether observed or planned – will extend across a certain region of time but may be quite transient.

(6) Since many motor schemas may initially be activated, constant processing is required to determine which actions are indeed executed – just as the "motor side" of eye control must execute a winner-take-all computation to select the next focus of attention.

(7) In VISIONS, an intermediate database bridges between the image and the short-term memory (STM, limited to the stock of schema instances linked to different regions of the scene). The state of this database depends both on visual input and top–down requests. As we extend our concern to dynamic scenes, we see that the state of the intermediate database will depend as much on the retention of earlier hypotheses as on the analysis of the current visual input. Indeed, much of that visual input will simply update the intermediate representation – e.g., adjusting the location of a segment or proto-object.

(8) The Salience, Vision, and Symbolic Schemas (SVSS) model explores the interaction of bottom–up salience and top–down hypotheses in building up a representation of a visual scene. Such interaction is crucial to understanding what "holds attention." Thus a static fixation may last long enough for us to categorize an object or decide it not of interest. Return may then be inhibited until either fading memory or relations with another object demands that it again receive attention. Movement itself is a strong salience cue. For a moving object of interest, "What will it do next?" may be enough to hold attention, but once attention shifts, inhibition of return will then apply to the region of its imminent trajectory.

(9) In SVSS, visual processing is no longer neutral but will be influenced by the current task (e.g., answering a question about a specific object or action). The Task Relevance Map (TRM) holds the locations in the scene currently deemed relevant to the task requirements. We extend the VISIONS-like STM to include instances of more symbol-like schemas which provide a qualitative summary of just a few aspects for as long as the entity is relevant. In terms of producing a description of a scene, we envision STM as holding cognitive structures (Cognitive Form; schema assemblages) from which some aspects are selected for conversion into semantic structures (hierarchical constituents expressing objects, actions, and relationships) which constitute a Semantic Form. Finally, ideas in the Semantic Form must be expressed in words whose markings and ordering provide a "phonological" structure, the Phonological Form.

(10) Just as recognizing a visual scene involves much more than recognizing an action, so does recognizing a sentence involve much more than recognizing individual words.

(11) A number of empirical studies have sought to link sentence perception and production to the eye movements of subjects looking at a visual display. However, these displays tend to be either formal arrangements of objects, or simple scenes which are exhausted by a single minimal subscene, thus avoiding a number of the important considerations which structure our analysis of complex scenes.

(12) Knott (2003) has suggested that, in linking sensorimotor sequences to sentences, the sensorimotor sequence of attention to the scene (the scanpath) is translated directly into the operations involved in constructing the syntactic tree for its description. However, we suggest that Knott's theory is more pertinent if we build scene description on the state of the Symbolic Working Memory, with items tagged for the "communicative salience" that is achieved following "apprehension of the subscene" rather than on the eye movements that went into the building of that representation.

(13) MSH suggests a view of the verb as having two distinct parts which play different roles in the syntax of sentences: one part (cf. the canonical neurons of F5 and the FARS model) focuses on the linkage of an action to the object of that action; the other part (cf. the mirror neurons of F5 and the part of the MNS model complementary to the FARS model) provides the means for the linkage of an action to the agent of that action. Diverse circuitry for the recognition of agents, actions, and objects must be linked to shared circuitry for the assemblage of words retrieved from the lexicon into well-formed sentences, and conversely for words to link to processes which modulate scene perception and ongoing action.

(14) Construction grammar provides a framework for specifying a variety of constructions which provide syntactic rules for a number of different ways of linking agents, actions, and objects. The approach to language via a large but finite inventory of constructions motivates a view of visual scene interpretation, compatible with the VISIONS framework, in which an inventory of "scene schemas" from the visual end may provide the linkage with constructions from the language end. Each constituent which expands a "slot" within a scene schema or verbal construction may be seen as a hierarchical structure in which extended attention to a given component of the scene extends the complexity of the constituents in the parse tree of the sentence.

(15) Finally, we revisited SVSS and then used Levelt's (1989, 2001) scheme for sentence production and understanding to provide the framework for the language processing system to be linked to systems for visual scene perception and action production.

MSH suggests that more than mere linkage of vision, action, and language is at play here. It postulates an evolutionary progression from manual actions via complex imitation and pantomime to protosign and protospeech. On this account, certain neural components of the language system have their evolutionary roots in the praxic system – so that recognition and production of words is viewed as employing mechanisms homologous to those involved in the recognition and production of manual actions. The attempt we have made here to unfold the interactions of task, attention, and visual perception is part of the effort required to lift MSH from the production and recognition of single actions (Fig. 9.13) to the understanding and planning of actions, agents, and objects as defining minimal episodes and their integration into the overall episodes which structure our lives.

References

Arbib, M. A., 1981. Perceptual structures and distributed motor control. In V. B. Brooks (ed.) *Handbook of Physiology*, Section 2, *The Nervous System*, vol. 2, *Motor Control*, Part 1. Bathesda, MD: American Physiological Society, pp. 1449–1480.

1989. *The Metaphorical Brain 2: Neural Networks and Beyond*. New York: Wiley-Interscience.

2001. The Mirror System Hypothesis for the language-ready brain. In A. Cangelosi & D. Parisi (eds.) *Computational Approaches to the Evolution of Language and Communication*. London: Springer-Verlag, pp. 229–254.

Arbib, M. A., and Bota, M., 2003. Language evolution: neural homologies and neuroinformatics. *Neur. Networks* **16**: 1237–1260.

Arbib, M. A., and Caplan, D., 1979. Neurolinguistics must be computational. *Behav. Brain Sci.* **2**: 449–483.

Arbib, M. A., and Didday, R. L., 1975. Eye-movements and visual perception: a two-visual system model. *Int. J. Man–Machine Stud.* **7**: 547–569.

Arbib, M. A., and Liaw, J.-S., 1995. Sensorimotor transformations in the worlds of frogs and robots. *Artif. Intell.* **72**: 53–79.

Arbib, M., and Rizzolatti, G., 1997. Neural expectations: a possible evolutionary path from manual skills to language. *Commun. Cogn.* **29**: 393–424.

Arbib, M. A., Érdi, P., and Szentágothai, J., 1998. *Neural Organization: Structure, Function, and Dynamics*. Cambridge, MA: MIT Press.

Arkin, R. C., 1998. *Behavior-Based Robotics*. Cambridge, MA: MIT Press.

Baddeley, A., 2003. Working memory: looking back and looking forward. *Nature Rev. Neurosci.* **4**: 829–839.

Baddeley, A. D., and Hitch, G. J., 1974. Working memory. In G. A. Bower (ed.) *The Psychology of Learning and Motivation*. New York: Academic Press, pp. 47–89.

Biederman, I., Teitelbaum, R. C., and Mczzanotte, R. J., 1983. Scene Perception: a failure to benefit from prior expectancy of familiarity. *J. Exp. Psychol. Learn. Mem. Cogn.* **9**: 411–429.

Bloom, P., 1997. Intentionality and word learning. *Trends Cogn. Sci.* **1**: 9–12.

Bullock, D., & Rhodes, B. J., 2003. Competitive queuing for planning and serial performance. in M. A. Arbib (ed.) *The Handbook of Brain Theory and Neural Networks*, 2 edn. Cambridge, MA: MIT Press, pp. 241–248.

Chambers, C. G., Tanenhaus, M. K., and Magnuson, J. S., 2004. Actions and affordances in syntactic ambiguity resolution. *J. Exp. Psychol. Learn. Mem. Cogn.* **30**: 687–696.

Chomsky, N., 1995. *The Minimalist Program*. Cambridge, MA: MIT Press.

Colby, C. L., and Goldberg, M. E., 1999. Space and attention in parietal cortex. *Annu. Rev. Neurosci.* **22**: 97–136.

Croft, W., 2001. *Radical Construction Grammar*. Oxford, UK: Oxford University Press.

Croft, W., and Cruse, D. A., 2004. *Cognitive Linguistics*. Cambridge, UK: Cambridge University Press.

Dominey, P. F., and Arbib, M. A., 1992. A cortico-subcortical model for generation of spatially accurate sequential saccades. *Cereb. Cortex.* **2**: 153–175.

Dominey, P. F., Arbib, M. A., and Joseph, J. P., 1995. A model of cortico-striatal plasticity for learning oculomotor associations and sequences. *J. Cogn. Neurosci.* **7**: 311–336.

Draper, B. A., Collins, R. T., Brolio, J., Hanson, A. R., and Riseman, E. M., 1989. The schema system. *Int. J. Comput. Vision.* **2**: 209–250.

Elsner, B., Hommel, B., Mentschel, C., *et al.*, 2002. Linking actions and their perceivable consequences in the human brain. *Neuroimage* **17**: 364–372.

Emmorey, K., 2004. The role of Broca's area in sign language. In Y. Grodzinsky and K. Amunts (eds.) *Broca's Region*. Oxford, UK: Oxford University Press, pp. 167–182.

Epstein, R., Stanley, D., Harris, A., and Kanwisher, N., 2000. The parahippocampal place area: perception, encoding, or memory retrieval? *Neuron* **23**: 115–125.

Fagg, A. H., and Arbib, M. A., 1998. Modeling parietal–premotor interactions in primate control of grasping. *Neur. Networks.* **11**: 1277–1303.

Fillmore, C. J., Kay, P., and O'Connor, M. K., 1988. Regularity and idiomaticity in grammatical constructions: the case of *let alone*. *Language* **64**: 501–538.

Fogassi, L., Ferrari, P. F., Gesierich, B., *et al.*, 2005. Parietal lobe: from action organization to intention understanding. *Science* **308**: 662–667.

Gandhi, S. P., Heeger, M. J., and Boyton, G.M., 1998. Spatial attention affects brain activity in human primary visual cortex. *Proc. Nat. Acad. Sci. USA* **96**: 3314–3319.

Gibson, J. J., 1979. *The Ecological Approach to Visual Perception*. Boston, MA: Houghton Mifflin.

Goldberg, A., 1995. *Constructions: A Construction Grammar Approach to Argument Structure*. Chicago, IL: University of Chicago Press
   2003. Constructions: a new theoretical approach to language. *Trends Cogn. Sci.* **7**: 219–224.

Goldin-Meadow, S., 1999. The role of gesture in communication and thinking. *Trends Cogn. Sci.* **3**: 419–429.

Griffin, Z., and Bock, K., 2000. What the eyes say about speaking. *Psychol. Sci.* **11**: 274–279.

Haruno, M., Wolpert, D. M., and Kawato, M., 2001. MOSAIC model for sensorimotor learning and control. *Neur. Comput.* **13**: 2201–2220.

Henderson, J. M., and Ferreira, F. (eds.), 2004. *Interface of Language, Vision, and Action: Eye Movements and the Visual World*. New York: Psychology Press.

Henderson, J. M., and Hollingworth, A., 1999. High-level scene perception. *Annu. Rev. Psychol.* **50**: 243–271.

Herzog, G., and Wazinski, P., 1994. Visual TRAnslator: linking perceptions and natural language descriptions. *Artif. Intell. Rev.* **8**: 175–187.

Hoff, B., and Arbib, M. A., 1993. Simulation of interaction of hand transport and preshape during visually guided reaching to perturbed targets. *J. Motor Behav.* **25**: 175–192.

Hollingworth, A., 2004. Constructing visual representations of natural scenes: the roles of short- and long-term visual memory. *J. Exp. Psychol. Hum. Percept. Perform.* **30**: 519–537.

Hollingworth, A., and Henderson, J. M., 1998. Does consistent scene context facilitate object perception? *J. Exp. Psychol. Gen.* **127**: 398–415.
   2002. Accurate visual memory for previously attended objects in natural scenes. *J. Exp. Psychol. Hum. Percept. Perform.* **28**: 113–136.
   2004. Sustained change blindness to incremental scene rotation: a dissociation between explicit change detection and visual memory. *Percept. Psychophys.* **66**: 800–807.

Indefrey, P. and Levelt, W. J. M., 2000. The neural correlates of language production. In M. Gazzaniga (ed.) *The New Cognitive Sciences*, 2nd edn. Cambridge, MA: MIT Press, pp. 845–865.

Irwin, D. E., and Andrews, R., 1996. Integration and accumulation of information across saccadic eye movements. In T. Inui and J. L. McClelland (eds.) *Attention and*

Performance, vol. 16, *Information Integration in Perception and Communication.* Cambridge, MA: MIT Press, pp. 125–155.

Irwin, D. E., and Zelinsky, G. J., 2002. Eye movements and scene perception: memory for things observed. *Percept. Psychophys.* **64**: 882–895.

Ito, M., and Gilbert, C. D., 1999. Attention modulates contextual influences in the primary visual cortex of alert monkeys. *Neuron* **22**: 593–604.

Itti, L., 2005. Quantifying the contribution of low-level saliency to human eye movements in dynamic scenes. *Visual Cogn.* **12**: 1093–1123.

Itti, L., and Koch, C., 2000. A saliency-based search mechanism for overt and covert shifts of visual attention. *Vision Res.* **40**: 1489–1506.

2001. Computational modeling of visual attention. *Nature Rev. Neurosci.* **2**: 194–203.

Itti, L., Gold, C., and Koch, C., 2001. Visual attention and target detection in cluttered natural scenes. *Opt. Engin.* **40**: 1784–1793.

Itti, L., Dhavale, N., and Pighin, F., 2003. Realistic avatar eye and head animation using a neurobiological model of visual attention. *Proceedings SPIE 48th Annual International Symposium on Optical Science and Technology*, pp. 64–78.

Jackendoff, R., 2002. *Foundations of Language.* Oxford, UK: Oxford University Press.

Jeannerod, M., 1997. *The Cognitive Neuroscience of Action.* Oxford, UK: Blackwell.

Jeannerod, M., and Biguer, B., 1982. Visuomotor mechanisms in reaching within extra-personal space. In D. J. Ingle, R. J. W. Mansfield, and M. A. Goodale (eds.) *Advances in the Analysis of Visual Behavior.* Cambridge, MA: MIT Press, pp. 387–409.

Jellema, T., Baker, C., Wicker, B., and Perrett, D., 2000. Neural representation for the perception of the intentionality of actions. *Brain Cogn.* **44**: 280–302.

Kayne, R. S., 1994. *The Antisymmetry of Syntax.* Cambridge, MA: MIT Press.

Knott, A., 2003. Grounding syntactic representations in an architecture for sensorimotor control. Available at http://www.cs.otago.ac.nz/trseries/oucs-2003-04.pdf

2004. Syntactic representations as side-effects of a sensorimotor mechanism. Abstract, *5th International Conference on Evolution of Language*, Leipzig, April, 2004.

Koopman, H., and Sportiche, D., 1991. The position of subjects. *Lingua* **85**: 211–258.

Lacquaniti, F., Perani, D., Guigon, E., *et al.*, 1997. Visuomotor transformations for reaching to memorized targets: a PET study. *Neuroimage*, **5**: 129–146.

Lesser, V. R., Fennel, R. D., Erman, L. D., and Reddy, D. R., 1975. Organization of the HEARSAY-II speech understanding system. *IEEE Trans. Acoust. Speech Signal Process.* **23**: 11–23.

Levelt, W. J. M., 1989. *Speaking.* Cambridge, MA: MIT Press.

2001. Spoken word production: a theory of lexical access. *Proc. Natl Acad. Sci. USA* **98**: 13464–13471.

Levelt, W. J. M., Roelofs, A., & Meyer, A. S., 1999. A theory of lexical access in speech production. *Behav. Brain Sci.* **22**: 1–75.

Luria, A. R., 1973. *The Working Brain.* New York: Penguin Books.

McCulloch, W. S., 1965. *Embodiments of Mind.* Cambridge, MA: The MIT Press.

McNeill, D. (ed.), 2000. *Language and Gesture.* Cambridge, UK: Cambridge University Press.

Medendorp, W. P., Goltz, H. C., Vilis, T., and Crawford, J. D., 2003. Gaze-centered updating of visual space in human parietal cortex. *J. Neurosci.* **23**: 6209–6214.

Megerdoomian, K., 2001. Event structure and complex predicates in Persian. *Can. J. Ling.–Rev. Can. Linguistique* **46**: 97–125.

Moran, J., and Desimone, R., 1985. Selective attention gates visual processing in the extrastriate cortex. *Science* **229**: 782–784.

Moreno, F. J., Reina, R., Luis, V., and Sabido, R., 2002. Visual search strategies in experienced and inexperienced gymnastic coaches. *Percept. Motor Skills* **95**: 901–902.

Navalpakkam, V., and Itti, I., 2005. Modeling the influence of task on attention. *Vision Res.* **45**: 205–231.

Nevatia, R., Zhao, T., and Hongeng, S., 2003. Hierarchical language-based representation of events in video streams. IEEE Workshop on Event Mining, Madison, WI, 2003.

Nodine, C. F., and Krupinski, E. A., 1998. Perceptual skill, radiology expertise, and visual test performance with NINA and WALDO. *Acad. Radiol.* **5**: 603–612.

Noton, D., and Stark, L., 1971. Scanpaths in eye movements during pattern perception. *Science(Washington)* **171**: 308–311.

Nunberg, G., Sag, I. A., and Wasow, T., 1994. Idioms. *Language* **70**: 491–538.

Oliva, A., and Schyns, P. G., 1997. Coarse blobs or fine edges? Evidence that information diagnosticity changes the perception of complex visual stimuli. *Cogn. Psychol.* **34**: 72 107.

2000. Diagnostic colors mediate scene recognition. *Cogn. Psychol.* **41**: 176–210.

O'Regan, J. K., 1992. Solving the "real" mysteries of visual perception: the world as an outside memory. *Can J. Psychol.* **46**: 461–488.

Oztop, E., and Arbib, M. A., 2002. Schema design and implementation of the grasp-related mirror neuron system. *Biol. Cybernet.* **87**: 116–140.

Oztop, E., Wolpert, D., and Kawato, M., 2005. Mental state inference using visual control parameters. *Cogn. Brain Res.* **22**: 129–151.

Perrett, D., Harries, M., Bevan, R., *et al.*, 1989. Frameworks of analysis for the neural representation of animate objects and actions. *J. Exp. Biol.* **146**: 87–113.

Pollock, J.-Y., 1989. Verb movement, universal grammar and the structure of IP. *Linguist. Inquiry* **20**: 365–424.

Potter, M. C., 1975. Meaning in visual search. *Science* **187**: 965–966.

Pulvermüller, F., Härle, M., and Hummel, F., 2001. Walking or talking? Behavioral and neurophysiological correlates of action verb processing. *Brain Lang.* **78**: 143–168.

Rensink R. A., 2000. Seeing, sensing, and scrutinizing. *Vision Res.* **40**: 1469–1487.

Rensink, R. A., O'Regan, J. K., and Clark, J. J., 1997. To see or not to see: the need for attention to perceive changes in scenes. *Psychol. Sci.* **8**: 368–373.

Rolls, E. T., and Arbib, M. A., 2003. Visual scene perception. In M. A. Arbib (ed.) *The Handbook of Brain Theory and Neural Networks*, 2nd edn. Cambridge, MA: MIT Press, pp. 1210–1215.

Rybak, I. A., Gusakova, V. I., Golovan, A. V., Podladchikova, L. N., and Shevtsova, N. A., 1998. A model of attention-guided visual perception and recognition. *Vision Res.* **38**: 2387–2400.

Sabbagh, M. A., and Baldwin, D. A., 2001. Learning words from knowledgeable versus ignorant speakers: links between preschoolers' theory of mind and semantic development. *Child Devel.* **72**: 1054–1070.

Sanocki, T., and Epstein, W., 1997. Priming spatial layout of scenes. *Psychol. Sci.* **8**: 374–378.

Savelsbergh, G. J., Williams, A. M., der Kamp, J. Van, and Ward, P., 2002. Visual search, anticipation and expertise in soccer goalkeepers. *J. Sports Sci.* **20**: 279–287.

Scalaidhe, S. P., Wilson, F. A., and Goldman-Rakic, P. S., 1999. Face-selective neurons during passive viewing and working memory performance of rhesus monkeys: evidence for intrinsic specialization of neuronal coding. *Cerebr. Cortex* **9**: 459–475.

Schill, K., Umkehrer, E., Beinlich, S., Krieger, G., and Zetzsche, C., 2001. Analysis with saccadic eye movements: top–down and bottom–up modeling. *J. Electr. Imag.*

Simons, D. J., 2000. Attentional capture and inattentional blindness. *Trends Cogn. Sci.* **4**: 147–155.

Tanenhaus, M. K., Chambers, C. G., and Hanna, J. E., 2004. Referential domains in spoken language comprehension: using eye movements to bridge the product and action traditions. In J. M. Henderson and F. Ferreira (eds.) *The Interface of Language, Vision, and Action: Eye Movements and the Visual World*. New York: Psychology Press, pp. 279–317.

Tipper, S., Lortie, C., and Baylis, G., 1992. Selective reaching: evidence for action-centred attention. *J. Exp. Psychol. Hum. Percept. Perform.* **18**: 891–905.

Tipper, S., Howard, L., and Houghton, G., 1998. Action-based mechanisms of attention. *Phil. Trans. Roy. Soc. London B* **353**: 1385–1393.

Torralba, A., 2003. Modeling global scene factors in attention. *J. Opt. Soc. America A, Opt. Image Sci. Vis.* **20**: 1407–1418.

Treisman, A., and Gelade, G., 1980. A feature integration theory of attention. *Cogn. Psychol.* **12**: 97–136.

Treue, S., and Martinez-Trujillo, J. C., 1999. Feature-based attention influences motion processing gain in macaque visual cortex. *Nature* **399**: 575–579.

Tversky, B., and Lee, P. U., 1998. How space structures language. In C. Freksa, C. Habel, and K. F. Wender (eds.) *Spatial Cognition: An Interdisciplinary Approach to Representing and Processing Spatial Knowledge*. Berlin: Springer-Verlag, pp. 157–175.

Weymouth, T. E., 1986. *Using Object Descriptions in a Schema Network for Machine Vision*, COINS Technical Report No. 86–24. Amherst, MA: Department of Computer and Information Science, University of Massachusetts.

Williams, E., 1995. Theta theory. In G. Webelhuth (ed.) *Government and Binding Theory and the Minimalist Program*. Oxford, UK: Blackwell, pp. 97–124.

Wolfe, J. M., 1994. Guided search 2.0: a revised model of visual search. *Psychonom. Bull. Rev.* **1**: 202–238.

Yarbus, A., 1967. *Eye Movements and Vision*. New York: Plenum Press.

Zacks, J., and Tversky, B., 2001. Event structure in perception and cognition. *Psychol. Bull.* **127**: 3–21.

Zhao, T., and Nevatia, R., 2004. Tracking multiple humans in complex situations. *IEEE Trans. Pattern Anal. Machine Intell.* **26**: 1208–1221.

10

Action verbs, argument structure constructions, and the mirror neuron system

David Kemmerer

10.1 Introduction

This chapter reviews recent evidence that the linguistic representation of action is grounded in the mirror neuron system. Section 10.2 summarizes the major semantic properties of action verbs and argument structure constructions, focusing on English but also considering cross-linguistic diversity. The theoretical framework is Construction Grammar, which maintains that the argument structure constructions in which action verbs occur constitute basic clausal patterns that express basic patterns of human experience. For example, the sentence *She sneezed the napkin off the table* exemplifies the Caused Motion Construction, which has the schematic meaning "X causes Y to move along path Z," and the sentence *She kissed him unconscious* exemplifies the Resultative Construction, which has the schematic meaning "X causes Y to become Z" (Goldberg, 1995).

Section 10.3 addresses the neuroanatomical substrates of action verbs and argument structure constructions. A number of neuroimaging and neuropsychological studies are described which suggest that different semantic properties of action verbs are implemented in different cortical components of the mirror neuron system, especially in the left hemisphere: (1) motoric aspects of verb meanings (e.g., the type of action program specified by *kick*) appear to depend on somatotopically mapped primary motor and premotor regions; (2) agent–patient spatial–interactive aspects of verb meanings (e.g., the type of object-directed path specified by *kick*) appear to depend on somatotopically mapped parietal regions; and (3) visual manner-of-motion aspects of verb meanings (e.g., the visual movement pattern specified by *kick*) appear to depend on posterior middle temporal regions. In addition, several neuropsychological studies are described which suggest that the meanings of argument structure constructions are implemented in left perisylvian cortical regions that are separate from, but adjacent to, those for verb meanings.

Action to Language via the Mirror Neuron System, ed. Michael A. Arbib. Published by Cambridge University Press. © Cambridge University Press 2006.

Finally, Section 10.4 broadens the discussion of action verbs and argument structure constructions by briefly considering the emergence of language during ontogeny and phylogeny.

10.2 Action verbs and argument structure constructions

10.2.1 Action verbs

Causal complexity

Although many verbs refer to abstract states and events (e.g., *exist, remain, increase, elapse*), the prototypical function of verbs in all languages is to denote physical actions, that is, situations in which an agent, such as a person or animal, engages in certain kinds of bodily movement (Croft, 1991). For this reason, and also because action is a prominent theme throughout the current volume, I will focus exclusively on action verbs. Most contemporary semantic theories assume that conceptions of space, force, and time constitute the foundation of the meanings of action verbs, and that an especially important dimension of semantic variation involves causal complexity (e.g., Pinker, 1989; Jackendoff, 1990; Van Valin and LaPolla, 1997; Croft, 1998; Rappaport Hovrav and Levin, 1998). The simplest verbs, which are usually intransitive, represent events in which an agent performs an activity that does not necessarily bring about any changes in other entities (e.g., *sing, laugh, wave, jog*). Other verbs, which are usually transitive, are more complex insofar as they express activities that do affect other entities in certain ways, such as by inducing a change of state (e.g., *slice, engrave, purify, kill*) or a change of location (e.g., *pour, twist, load, smear*), either through direct bodily contact or by means of a tool.

Semantic classes

In addition to the basic organizing factor of causal structure, the meanings of action verbs can be analyzed and compared in terms of the various semantic fields that they characterize. Levin (1993) sorted over 3000 English verbs (the majority of which are action verbs) into approximately 50 classes and 200 subclasses. Representative classes include verbs of throwing (e.g., *fling, hurl, lob, toss*), verbs of creation (e.g., *build, assemble, sculpt, weave*), and verbs of ingesting (e.g., *eat, gobble, devour, dine*). The verbs in a given class collectively provide a richly detailed categorization of the relevant semantic field by making distinctions, often of a remarkably fine-grained nature, along a number of different dimensions. For instance, verbs of destruction are distinguished by the composition of the entity to be destroyed (e.g., *tear* vs. *smash*), the degree of force (e.g., *tear* vs. *rip*), and the extent of deformation (e.g., *tear* vs. *shred*). The verbs in a given class are also organized according to principled semantic relations such as the following: synonymy, in which two verbs have nearly identical meanings (e.g., *shout* and *yell*); antonymy, in which two verbs have opposite meanings (e.g., *lengthen* and *shorten*); hyponymy, in which one verb is at a higher taxonomic level than another (e.g., *talk* and *lecture*); and cohyponymy, in which two verbs are at roughly the same taxonomic level (e.g., *bow* and *curtsey*) (Fellbaum, 1998).

Cross-linguistic diversity

Another important point about the meanings of action verbs is that they vary considerably across the 6000+ languages of the world. One manifestation of this variation involves the distinctions that are made within particular semantic fields or, as they are sometimes called, conceptual spaces. For example, Lakoff and Johnson (1999, p.576) describe several idiosyncratic differences in how languages carve up the conceptual space of hand actions:

- In Tamil, *thallu* and *ilu* correspond to English *push* and *pull*, except that they connote a sudden action as opposed to a smooth continuous force.
- In Farsi, *zadan* refers to a wide range of object manipulations involving quick motions – e.g., snatching a purse or strumming a guitar.
- In Cantonese, *mit* covers both pinching and tearing. It connotes forceful manipulation by two fingers, yet is also acceptable for tearing larger items when two full grasps are used.

A more systematic case of cross-linguistic variation in verb semantics derives from two different clausal patterns for encoding the manner and path components of motion events (Talmy, 1985). As shown below, in some languages (e.g., English, German, Russian, Swedish, and Chinese) manner is preferentially encoded by a verb and path by a preposition (or a similar grammatical category), whereas in other languages (e.g., French, Spanish, Japanese, Turkish, and Hindi) path is preferentially encoded by a verb and manner by an optional adverbial expression in a syntactically subordinate clause:

English: The dog ran$_{MANNER}$ into$_{PATH}$ the house.
French: Le chien est entré dans la maison en courant.
"The dog entered$_{PATH}$ the house by running$_{MANNER}$."

As a result of this fundamental difference, the inherently graded conceptual space of manner-of-motion is usually more intricately partitioned in languages of the former type than in languages of the latter type (Slobin, 2003). Thus, English distinguishes between *jump, leap, bound, spring*, and *skip*, but all of these verbs are translated into French as *bondir*; similarly, English distinguishes between *creep, glide, slide, slip*, and *slither*, but all of these verbs are translated into Spanish as *escabullirse*. Specialized manner verbs like these are not just dictionary entries, but are actively employed by English speakers in a variety of naturalistic and experimental contexts, including oral narrative, spontaneous conversation, creative writing, naming videoclips of motion events, and speeded fluency, i.e., listing as many manner verbs as possible in 1 minute (Slobin, 2003). In addition, recent research suggests that the cross-linguistic differences in semantic maps for manner-of-motion influence co-speech gesture (Kita and Özyürek, 2003; D. Kemmerer *et al.*, unpublished data) and lead to non-trivial differences in perceptual tuning and long-term memory for subtle manner details of motion events (A. W. Kersten *et al.*, unpublished data; Oh, 2003).

10.2.2 Argument structure constructions

The relation between verbs and constructions

Many action verbs occur in a wide range of argument structure constructions. For example, even though *kick* is usually considered to be a prototypical transitive verb, it occurs in at least nine distinct active-voice constructions (Goldberg, 1995):

(1) Bill kicked the ball.
(2) Bill kicked the ball into the lake.
(3) Bill kicked at the ball.
(4) Bill kicked Bob the ball.
(5) Bill kicked Bob black and blue.
(6) Bill kicked Bob in the knee.
(7) Bill kicked his foot against the chair.
(8) Bill kicked his way through the crowd.
(9) Horses kick.

These sentences describe very different kinds of events: (1) simple volitional bodily action directed at an object, (2) causing an object to change location, (3) attempting to contact an object, (4) transferring possession of an object, (5) causing an object to change state, (6) inducing a feeling in a person by contacting a part of their body, (7) causing a part of one's own body to contact an object, (8) making progress along a path by moving in a particular manner, and (9) having a tendency to perform a certain action. According to Construction Grammar and related theories (e.g., Goldberg, 1995, 2003; Croft, 2001; Jackendoff, 2002; Croft and Cruse, 2004), argument structure constructions are clausal patterns that are directly associated with specific meanings, and the interpretation of a sentence is in large part the outcome of a division of labor between the meaning of the construction and the meaning of the verb.[1] For instance, the *X's way* construction consists of a particular syntactic structure – roughly "Subject Verb *X's way* Oblique" – that is paired with a particular semantic structure – roughly "X makes progress along a path by V-ing." Thus, in a sentence like *Bill kicked his way through the crowd*, the general concept of "motion of the subject referent along a path" comes from the *X's way* construction itself, and the more specific notion of "forceful leg action" comes from *kick*. Table 10.1 shows how each of the sentences with *kick* listed in (1)–(9) above instantiates a construction that designates an idealized event type.

The meanings of these constructions are quite abstract and hence bear an interesting resemblance to the "minimal scenes" described by Itti and Arbib (this volume). As Goldberg (1998, p.206) points out, "we do not expect to find distinct basic sentence types that have semantics such as something turning blue, someone becoming upset, something turning over." This is because specific events like these do not happen frequently enough to warrant incorporation into a language's morphosyntactic design.

---

[1] The process of integrating verbs and constructions is, not surprisingly, quite complex. See Goldberg (1995) for an introduction.

Table 10.1 *Examples of English argument structure constructions*

Construction	Form	Meaning	Example
1. Transitive	Subject Verb Object	X acts on Y	Bill kicked the ball.
2. Caused motion	Subject Verb Object Oblique	X causes Y to move along path Z	Bill kicked the ball into the lake.
3. Conative	Subject Verb Oblique$_{at}$	X attempts to contact Y	Bill kicked at the ball.
4. Ditransitive	Subject Verb Object$_1$ Object$_2$	X causes Y to receive Z	Bill kicked Bob the ball.
5. Resultative	Subject Verb Object Complement	X causes Y to become Z	Bill kicked Bob black and blue.
6. Possessor ascension	Subject Verb Object Oblique$_{in/on}$	X contacts Y in/on body-part Z	Bill kicked Bob in the knee.
7. Contact$_{against}$	Subject Verb Object Oblique$_{against}$	X causes Y to contact Z	Bill kicked his foot against the chair.
8. *X's way*	Subject Verb *X's way* Oblique	X makes progress by performing action	Bill kicked his way through the crowd.
9. Habitual	Subject Verb	X performs action habitually	Horses kick.

The Grammatically Relevant Semantic Subsystem Hypothesis (GRSSH)

Research along these lines has led to what Pinker (1989) calls the Grammatically Relevant Semantic Subsystem Hypothesis (GRSSH), which maintains that a distinction exists between two large-scale components of meaning: (1) a set of fairly abstract semantic features that are relevant to grammar insofar as they tend to be encoded by closed-class items as well as by morphosyntactic constructions; and (2) an open-ended set of fairly concrete semantic features that are not relevant to grammar but instead enable open-class items to express an unlimited variety of idiosyncratic concepts. This distinction is illustrated by the ditransitive construction, which at the highest level of schematicity means "X causes Y to receive Z," but which also (as is true of many argument structure constructions) has several additional semantic restrictions, one of which is that while verbs of instantaneous causation of ballistic motion are acceptable (e.g., *I kicked/ tossed/rolled/bounced him the ball*), verbs of continuous causation of accompanied motion are not (e.g., **I carried/hauled/lifted/dragged him the box*) (Pinker, 1989).[2] According to the GRSSH, the ditransitive construction is sensitive to the relatively coarse-grained contrast between the two sets of verbs, but is not sensitive to the more

---

[2] Pinker (1989, p.358) points out, however, that some speakers find the sentences with verbs of accompanied motion to be acceptable, which suggests that there are dialectal or idiolectal differences in dativizability.

fine-grained contrasts between the verbs within each set – i.e., between *kick, toss, roll*, and *bounce* on the one hand, and between *carry, haul, lift*, and *drag* on the other.

Morphosyntax

Since the main focus of this chapter is on semantics, I will only make a few points concerning the morphosyntactic aspects of argument structure constructions (for a more detailed exposition, see Croft, 2001). As shown in Table 10.1, the form of these constructions is specified in terms of phrasal units that are arrayed around a verb according to syntactic relations like subject, object, and oblique. These phrasal units are assumed to have internal hierarchical structures captured by other (families of) constructions, such as the NP (noun-phrase) construction, the PP (prepositional phrase) construction, and so forth. With respect to grammatical categories like noun, verb, and adjective, the fact that the members of each category exhibit widespread distributional mismatches across constructions suggests that they fractionate into subclasses that comprise a vast multidimensional network or inheritance hierarchy with broad categories at the top and narrow ones at the bottom. For example, all English nouns can serve as the subject NP of a sentence, but "subject NP" is just one construction, and further investigation leads to a proliferation of subclasses of nouns with varying constructional distributions, such as pronouns, proper nouns, count nouns, and mass nouns, each of which breaks down into even smaller and quirkier groupings – e.g., proper nouns for days of the week and months of the year require different spatially based prepositions when used in expressions for temporal location (*on/*in Saturday, in/*on August*; cf. Kemmerer, 2005). As for English verbs, they all inflect for tense/aspect in main clauses, but again this is a construction-specific property justifying only the category that Croft (2001) calls "morphological verb," and closer scrutiny reveals that, as mentioned above, verbs display a tremendous range of distributional diversity, with approximately 50 classes and 200 subclasses based on combined semantic and syntactic factors (Levin, 1993). Finally, there are apparently no distributional criteria that justify a single overarching adjective category in English, as suggested by facts like the following. Some adjectives are both attributive and predicative (*the funny movie, that movie is funny*) whereas others are only attributive (*the main reason, *that reason is main*) or only predicative (**the asleep student, that student is asleep*). Moreover, when multiple adjectives occur prenominally, their linear order is determined primarily by which of several semantically and pragmatically defined subclasses they belong to, thus accounting for why it is grammatical to say *the other small inconspicuous carved jade idols* but not **the carved other inconspicuous jade small idols* (Kemmerer, 2000b; Kemmerer *et al.*, in press).

Three constructions

To further clarify the nature of argument structure constructions as well as the GRSSH, I will briefly describe two constructional alternations (the locative alternation and the body-part possessor ascension alternation) and one morphological construction (reversative *un-* prefixation). I will concentrate on semantic issues, and will return to all three constructions later, in the section on neuroanatomical substrates.

First, the locative alternation is illustrated by the constructions in (10) and (11):

(10) (a) Sam sprayed water on the flowers.
    (b) Sam dripped water on the flowers.
    (c) *Sam drenched water on the flowers.
(11) (a) Sam sprayed the flowers with water.
    (b) *Sam dripped the flowers with water.
    (c) Sam drenched the flowers with water.

Functionally, these two constructions encode different subjective construals of what is objectively the same type of event. Human cognition is remarkably flexible, and we are able to take multiple perspectives on events by allocating our attention to the entities in various ways (Tomasello, 1999). If I see Sam spraying water on some flowers, I can conceptualize the water as being most affected, since it changes location from being in a container to being on the flowers, or I can conceptualize the flowers as being most affected, since they change state from being dry to being wet. The construction in (10) captures the first kind of perspective since it has the schematic meaning "X causes Y to go to Z in some manner," whereas the construction in (11) captures the second kind of perspective since it has the schematic meaning "X causes Z to change state in some way by adding Y." Semiotically, the two constructions signal these different perspectives by taking advantage of a general principle that guides the mapping between syntax and semantics, namely the "affectedness principle," which states that the entity that is syntactically expressed as the direct object is interpreted as being most affected by the action (Gropen *et al.*, 1991). *Spray* can occur in both constructions because it encodes not only a particular manner of motion (a substance moves in a mist) but also a particular change of state (a surface becomes covered with a substance). However, *drip* and *drench* are in complementary distribution, largely because each constructional meaning is associated with a network of more restricted meanings that are essentially generalizations over verb classes (Pinker, 1989; Goldberg, 1995). One of the narrow-range meanings of the first construction is "X enables a mass Y to go to Z via the force of gravity," and this licenses expressions like *drip/dribble/pour/spill water on the flowers* and excludes expressions like *drench water on the flowers*. Similarly, one of the narrow-range meanings of the second construction is "X causes a solid or layer-like medium Z to have a mass Y distributed throughout it," and this licenses expressions like *drench/douse/soak/saturate the flowers with water* and excludes expressions like *drip the flowers with water*.

Second, the body-part possessor ascension alternation is illustrated by the constructions in (12) and (13):

(12) (a) Bill hit Bob's arm.
    (b) Bill broke Bob's arm.
(13) (a) Bill hit Bob on the arm.
    (b) *Bill broke Bob on the arm.

Like the locative alternation, the body-part possessor alternation provides two constructions for expressing different subjective construals of the same objective type of event – an event that basically involves something contacting part of someone's body. Also like the locative alternation, the body-part possessor alternation places restrictions on which verbs are acceptable, except that here the most interesting restrictions apply to just the second construction, which is traditionally called the ascension construction because the possessor NP – *Bob* in (13) – has "ascended" out of the modifier position in the complex NP of the first construction. Consider, for example, the following sentences: *Bill hit/bumped/tapped/whacked Bob on the arm* vs. **Bill broke/cracked/ fractured/shattered Bob on the arm*. There is not yet a completely satisfactory account of the precise semantic criteria that determine which verbs can occur in the construction, but one feature that appears to be relevant is "contact" (Kemmerer, 2003). Although both constructions make reference to physical contact, this feature is more prominent in the ascension construction. It is explicitly marked by a locative preposition (typically *on* or *in*) that introduces the body-part NP. Even more crucially, all of the verbs that can occur in the construction belong to classes that specify contact: verbs of touching (e.g., *caress, kiss, lick, pat, stroke*); verbs of contact by impact, which fractionate into three subclasses – "hit" verbs (e.g., *bump, kick, slap, smack, tap*), "swat" verbs (e.g., *bite, punch, scratch, slug, swipe*), and "spank" verbs (e.g., *bonk, clobber, flog, thrash, wallop*); verbs of poking, i.e., forceful contact by means of a sharp object (e.g., *jab, pierce, poke, prick, stick*); and verbs of cutting, i.e., forceful contact causing a linear separation in the object (e.g., *cut, hack, scrape, scratch, slash*) (Levin, 1993).[3] On the other hand, most of the verbs that cannot occur in the ascension construction belong to a class called "break" verbs (e.g., *break, rip, smash, splinter, tear*). These verbs do not necessarily entail contact but instead focus on just the change of state – the transformation of structural integrity – that an entity undergoes. Another important point is that although both of the constructions shown in (12) and (13) describe events involving bodily contact, they package the information differently. In accord with the affectedness principle mentioned above in the context of the locative alternation, the non-ascension construction focuses more on the body part than the person since the body part is mapped onto the direct object position (*Bill hit Bob's arm*), whereas the ascension construction focuses more on the person than the body part since the person is mapped onto this privileged syntactic position (*Bill hit Bob on the arm*). More specifically, the ascension construction conveys the impression that what is really being affected is the person as a sentient being – in other words, the person's inner sensations, thoughts, or feelings – and that this happens because a particular part of the person's body is contacted in some manner. Support for this aspect of the constructional meaning comes from the observation that the direct object of this construction must refer to an animate entity, as shown below (Wierzbicka, 1988):

---

[3] "Carve" verbs, which constitute a subclass of verbs of cutting (Levin, 1993), are somewhat problematic because although all of them denote contact, some do not occur naturally in the ascension construction (e.g., **The dentist drilled me in my tooth*). This issue is discussed briefly by Kemmerer (2003).

(14) (a') The puppy bit Sam on the leg.
    (a'') *The puppy bit the table on the leg.
    (b') Sam touched Kate on the arm.
    (b'') *Sam touched the library on the window.
    (c') A rock hit Sam on the head.
    (c'') *A rock hit the house on the roof.

The ascension construction therefore appears to have a schematic meaning that can be paraphrased rather loosely as follows: "X acts on person Y, causing Y to experience something, by contacting part Z of Y's body."

Third, reversative *un-* prefixation is a morphological construction that licenses some verbs but not others. Thus, one can *unlock* a door, *unwind* a string, *unwrap* a CD, *untwist* a wire, *untie* a shoe, and *unbutton* a shirt, but one cannot **unpress* a doorbell, **undangle* a bag, **unfluff* a pillow, **unhide* a present, or **unboil* a pot of water. According to a recent analysis by Kemmerer and Wright (2002), most of the verbs that allow reversative *un-* fall into two broad classes described by Levin (1993) as "combining/attaching" verbs (which further subdivide into five subclasses) and "putting" verbs (which further subdivide into seven subclasses). Both classes share the property of designating events in which an agent causes something to enter a constricted, potentially reversible spatial configuration. These semantic constraints are revealed in an especially striking way by the different uses of the verb *cross*: one can cross one's arms and then uncross them (because a constricted spatial configuration is created and then reversed), but if one crosses a street and then walks back again, it would be strange to say that one has uncrossed the street (because no constricted spatial configuration is involved). The reversative *un-* prefixation construction may therefore be restricted to verbs which have a schematic meaning something like "X causes Y to enter a constricted, potentially reversible spatial configuration relative to Z." No single verb encodes this idealized meaning, which is why the meaning is a purely constructional one. Yet many verbs have specific meanings that are compatible with this semantic template, and they are the ones that are allowed to occur in the construction.

Cross-linguistic diversity

As with action verbs, argument structure constructions vary greatly across languages in both form and meaning. Even a phenomenon as seemingly trivial as possessor ascension is manifested in such a wide variety of ways cross-linguistically that it has been the topic of book-length studies (Castillo, 1996; Chappell and McGregor, 1996). An especially instructive example of cross-linguistic constructional diversity involves causatives, broadly conceived. Many languages (including English; see Wierzbicka, 1998) have two or more causative constructions, and there is usually if not always a semantic difference between them involving at least one of the following nine parameters (Dixon, 2000):

- State/action: does a causative construction apply equally to state verbs and action verbs?
- Transitivity: does it apply equally to intransitive, transitive, and ditransitive verbs?
- Control: does the causee usually have or lack control of the induced activity?

- Volition: does the causee do it willingly or unwillingly?
- Affectedness: is the causee partially or completely affected?
- Directness: does the causer act directly or indirectly?
- Intention: does the causer achieve the result intentionally or accidentally?
- Naturalness: does the process happen fairly naturally (the causer only instigating it)?
- Involvement: is the causer (not just the causee) involved in the induced activity?

In addition, there are numerous language-specific restrictions (i.e., constraints unique to particular languages). For instance, in Nivkh a causer must be animate, so that one cannot say something like *The mist made us stay in the village*, but must instead resort to a non-causative construction such as *We stayed in the village because of the mist*.

10.3 Neuroanatomical substrates

In an illuminating discussion of the meanings of action verbs like *walk, jog, limp, strut*, and *shuffle*, Jackendoff (2002, p.350) argues that because they differ only in the manner of self-locomotion, the subtle semantic contrasts between them should probably be characterized directly in modality-specific visuospatial and motoric representational formats. In an effort to promote interdisciplinary cross-talk on issues like this, he then writes: "I hope researchers on vision and action might be persuaded to collaborate in the task." In this section I show that such collaboration is not only taking place, but is already yielding results that support Jackendoff's view. The section is organized in three parts. First, I summarize the Convergence Zone (CZ) theory and the Similarity-in-Topography (SIT) principle, which together comprise a model of the organization of conceptual knowledge in the brain. Second, I review research which suggests that (consistent with CZ theory and the SIT principle) the mirror neuron system contributes substantially to the representation of action concepts, including those encoded by verbs. Finally, I consider the neural correlates of the semantic and morphosyntactic aspects of argument structure constructions, concentrating on the following two findings from a series of neuropsychological studies that addressed the three constructions described above: focal brain damage can impair constructional meanings independently of verb meanings (consistent with the GRSSH), yet the lesion data suggest that constructional meanings are nevertheless implemented in cortical areas that are close to those that implement verb meanings (consistent with CZ theory and the SIT principle).

10.3.1 Convergence Zone (CZ) Theory and the Similarity-in-Topography (SIT) Principle

Several competing theories are currently available regarding the organization of conceptual knowledge in the brain (for reviews see Martin and Caramazza, 2003; Caramazza and Mahon, 2006). The framework that I adopt here is Convergence Zone theory (Damasio, 1989; Damasio *et al.*, 2004). It assumes that the various instances of a conceptual category are represented as fluctuating patterns of activation across modality-specific feature maps in primary and early sensory and motor cortices. These representations are experienced as explicit images, and they change continuously under the influence of external and internal inputs. For example, watching a dog run across a field generates transient activation patterns in multiple visual feature maps dedicated to coding information about shape, color, texture, size, orientation, distance, and motion. All of the instances of a conceptual category share certain properties that are neurally manifested as similar patterns of activation across feature maps. These commonalities are captured by "conjunctive neurons" (Simmons and Barsalou, 2003) in "convergence zones" (CZs: Damasio, 1989) that reside in higher-level association areas. CZs are reciprocally connected with feature maps, thereby enabling both recognition and recall. In addition, CZs exist at many hierarchical levels such that modality-specific CZs represent particular sensory and motor categories, while cross-modal CZs conjoin knowledge across modalities. Thus, to return to the example of watching a dog run across a field, the following stages of processing can be distinguished: first, activation patterns across visual feature maps are detected by modality-specific CZs that store purely visual knowledge about dogs; these modality-specific CZs then feed forward to a cross-modal CZ for the more general concept of a dog; next, the cross-modal CZ triggers the engagement of related modality-specific CZs in other knowledge domains; finally, the various modality-specific CZs may, depending on the task, generate explicit representations across the appropriate feature maps – e.g., auditory images of what dogs typically sound like, motor images of how one typically interacts with them (like reaching out and petting them), somatosensory images of how their fur feels, and so on. The evocation, whether conscious or unconscious, of some part of the large number of such neuronal patterns, over a brief lapse of time, constitutes activation of the conceptual knowledge pertaining to the category of entities at hand, namely dogs. More generally, CZ theory treats concept retrieval as a process of partial re-enactment or simulation of the sensorimotor states engendered by direct exposure to various instances of the given category (see also Barsalou, 2003; Barsalou *et al.*, 2003).

Functionally comparable CZs are neuroanatomically distributed within brain regions that are optimally situated for processing the given type of information. For example, research that has been guided by CZ theory suggests that CZs for animals – a conceptual domain that depends heavily on visual information – are implemented primarily in the right mesial–occipital/ventral–temporal region and the left mesial–occipital region (Tranel *et al.*, 1997, 2003a; Damasio *et al.*, 2004), whereas CZs for actions – a conceptual domain that depends heavily on motor programming and the perception of biological motion patterns – are implemented primarily in the left premotor/prefrontal region, the left inferior parietal region, and the left posterior middle temporal region, areas that participate in the mirror neuron system, as described in detail below (Damasio *et al.*, 2001; Tranel *et al.*, 2003b).

Simmons and Barsalou (2003) recently extended CZ theory by adding the Similarity-in-Topography (SIT) principle, which is a conjecture regarding the organization of conjunctive neurons in CZs: "The spatial proximity of two neurons in a CZ reflects the similarity of the features they conjoin. As two sets of conjoined features become more similar, the conjunctive neurons that link them lie closer together in the CZ's spatial topography." Simmons and Barsalou explore the implications of this principle for the neural representation of object concepts. In the next two sections, I show that it also has explanatory and predictive value in considering the neural representation of the kinds of action concepts that are encoded by verbs and argument structure constructions.

10.3.2 Action verbs

During the past few years, evidence has been accumulating for the view that the mirror neuron system supports not only non-linguistic action concepts in both monkeys and humans, but also the meanings[4] of language-specific action verbs. Here I provide a brief and selective review of this evidence. The discussion is organized around the relevant frontal, parietal, and temporal brain regions that together form a complex circuit (Keysers and Perrett, 2004); in addition, for each region the following three types of data are addressed in turn: monkey data, human non-linguistic data, and human linguistic data.

Frontal regions

Area F5 in the macaque brain contains neurons that represent a wide range of action types (Rizzolatti *et al.*, 1988). As Rizzolatti *et al.* (2000, p.542) put it:

> F5 is a store of motor schemas or . . . a "vocabulary" of actions. This motor vocabulary is constituted by "words", each of which is represented by a set of F5 neurons. Some words indicate the general goal of an action (e.g., grasping, holding, tearing); others indicate the way in which a specific action must be executed (e.g., precision grip or finger prehension); finally, other words are concerned with the temporal segmentation of the action into motor acts, each coding a specific phase of the grip (e.g., hand opening, hand closure).[5]

Crucially, many of these neurons discharge during both execution and observation of actions; they are called mirror neurons because the observed action seems to be reflected in the motor representation of the same action (Gallese *et al.*, 1996; Rizzolatti *et al.*, 1996a). Mirror neurons discharge even when the goal of the observed action is not visible but has recently been visible (Umiltà *et al.*, 2001). Moreover, some mirror neurons in the macaque discharge in response to presentation of either the sight or the sound of an action (Kohler *et al.*, 2002). In addition, there is evidence for mirror neurons representing oral actions (Ferrari *et al.*, 2003), but as yet no evidence for mirror neurons representing foot actions. Finally, a recent study found that neurons in the primary motor cortex of the macaque brain also have mirror properties (Raos *et al.*, 2004).

---

[4] Here and in what follows, only semantic structures are addressed, not phonological structures.

[5] Rizzolatti *et al.*'s vocabulary metaphor is quite provocative, but it should not be taken too literally. The cross-linguistic variation in verbs for hand actions described earlier (p. 349) suggests that the motor aspects of those verb meanings may be neurally implemented in language-specific CZs at a somewhat higher level of the motor hierarchy. See Gallese and Lakoff (2005) for an in-depth neurocognitive analysis of the concept of "grasping," with valuable discussion of pertinent linguistic issues, and see Mahon and Caramazza (2005) for a critique.

Turning to humans, a rapidly growing literature suggests that non-verbal action concepts depend on certain motor-related sectors of the frontal lobes. First, from the perspective of neurophysiology, experiments utilizing transcranial magnetic stimulation (TMS) have shown that motor evoked potentials recorded from a person's muscles are facilitated when the person observes either intransitive (non-object-directed) or transitive (object-directed) hand actions (Fadiga *et al.*, 1995). Second, from the perspective of functional neuroimaging, many important discoveries have recently been made, including the following:

- Broca's area, the human homologue of monkey F5 (Arbib and Bota, this volume), is activated during execution, observation, and imitation of hand actions, especially those involving complex finger movements (e.g., Rizzolatti *et al.*, 1996b; Decety *et al.*, 1997; Iacoboni *et al.*, 1999).
- The human mirror system is somatotopically organized – specifically, observation of both intransitive and transitive face, arm/hand, and leg/foot actions engages premotor areas in a somatotopic manner (e.g., Buccino *et al.*, 2001; Wheaton *et al.*, 2004), which is consistent with the SIT principle.
- The human mirror system has motor components only for types of actions that people are capable of performing (Buccino *et al.*, 2004).
- For action concepts that are within its repertoire, the human mirror system responds robustly even to degraded stimuli – e.g., point-light displays of people moving in particular ways (Saygin *et al.*, 2004a).

Third, from the perspective of neuropsychology, several lesion studies with large cohorts of brain-injured patients have found that damage in the left premotor/prefrontal region impairs conceptual knowledge of actions (Tranel *et al.*, 2003b; Saygin *et al.*, 2004b).

As with action concepts in general, there is increasing evidence that the meanings of action verbs – more precisely, the semantic features that specify the motoric aspects of the designated actions – are subserved by motor-related structures in the frontal lobes, especially in the left hemisphere (although the right hemisphere may also contribute; see Neininger and Pulvermüller, 2003). Studies employing high-density electroencephalography (e.g., Hauk and Pulvermüller, 2004), functional magnetic resonance imaging (fMRI) (Hauk *et al.*, 2004; Tettamanti *et al.*, 2005), magnetoencephalography (Pulvermüller *et al.*, 2005a), and TMS (Buccino *et al.*, 2005; Pulvermüller *et al.*, 2005b) indicate that verbs encoding face actions (e.g., *bite*), arm/hand actions (e.g., *punch*), and leg/foot actions (e.g., *kick*) differentially engage the corresponding inferior, dorsolateral, and dorsal–midline sectors of somatotopically mapped motor and premotor regions. These findings support the provocative notion that the motoric aspects of the meanings of action verbs are not part of an abstract symbolic representation in the brain (like the neural analogue of a dictionary entry), but are instead linked with the same frontal cortical structures that subserve action execution and observation. Further evidence for this view comes from studies indicating that damage in the left premotor/prefrontal disrupts knowledge of the meanings of action verbs (e.g., Bak *et al.*, 2001; Bak and Hodges, 2003; Kemmerer and Tranel, 2003). However, as yet no neuropsychological studies have directly tested the hypothesis that the meanings of verbs for face, arm/hand, and leg/foot actions should be differentially impaired by lesions affecting the pertinent somatotopically mapped motor regions (but see Kemmerer and Tranel (2000) for a preliminary investigation).

Parietal regions

In the macaque brain, mirror neurons in area F5 receive input from an inferior parietal area called PF (Rizzolatti and Luppino, 2001). This area contains neurons that are also mirror-like, firing during production as well as perception of particular types of transitive (object-directed) hand actions (Gallese *et al.*, 2002). These neurons appear to be especially sensitive to the intended goal of a complex action (Fogassi *et al.*, 2004).

The most impressive evidence that the human mirror system encompasses parietal structures comes from the frequently cited fMRI study by Buccino *et al.* (2001), which demonstrated that observation of transitive (object-directed) face, arm/hand, and leg/foot actions engages not only the frontal lobes but also the parietal lobes in a somatotopic manner. Additional evidence comes from the lesion study by Tranel *et al.* (2003b), which found that damage in the left inferior parietal cortex, especially the supramarginal gyrus (SMG), impairs knowledge of action concepts. Also relevant are lesion studies that have linked ideational apraxia – a disorder affecting the ideas or concepts underlying skilled movements – with damage in the left parieto-occipital junction (see Johnson-Frey (2004) for a review). Finally, mechanistic knowledge about the proper manipulation of tools has been associated with the left SMG (e.g., Boronat *et al.*, 2005).

Turning to action verbs, a recent fMRI study by Tettamanti *et al.* (2005) found that when subjects listened to transitive sentences describing object-directed face, arm/hand, and leg/foot actions, the left inferior parietal cortex was activated in a somatotopic manner, consistent with Buccino *et al.*'s (2001) findings. In addition, an earlier study employing positron emission tomography (PET) (Damasio *et al.*, 2001) found that activation in the left SMG was significantly greater for naming actions performed with tools (e.g., *write*) than without tools (e.g., *wave*), which fits the data mentioned above relating this region to conceptual knowledge for tool manipulation. Presumably all of the verbs in Damasio *et al.*'s (2001) tool condition were transitive, in line with the studies by Tettamanti *et al.* (2005) and Buccino *et al.* (2001); however, the authors did not indicate what proportion of the verbs in their non-tool condition were intransitive. Nevertheless, the data from all three functional neuroimaging studies lead to the intriguing prediction that lesions in the left inferior parietal cortex should impair semantic knowledge of transitive verbs (especially those encoding instrumental actions) to a significantly greater extent than semantic knowledge of intransitive verbs. While dissociations between these two large categories of verbs have been reported (e.g., Thompson *et al.*, 1997; Jonkers, 2000; Kemmerer and Tranel, 2000), the lesion correlates have not yet been carefully investigated. Finally, it is noteworthy that a recent fMRI study by Wu *et al.* (2004) found activation in the left inferior parietal cortex during attentive processing of the path component, as opposed to the manner component, of motion events. Although English preferentially encodes path information in prepositions like *into, out of, upward,* and *downward,* there are a few pure path verbs like *enter, exit, ascend,* and *descend* (all derived historically from Latin), and there are also some manner verbs that include spatial goal-oriented details – e.g., one subclass of "putting" verbs focuses on actions in which a substance is caused to move forcefully against the surface of an object in such a way that it ultimately has an idiosyncratic shape "on" the object (*smear, dab, streak, smudge,* etc.), whereas another subclass of "putting" verbs focuses on actions in which a flexible object extended in one dimension is caused to move along a circular path so that its final spatial configuration is "around" another object (*coil, wind, twirl, spin,* etc.) (Levin, 1993; Pinker, 1989). Thus, it is possible that the parietal activation reported by Wu *et al.* (2004) reflects these kinds of semantic features (see also Kemmerer and Tranel, 2003; Tranel and Kemmerer, 2004).

Temporal regions

In the dorsal stream of the macaque brain, the superior temporal sulcus (STS) receives input from area MST and projects to F5 via PF (Rizzolatti and Matelli, 2003). Within the STS, functionally specialized neurons respond to different types of face, limb, and whole-body motion not only when the stimuli are presented in full view but also when they are presented in point-light displays (see Puce and Perrett (2003) for a review). For example, one single-cell recording study identified a neuron that discharges vigorously during observation of a person dropping an object but not during observation of the same bodily movement without the object or during observation of the object motion alone (Keysers and Perrett, 2004; Barraclough *et al.*, 2005). Like F5 neurons, STS neurons continue to fire even when an agent has moved behind an occluder (Baker *et al.*, 2001) and also when the typical sounds associated with certain actions are detected (Barraclough *et al.*, 2005).

In humans, activation occurs in area MT and in areas along the superior temporal gyrus and STS during observation of biological motion patterns (again, see Puce and Perrett (2003) for a review). Activation in these regions can be elicited by point-light displays of real motion (Beauchamp *et al.*, 2003), by static images of implied motion (Kourtzi and Kanwisher, 2000; Senior *et al.*, 2000), and by motion-related sounds (Lewis *et al.*, 2004). Evidence that the left MT region contributes to conceptual knowledge of actions comes from Tranel *et al.*'s (2003a) lesion study, which found that damage to the white matter beneath this region severely disrupts that knowledge.

Regarding action verbs, numerous functional neuroimaging studies suggest that the visual motion patterns encoded by different verbs are represented in the left posterior middle temporal cortex anterior to area MT. This region is engaged by various tasks requiring the semantic processing of action verbs (see Martin *et al.* (2000) for a review; see also Damasio *et al.*, 2001; Kable *et al.*, 2002). In addition, area MT is activated significantly more when subjects use noun–verb homophones (e.g., *comb*) as verbs to name static pictures of implied actions than when subjects use them as nouns to name objects in the same pictures (Tranel *et al.*, 2005). Furthermore, the fMRI study by Wu *et al.* (2004) revealed activation in this cortical region during attentive processing of the manner component, as opposed to the path component, of motion events. From the perspective of neuropsychology, I am not aware of any studies that have directly tested the well-motivated hypothesis that knowledge of the visual motion patterns encoded by action verbs should be selectively impaired by damage anterior to area MT; however, Tranel *et al.*'s (2003b) lesion study strongly supports this possibility, since it showed that damage underneath MT disrupts non-verbal action concepts.

In this context, it is worth mentioning again that languages vary greatly in how they subdivide the conceptual space of manner-of-motion (see p. 349). In particular, it is intriguing to consider that CZ theory and the SIT principle together lead to the following hypothesis. Perhaps the cross-linguistic diversity in semantic distinctions is reflected, at least in part, in corresponding neuroanatomical diversity in the spatial arrangement of conjunctive neurons (or, more precisely, columns of such neurons) in some CZ within the mosaic of cortical areas extending from MT along the STG and STS – a CZ functionally dedicated to representing the visual motion patterns associated with language-specific verb meanings. According to this hypothesis, the topographical layout of the relevant conjunctive neurons is systematically different for English speakers compared to, say, Spanish speakers. For English speakers there are separate but tightly clustered conjunctive neurons for the closely related visual motion patterns encoded by *creep, glide, slide, slip,* and *slither;* however, for Spanish speakers such conjunctive neurons do not exist because (1) the Spanish manner-verb lexicon does not make any of those subtle semantic distinctions (the whole spectrum is covered by just one verb, *escabullirse*), and (2) there is no independent reason to expect those particular distinctions to be "natural" in the sense of being universally employed in the non-verbal categorization of motion events (see Slobin (2000, 2003) for relevant data and discussion from the perspective of language acquisition). As the spatial resolution of functional neuroimaging techniques continues to improve, it may eventually become feasible to test hypotheses of this nature, thereby shedding further light on the biological bases of the meanings of action verbs. For present purposes, the essential point is this: it may not be a coincidence that prominent theorists in both linguistic typology (e.g., Croft, 2001; Haspelmath, 2003) and cognitive neuroscience (e.g., Simmons and Barsalou, 2003) increasingly use the mapping metaphor in their characterizations of the organization of conceptual knowledge. Perhaps the metaphor is more appropriate than we have hitherto realized (see Kohonen and Hari (1999) for a review of pertinent neurocomputational modeling).

10.3.3 Argument structure constructions

Semantics

The nature of the meanings of argument structure constructions has attracted a great deal of attention in linguistics, but it has not inspired much work in cognitive neuroscience; hence the neuroanatomical substrates of these meanings remain, for the most part, *terra incognita*. Nevertheless, a few neuropsychological studies have yielded results that are consistent with predictions derived from the GRSSH, CZ theory, and the SIT principle.

According to the GRSSH as initially formulated by Pinker (1989) and further elaborated by other researchers (e.g., Mohanan and Wee, 1999), a fundamental division exists between constructional meanings and verb meanings. If this is true, then these two large-scale components of meaning are probably subserved by at least partially separate brain structures, perhaps involving different cortical areas implementing networks of CZs at different levels of abstraction. This in turn predicts that the two components of meaning could be impaired independently of each other by brain damage. I have been conducting a series of studies with aphasic patients to test this prediction, and have obtained results that support it.

The first study focused on the locative alternation (see p. 353) and documented the following double dissociation (Kemmerer, 2000a). Two patients performed well on a verb–picture matching test requiring discrimination between verbs that vary only with respect to subtle visual, affective, and motoric features that are grammatically irrelevant, e.g., *drip–pour–spray*, *coil–spin–roll*, and *decorate–adorn–embellish*. However, both patients failed a grammaticality judgement test requiring determination of the compatibility between, on the one hand, the meanings of the very same verbs that were used in the matching test and, on the other, the meanings of the constructions comprising the locative alternation – e.g., *drip* but not *drench* can occur in the construction with the (sub)meaning "X enables a mass Y to go to Z via the force of gravity" (cf. *Sam is dripping/*drenching water on the flowers*), whereas *drench* but not *drip* can occur in the construction with the (sub)meaning "X causes a solid or layer-like medium Z to have a mass Y distributed throughout it" (cf. *Sam is drenching/*dripping the flowers with water*). Strikingly, the patients treated as ungrammatical many sentences that are quite natural (e.g., *Sam is coiling the ribbon around the pole* and *Sam is decorating the pie with cream*), and treated as grammatical many sentences that are very odd (e.g., **Sam is coiling the pole with the ribbon* and **Sam is decorating cream onto the pie*). Their errors could not be attributed to an impairment of either syntactic processing or metalinguistic judgement ability because both patients passed another test that assessed the integrity of these capacities. The data therefore suggest that although the patients retained an impressive amount of knowledge regarding the semantic nuances of locative verbs, they suffered selective impairments of their appreciation of the more schematic meanings of locative constructions. Importantly, a third patient exhibited the opposite performance profile. She failed a significant number of items in the verb–picture matching test, but had no difficulty with the grammaticality judgement test. For example, even though she could not distinguish *coil* from *spin* and *roll*, she could correctly determine that **Sam is coiling the pole with the ribbon* is awkward and that *Sam is coiling the ribbon around the pole* is fine. In sum, the double dissociation identified by this study provides preliminary evidence that the neural substrates of constructional meanings are separate from those for verb meanings.

Two subsequent studies (Kemmerer and Wright, 2002; Kemmerer, 2003) focused on the body-part possessor alternation and the reversible *un-* prefixation construction (see pp. 353–5), using methods analogous to those employed in the study of the locative alternation. Both studies documented one-way dissociations involving preserved knowledge of verb meanings but impaired knowledge of constructional meanings. First, the study of the body-part possessor alternation found that three aphasic patients could discriminate between verbs that differ in idiosyncratic ways that are irrelevant to the ascension construction (e.g., *scratch–smack–spank* and *break–rip–fracture*), but could no longer make accurate judgements about which of these verbs could occur in the construction (e.g., *She scratched him on the arm* vs. **She broke him on the arm*). The patients' poor performances on the latter test were not due to purely syntactic disorders since they had no difficulty with a different test that evaluated their knowledge of the clausal organization of the ascension construction. Instead, the patients appeared to have deficits involving their knowledge and/or processing of the meaning of the ascension construction. Similarly, in the study of the reversative *un-* prefixation construction, two aphasic patients demonstrated intact knowledge of idiosyncratic aspects of verb meaning that are "invisible" to the construction (e.g., *wrap–buckle–zip* and *squeeze–press–push*), but were impaired at judging whether the very same verbs satisfy the semantic criteria of the construction (e.g., *unwrap* vs. **unsqueeze*). A separate test showed that the patients' errors were not due to an impaired understanding of the basic reversative meaning of *un-* or to problems with various task demands involving morphological analysis, but were instead most likely due to a selective disturbance of the knowledge and/or processing of the meaning of the construction.[6]

Although the three studies just described constitute only the initial cognitive–neuroscientific forays into the complex semantic territory of constructional meanings, the findings support the view that the neural correlates of these meanings are separate from those for verb meanings. This leads to the next question, which concerns the neuroanatomical localization of constructional meanings.

The meanings of argument structure constructions consist of abstract event schemas that often constitute semantic generalizations across various verb classes; hence they occupy a very high level of the action concept hierarchy. Given this background, CZ theory and the SIT principle together predict that the brain regions that operate as CZs for constructional meanings should be anatomically adjacent to those that operate as CZs for verb meanings – that is, they should be near or perhaps still within the fronto-parietotemporal circuitry underlying the human mirror system. It is difficult to predict exactly what regions these might be, but it is reasonable to suppose that they may include part of the left perisylvian cortex, because then the CZs for constructional meanings would also be close to the networks subserving morphosyntactic structures and processes (these networks are discussed briefly below). The three neuropsychological studies summarized above provide preliminary data that are consistent with this prediction, since the aphasic patients with selective impairments of constructional meanings had the greatest lesion

---

[6] The three studies just described, as well as the study involving prenominal adjective order reported by Kemmerer (2000b), employed some of the same brain-damaged subjects. It is noteworthy, however, that these subjects exhibited different performance profiles across the studies. For instance, subject 1962RR was impaired across the entire range of constructions, but subject 1978JB was impaired on just the three verb-based constructions, performing normally on the adjective-based construction. The dissociation exhibited by 1978JB raises the possibility that constructional meanings are organized in principled ways in the brain.

overlap in the left inferior premotor/prefrontal region and the left anterior SMG. These neuroanatomical results should be interpreted with caution, however, because they represent data from only a few patients. Hopefully, though, other researchers will soon be inspired to investigate this topic in greater depth by employing not only the lesion method but also hemodynamic methods which have much more precise spatial resolution.

Morphosyntax

The literature on the neural correlates of the morphosyntactic aspects of argument structure constructions is rapidly growing, but is still influenced much more by the Chomskian generative grammar framework than by the new constructionist approach. Here I will only mention a few salient findings. Grammatical categories like noun and verb – which, as noted earlier (see p. 352), fractionate into clusters of subcategories – may be supported by the cortex in and around Broca's area (see Caramazza and Shapiro (2004) for a review). In addition, Broca's area may contribute to the assembly of argument structure constructions during sentence production (Indefrey *et al.*, 2001, 2004). With respect to the parsing of argument structure constructions during sentence comprehension, many regions distributed throughout the left perisylvian cortex have been implicated (see Friederici (2004) for a review), but the anterior sector of the superior temporal gyrus appears to play a special role in processing morphosyntactic information (Dronkers *et al.*, 2004).

10.4 Discussion

In this concluding section I would like to broaden the discussion of action verbs and argument structure constructions by making a few remarks about the emergence of language during ontogeny and phylogeny. I will highlight the views of Michael Tomasello, since he is one of the leading advocates of the constructionist approach in these areas of inquiry.

10.4.1 Ontogeny

The best-known answer to the question of how children acquire language is that – as first proposed by Chomsky (1959) and later popularized by Pinker (1994), Jackendoff (1994), and others – they are guided in this seemingly monumental task by an evolutionarily specialized, genetically programmed, neurocognitive adaptation called Universal Grammar which includes a kind of blueprint of the basic design characteristics of all natural human languages. This orthodox view has been challenged, however, by a growing body of research motivated by the constructionist framework (see Tomasello (2003a) for the most well-articulated alternative theory). The most important empirical discovery is that virtually all of children's early linguistic competence is item-based in the sense of being organized around particular words and phrases, not around any system-wide innate categories. This was demonstrated by Tomasello (1992; see also Hill, 1983) in a detailed diary study of his own daughter's early language development. During exactly the same time period, this child used some verbs in only one type of very simple pivot construction or schema (e.g., *Cut ___*), but used other verbs in more complex frames of different types (e.g., *Draw ___, Draw___ on ___, I draw with ___, Draw ___ for ___, ___draw on ___*). For each individual verb, however, there was great continuity, such that new uses almost always replicated previous uses with only one small change (e.g., the addition of a new participant role). In fact, as Tomasello (2000, p.157) emphasizes, "by far the best predictor of this child's use of a given verb on a given day was *not* her use of other verbs on that same day, but rather her use of that same verb on immediately preceding days; there appeared to be no transfer of structure across verbs." These findings led to the Verb Island Hypothesis, which maintains that children's early linguistic competence consists almost entirely of an inventory of linguistic constructions like those just described – specific verbs with slots for narrowly defined participants such as "drawer" and "thing drawn," as opposed to subject and object (see Arbib and Hill (1988) and Culicover (1999) for similar proposals from rather different perspectives).

Of course, children eventually go beyond these early item-based constructions, and they appear to do so by first recognizing similarities across constructional schemas and then creatively combining these schemas to form novel utterances. For example, among the first three-word utterances creatively produced by Tomasello's daughter was *See Daddy's car*. She had previously said things like *See ball* and *See Mommy*, on the one hand, and things like *Daddy's shirt* and *Daddy's pen*, on the other. So the novel utterance may reflect the combination of a "*See ___*" schema with a "*Daddy's ___*" schema. Note that to accomplish this she had to understand that the complex expression *Daddy's car* was functionally equivalent to the simpler expressions that she had previously included in the slot for the "*See ___*" schema. More generally, the key idea is that through cognitive processes of this nature (which are not restricted to the linguistic domain), children gradually acquire increasingly abstract and adult-like argument structure constructions such as the following (Tomasello, 1999, p.141):

- Imperatives (*Roll it! Smile! Push me!*)
- Simple transitives (*Ernie kissed her; He kicked the ball*)
- Simple intransitives (*She's smiling; It's rolling*)
- Locatives (*I put it on the table; She took her book to school*)
- Resultatives (*He wiped the table clean; She knocked him silly*)
- Ditransitives (*Ernie gave it to her; She threw him a kiss*)
- Passives (*I got hurt; He got kicked by the elephant*)
- Attributives and identificationals (*It's pretty; She's my mommy*)

This overall approach to accounting for early language development has recently been supported by research using both naturalistic and experimental methods to study many different children acquiring many different languages (see Tomasello (2003a) for a review), and has also been successfully extended to the investigation of later language development (Diessel, 2004). A corresponding neurobiological theory of language development, with explicit links to the mirror neuron system, can be found in the work of Elizabeth Bates and her colleagues (Dick *et al.*, 2004).

10.4.2 Phylogeny

Tomasello (2003a, p.1) begins his book by pointing out that from an ethological perspective one of the most bizarre traits of *Homo sapiens* is that "whereas the individuals of all nonhuman species can communicate effectively with all of their conspecifics, human beings can communicate effectively only with other persons who have grown up in the same linguistic community – typically, in the same geographical region." Then, in direct opposition to the Chomskyan Universal Grammar framework which has dominated linguistics for over 40 years, but in complete agreement with the strongest version of the increasingly influential constructionist approach, he states that "one immediate outcome is that, unlike most other animal species, human beings cannot be born with any specific set of communicative behaviors." The central claim of the position that he subsequently defends is that the evolution of the capacity to communicate symbolically was sufficient for all natural human languages to arise; no independent adaptations for morphosyntax were necessary, the reason being that, as described above, most morphosyntactic constructions are actually symbolic devices that just happen to be more complex and schematic than words, and that develop very gradually on a historical timescale (see also Tomasello, 2003b; cf. Arbib, 2005, and Deacon, 1997, for similar proposals). Symbolic communication may have co-evolved with a number of other uniquely human neurocognitive adaptations, perhaps the most important of which was the ability to interpret and share intentions, since this is what enables human communities to establish social conventions for the referential use of arbitrary signs (Tomasello, 1999, 2005; see Frith and Wolpert, 2003, for pertinent neurobiological research; see also Stanford, this volume). Other relevant adaptations may include intuitive theories about various domains of the world, such as objects, forces, paths, places, manners, states, and substances – domains that are routinely encoded by words as well as constructions in languages worldwide. Pinker (2003) has suggested that all of these distinctively human skills – language, hypersociality, and sophisticated causal reasoning abilities – are adaptations to "the cognitive niche," i.e., a complex set of physical and social conditions that created selection pressures for acquiring and sharing information. I agree with this general hypothesis about the ancestral environment in which human symbolic capacity evolved; however, like Tomasello I doubt if there is currently enough evidence to support Pinker's more specific view – one that is also endorsed by Jackendoff (cf. Pinker and Jackendoff, 2005) – that full-blown human language depends on additional adaptations for morphosyntax. The resolution of this debate will hinge on future research in all of the disciplines that contribute to studying the evolution of language.

Acknowledgments

I thank Michael Arbib, Natalya Kaganovich, and three anonymous referees for comments on previous versions of this chapter.

References

Arbib, M. A., 2005. From monkey-like action recognition to human language: an evolutionary framework for neurolinguistics. *Behav. Brain Sci.* **28**: 105–167.

Arbib, M. A., and Hill, J. C., 1988. Language acquisition: schemas replace universal grammar. In J. A. Hawkins (ed.) *Explaining Language Universals.* Oxford, UK: Blackwell, pp. 56–72.

Bak, T. H., and Hodges, J. R., 2003. "Kissing and dancing" – a test to distinguish the lexical and conceptual contributions to noun/verb and object/action dissociations: preliminary results in patients with frontotemporal dementia. *J. Neuroling.* **16**: 169–181.

Bak, T. H., O'Donovan, D. G., Xuereb, J. H., Boniface, S., and Hodges, J. R., 2001. Selective impairment of verb processing associated with pathological changes in Brodmann areas 44 and 45 in the motor neurone disease–dementia–aphasia syndrome. *Brain* **124**: 103–130.

Baker, C. I., Keysers, C., Jallema, T., Wicker, B., and Perrett, D. I., 2001. Neuronal representation of disappearing and hidden objects in temporal cortex of the macaque. *Exp. Brain Res.* **140**: 375–381.

Barraclough, N. E., Xiao, D., Baker, C. I., Oram, M. W., and Perrett, D. I., 2005. Integration of visual and auditory information by STS neurons responsive to the sight of actions. *J. Cogn. Neurosci.* **17**: 377–391.

Barsalou, L. W., 2003. Situated simulation in the human conceptual system. *Lang. Cogn. Proc.* **18**: 513–562.

Barsalou, L. W., Simmons, W. K., Barbey, A., and Wilson, C. D., 2003. Grounding conceptual knowledge in modality-specific systems. *Trends Cogn. Sci.* **7**: 84–91.

Beauchamp, M. S., Lee, K. E., Haxby, J. V., and Martin, A., 2003. fMRI responses to video and point-light displays of moving humans and manipulable objects. *J. Cogn. Neurosci.* **15**: 991–1001.

Buccino, G., Binkofski, F., Fink, G. R., *et al.*, 2001. Action observation activates premotor and parietal areas in a somatotopic manner: an fMRI study. *Eur. J. Neurosci.* **13**: 400–404.

Buccino, G., Lui, F., Canessa, N., *et al.*, 2004. Neural circuits involved in the recognition of actions performed by non-conspecifics: an fMRI study. *J. Cogn. Neurosci.* **16**: 114–126.

Buccino, G., Riggio, L., Melli, G., *et al.*, 2004. Listening to action-related sentences modulates the activity of the motor system: a combined TMS and behavioral study. *Cogn. Brain Res.* **24**: 355–363.

Castillo, M. V., 1996. *The Grammar of Possession: Inalienability and Possessor Ascension in Guarani.* Amsterdam, Netherlands: John Benjamins.

Caramazza, A., and Mahon, B. Z., 2006. The organization of conceptual knowledge in the brain: the future's past and the some future directions. *Cogn. Neuropsychol.* **23**: 13–38.

Caramazza, A., and Shapiro, K., 2004. The organization of lexical knowledge in the brain: the grammatical dimension. In M. Gazzaniga (ed.) *The Cognitive Neurosciences*, vol. 3. Cambridge, MA: MIT Press, pp. 803–814.

Chappell, H., and McGregor, W. (eds.), 1996. *The Grammar of Inalienability*. Berlin, Germany: Mouton de Gruyter.

Chomsky, N., 1959. A review of B. F. Skinner's *Verbal Behavior*. *Language* **35**: 26–58.

Croft, W., 1991. *Syntactic Categories and Grammatical Relations*. Chicago, IL: University of Chicago Press.

  1998. Event structure in argument linking. In M. Butt and W. Geuder (eds.) *The Projection of Arguments*. Stanford, CA: CSLI, pp. 21–64.

  2001. *Radical Construction Grammar*. Oxford, UK: Oxford University Press.

Croft, W., and Cruse, D. A., 2004. *Cognitive Linguistics*. Cambridge, UK: Cambridge University Press.

Culicover, P. W., 1999. *Syntactic Nuts*. Oxford, UK: Oxford University Press.

Damasio, A. R., 1989. Time-locked multiregional retroactivation: a systems-level proposal for the neural substrates of recall and recognition. *Cognition* **33**: 25–62.

Damasio, H., Grabowski, T. J., Tranel, D., *et al.*, 2001. Neural correlates of naming actions and of naming spatial relations. *Neuroimage* **13**: 1053–1064.

Damasio, H., Tranel, D., Grabowski, T. J., Adolphs, R., and Damasio, A. R., 2004. Neural systems behind word and concept retrieval. *Cognition* **92**: 179–229.

Deacon, T. W., 1997. *The Symbolic Species*. New York: Norton.

Decety, J., Grezes, J., Costes, N., *et al.*, 1997. Brain activity during observation of actions: influence of action content and subject's strategy. *Brain* **120**: 1763–1777.

Dick, F., Dronkers, N., Pizzamiglio, L., *et al.*, 2004. Language and the brain. In M. Tomasello and D. Slobin (eds.) *Beyond Nature–Nurture: Essays in Honor of Elizabeth Bates*. Mahwah, NJ: Lawrence Erlbaum, pp. 237–261.

Diessel, H., 2004. *The Acquisition of Complex Sentences*. Cambridge, UK: Cambridge University Press.

Dixon, R. M. W., 2000. A typology of causatives: form, syntax, and meaning. In R. M. W. Dixon and A. Y. Aikhenvald (eds.) *Changing Valency*. Cambridge, UK: Cambridge University Press, pp. 30–83.

Dronkers, N. F., Wilkins, D. P., Van Valin, Jr., R. D., Redfern, B. B., and Jaeger, J. J., 2004. Lesion analysis of the brain areas involved in language comprehension using a new method of lesion analysis. *Cognition* **92**: 145–177.

Fadiga, L., Fogassi, L., Pavesi, G., and Rizzolatti, G., 1995. Motor facilitation during action observation: a magnetic stimulation study. *J. Neurophys.* **73**: 2608–2611.

Fellbaum, C., 1998. A semantic network of English verbs. In C. Fellbaum (ed.) *Wordnet*. Cambridge, MA: MIT Press, pp. 69–104.

Ferrari, P. F., Gallese, V., Rizzolatti, G., and Fogassi, L., 2003. Mirror neurons responding to the observation of ingestive and communicative mouth actions in the monkey ventral premotor cortex. *Eur. J. Neurosci.* **17**: 1703–1714.

Fogassi, L., Ferrari, P. F., Gesierich, B., Rozzi, S., Chersim, F., and Rizzolatti, G. 2005. Parietal lobe: from action organization to intention understanding. *Science* **308**: 662–667.

Friederici, A. D., 2004. The neural basis of syntactic processes. In M. Gazzaniga (ed.) *The Cognitive Neurosciences*, vol. 3. Cambridge, MA: MIT Press, pp. 803–814.

Frith, C., and Wolpert, D. (eds.), 2003. *The Neuroscience of Social Interaction*. Oxford, UK: Oxford University Press.

Gallese, V., and Lakoff, G., 2005. The brain's concepts: the role of the sensory-motor system in conceptual knowledge. *Cogn. Neuropsych.* **22**: 455–479.

Gallese, V., Fadiga, L., Fogassi, L., and Rizzolatti, G., 1996. Action recognition in the premotor cortex. *Brain* **119**: 593–609.

Gallese, V., Fogassi, L., Fadiga, L., and Rizzolatti, G., 2002. Action representation and the inferior parietal lobule. In W. Prinz and B. Hommel (eds.) *Attention and Performance, vol. 19, Common Mechanisms in Perception and Action.* Oxford, UK: Oxford University Press, pp. 334–355.

Goldberg, A., 1995. *Constructions: A Construction Grammar Approach to Argument Structure.* Chicago, IL: Univerisity of Chicago Press.

    1998. Patterns of experience in patterns of language. In M. Tomasello (ed.) *The New Psychology of Language*, vol. 1. Mahwah, NJ: Lawrence Erlbaum, pp. 203–220.

    2003. Constructions: a new theoretical approach to language. *Trends Cogn. Sci.* **7**: 219–224.

Gropen, J., Pinker, S., Hollander, M., and Goldberg, R., 1991. Affectedness and direct objects: the role of semantics in the acquisition of verb argument structure. *Cognition* **41**: 143–195.

Haspelmath, M., 2003. The geometry of grammatical meaning: semantic maps and cross-linguistic comparison. In M. Tomasello (ed.) *The New Psychology of Language*, vol. 2. Mahwah, NJ: Lawrence Erlbaum, pp. 211–242.

Hauk, O., and Pulvermüller, F., 2004. Neurophysiological distinction of action words in the fronto-central cortex. *Hum. Brain Map.* **21**: 191–201.

Hauk, O., Johnsrude, I., and Pulvermüller, F., 2004. Somatotopic representation of action words in human motor and premotor cortex. *Neuron* **41**: 301–307.

Hill, J. C., 1983. A computational model of language acquisition in the two-year-old. *Cogn. Brain Theory* **6**: 287–317.

Iacoboni, M., Woods, R. P., Brass, M., *et al.*, 1999. Cortical mechanisms of human imitation. *Science* **286**: 2526–2528.

Indefrey, P., Brown, C. M., Hellwig, F., *et al.*, 2001. A neural correlate of syntactic encoding during speech production. *Proc. Natl Acad. Sci. USA*, **98**: 5933–5936.

Indefrey, P., Hellwig, F., Herzog, H., Seitz, R. J., and Hagoort, P., 2004. Neural responses to the production and comprehension of syntax in identical utterances. *Brain Lang.* **89**: 312–319.

Jackendoff, R., 1990. *Semantic Structures.* Cambridge, MA: MIT Press.

    1994. *Patterns in the Mind.* New York: Basic Books.

    2002. *Foundations of Language.* Oxford, UK: Oxford University Press.

Johnson-Frey, S. H., 2004. The neural bases of complex tool use in humans. *Trends Cogn. Sci.* **8**: 71–78.

Jonkers, R., 2000. Verb finding problems in Broca's aphasics: the influence of transitivity. In R. Bastiaanse and Y. Grodzinsky (eds.) *Grammatical Disorders in Aphasia.* London: Whurr, pp. 105–122.

Kable, J. W., Lease-Spellmeyer, J., and Chatterjee, A., 2002. Neural substrates of action event knowledge. *J. Cogn. Neurosci.* **14**: 795–805.

Kemmerer, D., 2000a. Grammatically relevant and grammatically irrelevant features of verb meaning can be independently impaired. *Aphasiology* **14**: 997–1020.

    2000b. Selective impairment of knowledge underlying prenominal adjective order: evidence for the autonomy of grammatical semantics. *J. Neuroling.* **13**: 57–82.

2003. Why can you hit someone on the arm but not break someone on the arm? A neuropsychological investigation of the English body-part possessor ascension construction. *J. Neuroling*. **16**: 13–36.

2005. The spatial and temporal meanings of English prepositions can be independently impaired. *Neuropsychologia* **43**: 795–806.

Kemmerer, D., and Tranel, D., 2000. Verb retrieval in brain-damaged subjects. I. Analysis of stimulus, lexical, and conceptual factors. *Brain Lang*. **73**: 347–392.

2003. A double dissociation between the meanings of action verbs and locative prepositions. *Neurocase* **9**: 421–435.

Kemmerer, D., and Wright, S. K., 2002. Selective impairment of knowledge underlying *un*-prefixation: further evidence for the autonomy of grammatical semantics. *J. Neuroling*. **15**: 403–432.

Kemmerer, D., Weber-Fox, C., Price, K., Zdansczk, C., and Way, H. (in press). *Big brown dog or Brown big dog?* An electrophysiological study of semantic constraints on prenominal adjective order. *Brain and Language*.

Keysers, C., and Perrett, D. I., 2004. Demystifying social cognition: a Hebbian perspective. *Trends Cogn. Sci*. **8**: 501–507.

Kita, S., and Özyürek, A., 2003. What does cross-linguistic variation in semantic coordination of speech and gesture reveal? Evidence for an interface representation of spatial thinking and speaking. *J. Mem. Lang*. **48**: 16–32.

Kohler, E., Keysers, C., Umiltà, M. A., *et al*., 2002. Hearing sounds, understanding actions: action representation in mirror neurons. *Science* **297**: 846–848.

Kohonen, T., and Hari, R., 1999. Where the abstract feature maps of the brain might come from. *Trends Neurosci*. **22**: 135–139.

Kourtzi, Z., and Kanwisher, N., 2000. Activation in human MT/MST by static images with implied motion. *J. Cogn. Neurosci*. **12**: 48–55.

Lakoff, G., and Johnson, M., 1999. *Philosophy in the Flesh*. Chicago, IL: University of Chicago Press.

Levin, D., 1993. *English Verb Classes and Alternations*. Chicago, IL: University of Chicago Press.

Lewis, J. W., Wightman, F. L., Brefczynski, J. A., *et al*., 2004. Human brain regions involved in recognizing environmental sounds. *Cereb. Cortex* **14**: 1008–1021.

Mahon, B., and Caramazza, A., 2005. The orchestration of the sensory-motor systems: clues from neuropsychology. *Cogn. Neuropsych*. **22**: 480–494.

Martin, A., and Caramazza, A. (eds.) 2003. *The Organization of Conceptual Knowledge in the Brain*. Philadelphia, PA: Psychology Press.

Martin, A., Ungerleider, L. G., and Haxby, J. V., 2000. Category specificity and the brain: the sensory/motor model of semantic representations of objects. In M. S. Gazzaniga (ed.) *The New Cognitive Neurosciences*. Cambridge, MA: MIT Press, pp. 1023–1036.

Monahen, T., and Wee, L. (eds.) 1999. *Grammatical Semantics*. Stanford, CA: CLSI.

Neininger, B., and Pulvermüller, F., 2003. Word-category specific deficits after lesions in the right hemisphere. *Neuropsychologia* **41**: 53–70.

Oh, K., 2003. Language, cognition, and development: motion events in English and Korean. Ph. D. dissertation, University of California, Berkeley, CA.

Pinker, S., 1989. *Learnability and Cognition*. Cambridge, MA: MIT Press.

2003. Language as an adaptation to the cognitive niche. In M. H. Christiansen and S. Kirby (eds.) *Language Evolution*. Oxford, UK: Oxford University Press, pp. 16–37.

Pinker, S., and Jackendoff, R., 2005. The faculty of language: what's special about it? *Cognition* **95**: 201–236.

Puce, A., and Perrett, D., 2003. Electrophysiology and brain imaging of biological motion. In C. Frith and D. Wolpert (eds.) *The Neuroscience of Social Interaction*. Oxford, UK: Oxford University Press, pp. 1–22.

Pulvermüller, F., Shtyrov, Y., and Ilmoniemi, R., in press-a. Brain signatures of meaning access in action word recognition. *J. Cogn. Neurosci.*

Pulvermüller, F., Hauk, O., Nikulin, V., and Ilmoniemi, R., in press-b. Functional links between motor and language systems. *Eur. J. Neurosci.*

Raos, V., Evangeliou, M. N., and Savaki, H. E., 2004. Observation of action: grasping with the mind's hand. *Neuroimage*, **23**: 193–201.

Rappaport Hovrav, M., and Levin, B., 1998. Building verb meanings. In M. Butt and W. Geuder (eds.) *The Projection of Arguments*. Stanford, CA : CSLI, pp. 97–134.

Rizzolatti, G., and Luppino, G., 2001. The cortical motor system. *Neuron* **31**: 889–901.

Rizzolatti, G., and Matelli, M., 2003. Two different streams form the dorsal visual system: anatomy and functions. *Exp. Brain Res.* **153**: 146–157.

Rizzolatti, G., Camarda, R., Fogassi, L., *et al.*, 1988. Functional organization of inferior area 6 in the macaque monkey. II. Area F5 and the control of distal movements. *Exp. Brain Res.* **71**: 491–507.

Rizzolatti, G., Fadiga, L., Gallese, V., and Fogassi, L., 1996a. Premotor cortex and the recognition of motor actions. *Cogn. Brain Res.* **3**: 131–141.

Rizzolatti, G., Fadiga, L., Matelli, M., *et al.*, 1996b. Localization of grasp representations in humans by PET. I. Observation vs. execution. *Exp. Brain Res.* **111**: 246–252.

Rizzolatti, G., Fogassi, L., and Gallese, V., 2000. Cortical mechanisms subserving object grasping and action recognition: a new view on the corticala motor functions. In M. S. Gazzaniga (ed.) *The New Cognitive Neurosciences*. Cambridge, MA: MIT Press, pp. 539–552.

Saygin, A. P., Wilson, S. M., Hagler, D. J., Bates, E., and Sereno, M. I., 2004a. Point-light biological motion perception activates human premotor cortex. *J. Neurosci.* **24**: 6181–6188.

Saygin, A. P., Wilson, S. M., Dronkers, N. F., and Bates, E., 2004b. Action comprehension in aphasia: linguistic and non-linguistic deficits and their lesion correlates. *Neuropsychologia*, **42**: 1788–1804.

Senior, C., Barnes, J., Giampietro, V., *et al.*, 2000. The functional neuroanatomy of implicit motion perception or "representational momentum." *Curr. Biol.* **10**: 16–22.

Simmons, K., and Barsalou, L. W., 2003. The similarity-in-topography principle: reconciling theories of conceptual deficits. *Cogn. Neuropsych.* **20**: 451–486.

Slobin, D. I., 2000. Verbalized events: a dynamic approach to linguistic relativity and determinism. In S. Niemeier and R. Dirven (eds.) *Evidence for Linguistic Relativity*. Amsterdam, Netherlands: Benjamin, pp. 107–138.

2003. Language and thought online: cognitive consequences of linguistic relativity. In D. Gentner and S. Goldin-Meadow (eds.) *Language in Mind: Advances in the Study of Language and Thought*. Cambridge, MA: MIT Press, pp. 157–192.

Talmy, L., 1985. Lexicalization patterns: semantic structure in lexical forms. In T. Shopen (ed.) *Language Typology and Syntactic Description*, vol. 3. Cambridge, UK: Cambridge University Press, pp. 136–149.

Tettamanti, M., Buccino, G., Saccuman, M. C., *et al.*, 2005. Listening to action-related sentences activates fronto-parietal motor circuits. *J. Cogn. Neurosci.* **17**: 273–281.

Tomasello, M., 1999. *The Cultural Origins of Human Cognition*. Cambridge, MA: Harvard University Press.

　　2000. The item-based nature of children's early syntactic development. *Trends Cogn. Sci.* **4**: 156–163.

　　2003a. *Constructing a Language*. Cambridge, MA: Harvard University Press.

　　2003b. On the different origins of symbols and grammar. In M. H. Christiansen and S. Kirby (eds.) *Language Evolution*. Oxford, UK: Oxford University Press, pp. 94–110.

　　(in press). Understanding and sharing intentions: the origins of cultural cognition. *Behav. Brain Sci.*

Tranel, D., and Kemmerer, D., 2004. Neuroanatomical correlates of locative prepositions. *Cogn. Neuropsychol.* **21**: 719–749.

Tranel, D., Damasio, H., and Damasio, A. R., 1997. A neural basis for the retrieval of conceptual knowledge. *Neuropsychologia* **35**: 1319–1327.

Tranel, D., Damasio, H., Eichhorn, G., et al., 2003a. Neural correlates of naming animals from their characteristic sounds. *Neuropsychologia* **41**: 847–854.

Tranel, D., Kemmerer, D., Adolphs, R., Damasio, H., and Damasio, A. N., 2003b. Neural correlates of conceptual knowledge of actions. *Cogn. Neuropsychol.* **20**: 409–432.

Umiltà, M. A., Kohler, E., Gallese, V., et al., 2001. "I know what you are doing": a neurophysiological study. *Neuron*, **32**: 91–101.

Van Valin, R., and LaPolla, R., 1997. *Syntax*. Cambridge, UK: Cambridge University Press.

Wheaton, K. J., Thompson, J. C., Syngeniotis, A., Abbott, D. F., and Puce, A., 2004. Viewing the motion of human body parts activates different regions of premotor, temporal, and parietal cortex. *Neuroimage* **22**: 277–288.

Wierzbicka, A., 1988. *The Semantics of Grammar*. Amsterdam, Netherlands: Benjamin.

　　1998. The semantics of English causative constructions, in a universal-typological perspective. In M. Tomasello (ed.) *The New Psychology of Language*, vol. 1. Mahwah, NJ: Lawrence Erlbaum, pp. 113–153.

Wu, D. H., Morganti, A., and Chatterjee, A., 2004. Neural substrates of the path and manner of movement. *Hum. Brain Map. Abstracts*, Poster No. TU 332.

11

Language evidence for changes in a Theory of Mind

Andrew S. Gordon

11.1 Introduction

One topic that is strikingly pervasive across the cognitive sciences is that of Theory of Mind, referring to the abilities that people have in reasoning about their own mental states and those of others. It is the set of Theory of Mind abilities that enable people to reflect introspectively on their own reasoning, to empathize with other people by imagining what it would be like to be in their position, and to generate reasonable expectations and inferences about mental states and processes.

Although there are inherent difficulties involved in investigating behavior that is largely unobservable, a relatively sophisticated understanding of Theory of Mind abilities has emerged through the synthesis of widely disparate sources of evidence. This evidence suggests that Theory of Mind abilities progressively develop in children and adults (Happé *et al.*, 1998; Wellman and Lagattuata, 2000), are degraded in people diagnosed with the illness of autism (Baron-Cohen, 2000), have a relationship to localized brain regions (Happé *et al.*, 1999; Frith and Frith, 2000), and are a uniquely human cognitive faculty not available to other primates, e.g. chimpanzees and orangutans (Call and Tomasello, 1999). This last contribution to our understanding of Theory of Mind suggests that these abilities must have arisen in the human lineage only after a split from that of chimpanzees some 6–8 million years ago (but see Stanford, this volume, for a review of dissenting opinions).

Although it may be reasonable to assume that Theory of Mind abilities emerged in humans through a combination of natural evolution and cultural evolution, the relative importance that one ascribes to either of these two forces can radically change one's conception of the mental lives of humans that are contemporary on a genetic timescale and early on a cultural timescale. If genetic evolution is the prime contributor to human Theory of Mind abilities, then we could imagine that human beings tens of thousands of years ago reflected introspectively on their own reasoning, empathized with other humans they had contact with, and were able to generate expectations and inferences about mental

Action to Language via the Mirror Neuron System, ed. Michael A. Arbib. Published by Cambridge University Press. © Cambridge University Press 2006.

states and processes. If instead culture is seen as the prime contributor, then human beings at some stage perhaps tens of thousands of years ago would not have been capable of these behaviors.

In attempting to determine whether the emergence of Theory of Mind abilities is genetic or cultural, researchers are immediately faced with the problem of evidence. Drawing comparisons between Theory of Mind abilities is an extremely difficult task, even between people who participate in controlled psychological experiments, let alone across cultures separated by distance and/or time. Several researchers have thought that the strongest evidence for a cultural Theory of Mind would be the discovery of significantly different Theory of Mind abilities across contemporary cultures. Lillard's (1998) review of research in cultural variation in Theory of Mind suggests that meaningful variations may exist among the peoples of the world, but argues that there is little evidence available to draw firm conclusions, and that the methodologies employed in the past to study mental representations in other cultures have been problematic.

A second type of evidence for a cultural Theory of Mind has looked for significant variation within a culture across time. One of the more provocative of these historical analyses was that of Julian Jaynes (1976) in support of his ideas on the emergence of consciousness. In this work, Jaynes examines references to psychological concepts as they appear in a variety of early narratives, including the *Iliad* and the *Odyssey*. By comparing how these texts and others use terms such as *thumos, phrenes, kradie, etor, noos*, and *psyche*, Jaynes advances his claim that there was a shift in the way the people of ancient Greece thought about consciousness.

Although Jaynes successfully argues that there was a shift in the way that mental concepts were referenced in these early texts, one could argue that these changes can be attributed solely to changes in linguistic convention, rather than to changes in the underlying semantics of the language or the cognitive abilities that are based on these representations. For this criticism, Jaynes makes the controversial retort, "Let no one think these are *just* word changes. Word changes are concept changes and concept changes are behavioral changes." (p.292, original emphasis) While this comment raises a much larger philosophical debate concerning mental representation, arguments made on either side would be improved if the evidence were stronger. Among other points, it is tempting to argue that the changes exhibited in Jaynes' sampling of early texts are not representative of the cultural environment in which these texts were produced.

A third type of evidence for a cultural Theory of Mind has considered whether the development of Theory of Mind reasoning abilities in young children is predicated on the acquisition of culturally shared concepts. By one account (Bartsch and Wellman, 1995), children's Theory of Mind reasoning abilities are significantly the product of conceptual changes that occur during development, as opposed to a maturational change in cognitive function (e.g., Scholl and Leslie, 2001). The success of conceptual change accounts of children's developing Theory of Mind abilities open the door for arguments in favor of the cultural acquisition of a Theory of Mind, where children's Theory of Mind concepts are developed through the experiences that they have in rich social and linguistic environments. One strategy for investigating this possibility is to look directly at the linguistic environment and productions of young children as they are acquiring competency in Theory of Mind tasks. A strong relationship between language acquisition and Theory of Mind reasoning abilities would further support arguments in favor of a cultural Theory of Mind.

This chapter explores the role that language evidence can play in answering questions concerning the cultural development of a Theory of Mind, and its implications for how we are to conceive of the mental lives of early humans. The primary tool that is used to explore these issues is that of automated corpus analysis, where computer programs are constructed to gather statistics about the linguistic elements found in large collections of electronic text documents. Automated techniques for corpus analysis enable researchers to study text collections that are larger than could be reasonably processed by human analysts, but the trade-off is that effort must be directed toward the development of the tool. The next section describes an effort aimed at constructing corpus analysis tools capable of identifying references to Theory of Mind concepts in English text in a robust manner, and the application of these tools toward investigations of the cultural development of a Theory of Mind.

11.2 The Theory of Mind in language

Everyday language-use is rich with words and phrases that refer to Theory of Mind concepts. Examples of these references include the open-class English words *belief, assumption, intend, prefer, inadvertently, carefully, reminiscent*, and *passionate*, as well as common phrases such as to *get one's head around something*, and to *come to terms with something*. While some of these examples go beyond topics that have been traditionally associated with Theory of Mind studies (often restricted to notions of belief and intent), each are related to commonsense notions of human psychology that more broadly include the faculties of memory, imagination, explanations, predictions, emotions, planning, and control, among others. In the cognitivist tradition, it is assumed that these linguistic expressions have a correspondence with non-linguistic concepts that entail their meaning, concepts that are manipulated in other non-linguistic reasoning processes. In short, these words are associated with their meaning for a speaker or a listener. Collectively, the conceptual meaning of words and phrases associated with mental states and processes contribute to a mental model of how people think, a representational Theory of Mind.

For the purpose of investigating Theory of Mind through a study of language, it would be particularly helpful if it were possible to describe the associations between language and Theory of Mind concepts on a large scale. The immediate challenge that one faces, however, is that only one side of these associations (the words and phrases that people use) can be directly observed. As for identifying the concepts that these words are associated with, the most compelling attempts have been conducted within the computer science subfield of knowledge representation.

In my own research in knowledge representation (Gordon, 2004), I have found that large-scale analytic techniques can be effectively used to identify the breadth of concepts that must participate in a representational Theory of Mind. In this work, 372 strategies were collected from ten different real-world planning domains (including governance, warfare, and artistic performance). These strategies were then encoded as preformal definitions aimed at identifying the formal knowledge representations that will be required to enable the creation of cognitive models of human strategic reasoning. Of the 988 concepts that were required to author preformal definitions of these strategies, 635 were identified as pertaining to mental states and processes. In Gordon (2002) these 635 concepts are organized into 30 representational areas, a set which stands as the most comprehensive characterization of human commonsense psychological models available today.

The 30 representational areas that organize these 635 concepts are as follows: (1) Managing knowledge, (2) Similarity comparisons, (3) Memory retrieval, (4) Emotions, (5) Explanations, (6) World envisionment, (7) Execution Envisionment, (8) Causes of failure, (9) Managing expectations, (10) Other agent reasoning, (11) Threat detection, (12) Goals, (13) Goal themes, (14) Goal management, (15) Plans, (16) Plan elements, (17) Planning modalities, (18) Planning goals, (19) Plan construction, (20) Plan adaptation, (21) Design, (22) Decisions, (23) Scheduling, (24) Monitoring, (25) Execution modalities, (26) Execution control, (27) Repetitive execution, (28) Plan following, (29) Observation of execution, and (30) Body interaction.

Using this collection of 635 concepts as a starting point, my research group began a second large-scale effort to catalog the associations that exist between English words and phrases and concepts of mental states and processes. One of our central assumptions in this work was that there were synonymous ways of referring to a single Theory of Mind concept. The associations that exist between words and phrases in the English language and Theory of Mind concepts are largely many-to-one. For example, we assumed that each of the following sentences contained phrases whose meaning is the same, namely referencing the mental event of *reminding*, one of the terms from the representational area of Memory retrieval.

(1) He *was reminded of* the time he crashed his car.
(2) The broken headlight *made him think of* when he crashed his car.
(3) Every car horn *evoked memories* of that fateful day.
(4) The constant *remindings* of the tragedy were relentless.
(5) Talk of car crashes *brought back memories*.
(6) His *memories were triggered* by the sudden car horn.

This set of 635 concepts had the necessary breadth for describing the associations between language and Theory of Mind concepts, although there remained some room for improvement. Because these concepts contained only the terms necessary to adequately define the real-world strategies that were analyzed as part of that study, some redundancy and gaps were evident. For example, the representational area of Managing Expectations (dealing with the events that people expect to happen in future states) listed the term *Expectation violation*, referring to the mental event of being surprised by something that occurs, but does not include in the list of eight terms a corresponding concept for *Expectation confirmation*, referring to the mental event of realizing that one's expectations have been met.

To elaborate these 635 concepts as well as reduce the redundancy that was apparent, we decided to use natural language as additional empirical evidence for the commonsense psychological concepts that were necessary to manipulate in formal theories. Beginning in Summer 2002 and ending in Fall 2004, we conducted a large-scale effort to identify every possible way to express in the English language all of the concepts related to the 30 representational areas. This work was completed by first identifying multiple ways to express each of the 635 concepts in the English language. These examples were then used as a launching point for large-group brainstorming sessions aimed at eliciting even more examples (typically dozens for each concept). The resulting set of examples was then organized to determine cases where the existing concept set for a representational area lacked a concept that was expressible in language (a term needed to be added) and cases where language made no distinction between two existing concepts (two concepts needed to be combined into one). Graduate students in my research group (native English speakers) then identified full sets of synonymous expressions for each of the examples using thesauri, topic dictionaries, and existing phrase books. A final set of 528 Theory of Mind concepts with corresponding examples was produced.

To encode the associations that exist between this set of concepts and the words and phrases contained in their corresponding examples, our group constructed 528 finite-state transducers that could be applied algorithmically to an electronic text document in order to recognize and tag references to each of these concepts. We utilized the Intex Corpus Processor software (Silberztein, 1999), which allowed us to author finite-state transducers as linguistic patterns using a graphical user interface. To simplify the specification of linguistic patterns, we relied heavily on a large-coverage English dictionary compiled by Blandine Courtois, allowing us to specify components of our finite-state automata at a level that generalized over noun cardinality and verb inflections. For example, a single pattern for a memory retrieval expression can be described with a finite-state automaton of four successive transitions that handles both *made him think of* and *makes her think of* by generalizing over the verb and the pronoun. This single pattern can then be grouped with those for alternate phrasings of references to the concept of memory retrieval into a single finite-state transducer for tagging this specific concept. The 528 finite-state transducers that were authored typically contained dozens of linguistic patterns each for a given concept reference, generalizing over hundreds of surface forms in the English language. When we completed the authoring process in September 2004, we combined each of these transducers into a single massive transducer that could be applied to electronic text documents for the purpose of tagging references to all recognized Theory of Mind concepts. The compiled minimal finite-state transducer contained 9556 states and 85 153 transitions.

Gordon *et al.* (2003) evaluated the quality of the finite-state transducers that are produced using this methodology according to traditional information retrieval standards. The transducers for four of the representational areas (Managing knowledge, Memory retrieval, Explanations, and Similarity comparisons) were applied to sentences judged to contain references to each of these areas, elicited from volunteers who participated in a linguistic survey. Results indicated that this approach was effective at identifying 81.51% of the expressions associated with these representational areas in English written text (recall score), and that 95.15% of the identified expressions would be judged as appropriate examples in the representational area by a human rater (precision score).

In authoring a large-scale finite-state transducer for recognizing Theory of Mind concepts in English text, we have created a new tool that may have some applicability to the investigation of cultural, developmental, and evolutionary questions surrounding the relationship between Theory of Mind and language use. This tool has its limitations, in that it does not achieve perfect levels of precision or recall even when analyzing contemporary written text. Still, the real value of this tool is that statistics of references to Theory of Mind concepts can be quickly computed from analyses of extremely large text corpora. In the remainder of this chapter, two investigations are described that take advantage of this tool in ways that inform discussion of issues related to the historical evolution of language. In the first investigation, evidence is sought for cultural changes in Theory of Mind over time by tracking references to non-conscious desires across a century of English language novels. In the second investigation, evidence is sought in transcripts of children's speech for a relationship between linguistic competency for Theory of Mind concepts and the reasoning capacity that young children acquire for false-belief tasks.

11.3 Looking for a Freudian shift

Probably the strongest argument for recent, widespread cultural change in Theory of Mind involves our shared understanding of the relationship between conscious and unconscious thought. As Whyte (1978) argues, our modern conception of unconscious thought is an invention that was conceivable around 1700, topical around 1800, and effective around 1900. While scientific discussion of the unconscious became prevalent in the latter half of the nineteenth century (e.g., Hartmann, 1869), truly widespread socialization of concepts related to the unconscious can be largely attributed to the impact of the writings of Sigmund Freud. In *The Interpretation of Dreams*, first published in German in 1900 and translated into English in 1913, Freud outlined an understanding of human cognition that involved the influence of subconscious desires in behavior. Although Freud's use of symbolism in explaining dreams is often joked about in everyday conversation, the more central idea of a subconscious desire has become a fundamental part of the way people talk about their goals. It is uncontroversial when a person reports, "I must have wanted to leave my wife all along, but I only recently became aware of this subconscious desire of mine."

In the English language, there are, in fact, many ways of expressing the Freudian notion of a subconscious desire without using this term – or any special terminology at all. As part of our work in developing finite-state transducers for recognizing Theory of Mind expressions, we explored references to this concept within the representational area of Goals (one of the 30 representational areas that we completed). The following list gives some specific examples of the general linguistic forms that our hand-authored rules are able to identify and tag as a subconscious goal:

(1) He *realized he hoped* the plan would fail.
(2) He *guessed that what he really wished for* was something different entirely.
(3) He *didn't figure out what it was that he desired* until it was too late.
(4) I'm *not sure I want* to go down that route.
(5) He *wasn't aware of what he yearned* for.
(6) It was what I *actually wanted all along*.
(7) His *subconscious wish* now came to the surface.
(8) She later learned of his *unconscious desire*.

The ability to automatically identify and collect references to subconscious desires in English text allows us to explore whether this Freudian idea is a modern addition to our understanding of mental states and processes – the product of a cultural innovation and dissemination. If so, we hypothesize that we would not be able to find in English text documents any of the linguistic patterns listed above before the period in history when this Freudian idea was disseminated, outside of the writings of Freud's immediate predecessors. If instead we view this concept as one that was already well established in the pre-Freudian mind, then we should be able to find some variations of the above expressions before this time among representative writing samples.

To explore the way that people use language related to the Freudian notion of a subconscious goal, it was necessary to construct a very large text corpus of uniform genre that spanned the relevant years in history. For this, we quickly settled on the genre of the novel as the most appropriate data source. First, electronic texts of novels (necessary for automated analysis) are readily available in the public domain[1]. Second, the mix of narration and dialogue that is found in the genre makes it likely that expression related to desires of all sorts will occur in abundance, largely because of an emphasis on character development. Third, the genre of the novel has remained remarkably constant over the course of the last several hundred years in format, aiding in the comparison of novels over time.

The main downside of the novel as a genre of analysis is that US copyright law has made it difficult to obtain electronic texts of works published after 1922. Freud's *The Interpretation of Dreams* was certainly well known and discussed among English-language writers by this date, but we imagine that the cultural impact of this work would

---

[1] Particularly from Project Gutenberg online resource: http://www.gutenberg.net

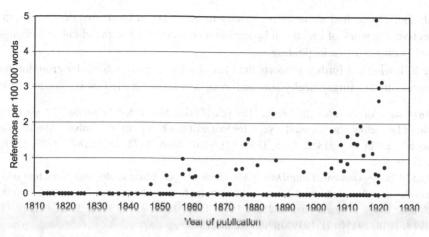

Figure 11.1 Subconscious goal references per 100 000 words in 176 English-language novels published between 1813 and 1922.

have continued to grow after this time. Still, we felt that this cut-off date for novel data was sufficiently late in history to answer the fundamental question.

We assembled a corpus of 176 English-language novels with publication dates spanning the years between 1813 and 1922. This corpus consisted primarily of American and British novels, all of which appeared on at least one "great books list" of the nineteenth and twentieth centuries and which were available electronically from Project Gutenberg. With an average of 118 000 words per novel, our corpus consisted of over 20 million words.

To analyze this corpus, we applied our hand-authored finite-state transducers for subconscious desires to each novel, collecting each sentence from the corpus that matched any of the linguistic patterns for this concept. Surprisingly, only 69 occurrences were identified.

Figure 11.1 displays the distribution of the occurrences that were identified. To normalize this data across novels of varying length, each data point on the graph indicates the number of references to subconscious goals found in a single novel divided by the word count of the novel, plotted as references per 100 000 words. While this normalization permits us to compare the frequency of references in a more valid manner, the relatively constant size of novels in our corpus yields a normalization that has little effect on the shape of the graph.

The evidence from this corpus supports neither of our original two hypotheses. There is certainly a shift in the frequency of references to subconscious goals within our data set, but this shift is not an immediate one following the widespread cultural dissemination of Sigmund Freud's ideas. Instead, we find a gradual shift beginning in the middle of the nineteenth century that continues to grow into the twentieth century through the last of our data points. We believe this evidence argues for a "Pre-Freudian Shift" in the mental models that people had about their capacity to be aware of their own desires. From this perspective, the work of Freud can be seen in a context of widespread cultural change, as an effect rather than an instigating force.

The following list further supports the idea of a Pre-Freudian Shift by grouping the 69 references into ten linguistically similar forms, and showing the first occurrence of each.

(1) 1814 Jane Austen, *Mansfield Park*: "Her year!" cried Mrs. Price; "*I am sure I hope* I shall be rid of her before she has staid a year, for that will not be up till November." Similar usage is found later in 1847, 1852, 1853, 1857(×2), 1861, 1868, 1872, 1876, 1877, 1886, 1887(×2), and 1905.

(2) 1857 Charles Dickens, *Little Dorrit*: Mr. Clennam got it him to do, and gives him odd jobs besides in at the Works next door – makes 'em for him, in short, when *he knows he wants* 'em. Similar usage is found later in 1857, 1877, 1878, 1886, 1898, 1906, 1908(×2), 1910(×2), 1913, 1914, 1915(×2), 1919, 1920(×4), and 1922.

(3) 1859 Charles Dickens, *A Tale of Two Cities*: He'd never have no good of it; *he'd want all along* to be out of the line, if he, could see his way out, being once in – even if it wos so. Similar usage is found later in 1903 and 1914.

(4) 1860 George Eliot, *The Mill on the Floss*: Maggie, in her brown frock, with her eyes reddened and her heavy hair pushed back, looking from the bed where her father lay to the dull walls of this sad chamber which was the centre of her world, was a creature full of eager, passionate longings for all that was beautiful and glad; thirsty for all knowledge; with an ear straining after dreamy music that died away and would not come near to her; with a blind, *unconscious yearning* for something that would link together the wonderful impressions of this mysterious life, and give her soul a sense of home in it. Similar usage is found later in 1874 (Thomas Hardy's *Far from the Madding Crowd*).

(5) 1881 Henry James, *The Portrait of a Lady*: "It's not only that," said Isabel; "but *I'm not sure I wish* to marry any one."

(6) 1896 Thomas Hardy, *Jude the Obscure*: "I don't care to go into them," she replied evasively. "I make a very good living, and *I don't know that I want* your company." Similar usage is found later in 1903, 1905, 1909, 1910, 1912(×3), 1913(×2), 1914, 1915, 1919(×4), 1920(×3), and 1921(×2).

(7) 1913 D. H. Lawrence, *Sons and Lovers*: "*You are sure you want* me?" he asked, as if a cold shadow had come over him.

(8) 1914 Theodore Dreiser, *The Titan*: "Oh, I don't know," replied Cowperwood, easily; "*I guess I want* you as much as ever." Similar usage is found later in 1918 (Willa Cather's *My Antonia*).

(9) 1917 Edith Wharton, *Summer*: "What's all this about wanting?" he said as she paused. "Do *you know what you really want*?"

(10) 1920 James Joyce, *Ulysses*: . . . ah yes *I know I hope* the old press doesn't creak ah I knew it would . . .

The first form in this list, "I am sure I hope," accounts for most instances in the latter half of the nineteenth century and could be viewed as a verbose way of expressing "I hope." Henry James' negation of this phrase in 1881 begins to connote a sense of doubt about one's desires, and gives rise to the twentieth-century usage first seen in 1896, "I don't know that I want." This trend becomes fully developed in 1913 when the certainty of someone else's desires is questioned, and in 1914 when one's own desires must be guessed. Most striking in this list is the direct references to *unconscious yearning* by George Eliot (pseudonym of Mary Anne Evans) in 1860 and similarly to *unconscious desire* by Thomas Hardy in 1874: "There was no necessity for any continuance of speech, and the fact that she did add more seemed to proceed from an *unconscious desire* to show unconcern by making a remark, which is noticeable in the ingenuous when they are acting by stealth."

The appearance of a Pre-Freudian Shift in the way that people refer to the idea of a subconscious goal does not, in itself, prove that Theory of Mind abilities are the result of cultural development. Nor does this shift provide direct evidence for a change in the abilities that nineteenth- and twentieth-century English-speaking people had in reasoning about their own goals or the goals of other people. The evidence that is presented is simply that people have changed the way they refer to goals in English over this period in history. The argument for the cultural development of a Theory of Mind, however, is that this shift is indicative of a significant cultural change in the mental models of psychology that people hold – one that now includes the concept of a subconscious goal. If we believe that our Theory of Mind abilities are predicated on tacit representational models of mental states and processes, then we would expect that some change in our Theory of Mind abilities might have also occurred. If so, it could be viewed as just one recent change in a long history of cultural innovations that would have included the more fundamental commonsense psychology concepts of memories, beliefs, emotions, goals, and plans, among others that would not have participated in the mental lives of early man.

11.4 Looking for a false-belief shift

Within the research area of developmental psychology, Theory of Mind has been studied as a set of cognitive abilities that progressively emerge in children. A prevalent experimental instrument for studying children's Theory of Mind abilities is the false-belief task. In a standard version of this task (Wimmer and Perner, 1983), the child is introduced to two characters, Maxi and his mother. Maxi places an object of interest into a cupboard, and then leaves the scene. While he is away, his mother removes the object from the cupboard and places it in a drawer. The child is then asked to predict where Maxi will look for the object when he returns to the scene. Success on this task has been criticized as neither entirely dependent on Theory of Mind abilities nor broadly representative of them (Bloom and German, 2000); however, its utility has been in reliably demonstrating a developmental shift. Wellman *et al.* (2001) analyzed 178 separate studies that employed a version of this task, finding that 3-year-olds will consistently fail this task on the majority of trials by indicating that Maxi will look for the object in the location to which his mother has moved it; 4-year-olds will succeed on half the trials, while 5-year-olds will succeed on the majority of trials. Call and Tomasello (1999) demonstrate that these results are consistent across verbal and non-verbal versions of this task.

Children's developing performance on the false-belief task is particularly interesting when considered within the larger debate concerning maturation and conceptual change in cognitive development. Like every other cognitive ability that emerges in childhood, performance on Theory of Mind tasks is likely due to a complex combination of maturing innate abilities and conceptual knowledge learned through experience. Still, understanding the relative importance of these two factors significantly impacts how we are to conceive of the mental lives of early humans that are contemporary on a genetic time-scale, but early on a cultural timescale. If maturational factors are the prime contributors to children's developing Theory of Mind abilities, then we can argue that these abilities would have been a well-established part of the mental lives of early humans tens of thousands of years ago and earlier. If instead it is the acquisition of conceptual knowledge (a representational Theory of Mind) that is the prime contributing factor (e.g., Bartsch and Wellman, 1995), then the possibility exists that enculturation might play some important role, and that early humans would have had impoverished Theory of Mind abilities.

One approach to investigating this issue is to look for evidence of the acquisition of a representational Theory of Mind in the language that children use in everyday conversation. The contemporary cognitivist view of natural language understanding and generation presupposes that the meaning of verbal expressions are representational in nature, and that these underlying representations are the same ones that would be manipulated for the purposes of inference (e.g., explanation and prediction). By tracking the production of children's speech that references Theory of Mind concepts, we can look for some correlation between linguistic competency with these concepts and emerging Theory of Mind abilities. Of particular interest is children's acquisition of a linguistic competency for concepts related to knowledge and beliefs, as they are the most related to the false-belief task. By examining the correspondence between these linguistic competencies and the ages in which children acquire cognitive competencies in Theory of Mind tasks, our aim is to provide an additional point of evidence that can be used in arguing for or against the competing developmental models.

Among those 528 finite-state transducers that we authored for recognizing expressions related to Theory of Mind concepts are 37 related to concepts specifically dealing with knowledge and belief, including assumptions, contradictions, justifications, logical consequences, truth, falsehood, and the mental processes associated with these commonsense psychological entities. An evaluation described in Gordon *et al.* (2003) indicates that these finite-state transducers are effective at identifying 83.92% of the expressions related to knowledge and belief in English written text (recall score), and that 92.15% of the tagged expressions would be judged as correct by a human rater (precision score).

Examples of the expressions that are tagged with these 37 concepts are as follows: He's got a logical mind (*managing-knowledge-ability*). She's very gullible (*bias-toward-belief*). He's skeptical by nature (*bias-toward-disbelief*). It is the truth (*true*). That is completely false (*false*). We need to know whether it is true or false (*truth-value*). His claim was bizarre (*proposition*). I believe what you are saying (*belief*). I didn't know about that (*unknown*). I used to think like you do (*revealed-incorrect-belief*). The assumption was widespread (*assumption*). There is no reason to think that (*unjustified-proposition*). There is some evidence you are right (*partially-justified-proposition*). The fact is well established (*justified-proposition*). As a rule, students are generally bright (*inference*). The conclusion could not be otherwise (*consequence*). What was the reason for your suspicion (*justification*)? That isn't a good reason (*poor-justification*). Your argument is circular (*circular-justification*). One of these things must be false (*contradiction*). His wisdom is vast (*knowledge*). He knew all about history (*knowledge-domain*). I know something about plumbing (*partial-knowledge-domain*). He's got a lot of real-world experience (*world-knowledge*). He understands the theory behind it (*world-model-knowledge*). That is just common sense (*shared-knowledge*). I'm willing to believe that (*add-belief*). I stopped believing it after a while (*remove-belief*). I assumed you were coming (*add-assumption*). You can't make that assumption here (*remove-assumption*). Let's see what follows from that (*check-inferences*). Disregard the consequences of the assumption (*ignore-inference*). I tried not to think about it (*suppress-inferences*). I concluded that one of them must be wrong (*realize-contradiction*). I realized he must have been there (*realize*). I can't think straight (*knowledge-management-failure*). It just confirms what I knew all along (*reaffirm-belief*).

As a corpus of analysis, we utilized the CHILDES database of children's speech (MacWhinney, 2000), a collection of transcripts from a wide variety of psycholinguistic studies conducted largely in the 1980s. Specifically, we analyzed the transcripts from the 42 research studies that contributed data of normally developing monolingual English-learning children. To facilitate the analysis of this data set according to the age of the children, individual files were generated containing only the transcripts of speech produced by a single child for each of the transcript files (a total of 3001 individual files). The total number of words in each file was calculated and the age of the child (in months) was recorded. There were 3 347 340 words transcribed in these files from children ranging in age from 11 to 87 months.

Figure 11.2 presents a histogram of the number of words in the files associated with each age of the children. The notable spike that appears in this figure is due to a large data set that exists within the CHILDES database contributed from a study by Hall *et al.* (1984). There, the groups of children were collectively identified only as being between the ages of 54 and 60 months without differentiation, so all of this data set was used for evidence at the low end of this range. More significantly, Fig. 11.2 reveals that comparatively few data exist within the CHILDES corpus of normally developing monolingual English-learning children after the age of 5 years (60 months). Although the available data should allow for the observation of some interesting trends throughout the age range of the corpus, some caution is necessary when drawing strong conclusions about children older than 60 months.

In order to enable comparisons between children and adults, each of the analyses were also conducted on the CALLHOME American English Speech data collected and transcribed by the Linguistic Data Consortium (1997). The CALLHOME database

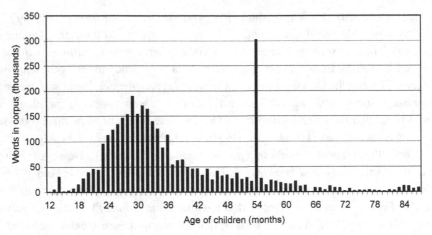

Figure 11.2 Number of words in corpus by age of children in months.

consists of transcripts of 120 unscripted telephone conversations (302 083 words) between native speakers of English, where the callers average 38.875 ($\sigma = 16.14$) years of age.

Given these text corpora, two analyses were conducted. Both of these analyses involved the use of tag frequency as data points. To compute tag frequency, the 37 finite-state transducers described early were applied to the corpora in order to find references to concepts related to knowledge and belief. The number of tagged expressions was then divided by the number of words that were searched to compute each frequency data point. In the first analysis, the frequency of all expressions of the 37 concepts related to knowledge and belief were tabulated for each of the data sets. In the second analysis, the frequencies of expressions related to each individual concept (of the 37 total) were tabulated.

No attempt was made to filter the results of the application of the finite-state transducers to improve precision, and no evaluation was conducted to estimate the recall rate on these corpora. However, after reviewing the resulting tags we believe that the precision and recall scores obtained on these corpora are only marginally less than was achieved in the evaluation of written text tagging conducted by Gordon *et al.* (2003) using the same set of local grammars.

11.4.1 Frequency of all expressions related to knowledge

The first analysis that we conducted was to apply all of the local grammars for the 37 concepts related to knowledge and belief to each of the data files corresponding to children of different ages. In all, there were 18 283 tags produced through the application of these local grammars, with only 19 of the 37 tags appearing in the data. Nearly half of these tags were for the concept of a justified proposition (9113 tags), while the remaining

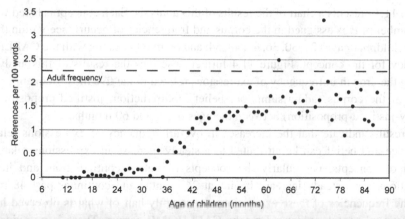

Figure 11.3 Frequency of references to all concepts related to knowledge and belief by age of children.

half consisted of belief (3150 tags), contradictions (3485 tags), and partially justified propositions (1483 tags).

Applying the full set of local grammars to the CALLHOME data set produced 6775 tags, yielding a frequency of 2.24 references per 100 words of speech. Of the 37 tags, 21 were assigned to these data, with the highest frequencies going to the concepts of justified proposition (3172 tags), contradiction (1551 tags), belief (1000 tags), and partially justified proposition (493 tags).

Figure 11.3 presents a graph of the frequency per 100 words of speech for all expressions related to the concepts of knowledge and belief based on the age of the children (in months) of the analyzed data. As a point of comparison, the frequency for the CALL-HOME data (2.24) is also indicated on the graph as a dashed horizontal line. The data on the graph can be described by the linear function $y = 0.0281x - 0.3914$, where the correlation statistic (r^2) is 0.7021.

The results indicate that expressions related to knowledge and belief do not appear at the beginning of children's speech production, but increase in frequency in a strongly linear manner from 30 months (2.5 years) until 48 months (4 years), when the frequencies of these expressions are roughly half of what is observed in adult conversational speech.

11.4.2 Frequency of expressions of individual concepts

In the second analysis, we individually applied each of the 19 local grammars that produced at least one tag in the corpus to each of the transcript data for children of different ages. The primary purpose of this analysis was to track the relative increase in frequency for each concept over the developmental period where a change in Theory of Mind abilities is evident (between 36 and 60 months of age).

Table 11.1 presents a chart of the results of this analysis. Each concept is listed with the total number of tags assigned in the corpus and frequencies of occurrence within the data sets for children of ages 24, 30, 36, 42, 48, 54, and 60 months, along with the CALLHOME frequency for the concept. Figure 11.4 further describes the results of this analysis by charting the growth in frequency of expressions related to the five most frequent concepts tagged in the corpus (add-assumption, belief, contradiction, justified-proposition, and partially-justified-proposition) between the ages of 24 and 60 months.

The results indicate that the increases in overall frequency of expressions related to knowledge and belief can be attributed to a steady increase in expressions related to a handful of concepts, particularly the concepts of belief, contradiction, and justified-propositions. This steady increase begins at 24 months and continues past 48 months, when the frequencies of these expressions are roughly half of what is observed in adult conversational speech evidenced by the CALLHOME corpus. There is no evidence of any qualitative change in the sorts of concepts related to knowledge and belief that are expressed by children of different ages.

11.4.3 Absence of a false-belief shift

The overall purpose of this study was to determine the relationship between linguistic competency in the production of expressions related to knowledge and belief and children's developing Theory of Mind abilities, particularly during the age range where children acquire competency on the false-belief task (between 3 and 5 years in age). In this section, we will consider the results of our analysis with respect to this purpose.

First, there is no evidence to suggest a qualitative change in the frequencies that children express concepts related to knowledge and belief between the ages of 3 and 5. Looking first only at the frequency of all expressions related to knowledge and belief we see that children between the ages of 3 and 5 are continuing a steady increase in frequency that started at the beginning of their speech production. The sparse data that we have for children older than 60 months suggests that this gradual increase begins to level off after this point. If we had seen a non-linear shift in the frequencies of expression between 3 and 5, then an argument could have been made relating linguistic competency to Theory of Mind abilities. Finding no such shift, one could reasonably infer that the developing linguistic competencies that children have for expressions related to knowledge and belief are not strongly tied to their reasoning abilities in Theory of Mind tasks.

Second, there is little evidence to suggest a qualitative change in the concepts that children express related to knowledge and belief between the ages of 3 and 5. Looking at the individual frequencies for each of the 19 concept tags that were assigned to the corpus we see that the handful of concepts that account for the vast majority of tags increase in frequency at a constant rate from the very beginning of children's speech production. Very few expressions appear in these data related to other concepts that appear with slightly higher frequencies in adult discourse, and there is no evidence that linguistic competency is acquired for these concepts during this period of time either. If we had seen

Table 11.1 *Frequency of expressions of individual concepts related to knowledge and belief (tags per 10 000 words)*

Conceptual tag	Total tags	24 mo.	30 mo.	36 mo.	42 mo.	48 mo.	54 mo.	60 mo.	Adult
add-assumption	499	0.27	0.65	1.34	3.31	0.91	4.22	2.59	5.16
assumption	22	0	0	0	0	0	0.17	0	0.07
belief	3150	0.72	3.21	3.67	17.67	18.71	12.39	24.63	33.1
bias-toward-disbelief	26	0	0	0	0	0	0.17	0.65	1.22
check-inferences	4	0	0	0	0	0	0	0.65	0.23
consequence	1	0	0	0	0	0	0	0	0.26
contradiction	3485	0	0	0	22.09	29.27	30.43	27.22	51.34
false	69	0	0	0	1.1	0.6	0.37	0.65	0.26
ignore-inference	1	0	0	0	0	0	0.03	0	0
justified-proposition	9113	2.07	13.02	28.35	51.46	45.87	61.42	73.23	105
knowledge	5	0.09	0	0	0	0	0	0	0.53
managing-knowledge	26	0	0	0	0	0	0.33	0	0.36
partial-domain-knowledge	23	0	0	0	0.22	0	0.2	0	0.6
partially-justified-proposition	1483	0.27	0	0	6.85	6.04	7.77	8.43	16.32
reaffirm-belief	19	0	0.26	0.09	0	0	0.13	0	0.36
realize	229	0.18	0.13	0.09	0.22	1.81	1.3	2.59	5.63
true	67	0	0.07	0	0	0	0.47	0	2.02
unjustified-proposition	51	0	0	0	1.1	0	0.07	0	0.93
world-model-knowledge	10	0	0	0	0	0	0	0	0.73

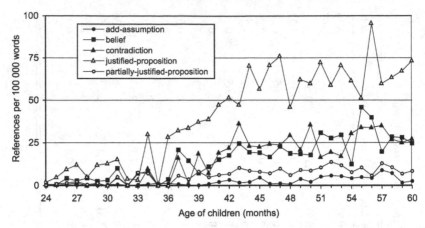

Figure 11.4 Frequency of references to the five most frequent individual concepts by age of children.

a change in the concepts that were being expressed between the ages of 3 and 5, then a different argument could have been made relating linguistic competency to Theory of Mind abilities. Finding no such change one could again reasonably infer the lack of a direct relationship between language-use and acquired reasoning abilities.

Together these two points argue against a strong relationship between linguistic competencies for expressions related to knowledge and belief and children's developing Theory of Mind abilities. This argument is particularly important in evaluating cognitive models that assume that Theory of Mind abilities and language abilities are enabled by representational mental models of the same type. If we assume that the sophistication of children's representational theories of knowledge and belief is closely related to the way that children express these concepts in language, then there is little evidence to suggest that these representational theories change at all between the ages of 3 and 5, when competency on the false-belief task develops. Accepting this assumption, the evidence in this section would argue against any strong conceptual change account of Theory of Mind abilities where competency on the false-belief task is due solely to the acquisition of more sophisticated representational mental models. This evidence would argue instead for a maturational account, where competency on the false-belief task can be attributed to the development of new cognitive abilities for taking the perspective of other people or in the monitoring of one's own mental state between the ages of 3 and 5. One strong counterargument that could be made against this line of reasoning concerns the differences in the linguistic competencies between language production and language understanding. In analyzing transcript data consisting of words uttered by children, this study can make no claims regarding the linguistic competency that these children might have for understanding expressions related to knowledge and belief during the relevant periods of development.

11.5 Discussion

An inherent problem in the study of language evolution is that it will never be possible to examine the way early humans first used language. Instead, the best that we can do is interpret the evidence that can be gleaned from examining contemporary language use and the written record. In this chapter, it is argued that automated corpus analysis techniques can be applied to large collections of contemporary language use in order to inform debate on language evolution. In the two sets of analyses that are described, two broad claims can be put forth as interpretations of this data, namely:

(a) People in a culture change the way they talk about mind over time.
(b) The way that people talk about the mind is not strongly tied to their Theory of Mind reasoning abilities.

The first of these two claims is evidenced in the way that English-speaking people (specifically novelists) changed the way that they referenced unconscious goals during the middle part of the nineteenth century. The "Pre-Freudian Shift" represents a significant change in the way that English-speaking people talked about the mind over time. The introduction of expressions of non-conscious goals into narrative text is likely only one example of a long history of additional concepts that can be referenced by the language that a culture employs.

The second of these two claims is evidenced by the way that English-learning children speak about knowledge and belief during a developmental stage where a dramatic shift in Theory of Mind reasoning abilities is apparent (performance on false-belief tasks). Children begin expressing several concepts related to knowledge and belief from the very beginning of language production, and steadily increase the frequencies of these expressions through-out childhood without adding expressions related to other knowledge and belief concepts to their repertoire to any substantial degree. The sorts of Theory of Mind reasoning tasks that they can complete do not appear to be strongly tied to the way they talk about the mind.

When paired together, these two claims present two interesting and competing possibilities, either:

(c) People's Theory of Mind reasoning abilities are subject to cultural changes that are not strongly tied to the way people talk about the mind.
    or:
(d) Theory of Mind reasoning abilities are not subject to cultural changes.

The implications for the mental lives of early humans are as follows. If the first of these last two hypotheses is correct, then we would not expect the Theory of Mind reasoning abilities of early humans to be strongly tied to the evolution of language. It would be reasonable to suppose that other cultural factors (changing population densities, the development of clans and chiefdoms, the invention of agriculture) may have played a more significant role in the development of a representational Theory of Mind, enabling new reasoning abilities. Insomuch as language was available to early humans during the cultural development of Theory of Mind abilities, it played a supporting rather than primary role. Still, elaborating how language could support emerging reasoning abilities among groups of early humans (as well as in child development) remains an important research direction.

If the second of these hypotheses is correct, then we would expect that Theory of Mind reasoning abilities in humans change only on a genetic timescale. These abilities would have been available to early humans throughout the evolution of language, and likely would have played an important role in enabling language use in the first place. Accordingly, it would be possible that early language use included language behavior that require speakers to reflect introspectively on their own reasoning, to empathize with other people by imagining what it would be like to be in their position, or to generate reasonable expectations and inferences about mental states and processes. It remains to understand the cultural factors that lead to changes in the way that people talk about the mind over time, even as their underlying Theory of Mind reasoning capabilities remain constant.

Acknowledgments

Several students at the University of Southern California contributed to research efforts described in this chapter, including Ryan Kashfian, Abe Kazemzadeh, Ashish Mestry, Anish Nair, Milena Petrova, and Reid Swanson. The project or effort depicted was or is sponsored by the US Army Research, Development, and Engineering Command (RDECOM), and the content or information does not necessarily reflect the position or the policy of the Government, and no official endorsement should be inferred.

References

Baron-Cohen, S., 2000. Theory of mind and autism: a fifteen year review. In S. Baron-Cohen, H. Tager-Flusberg, and D. Cohen (eds.) *Understanding Other Minds: Perspectives from Developmental Cognitive Neuroscience*, 2nd edn. Oxford, UK: Oxford University Press, pp. 3–20.

Bartsch, K., and Wellman, H., 1995. *Children Talk about the Mind*. New York: Oxford University Press.

Bloom, P., and German, T., 2000. Two reasons to abandon the false belief task as a test of theory of mind. *Cognition* **77**: B25–B31.

Call, J., and Tomasello, M., 1999. A non-verbal false-belief task: the performance of children and great apes. *Child Devel.* **70**: 381–395.

Frith, C., and Frith, U., 2000. The physiological basis of theory of mind: functional neuroimaging studies. In S. Baron-Cohen, H. Tager-Flusberg, and D. Cohen (eds.) *Understanding Other Minds: Perspectives from Developmental Cognitive Neuroscience*, 2nd edn. Oxford, UK: Oxford University Press, pp. 334–356.

Gordon, A., 2002. The theory of mind in strategy representations. In W. Gray and C. Schunn (eds.) *Proceedings of the 24th Annual Conference of the Cognitive Science Society*, Mahwah, NJ: Lawrence Erlbaum, pp. 375–380.

2004. *Strategy Representation: An Analysis of Planning Knowledge*. Mahwah, NJ: Lawrence Erlbaum.

Gordon, A., Kazemzadeh, A., Nair, A., and Petrova, M., 2003. Recognizing expressions of commonsense psychology in English text. *Proceedings 41st Annual Meeting of the Association for Computational Linguistics*, Sapporo, Japan, pp. 208–215.

Hall, W., Nagy, W., and Linn, R., 1984. *Spoken Words: Effects of Situation and Social Group on Oral Word Usage and Frequency*. Mahwah, NJ: Lawrence Erlbaum.

Happé, F., Brownell, H., and Winner, E., 1998. The getting of wisdom: theory of mind in old age. *Devel. Psychol.* **34**: 358–362.

1999. Acquired theory of mind impairments following stroke. *Cognition* **70**: 211–240.

Hartmann, E., 1869. *Philosophy of the Unconscious: Speculative Results according to the Inductive Method of Physical Science*. Republished 1931, New York: Harcourt Brace.

Jaynes, J., 1976. *The Origin of Consciousness in the Breakdown of the Bicameral Mind*. Boston, MA: Mariner.

Lillard, A., 1998. Ethnopsychologies: cultural variations in theories of mind. *Psychol. Bull.* **123**: 3–32.

Linguistic Data Consortium, 1997. *CALLHOME American English Transcripts*, LDC catalog no. LDC97T14. Available at http://www.ldc.upenn.edu

MacWhinney, B., 2000. *The CHILDES Project: Tools for Analyzing Talk*, vols. 1 and 2. Mahwah, NJ: Lawrence Erlbaum.

Scholl, B., and Leslie, A., 2001. Minds, modules, and meta-analysis. *Child Devel.* **72**: 696–701.

Silberztein, M., 1999. Text indexing with INTEX. *Comput Humanities* **33**: 265–280.

Wellman, H. M., and Lagattuta, K. H., 2000. Developing understandings of mind. In S. Baron-Cohen, H. Tager-Flusberg and D. Cohen (eds.) *Understanding Other Minds: Perspectives from Developmental Cognitive Neuroscience*, 2nd edn. Oxford, UK: Oxford University Press, pp. 21–49.

Wellman, H., Cross, D., and Watson, J., 2001. Meta-analysis of theory-of-mind development: the truth about false belief. *Child Devel.* **72**: 655–684.

Whyte, L., 1978. *The Unconscious before Freud*. New York: St. Martin's Press.

Wimmer, H., and Perner, J., 1983. Beliefs about beliefs: representation and constraining function of wrong beliefs in young children's understanding of deception. *Cognition* **13**: 103–128.

Part V

Development of action and language

12

The development of grasping and the mirror system

Erhan Oztop, Michael A. Arbib, and Nina Bradley

12.1 Introduction: a mirror system perspective on grasp development

Neonates and young infants are innately compelled to move their arms, the range of possible spontaneous movements being biologically constrained by anatomy, environmental forces, and social opportunity. Over the first 9 postnatal months, reaching movements are transformed as infants establish an array of goal-directed behaviors, master basic sensorimotor skills to act on those goals, and acquire sufficient knowledge of interesting objects to preplan goal-directed grasping. In monkeys, it appears that the neural circuit for control of grasping also functions to understand the manual actions of other primates and humans (Arbib, Chapter 1, this volume). Within the grasp circuitry, "mirror neuron" activity encodes both the manual actions executed by the monkey and the observed goal-directed actions of others. Recent imaging studies on humans indicate that a mirror neuron network may exist in humans linking observation and execution functions. However, the link between grasp development and mirror system development is widely unexplored. To address this, we will build models based both on behavioral data concerning the course of development of reaching in human infants and on neurophysiological data concerning mirror neurons and related circuitry in macaque monkeys.

In humans, the foundation for reaching may begin as early as 10–15 weeks of fetal development when fetuses make hand contact with the face and exhibit preferential sucking of the right thumb (de Vries *et al.*, 1982; Hepper *et al.*, 1991). Soon after birth, neonates lie quietly on their backs and preferentially place their hands within their visual field (van der Meer *et al.*, 1995). By 2–3 postnatal months infants enthusiastically fling their arms and direct their hands in paths influenced by the direction of movement of objects (von Hofsten, 1991), and by 3–4 months, they acquire sufficient skill to aim and make object contact (von Hofsten and Rönnqvist, 1988).

Though vision informs and increasingly biases limb trajectories, it will take another 2–5 months of experience to implement visually informed feedforward control of the

Action to Language via the Mirror Neuron System, ed. Michael A. Arbib. Published by Cambridge University Press.
© Cambridge University Press 2006.

hand to preshape it to grasp a desired object. In fact, during this interim period infants do not appear to require vision of their hands for initiating reaches or contacting and grasping an object. Nor do they appear to use vision to guide hand trajectory or to orient the hand toward the object prior to initial contact. At 4–5 months, reaches are as good with vision available during the reach as when vision is removed after onset of the reach (Clifton et al., 1993). The apparent lack of visual guidance during a reach has been interpreted as indicating that younger infants initiate reaching using a ballistic strategy to aim the hand toward the target and then employ tactile-based feedback to make corrective movements for grasping once the object is contacted (Lasky, 1977; Lockman et al., 1984; Wishart et al., 1978).

Infants use a ballistic reaching strategy well into childhood, but beyond 5 months of age, infants begin to attend to visual information during reaching. By 5 months of age infants adjust their gaze, anticipating the future trajectory of an object (van der Meer et al., 1994). Given that infants begin looking at their hands early in postnatal develop-ment, the question arises as to why it takes them approximately 9 months to acquire visually guided reaching. We favor a motor learning view of this developmental delay, learning that refines an inherent action-recognition system supporting visually guided reaching. In support of this view, initially random, spontaneous neonatal arm movements are rapidly transformed. Two-month-old infants reliably engage their hands in mutual fingering when both hands are at midline, and they manipulate or play with objects put in their hands (Bayley, 1969). Play and manipulation are fundamental to discovery and exploration of hand space for establishing a rudimentary array of biomechanically feasible reach configurations such as those stored in the wrist rotation layer of the Infant Learning to Grasp Model (ILGM) presented later in this chapter. Play and manipulation will also lead to development of attention to manipulated items, a precursor to extracting object affordances.[1] By 3 months of age, if a hand contacts a glowing or sounding object presented in the dark, infants attempt to grasp it (Clifton et al., 1993). Although infants at 5 months take little notice of the hand during flight, by 6 to 7 months, blocking vision of the hand as infants reach for a virtual (mirror image) object frequently disrupts or impairs reach performance (Lasky, 1977; Wishart et al., 1978). By 9 months, infants can correct errors in reach trajectory prior to object contact (von Hofsten, 1979) and orient the hand to match the object's orientation.

However, application of a ballistic–feedback dichotomy during reaching in adults has been questioned, and we note the issues briefly here as a challenge for future develop-mental analysis. Arbib and Hoff (1994) proposed a model of adult reaching that predicts the internal representation of a reach is sufficient to direct the hand towards a visible object with little error so that it *appears* ballistic. Further, their modeling of perturbation experiments supported the interpretation that error monitoring in adults is always "on"

---

[1] We recall from Chapter 1 that a successful grasp requires visual analysis of the *affordances* (opportunities for grasping) of the object to guide the selection of an appropriate grasp by the (pre)motor system. Our developmental account must then address not only the acquisition of a repertoire of grasps but also the development of the visual skill to recognize affordances and the visuomotor coordination that links this recognition to the selection and control of movement.

and discrepancies between trajectory and target will be compensated for in about 100 ms. Veering angles, curves in hand path directed toward an object, occurring within a movement unit formed by one acceleration and one deceleration, also suggest continuous updating occurs within a reach plan, though some have suggested veering in infant reaching is a byproduct of the mechanics of the developing infant limb rather than a cortical plan (Fetters and Todd, 1987). Mathew and Cook (1990) reported veering occurred independent of speed valleys, the deceleration of one movement unit and acceleration of the next unit, as early as 6 months of age. This veering motivates us to ask whether internal reach representations have achieved a level of refinement by 6 months that in some way facilitates emergence of on-line monitoring and emergence of supination/pronation corrections in hand orientation at 75% of reach distance by 7 months of age (McCarty *et al.*, 2001).

But how does the child acquire the ability to grasp objects appropriately? We propose that a positive reward stimulus generated by the tactile feedback of a successful grasp, which we refer to as *Joy of Grasping*, motivates infants to explore and learn actions that lead to grasp-like experiences. Sporns and Edelman (1993) have proposed a similar evaluation mechanism for sensorimotor learning which they refer to as the *adaptive value* of an action. Presumably the infant requires a substantial number of grasp experiences before she can knowingly select the subset of motor commands for reaching most compatible with online visual information on the state of the arm, hand, and object. We propose that the amount of time the object is held in the hand increases with the stability of the grasp and thus provides the reinforcement signal for a system which extracts from the vast space of possible handshapes those that have yielded successful grasping. We will say more about this evaluation function when we discuss ILGM below.

In contrast to a motor learning focus such as that above, delays in fine motor skills like reach to grasp have to varying degrees also been ascribed to the postnatal myelination of corticospinal pathways (McGraw, 1945; Lawrence and Hopkins, 1976; Olivier *et al.*, 1997; Forrsberg, 1998). While these may play some role in the refinement of reach to grasp skills, all motor skills are indeed dependent on the dynamic interactions of multiple neurological, biomechanical, environmental, and social contexts of movement. Thus delays in any contributing system can delay skill development (Thelen and Ulrich, 1991). For example, the delayed acquisition of binocular coordination for visual depth perception may delay not only the development of reach to grasp skills (von Hofsten, 1990) but also the ability to assess an object's affordances for grasp planning. Rate-limiting visuomotor control may extend the amount of reach and grasp practice needed before attention can be effectively directed to determine object affordances. Delays in visual development may in turn favor greater weighting of tactile/manipulative modalities during initial postnatal months. This modality bias may motivate infants to adapt grasp configuration and one/two-hand reaching upon manual contact with an object (Newell *et al.*, 1989, 1993; Siddiqui, 1995). Further, primary dependence on a power grasp during the first postnatal year, though affording greater probability of successfully securing an object, is likely to constrain manipulative configurations for tactile exploration of object affordances. *Joy of Grasping* may contribute a counter dynamic to ensure that manual exploration is not so constrained by successful power grasping as to limit discovery of new manual possibilities and object affordances for grasping. As the frequency and duration of positive reward signals increase, *Joy of Grasping* may also motivate infants to modify their grasp to explore new bimanual possibilities, for shortly after infants become adept at bimanually securing a dangling ring around 4 months of age, they begin to release the grasp of one hand and transfer the object between hands (Bayley, 1969). This might be accomplished by reward signal content that conveys the risk of losing control of the object is low during select actions. Bimanual transfers of objects offer new self-motivated experiences that may increasingly direct the infant's attention to select attributes of the object and increase affordance recognition. However, the study of bimanual coordination is beyond the reach of this chapter.

With this we turn to the issue of the existence and development of a mirror system for grasping in humans. Imaging studies indicate that a mirror neuron network may exist in humans linking observation and execution functions (e.g., Grafton *et al.*, 1996; Rizzolatti *et al.*, 1996; Hari *et al.*, 1998; Nishitani and Hari, 2000; Buccino *et al.*, 2001). However, the link between grasp development and development of a possible mirror system has remained empirically unexplored. Pursuit of this link is critical because it may shed light on imitation, pretend play and pantomime understanding, as the ancestral mirror system for grasping might be the evolutionary precursor of the brain areas that implement these functions. Study of early reaching behavior in the infant may present a unique window of access for identification and comparative study of the mirror neuron system akin to the search for evidence of a locomotor pattern generator in studies of infant stepping (Bradley, 2003).

There are several reasons to propose that the Mirror Neuron System (MNS) is not fully developed before the end of the first postnatal year in humans.[2] The basic assumption behind this reasoning is that mirror neuron activity and grasp execution in the human are likely to be mediated by overlapping circuits of neurons involved in both execution and observation, as found in monkey. Thus, any delays in grasp circuitry development are likely to include delays in shared circuitry. During the first postnatal year infants largely employ a power grasp during reaches, though they are physically able to match hand configuration to an object after contact as early as 5 months of age (Newell *et al.*, 1993), and rudimentary finger pincer skills begin to emerge between 6 and 12 months (Bayley, 1969). The apparent dependence on a power grip suggests that visual assessment of object affordances plays a minimal role in execution of the grasp before 9 months (Newell *et al.*, 1993). We note that infants lack the ability to dampen self-generated perturbations during

---

[2] In the monkey, initial work on the F5 mirror neuron system emphasized a vision-based system that serves the analysis of hand and object interaction during "standard" precision and power grasps. More recent research has shown that some of these neurons are auditorily driven (Keysers *et al.*, 2003; Kohler *et al.*, 2002), while exploration of the adjacent orofacial area of F5 has revealed mirror-like neurons associated with orofacial movements that may bridge between ingestion and communication (Ferrari *et al.*, 2003). More to the point here is that these studies have included skills such as breaking peanuts and tearing paper that suggest that the mirror system is plastic, thus suggesting that its involvement in even "basic" grasps may involve learning rather than the expression of an innate repertoire.

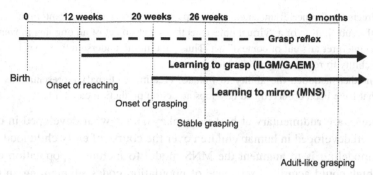

Figure 12.1 The developmental course of grasp learning and mirror neuron formation. GAEM, Grasp Affordance Emergence Model; ILGM, Infant Learning to Grasp Model; MNS, Mirror Neuron System.

reach to grasp, many of which can actually drag the arm away from an object (Thelen *et al.*, 1993; Corbetta *et al.*, 2000). As a result, the quality of all sensory inputs, including vision, may be so noisy during early months that the action recognition system cannot effectively extrapolate reliable associations, thus delaying maturation of the mirror neuron system. We propose that a ready-to-learn manipulation/mirror neuron system is gradually elaborated, and that it may take the first postnatal year for infants to acquire sufficient experiences before they can reliably discriminate among hand manipulation options.

More specifically, we propose that the system develops in four stages:

1. The system initially provides the basic visual frame for reaching toward objects (we present a classic model proposed by Kuperstein).
2. The system gradually develops a repertoire of successful grasps (we present ILGM for this).
3. The system comes to recognize those object affordances associated with successful grasps and develops the mapping from affordances to grasps (we present the Grasp Affordance Emergence Model (GAEM) for this process).
4. The system establishes associations between the infant's object–grasp repertoire and the corresponding visual repertoire of the object–hand trajectories. We will present the MNS model for this: "canonical" neurons[3] encoding grasps already in the infant's repertoire provide training signals for mirror neurons to learn to recognize these grasps from the sight of appropriate trajectories relating hand and object. The developmental course that is proposed by the four stages for the human (and macaque) infant is summarized in Fig. 12.1.

We assume that stages (1) through (4) are much the same in monkey and human, and that they ground two further stages:

5. The human infant (with maturation of visual acuity) gains the ability to map another individual's actions into his internal motor representation. This is possible when the actions are represented in

---

[3] Canonical neurons are the grasp-related neurons of premotor area F5 of the macaque which do *not* have the mirror property; i.e., they are active during object vision but not for action observation per se.

a goal-directed reference frame, i.e., with respect to a target object. Pantomime recognition requires the substitution of a "virtual object" as the target object of an ongoing movement in an attempt to interpret an "out of context" act. Thus pantomime understanding goes beyond mirror neurons' ability to recognize goal directed actions.

6. The infant then acquires the ability to imitate, creating (internal) representations for novel actions that have been observed and develops an action prediction capability.

These stages are rudimentary at best in monkeys, somewhat developed in chimpanzees, and well developed in human children over the course of early childhood. A target for future modeling is to augment the MNS model to include a population of mirror neurons which could acquire a sequence of population codes segmenting an observed action not yet in the repertoire of self-actions, then in stage (6) the sequence of mirror neuron activity would provide training for the canonical neurons, reversing the information flow seen in the MNS model. We note that this raises the question whether human infants come to recognize movements completely exclusive of their current repertoire (e.g., movements outside their anatomical abilities) using the mirror system or a separate system possibly implemented along the ventral pathway. In this case, the cumulative development of action recognition may proceed to increase the breadth and subtlety of the range of actions that are recognizable but cannot be performed by children. Presumably, the ILGM mechanisms (2) remain operative to continually refine our skills via proprioceptive and somatosensory experience, but rather than acquiring the repertoire through trial-and-error as the infant does, the older child and adult can use observation to achieve what Arbib (2002; Chapter 1, this volume) calls *complex imitation* – gaining the first approximation of a new skill by recognizing it as an assemblage of variants of a known action (cf. Byrne, 2003). More practice or less may be required to hone this approximation both by increasing the skilled perceptuomotor integration within each part of the motion and by mastering the graceful phasing in and out of the components.

12.2 Acquiring the visuomotor frame

Human fetuses initiate hand-to-face contact by 10 weeks of gestation (de Vries *et al.*, 1982) and within days after birth presentation of a visual stimulus triggers reaching and hand trajectories anchored around the stimulus. The directness of neonatal reaching triggered by a visual stimulus has been interpreted as evidence of a genetically established neural coordinate system that links the trajectories of the hand with the face and direction of gaze toward visually fixated objects within reach distance (von Hofsten, 1982). For example, in a study of 3-day-old newborns supported in a reclining infant seat, presentation of a stimulus moving in an arc 12 cm from the eyes yielded forward-extending arm movements that were aimed much closer to the target while infants visually fixated it compared to arm movements while they looked elsewhere or closed their eyes (von Hofsten, 1982). Further, observation that infants overcome weighted perturbations to maintain view of their hands when lying supine has been interpreted as evidence that infants purposefully move their arms within visual range to construct a reference frame for action (van der Meer *et al.*, 1995).

Kuperstein (1988) developed a model which addresses the problem of visuomotor association. A neural network monitors the tensions of the extraocular eye muscles during foveation of the tip of the hand. Using a distributed representation of muscle length information, an association is built up between the activity of the extraocular muscles and that of the muscles of the arm in its postural state. Later, when an object of interest is foveated, the correct arm muscle tensions are recalled in order to position the arm to grasp the target. The point of the computational model is that problems in stereo vision and arm kinematics can be solved through a single computational stage, without a priori knowledge of the geometry of the system. The solution is stored in an associative-like memory, built through experience. Note that issues in arm dynamics and trajectory generation are not addressed; only the terminal posture is learned. In other words the trajectory through which the arm is brought to the target posture and the applied forces are not specified by this model of learning.

12.3 The Infant Learning to Grasp Model (ILGM)

We have recently presented a computational model of infant grasp learning (ILGM) based on the initial open-loop reaching strategies, subsequent feedback adjustments, and emergent feedforward strategies employed by infants in the first postnatal year (Oztop *et al.*, 2004) (Fig. 12.2). The work on ILGM assumes that the initial stage of grasp acquisition is common to monkey and human infants. The model draws upon the literature (sampled in Section 12.1) examining how human infants develop grasping skills. The ILGM is anchored in the knowledge that from the beginning the infant begins discovering through exploratory behavior its own possibilities for action in the environment and the affordances of objects (the "affordances of objects" in this context refers to the set of grasps that can be applied to secure an object). By 2–3 months, infants are exploring their bodies, touching themselves, staring at their hands, and babbling (Bayley, 1969). Infants quickly progress from a crude reaching ability at birth to finer reaching and grasping abilities around 4 months of age. Infants learn to overcome problems associated with reaching and grasping by interactive searching (von Hofsten and Ronnqvist, 1993; Berthier *et al.*, 1996). To grasp successfully infants have to learn how to control their arms and to match the abilities of their limbs with affordances presented by the environment (Thelen, 2000). At first, the poorly controlled arm, trunk, and postural movements make it very difficult for the young infant to generate consistent feedback to form stable links between perceptual and motor schemas. Nonetheless, the accumulation of experiences eventually yields a well-established set of grasps, including the precision grip, with preshaping to visual affordances by 12–18 months of age (Berthier *et al.*, 1999).

The model assumes initial reaches are opportunistic, a function of the neonate's innate bias to move, and that movements are inherently variable, while also biologically constrained. The model proposes that sensory feedback arising from successful object

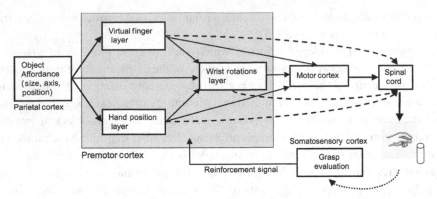

Figure 12.2 The structure of the Infant Learning to Grasp Model (ILGM) of Oztop *et al.* (2004). The individual layers inside the gray box are trained based on a *Joy of Grasping* reinforcement signal arising from somatosensory feedback concerning the stability of the current grasp.

contact generates a positive motivating signal, *Joy of Grasping*, which nurtures further exploratory reaches to grasp. The requirements of the model are that infants will (1) execute a reflex grasp when triggered by contact of the inside of the hand with an object encountered while reaching, and can (2) sense the effects of their actions and (3) use feedback to adjust their movement planning parameters. The model proposes a process by which infants may learn to generate grasps to match an object's affordances.

The ILGM is a systems-level model based on the broad organization of primate visuomotor control with visual features extracted by the parietal cortex and used by the premotor cortex to generate high-level motor signals that drive lower motor centers thereby generating movement. The feedback arising from object contact is used to modify grasp generation mechanisms within the premotor cortex. The model consists of four modules: the input module (parietal cortex), the grasp learning module (premotor cortex), the movement generation module (motor cortex and spinal cord), and the grasp evaluation module (somatosensory cortex). In addition, the grasp learning module contains three key computational layers: *Virtual Finger, Hand Position*, and *Wrist Rotations.*

While the behavior of the ILGM is designed to learn grasping in a manner consistent with data from human infants, the layers of the model are associated with general brain areas based on monkey neurophysiological data. The input module is located in the parietal lobe and it functions to extract object affordances that it then relays to the grasp learning module. The grasp learning module is assumed to be located in premotor cortex, the cortical area implicated in grasp programming. The computational layers of the grasp learning module are based on the *Preshape, Approach Vector* and *Orient* grasping schemas proposed by Iberall and Arbib (1990). These layers encode a minimal set of kinematic parameters specifying basic grasp actions. Thus, the grasp learning module formulates a grasp plan and instructs the movement generation module, located in both the spinal cord and motor cortex. The movement generation module completes task execution (Jeannerod *et al.*, 1995). The sensory stimuli generated by the execution of the plan are then integrated by the movement evaluation module, located in the primary somatosensory cortex. Output of the somatosensory cortex, the reinforcement signal, is used to adapt the connection strength between parietal–premotor connections (Fig. 12.2). Implementation of the reinforcement signal is defined simply in terms of grasp stability: a grasp attempt that misses the target or yields an inappropriate object contact produces a negative reinforcement signal. For the sake of simplicity, the brain areas that relay the output of the somatosensory cortex to the parietal and premotor areas have not been modeled. However it is known that primate orbitofrontal cortex receives multimodal sensory signals (conveying both pleasant and non-pleasant sensations), including touch from the somatosensory cortex (Rolls, 2004). Furthermore there is evidence that orbito-frontal cortex may be involved in stimulus-reinforcement learning, affecting behavior through the basal ganglia (Rolls, 2004).

ILGM emulates infant learning at various levels of (limited) capability for affordance extraction. At one extreme it can learn to generate movements based on having no affordance information, capturing the probability distribution of grasp plans afforded by the given context without any information on the presented object other than its location. At the other, it can incorporate some of the affordances of a presented object (e.g., the orientation), and produce grasp plan distributions as a function of the encoded affordance. Whichever the situation, ILGM will generate exploratory movements as well as movements based on its internal parameters. The exploration is necessary for learning, yielding neural structures that can guide behavior satisfactorily – unless and until changed circumstances neccesitate further learning.

During early visually elicited reaching while infants are reaching towards visual or auditory targets, they explore the space around the object and occasionally touch the object (Clifton *et al.*, 1993). ILGM models the process of grasp learning starting from this stage. We represent infants' early reaches using an object-centered reference frame and include random disturbances during initial reach trials by ILGM so that the space around the object is explored. We posit that the *Hand Position* layer specifies the hand approach direction relative to the object, restricting the angles from which the object can be grasped and touched on the basis, e.g., of the location of the object relative to the infant's shoulder.

Wrist orientation depends crucially on where the object is located relative to the shoulder as well as on how the reach is directed towards the object. Thus, the task of ILGM is to discover which orientations and approach directions are appropriate for a given object at a certain location, emphasizing that both an object's intrinsic properties and also its location create a context in which the grasp can be planned and executed. Thus, the *Wrist Rotation* layer learns the possible wrist orientations given the approach direction specified by the *Hand Position* layer. The *Wrist Rotation* layer also receives projections from the *Affordance* layer because, in general, different objects afford different paired sets of approach-direction and wrist-rotation solutions. The parameters generated by this layer determine the movements of hand extension–flexion, hand supination–pronation, and ulnar and radial deviation.

The *Virtual Finger* layer indicates which fingers will move together as a unit given an input (Arbib *et al.*, 1985). This layer's functionality is fully utilized in adult grasping but in the current simulations is engaged only in learning the synergistic enclosure rate of the hand. This layer can account for the developmental transition from synergistic finger coordination during grasping (Lantz *et al.*, 1996) to selective digit control for matching hand shape to object shape (Newell *et al.*, 1993). In the simulations presented by Oztop *et al.* (2004) the output of the *Virtual Finger* layer is a scalar value determining the rate of enclosure of the fingers during the transport phase. However, this does not restrain the model from reproducing infant behavior and generating testable predictions.

For human infants, the advent of voluntary grasping of objects is preceded by several weeks in which the infant engages in arm movements and fisted swipes in the presence of visible objects (von Hofsten and Fazel-Zandy, 1984). An infant, once contacting an object, will occasionally try to grasp it. Within a few weeks, infants acquire the ability to sculpt hand movements in the presence of a palmar stimulus – cutaneous inputs initially aid or increase the probability in securing grasp of an object, such as the red ring used in Bayley Motor Scales (Bayley, 1969), and infants come to selectively use them to assist their movements. By around 7 months infants are able to stabilize the grasp (Clifton *et al.*, 1993). However, infants do not readily demonstrate control over fractionated finger movements before the end of the first year. Newell *et al.* (1989) identified rudimentary handshaping after contact starting at 4 to 6 months, whereas 7- to 8-month-olds did not appear to need contact to initiate shaping. Von Hofsten and Ronnqvist (1988) found that children would start shaping the hand midreach by 9–13 months. It appears that in early infancy the fractionated control of fingers is mainly driven by somatosensory feedback. Newell *et al.* (1989) found that the older infants' visually programmed and younger infants' haptically adjusted grasp configurations are very similar. This strongly suggests that the earlier haptic phase grasp configurations serve as target configurations (i.e., the teaching pattern in supervised learning terminology) to be associated with visual representation of the objects mediating visual grasp planning in the infant brain.

Such data constrained the design of ILGM, and enabled us to evaluate its relevance to infant learning through explicit comparisons. The model interacts with its environment (plans and executes grasp actions) and observes the consequences of its actions (grasp feedback) and modifies its internal parameters (corresponding to neural connections) in such a way that certain patterns (grasp plans) are selected and refined amongst many other possibilities.

Simulations with the ILGM model explained the development of units with properties similar to F5 canonical neurons. Examples are shown in Fig. 12.3. The important point is that ILGM discovered the appropriate way to orient the hand towards the object as well as the disposition of the fingers to grasp it. Such results provide computational support for the following three hypotheses:

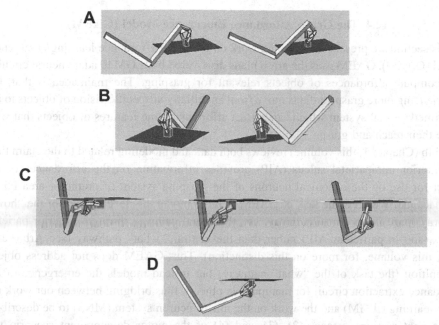

Figure 12.3 (A, B) With an object placed atop a table, ILGM learned a "menu" of precision grips with the common property that the wrist was placed well above the object. The orientation of the hand and the contact points on the object showed some variability. In addition to precision grips where the thumb opposes more than one finger (A), ILGM also acquired precision grips where the thumb opposes only the index finger (B). (C, D) Underarm grasping was learned for an object placed underneath a table. Most of the learned grasps were precision grips (C). Side or lateral grips (thumb opposing the side of the index finger) were also observed, as in panel D. The wrist rotations were determined implicitly by the location of the target object. The larger wrist flexion found in A and B was due to the fact the table afforded more constrained reaching from the top compared to reaching from the bottom. (Adapted from Oztop *et al.*, 2004.)

(H1) In the early postnatal period infants acquire the skill to orient their hand towards a target, rather than innately possessing it.

(H2) Infants are able to acquire grasping skills before they develop an elaborate adult-like object visual analysis capability.

(H3) Task constraints due to environmental context (or the action opportunities afforded by the environment) are factors shaping infant grasp development.

To this we add a fourth hypothesis, to be explored in the next section.

(H4) Grasping performance in the absence of object affordance knowledge, "affordance-off stage", mediates the development of a visual affordance extraction circuit.

Of course, ILGM says nothing about the mirror system – rather it shows how the infant brain may acquire the basic repertoire of grasps that "gets the mirror neurons started" along the lines delineated in the MNS model to be described below.

12.4 The Grasp Affordance Emergence Model (GAEM)

In this section we present our ongoing work on the grasp affordance learning/emergence model (GAEM). GAEM uses the grasp plans discovered by ILGM to adapt neural circuits that compute affordances of objects relevant for grasping. The main idea is that, by experiencing more grasps, infants move from an initially unspecific vision of objects to an "informed" visual system which can extract affordances, the features of objects that will guide their reach and grasp.

Arbib (Chapter 1, this volume) reviews both data and modeling related to the claim that the anterior intraparietal sulcus (AIP) provides information on the affordances of an object for use by the canonical neurons of the grasping system of premotor area F5 in the macaque. We view the task of GAEM to be to model the development of the "how" pathway (part of the dorsal pathway, i.e., the visual pathway through posterior parietal cortex, and in particular AIP) rather than the ventral, "what" pathway (see Arbib and Bota, this volume, for more on this distinction). Thus GAEM does not address object recognition (the task of the "what" pathway) but instead models the emergence of an affordance extraction circuit for manipulable objects, thus bridging between our work on grasp learning (ILGM) and the work on the mirror neuron system (MNS) to be described in the next section – stages (2), (3), and (4) of the motor development scenario we presented in the first section.

During infant motor development, it is likely that ILGM and GAEM learning coexists in a recursive manner. In this setting, AIP units that are being shaped by GAEM learning provide "better" affordance input for ILGM. In turn ILGM expands the class of grasp actions (learns new ways to grasp) providing more data points for GAEM learning. When this dual learning system stabilizes, GAEM and ILGM will be endowed with a set of affordance extraction (visual cortex→AIP weights) and robust grasp planning (ILGM weights) mechanisms. However note that there is no evidence to reject a *staged learning* (several iterations of ILGM adaptation followed by GAEM adaptation) in favor of the recursive learning speculated above. Here we adopt a staged learning approach solely for ease in implementing GAEM learning. In other words, in the simulations presented GAEM adaptation is based on the successful set of grasps generated by ILGM in a stabilized state (ILGM learning is complete and weights are fixed).

To model the unprocessed/unspecific raw visual input, the model represents the visual sensation of object presentation as a matrix which can be considered a simple pattern of retinal stimulation. Currently we do not model stereo vision, thus reducing computational load. The ILGM model showed how grasp plans could be represented by F5 canonical neurons without competition among the learned grasps. The GAEM model proposes that the successful grasp plans (and associated grasp configurations) represented by F5 canonical neurons could form a self-organizing map where the grasp plans are clustered (see Fig. 12.4) into sets of grasps with similar kinematics (e.g., common grip types and fixed aperture range). A noteworthy point here is that the clusters are not specified manually but are formed via self-organization. Parallel to the self-organization, a vertical

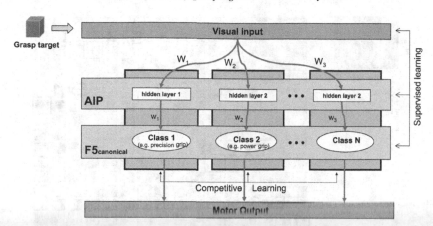

Figure 12.4 The Grasp Affordance Learning/emergence Model (GAEM) derives its training inputs from ILGM. The successful grasping movements provide the basis for learning by GAEM: vertical learning, supervised; horizontal learning, competitive. See text for details.

supervised learning takes place that aims at learning the mapping from the visual input to the cluster (see Fig. 12.4), modeling learning along the dorsal pathway, namely visual cortex → AIP → F5. Note that a single object may afford more than one grasp plan and thus each cluster forms an independent subnetwork. In this sense, the self-organization could be seen as the competition of subnetworks to learn the data point available to GAEM at a given instant.

The critical test for the model is the evaluation of the activities of the hidden units in the supervised learning network which models the AIP neural units during an object presentation. Our results to date show that the modeled AIP units in the hidden layers emergently represent features of the visual scene relevant to the grasping movements, consistent with the GAEM model. In these simulations the visual input was modeled as a 32×32 depth map centered on the target object. The motor command was modeled as the joint configuration of the hand (7 joints: 3 joints for the thumb, 2 joints each for the index and middle fingers.) The number of subnetworks was 16, and each had 16 hidden (AIP) units. For the self-organizing map, a two-dimensional mesh was used. The number of the nodes was equal to the number of subnetworks (16), and the input dimension was equal to the motor output (7). For input, simple objects with variable dimensions were used. The objects were: a rectangular prism which could change size in two dimensions; a vertical cylinder that could change in height and diameter; and a horizontal cylinder with variable height and diameter.

After GAEM learning, objects with varying dimensions were presented and resulting AIP unit responses were recorded (e.g., for the cylinders the diameter and the height were varied). Thus for a given object a two-dimensional response map was obtained for each unit. The response map is intensity-coded such that lighter areas indicate maximal

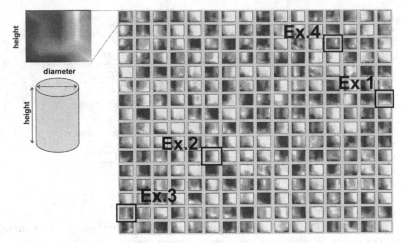

Figure 12.5 The tuning curves of the modeled AIP units to diameter and height of a vertical cylinder are shown. Each row corresponds to a subnetwork. The zoom-in illustration on the left describes the relation between cylinder features and the plot axes. In the plots, white and black colors indicate maximal response and zero response respectively.

response and darker regions indicate no response (Fig. 12.5). A vertical gradation in intensity indicates that the unit encodes the height of the cylinder (e.g., unit Ex. 1), whereas a horizontal gradation indicates diameter encoding (e.g., unit Ex. 2). There are also units which encode a range of width or height values (e.g., unit marked with Ex. 3 prefers a certain diameter range regardless of height). These types of units are rarer than the width and height encoding units. Finally there are a few units that prefer certain width and height combinations such as the one marked with Ex. 4.

12.5 The Mirror Neuron System model (MNS)

Many people view the mirror system as a social action-of-others recognition system. But what is the basis for the development of such a capability? Our view is that the primary basis for the mirror system is its use to adapt the infant's grasps relative to object affordances. There are two alternatives within this hypothesis. According to the first alternative, mirror neurons represent the *visual error* of a manipulative movement, and hence the observation of both the self and others' movement could generate similar error patterns (thus similar mirror neuron firing patterns) if suitable invariances are incorporated in the visuomanual control (visuomotor control for hand actions). According to the second alternative, the mirror neurons implement a set of *sensory predictors* (i.e., forward models) to help visuomanual control by compensating delays in the sensorimotor loop. We shall say more about forward (and inverse) models in Section 12.6.

However, the MNS model (Oztop and Arbib, 2002) to be presented here does not explicitly show the adaptive use of the mirror system. Rather it offers a notion of "hand state," trajectories linking the hand's movement to the location and affordances of the object in an object-centered representation. By engaging hand state, the infant can recognize grasps already within its own repertoire in such a way that this recognition readily generalizes to the grasps performed by others. It is this linkage between the infant's knowledge of her own repertoire and grasp actions of others that may be exploited for analyzing and understanding another's movements. Although our focus in this chapter is developmental rather than evolutionary, we note that this approach supports the view that the mirror-system-as-part-of-motor-control is the evolutionary precursor of mirror-system-for-recognition-of-other's actions.

By extension, the hypothesis implies that while an infant is learning to extract object affordances and their associations to grasping movements, she is also learning the task-specific features that will enable her to better control her own actions, and that these features in turn will help her understand the actions of others. The proposed learning will enable the retrieval of the corresponding motor representation in the mirror neuron system during an action observation, provided that the features extracted for the self-hand-object and other's-hand-object have certain invariance properties as formulated in the "hand state."

Of course, it is possible to envision an action recognition or imitation system that directly associates another's actions with internal motor representations. To do so an infant needs a binding element to set up the link between the external action and the internally represented motor program. One such link could be the *goal* of the movement (i.e., the object being grasped). However, the direct association (non-self-observation) learning case requires a high-level processing mechanism, which may require other processes such as object recognition and retrieval of past memories, processes that may not be available to the infant at early ages. In contrast, our hypothesis implies that learning from self-observation does not require excess processing to generate a signal for mediating learning. We favor the view that learning can be accomplished by using the internal signals produced during self-action, such as motor commands (corollary discharge) and cutaneous or proprioceptive sensations experienced during grasping and holding (i.e., the developmental continuation of *Joy of Grasping* signal introduced in the ILGM) to structure learning.

With this, we turn to a brief exposition of the MNS model (Oztop and Arbib, 2002) shown in Fig. 12.6. First, we look at those elements involved when the monkey itself reaches for an object. Areas IT and cIPS provide visual input concerning the nature of the observed object and the position and orientation of the object's surfaces, respectively, to AIP. The job of AIP is then to extract the affordances the object offers for grasping. The upper diagonal in Fig. 12.6 corresponds to the basic pathway AIP $\rightarrow$ F5$_{canonical}$ $\rightarrow$ M1 (primary motor cortex) of the FARS model (Arbib, Chapter 1, this volume), but Fig. 12.6 does not include the important role of prefrontal cortex in action selection.

The lower right diagonal (MIP/LIP/VIP $\rightarrow$ F4) of Fig. 12.6 completes the "canonical" portion of the MNS model, since motor cortex must not only instruct the hand muscles

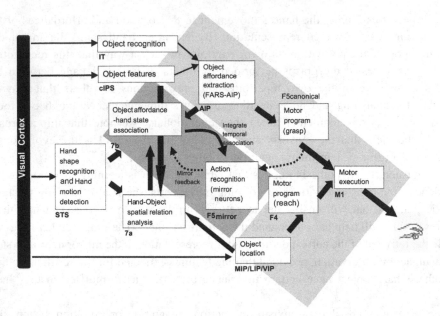

Figure 12.6 A schematic view of the Mirror Neuron System. The Mirror Neuron System (MNS) model (Oztop and Arbib, 2002) focuses on the circuitry highlighted by the gray diagonal rectangle.

how to grasp but also (via various intermediaries) the arm muscles how to reach, transporting the hand to the object.

The rest of Fig. 12.6 (the gray diagonal rectangle) presents the core elements of the mirror system. Mirror neurons do not fire when the monkey sees the hand movement or the object in isolation – it is the sight of the hand moving appropriately to grasp or otherwise manipulate a seen (or recently seen) object that is required for the mirror neurons attuned to the given action to fire. This requires schemas for the recognition of both the shape of the hand and analysis of its motion (ascribed in the figure to the superior temporal sulcus, STS), and for analysis of the relation of these hand parameters to the location and affordance of the object (7a and 7b in the figure; we identify 7b with paristal area PF in what follows).

In the MNS model, the *hand state* was accordingly defined as a vector whose components represented the movement of the wrist relative to the location of the object and of the hand shape relative to the affordances of the object. For example, one of the features captured in hand state is the orientation difference between the opposition axis for a precision pinch (vector connecting the index finger tip to the thumb tip) and the target object's grasp axis (e.g., the vector passing through the centers of the circular faces of a short cylinder). This representation is invariant, whether the action is executed by the self or by others. Oztop and Arbib (2002) showed that an artificial neural network corresponding to PF and $F5_{\text{mirror}}$ could be trained to recognize the grasp type from the *hand state trajectory*, with correct classification often achieved well

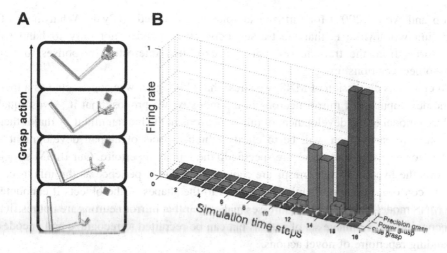

Figure 12.7 Power and precision grasp resolution. (A) The frames illustrate the movement of the hand starting from the initial configuration to the final configuration. (B) The distinctive feature of this trajectory is that the hand initially opens wide to accommodate the length of the object, but then thumb and forefinger move into position for a precision grip. Even though the model has been trained only on precision grips and power grips separately, it reflects the ambiguities of this novel trajectory – the curves for power and precision cross towards the end of the action, showing the resolution of the initial ambiguity by the network. (Adapted from Figure 14 of Oztop and Arbib, 2002.)

before the hand reached the object. The modeling assumed that the neural equivalent of a grasp in the monkey's repertoire is a pattern of activity in F5 canonical neurons that commands the grasp. During training, the grasp code produced by F5 canonical neurons for grasp execution was used as the training signal for the F5 mirror neurons, enabling them to learn which hand–object trajectories corresponded to the canonically encoded grasps. Moreover, the input to the F5 mirror neurons encodes the trajectory of the hand to the object rather than the visual appearance of the hand in the visual field. As a result of this training, the appropriate mirror neurons come to fire in response to the appropriate trajectories even when the trajectory is not accompanied by F5 canonical firing.

Training prepares the F5 mirror neurons to respond to hand–object relational trajectories even when the hand belongs to the "other" rather than the "self" because the hand state is based on the movement of a hand relative to the object. Thus F5 responses are directly related to relations between hand and object and only *indirectly* to the retinal input of seeing hand and object, views which can differ greatly between observation of self and other.

Despite the use of a non-physiological neural network, simulations with the model revealed a range of putative properties of mirror neurons that suggest new neurophysiological experiments. Figure 12.7 gives just one example; the reader is referred to

Oztop and Arbib (2002) for further examples and detailed analysis. What makes the modeling worthwhile is that the trained network responded not only to hand state trajectories from the training set, but also exhibited interesting responses to novel hand–object relationships.

To close this discussion of MNS, we stress that although it was constructed as a model of the development of mirror neurons in the monkey, we propose that it serves equally well as a model of the development of mirror neurons in the human infant. A major theme for future research, then, will be to clarify which aspects of human development are generic for primates, and which are specific to the human repertoire. Our ILGM strongly supports the hypothesis that grasps are acquired through experience as the infant learns how to conform the biomechanics of its hand to the shapes of the objects it encounters. The MNS model and the data it addresses make clear that mirror neurons are not restricted to recognition of an innate set of actions but can be recruited to recognize and encode an expanding repertoire of novel actions.

12.6 Forward and inverse models

The present section introduces inverse and forward models of motor control and shows how they have been related to the mirror system for grasping. Skipper, Nusbaum, and Small (this volume) employ these concepts in relating the mirror system to the production and perception of audiovisual speech.

Jordan and Rumelhart (1992) considered the problem of a human infant learning to pronounce words. The Neural Controller for the Motor Program (MP in Fig. 12.8) does not have direct access to the sound of the word or the articulations that produce a sound. Instead, it must go from the neural code for a heard word to the neural code for its articulation. The challenge is to do this in such a way that the sounds the child hears and the sounds it produces become recognizably similar. Let, then, X be the set of neural motor commands and Y the set of neural patterns produced by auditory inputs. Then the physical system

$$\text{Muscles} + \text{Vocal Tract} + \text{Sound Waves} + \text{Ear} + \text{Auditory System}$$

defines a function $f: X \rightarrow Y$ such that $f(x)$ is the neural code for what is heard when neural command x is sent to the articulators. (For more on these articulators, see Goldstein, Byrd, and Saltzman, this volume.)

The challenge is to train MP to yield successful imitation of an auditory target – the catch being that neither signal set X nor Y is accessible to the teacher. The answer offered by Jordan and Rumelhart (1992) is to "backpropagate through the world" to learn a "mental model" (MM) of the physics. Backpropagation (Rumelhart *et al.*, 1986) is a way of training a network across many trials so that it comes to match input patterns to desired output patterns. However, there is nothing in this discussion that rests on

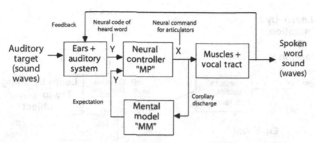

Figure 12.8 A recasting of the Jordan–Rumelhart model of imitation: train a *direct (forward) model* MM (mental model) of the neural correlates of the transformation of motor commands to auditory output, and then train an *inverse model* MP (motor program) that generate motor commands which match auditory inputs.

the training method used; as long as $X \rightarrow Y$ is a well-defined function (i.e., X uniquely determines Y) a general function approximator would do. For example the results described here have also been attained with reinforcement learning (Barto and Jordan, 1987).

- First learn the "mental model" MM, a network which provides a *direct* model of the effect of commands – MP models the "physics" $f: X \rightarrow Y$, predicting the sensory effects in Y of a wide set of commands in X.
- Once MM is built, it provides a stable reference for training MP by adjusting the total network [MP $\rightarrow$ MM] so that input sounds are reproduced fairly accurately.

MP thus comes to implement a function

$$g : Y \rightarrow X \text{ with the property } f(g(y)) \approx y$$

for each desired sensory situation y, generating the motor program for a desired response. That is to say, MP becomes an *inverse model* of the effect of commands: given the neural code for an auditory target y, MP will produce the neural code for a motor command which will generate a spoken sound that approximates y. One remark is in order here. In general, f may be non-invertible and hence g cannot be taken as simply the inverse of f. This situation is severe in the control of redundant systems (e.g., a 7 degrees-of-freedom arm). However we can continue our disposition without loss of generality since there are techniques that can deal with this situation (e.g., restricting the range of g such that g is one-to-one and for all y, $g(y) \in f^{-1}(y)$; or modeling the relation between X and Y as a joint probability distribution $p(X, Y)$).

Arbib and Rizzolatti (1997) built on Jordan and Rumelhart's (1992) account of a human infant learning to pronounce words in terms of a *forward* "mental model" MM of the effect of commands and the motor program MP viewed as an *inverse model* of the effect of commands. Fig. 12.9 presents their conceptual framework:

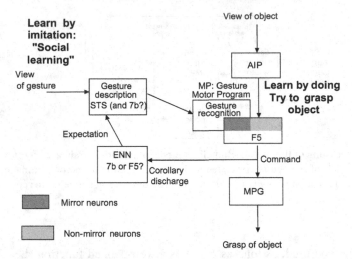

Figure 12.9 A conceptual framework analyzing the role of F5 in grasping (Arbib and Rizzolatti, 1997). The path along the right side of the page provides mechanisms for grasping a seen object. The loop on the left provides mechanisms for imitating observed gestures in such a way as to create expectations which enable the visual feedback loop to serve both for (delayed) error correction during one's own actions and for "social learning" through imitation of the actions of others.

(1) The vertical path to the right is the *execution system* encompassing "view of object" via parietal area AIP (visual recognition of affordances – possible ways to grasp an object) and F5 (motor schemas) and motor cortex to grasp an object. This pathway (and the way in which prefrontal cortex may modulate it) has been analyzed in the Fagg–Arbib–Rizzolatti–Sakata (FARS) model (Fagg and Arbib, 1998; see Arbib, Chapter 1, this volume), and ILGM (see Section 12.3) and demonstrated that learning is possible even with limited affordance input.

(2) The loop on the left provides both the *observation matching system* from "view of gesture" via gesture description (posited to be in STS) and gesture recognition (mirror neurons in F5 or area 7b) to a representation of the "command" for such a gesture, and the *expectation system* from an F5 command via the expectation neural network (ENN) to MP, the motor program for generating a given gesture. The latter path may provide visual feedback comparing "expected gesture" and "observed gesture" for monkey's self-generated movements, and also create expectations which enable the visual feedback loop to serve for learning an action through imitation of the actions of others. In fact, Arbib and Rizzolatti (1997) conflate three separate roles of the mirror system in the loop (2):

a. use of the mirror system to provide visual feedback appropriate to dexterous action;

b. ability to recognize the actions of others (cf. the MNS model, Section 12.5); and

c. ability to use that recognition as the basis for imitation.

   Arbib (Chapter 1, this volume) suggests that (a) and a simple form of (b) were already present in the common ancestor of human, chimpanzee, and monkey; that (a) and simple forms of (b) and (c) were already present in the common ancestor of human and chimpanzee; but that the extended capacity for (b) and (c) – complex action analysis and complex imitation – evolved only along the hominid line (see also the chapters by

Stanford and Greenfield, this volume). But since we are dealing here with the relation between action and language in the human brain, we may proceed with the diagram as labeled, but with F5 now interpreted loosely as being in or near Broca's area.

The integrated model of Fig. 12.9 thus relates the "grasp an object" system to the "view a gesture" system. The expectation network is driven by F5 irrespective of whether the motor command is "object-driven" (via AIP) or "gesture-driven." It thus creates expectations both for what a hand movement will look like when "object-driven" (an instrumental action directed towards a goal) or "gesture-driven" (a "social action" aimed at making a self-generated movement approximate an observed movement). The right-hand path of Fig. 12.9 exemplifies "learning by doing," which is modeled in detail with ILGM (Section 12.3). The left-hand path of Fig. 12.9 exemplifies a complementary mode of learning which creates expectations about gestures as well as exemplifying "social learning" based on imitation of gestures made by others.

Arbib and Rizzolatti (1997) discussed two main possibilities. The first is that MP is located along the path leading from STS to F5 via 7b. The reciprocal path from F5 to STS would provide the direct model (i.e., forward model), ENN. An alternative would be that both ENN and MP are located in F5. The MNS model of MP (Section 12.5) was developed on the former basis: signals from STS and 7b encoding the relation between hand and object become correlated with the F5 codes for a variety of grasps. Iacoboni *et al.* (2001; Iacoboni, 2004) analyzed human brain imaging studies to support the view that STS is responsible for the visual representation of observed actions with (i) the connections from STS to PF and onwards to the mirror cells in F5 forming an inverse model, converting this visual representation into a motor plan, and (ii) connections from mirror cells in F5 to PF and back to STS forming a forward model converting the motor plan back into a predicted visual representation (a sensory outcome of action). These views implicate the STS as the comparison buffer for the observed and the executed action. Miall (2003) has offered a brief essay on challenges which might guide new modeling studies inspired by this viewpoint.

We close by stressing that the inverse and forward models we have discussed here may in reality be seen as encompassing a whole family of inverse and forward models. Thus, in recognizing an action (as in the MNS model), we are not so much employing "the" inverse model, one of multiple inverse models best matching the observed interaction of hand and object. Rather, the system may recognize that the current action can better be viewed as a combination of actions already within the repertoire. One may relate this case (recognition of trajectories extended over time) to the issue of modular and hierarchical systems in which the learner must not only acquire appropriate models or pattern recognition systems but must also learn how to partition the input space into regions where a given model provides the best analysis of the data. Jacobs *et al.* (1991) developed learning algorithms that compute a set of posterior probabilities estimating the utility of each network module across the input space. Jordan and Jacobs (1994) extended the modular system to a hierarchical system, made links to the statistical literature on classification and regression trees and developed an Expectation-Maximization (EM) algorithm for the architecture. Haruno *et al.* (2001) also adopted a modular methodology and developed a model composed of multiple paired forward–inverse models they call the MOSAIC architecture for human motor control. The distinctive feature of the MOSAIC model is that the forward models are used independently to compute suitableness (responsibility signals) of the forward–inverse pairs for control. This is in contrast to other mixtures of expert systems where a central neural network decides which module should become responsible in a given input or a state space. The responsibility signals determine the amount of the contribution of each inverse model to the global control. The responsibility signal is computed by comparing the forward prediction and the actual sensory feedback: a good match renders the corresponding inverse model more influential in the overall control, whereas a bad match lowers the effect of the corresponding inverse model.

Some of these ideas have been applied by Oztop *et al.* (2005), developing the "sensory predictor" hypothesis of mirror neurons which views *mental simulation* as the substrate for inferring others' intentions. An observer may "guess" the intentions of an actor, and generate task-appropriate motor commands just as if it were executed in an on-line feedback control mode. The similarity of the expected sensory outcome associated with the simulated movement and the expected sensory outcome associated with the observation would provide a means for evaluating the accuracy of the guess. A similar mechanism may mediate imitation with an addition of a memory system that can store the signals generated by mirror neurons for comparison. An imitator may execute movement X so as to match the mirror signals generated when she observed the movements of the demonstrator. The latter assumes that although the mirror neurons are not tuned for the entirety of the observed action, they respond to segments of a movement generating a sequence of neural codes that could be used to replicate an observed movement.

12.7 Discussion

To summarize, mirror neurons likely provide dexterity in task execution (by either providing *sensory predictions* or *kinematic error signals*), possibly mediate understanding of other's intentions, and may support learning by imitation. According to the "sensory predictor" hypothesis of mirror neurons, *mental simulation* could be the substrate for inferring others' intentions (Oztop *et al.*, 2005; for more on the simulation theory see Gallese and Goldman, 1998; Jeannerod, 2004). An important stage in the evolution of language according to the form of the Mirror System Hypothesis presented by Arbib (Chapter 1, this volume) is pantomime, seen as the intermediate between the imitation of praxic actions and the development of protosign as a conventionalized form of manual communication. This raises the developmental issue of how children come to recognize pantomime. Since activation of mirror neurons in monkeys requires a recent view of the object or a working memory of its presence (Umiltà *et al.*, 2001), it seems that the mirror neuron system in the monkey is not sufficiently elaborated to support pantomime. Interestingly, younger children, 3 to 5 years of age, seem to have difficulty understanding pantomimes of actions towards objects when the objects are missing, and are better able to comprehend the pantomime if the demonstrator substitutes a body part for the imagined object (O'Reilly, 1995). The transitional requirement of an object may indicate that the maturing human mirror system can be driven by a recognized substitute, given understanding of the movement goal. The substitute then is no longer required once object affordances relative to the desired goal are so well learned that a "virtual object" is readily imagined. Both levels of abstraction, body substitution and virtual object, would suggest the existence of a visuomanual feedback circuit that distinguishes the evolutionary transformations between monkey and human of the mirror neuron system.

We suggest that as the infant expands her database of self-observations and associated motor commands, she begins to recognize common elements shared by her actions and those of another. Development of this recognition raises the question of whether self-observation and the establishment of a mirror system in humans are at some level also associated with the development of imitation learning and social development. Infants can be stimulated to imitate facial and manual gestures soon after birth (Meltzoff and Moore, 1977) and imitate object manipulation by 14 months of age (Meltzoff, 1988). However, consistent with the data that grounded our work on ILGM, we would view the "imitation" of the neonate as different from the "real" imitation that begins to be seen in the second year. The beginnings of this transition are perhaps seen in the observation that infants exposed to parents using American Sign Language (ASL) begin to make arm movements suggestive of ASL motions by 6 months (Petitto *et al.*, 2001), at approximately the same time infants expect to see their own hands during reaching. The concurrence of these last two developmental events suggests they may not only share common neural elements, but that these elements may be critical to the maturation of both systems, linking them for essential sensorimotor and language development. Further, given that both developmental areas are embedded in social context, this particular linkage between motor control and mirror systems may also serve a social intention recognition system motivated by a social recognition reward signal, the "joy of being involved."

References

Arbib, M. A., 2002. The mirror system, imitation, and the evolution of language. In C. Nehaniv and K. Daurenhahn (eds.) *Imitation in Animals and Artifacts.* Cambridge, MA: MIT Press, pp. 229–280.

Arbib, M. A., and Hoff, B., 1994. Trends in neural modeling for reach to grasp. In K. M. B. Bennett and U. Castiello (eds.) *Insights into the Reach to Grasp Movement.* Amsterdam, Netherlands: North-Holland, pp. 311–344.

Arbib, M., and Rizzolatti, G., 1997. Neural expectations: a possible evolutionary path from manual skills to language. *Commun. Cogn.* **29**: 393–424.

Arbib, M. A., Iberall, T., and Lyons, D., 1985. Coordinated control program for movements of the hand. *Exp. Brain Res. Suppl.* **10**: 111–129.

Barto, A. G., and Jordan M. I., 1987. Gradient following without back-propagation in layered networks. *Proceedings 1st IEEE Annual Conference on Neural Networks*, San Diego, CA, pp. II-629–II-636.

Bayley, N., 1969. *Bayley Scales of Infant Development*. New York: Psychological Corporation.

Berthier, N. E., Clifton, R. K., Gullapalli, V., McCall, D. D., and Robin, D. J., 1996. Visual information and object size in the control of reaching. *J. Mot. Behav.* **28**: 187–197.

Berthier, N. E., Clifton, R. K., McCall, D. D., and Robin, D. J., 1999. Proximodistal structure of early reaching in human infants. *Exp. Brain Res.* **127**: 259–269.

Bradley, N. S., 2003. Connecting the dots between animal and human studies of locomotion: focus on "Infants adapt their stepping to repeated trip-inducing stimuli." *J. Neurophysiol.* **90**: 2088–2089.

Buccino, G., Binkofski, F., Fink, G. R., *et al.*, 2001. Action observation activates premotor and parietal areas in a somatotopic manner: an fMRI study. *Eur. J. Neurosci.* **13**: 400–404.

Byrne, R. W., 2003. Imitation as behaviour parsing. *Phil. Trans. Roy. Soc. London. B* **358**: 529–536.

Clifton, R. K., Muir, D. W., Ashmead, D. H., and Clarkson, M. G., 1993. Is visually guided reaching in early infancy a myth. *Child Devel.* **64**: 1099–1110.

Corbetta, D., Thelen, E., and Johnson, K., 2000. Motor constraints on the development of perception–action matching in infant reaching. *Infant Behav. Devel.* **23**: 351–374.

de Vries, J. I. P., Visser, G. H. A., and Prechtl, H. F. R., 1982. The emergence of fetal behavior. I. Qualitative aspects. *Early Hum. Devel.* **7**: 301–322.

Fagg, A. H., and Arbib, M. A., 1998. Modeling parietal–premotor interactions in primate control of grasping. *Neur. Networks* **11**: 1277–1303.

Ferrari, P. F., Gallese, V., Rizzolattia, G., and Fogassi, L., 2003. Mirror neurons responding to the observation of ingestive and communicative mouth actions in the monkey ventral premotor cortex. *Eur. J. Neurosci.* **17**: 1703–1714.

Fetters, L., and Todd, J., 1987. Quantitative assessment of infant reaching movements. *J. Mot. Behav.* **19**: 147–166.

Forrsberg, H., 1998. The neurophysiology of manual skill development. *Clin. Devel. Med.* **147**: 97–122.

Gallese, V., and Goldman, A., 1998. Mirror neurons and the simulation theory of mind-reading. *Trends Cogn. Sci.* **2**: 493–501.

Grafton, S. T., Arbib, M. A., Fadiga, L., and Rizzolatti, G., 1996. Localization of grasp representations in humans by PET. II. Observation compared with imagination. *Exp. Brain Res.* **112**: 103–111.

Hari, R., Forss, N., Avikainen, S., *et al.*, 1998. Activation of human primary motor cortex during action observation: a neuromagnetic study. *Proc. Natl Acad. Sci. USA* **95**: 15061–15065.

Haruno, M., Wolpert, D. M., and Kawato, M., 2001. MOSAIC model for sensorimotor learning and control. *Neur. Comput.* **13**: 2201–2220.

Hepper, P. G., Shahidullah, S., and White, R., 1991. Handedness in the human fetus. *Neuropsychologia* **29**: 1107–1111.

Iacoboni, M., 2004. Understanding others: imitation, language, empathy. In S. Hurley and N. Chater (eds.) *Perspectives on Imitation: From Cognitive Neuroscience to Social*

Science, vol. 1, Mechanisms of Imitation and Imitation in Animals. Cambridge, MA: MIT Press, pp. 77–99.

Iacoboni, M., Koski, L. M., Brass, M., *et al.*, 2001. Reafferent copies of imitated actions in the right superior temporal cortex. *Proc. Natl Acad. Sci. USA* **98**: 13995–13999.

Iberall, T., and Arbib, M. A., 1990. Schemas for the control of hand movements: an essay on cortical localization. In M. A. Goodale (ed.) *Vision and Action: The Control of Grasping.* Norwood, NJ: Ablex, pp. 204–242.

Jacobs, R. A, Jordan, M. I., Nowlan, S. J., and Hinton, G. E., 1991. Adaptive mixtures of local experts. *Neur. Comput.* **3**: 79–87.

Jeannerod, M., 2004. How do we decipher other's minds? In J.-M. Fellous and M. A. Arbib (eds.) *Who Needs Emotion? The Brain Meets the Robot.* New York: Oxford University Press, pp. 147–169.

Jeannerod, M., Arbib, M. A., Rizzolatti, G., and Sakata, H., 1995. Grasping objects: the cortical mechanisms of visuomotor transformation. *Trends Neurosci.* **18**: 314–320.

Jordan, M. I., and Jacobs, R. A., 1994. Hierarchical mixtures of experts and the EM algorithm. *Neur. Comput.* **6**: 181–214.

Jordan, M. I., and Rumelhart, D. E., 1992. Forward models: supervised learning with a distal teacher. *Cogn. Sci.* **16**: 307–354.

Keysers, C., Kohler, E., Umiltà, M. A., *et al.*, 2003. Audiovisual mirror neurons and action recognition. *Exp. Brain Res.* **153**: 628– 636.

Kohler, E., Keysers, C., Umiltà M. A., *et al.*, 2002. Hearing sounds, understanding actions: action representation in mirror neurons. *Science* **297**: 846–848.

Kuperstein, M., 1988. Neural model of adaptive hand–eye coordination for single postures. *Science* **239**: 1308–1311.

Lantz, C., Melen, K., and Forssberg, H., 1996. Early infant grasping involves radial fingers. *Devel. Med. Child Neurol.* **38**: 668–674.

Lasky, R. E., 1977. The effect of visual feedback of the hand on the reaching and retrieval behavior of young infants. *Child Devel.* **48**: 112–117.

Lawrence, D. G., and Hopkins, D. A., 1976. The development of motor control in the rhesus monkey: evidence concerning the role of corticomotoneuronal connections. *Brain* **99**: 235–254.

Lockman, J., Ashmead, D. H., and Bushnell, E. W., 1984. The development of anticipatory hand orientation during infancy. *J. Exp. Child Psychol.* **37**: 176–186.

Mathew, A., and Cook, M., 1990. The control of reaching movements by young infants. *Child Devel.* **61**: 1238–1258.

McCarty, M. K., Clifton, R. K., Ashmead, D. H., Lee, P., and Goulet, N., 2001. How infants use vision for grasping objects. *Child Devel.* **72**: 973–987.

McGraw, M. B., 1945. *The Neuromuscular Maturation of the Human Infant.* New York: Hafner.

Meltzoff, A. N., 1988. Infant imitation and memory: nine-month-olds in immediate and deferred tests. *Child Devel.* **59**: 217–225.

Meltzoff, A. N., and Moore, M. K., 1977. Imitation of facial and manual gestures by human neonates. *Science* **198**: 74–78.

Miall, R. C., 2003. Connecting mirror neurons and forward models. *Neuroreport* **14**: 2135–2137.

Newell, K. M., Scully, D. M., McDonald, P. V., and Baillargeon, R., 1989. Task constraints and infant grip configurations. *Devel. Psychobiol.* **22**: 817–831.

Newell, K. M., McDonald, P. V., and Baillargeon, R., 1993. Body scale and infant grip configurations. *Devel. Psychobiol.* **26**: 195–205.

Nishitani, N., and Hari, R., 2000. Temporal dynamics of cortical representation for action. *Proc. Natl Acad. Sci. USA* **97**: 913–918.

Olivier, E., Edgley, S. A., Armand, J., and Lemon, R. N., 1997. An electrophysiological study of the postnatal development of the corticospinal system in the macaque monkey. *J. Neurosci.* **17**: 267–276.

O'Reilly, W., 1995. Using representations: comprehension and production of actions with imagined objects. *Child Devel.* **66**: 999–1010.

Oztop, E., and Arbib, M. A., 2002. Schema design and implementation of the grasp-related mirror neuron system. *Biol. Cybernet.* **87**: 116–140.

Oztop, E., Bradley, N. S., and Arbib, M. A., 2004. Infant grasp learning: a computational model. *Exp. Brain Res.* **158**: 480–503.

Oztop, E., Wolpert, D., Kawato, M., 2005. Mental state inference using visual control parameters. *Brain Res. Cogn. Brain Res.* **22**: 129–151.

Petitto, L. A., Katerelos, M., Levy, B. G., *et al.*, 2001. Bilingual signed and spoken language acquisition from birth: implications for the mechanisms underlying early bilingual language acquisition. *J. Child Lang.* **28**: 453–496.

Rizzolatti, G., Fadiga, L., Matelli, M., *et al.*, 1996. Localization of grasp representations in humans by positron emission tomography. I. Observation versus execution. *Exp. Brain Res.* **111**: 246–252.

Rolls, E. T., 2004. Convergence of sensory systems in the orbitofrontal cortex in primates and brain design for emotion. *Anat. Rec.* **281A**: 1212–1225.

Rumelhart, D. E., Hinton, G. E., and Williams, R. J., 1986. Learning internal representations by error propagation. In D. Rumelhart and J. McClelland (eds.) *Parallel Distributed Processing: Explorations in the Microstructure of Cognition* vol. 1. Cambridge, MA: MIT Press, pp. 318–362.

Siddiqui, A., 1995. Object size as a determinant of grasping in infancy. *J. Genet. Psychol.* **156**: 345–358.

Sporns, O., and Edelman, G. M., 1993. Solving Bernstein's problem: a proposal for the development of coordinated movement by selection. *Child Devel.* **64**: 960–981.

Thelen, E., 2000. Motor development as foundation and future of developmental psychology. *Int. J. Behav. Devel.* **24**: 385–397.

Thelen, E., and Ulrich, B. D., 1991. Hidden skills: a dynamic systems analysis of treadmill stepping during the first year. *Monogr. Soc. Res. Child Devel.* **56**: 1–98.

Thelen, E., Corbetta, D., Kamm, K., and Spencer, J. P., 1993. The transition to reaching: mapping intension and intrinsic dynamics. *Child Devel.* **64**: 1058–1098.

Umiltà, M. A., Kohler, E., Gallese, V., *et al.*, 2001. I know what you are doing: a neurophysiological study. *Neuron* **31**: 155–165.

van der Meer, A. L. H., van der Weel, F. R., and Lee, D. N., 1994. Prospective control in catching by infants: prospective control in catching by infants. *Perception* **23**: 287–302.

1995. The functional significance of arm movements in neonates. *Science* **267**: 693–695.

von Hofsten, C., 1979. Development of visually guided reaching: the approach phase. *J. Hum. Movt Stud.* **5**: 160–178.

1982. Eye–hand coordination in the newborn. *Devel. Psychol.* **18**: 450–461.

1990. A perception–action perspective on the development of manual movements. In M. Jeannerod (ed.) *Attention and Performance*, vol. 13. Hillsdale, NJ: Lawrence Erlbaum, pp. 739–762.

1991. Structuring of early reaching movements: a longitudinal study. *J. Mot. Behav.* **23**: 280–292.

Von Hofsten, C., and Fazel-Zandy, S., 1984. Development of visually guided hand oriented in reaching. *J. Exp. Child Psychol.* **38**: 208–219.

von Hofsten, C., and Ronnqvist, L., 1988. Preparation for grasping an object: a developmental study. *J. Exp. Psychol. Hum. Percept. Perform.* **14**: 610–621.

1993. The Structuring of neonatal arm movements. *Child Devel.* **64**: 1046–1057.

Wishart, J. G., Bower, T. G. R., and Dunkeld, J., 1978. Reaching in the dark. *Perception* **7**: 507–512.

13

Development of goal-directed imitation, object manipulation, and language in humans and robots

Ioana D. Goga and Aude Billard

13.1 Introduction

The aim of the present volume is to enrich human language dimensions by seeking to understand how the use of language may be situated with respect to other systems for action and perception. There is strong evidence that higher human cognitive functions, such as imitation and language, emerged from or co-evolved with the ability for compositionality of actions, already present in our ancestors (Rizzolatti and Arbib, 1998; Lieberman, 2000; Arbib, 2003; Arbib, Chapter 1, this volume). Corroborating evidence from psychology (Greenfield *et al.*, 1972; Iverson and Thelen, 1999; Glenberg and Kaschak, 2002), neurobiology (Pulvermüller, 2003) and cognitive sciences (Siskind, 2001; Reilly, 2002) strongly support a close relationship between language, perception, and action. Social abilities, such as imitation, turn-taking, joint attention and intended body communication, are fundamental for the development of language and human cognition. Together with the capacity for symbolization, they form the basis of *language readiness* (Rizzolatti and Arbib, 1998; Arbib, 2003).

The work presented in this chapter takes inspiration from this body of experimental evidence in building a composite model of the human's cognitive correlates to action, imitation and language. The model will contribute to develop a better understanding of the common mechanisms underlying the development of these skills in human infants, and will set the stage for reproducing these in robots and simulated agents.

A recent trend of robotics research follows such views, by equipping artifacts with social capabilities. The rationale is that such abilities would enable the artifact to communicate with humans using "natural" means of communication (Schaal, 1999; Breazeal and Scassellati, 2002; Billard and Mataric, 2001; Kozima and Yano, 2001; Demiris and Hayes, 2002). We follow this trend and investigate the role that imitation and joint attention play in early language acquisition. This work builds upon other work of ours that investigate the basics components of imitation learning, such as the ability to extract the important features of a task (*what to imitate*) (Billard *et al.*, 2003; Calinon and Billard,

Action to Language via the Mirror Neuron System, ed. Michael A. Arbib. Published by Cambridge University Press.
© Cambridge University Press 2006.

in press), and the ability to map motion of others into one's own repertoire (*how to imitate*) (Sauser and Billard, 2005). See the chapters by Greenfield and by Zukow-Goldring (this volume) for complementary work on the role of the caregiver in the child's acquisition of these abilities.

We now briefly outline the methodological approach followed in this work.

13.1.1 Outline of the methodological approach

A considerable body of cognitive and robotic theories point to at least three conditions to be met by a system, in order for the system to develop human-like cognition: (a) *sociocultural situatedness*, understood as the ability to engage in acts of communication and participate in social practices and language games within a community; (b) *naturalistic embodiment*, that is, the possession of bodily structures to experience the world directly; (c) *epigenetic development*: the development of physical, social, and linguistic skills in an incremental, step-wise manner (Harnad, 1990; Clark, 1997; Brooks *et al.*, 1998; Zlatev and Balkenius, 2001; Steels, 2003).

Embodiment

Embodiment allows artificial systems to ground symbolic representations in behavioral interactions with the environment in such a way that the agent's behaviors, as well as its internal representations, are intrinsic and meaningful to itself (Harnad, 1990; Ziemke, 1999). Symbols are grounded in the capacity to discriminate and identify the objects, events, and states of affairs that they stand for, from their sensory projections (Regier, 1995). In addition, by enabling the agent to act upon its environment, the transduction mechanism develops a functional value for the agent, and can be considered meaningful to itself (Bailey, 1997; Brooks *et al.*, 1998).

Embodiment is at the core of our methodology. In previous work, we took the stance that the robot's body was fundamental to convey and ground meaning in words, transmitted and taught by another teacher agent (Billard, 2002). Meaning was, thus, grounded in the learner robot's perceptions (i.e., sensor measurements and motor states).

Development

A second core stance of our methodology stresses the role that development plays in the acquisition of compositional skills, such as language and imitation. Development represents a framework through which humans acquire increasingly more complex structures and competencies (Piaget, 1970). A developmental process starting with a simple system that gradually becomes more complex allows efficient learning throughout the whole process, and makes learning easier and more robust. A developmental framework also supports the system capacity to organize words and concepts in a hierarchical, recursive, and compositional fashion. The epigenetic developmental approach is increasingly exploited in the artificial intelligence (AI) field to account for the building of complex cognitive structures from low-level sensorimotor schemas (Metta *et al.*, 1999; Weng *et al.*, 2001; Zlatev and Balkenius, 2001; Reilly and Marian, 2002).

The starting point of our approach to action, imitation, and language is the definition of a developmental benchmark, against which modeling can be compared. This benchmark has to meet several criteria: (a) it must be grounded in the observation of infants' behavior in a complex social scenario, which involves social interaction, imitation, object manipulation, and language understanding and production; (b) it must permit the characterization of developmental stages in the acquisition of the skills under observation; (c) it should be sufficiently realistic, so that it could be replicated experimentally; (d) the infants' behavior under observation must be such that they can be modeled and implemented in artificial systems. The *seriated nesting cups* task (Greenfield *et al.*, 1972; see also Greenfield, 1991, and this volume) is our benchmark.

The role of imitation

We argue that the abilities to imitate and to manipulate objects lay at the foundation of language development in humans. The capacity to imitate goal-directed actions plays an important role in coordinating different behaviors. Billard and Dautenhahn (2000) showed that imitation allows sharing of a similar perceptual context, a prerequisite for symbolic communication to develop. In the work reported here, imitation will be studied as exploiting the capacity to recognize and extract others' actions (where mirror neurons have been shown to play a central function), to memorize, learn, and reproduce the demonstrated behavior. Our working hypothesis is that imitation requires the ability to extract meaning, by inferring, and furthermore, by understanding the demonstrator's goal and intention (Byrne and Russon, 1998). Moreover, the ability to infer others' intention, to imitate novel behavior and the capacity for language is investigated as a process that follows a common developmental path, and which may have a common underlying neural process.

Thus, essential to our approach is the Mirror System Hypothesis (Arbib, Chapter 1 this volume), that is, the idea that the communication system develops atop an action system capable of object manipulation, with its capacity to generate and recognize a set of actions. This assumption represents a starting point for the present approach, whose goal is to investigate the computational means by which interaction of the action and language systems develops on a common neural substrate. The model we develop here follows from and complements other works by providing a more detailed description of the role of imitation and joint attention in the acquisition of sequential manipulation of objects.

Modeling

The core of our methodological approach, outlined in Fig. 13.1, resides in the computational modeling of the developmental path that human infants follow in developing the capacity to create assemblages of objects and to generate well-formed sentences. Our model is constrained by evidence from neuroscience and developmental psychology. The

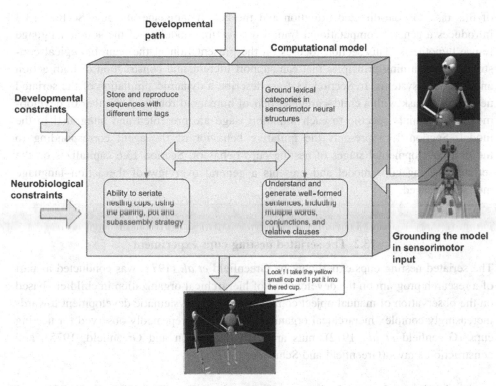

Figure 13.1 A neurobiologically and developmentally constrained approach to action–language modeling task. The developmental path envisaged has four hypothesized stages: (a) development of the ability to learn and reproduce sequences with different time lags and complexity; (b) acquisition of the capacity to seriate nesting cups, through an epigenetic process that replicates the strategic rule-bound behavior observed in human infants; (c) learning of a number of lexical categories grounded on the sensorimotor neural structures developed previously; (d) development of the capacity to generate well-formed, meaningful sentences that describe the actions performed, by following the same linguistic developmental path that human infants follow.

acquired knowledge is grounded in the sensorimotor interaction of the simulated agent with the environment.

We start with a simple system capable first of imitating short sequences of actions, that gradually develops afterwards to accommodate more complex behaviors. As learning and development proceed, prior structures get integrated and provide competencies that can be reused (the vertical flow in Fig. 13.1). Previously acquired behavioral manifestations put constraints on the later structures and proficiencies (the horizontal flow in Fig. 13.1). To facilitate learning, the gradual increase in internal complexity is accompanied by a gradual increase in the complexity of the external world to which the "infant" is exposed (see more on the language modeling framework in Section 13.6).

The rest of this chapter is structured as follows. Section 13.2 describes the original experiment of seriated nesting cups (Greenfield *et al.*, 1972) and discusses the relevance of this task to goal-directed imitation and the action–language analogy. Section 13.3 introduces a general computational framework for the modeling of the action–language re-use hypothesis. This section focuses on the presentation of the neurobiological constraints and learning principles that can support incremental construction of both action and language systems. In Section 13.4 we describe a dynamic simulation of the seriated nesting cups task with a child–caregiver pair of humanoid robots. A number of developmental constraints specific to each imitation stage are presented and integrated in the model. Section 13.5 presents the imitative behavior of the agent corresponding to the first developmental stages of nesting cups behavior. Section 13.6 capitalizes on the seriated nesting cups model and presents a general overview of the action–language model envisaged.

13.2 The seriated nesting cups experiment

The seriated nesting cups experiment by Greenfield *et al.* (1972) was conducted as part of a research program on the development of hierarchical organization in children, based on the observation of manual object combination tasks. Systematic development towards increasingly complex hierarchical organization has been repeatedly observed for nesting cups (Greenfield *et al.*, 1972), nuts and bolts (Goodson and Greenfield, 1975), and construction straws (Greenfield and Schneider, 1977).

13.2.1 Development of rule-bound strategies for manipulating seriated cups

Greenfield *et al.* (1972) report that children between 11 and 36 months of age exhibit different strategies, correlated to their developmental age, for combining cups of different sizes. The seriated nesting cups experiment consists first of a demonstration phase, during which the experimenter manipulates the cups to form a nest (i.e., insert all cups into one another using the most advanced (subassembly) strategy (see Fig. 13.2c)), followed by a spontaneous imitation phase, during which the child is left free to play with the cups.

Three manipulative strategies were identified and analyzed: (1) *the pairing method*, when a single cup is placed in/on a second cup; (2) *the pot method*, when two or more cups are placed in/on another cup; (3) *the subassembly method*, when a previously constructed structure consisting of two or more cups is moved as a unit in/on another cup or cup structure (see Fig. 13.2).

The child's choice of the acting/acted upon cups seems to be based on one of three criteria: size, proximity, and contiguity. During the first developmental stage, children typically use the proximity criterion (i.e., same side of the table with the moving hand) for pairing cups. Children also operate as though size is a binary concept, with one cup treated as the "biggest" while all the others belong to the category "little." A common phenomenon is the transfer of a single moving cup from one stationary cup to another without letting go of the original cup. Intermediate constructions between stage 1 and

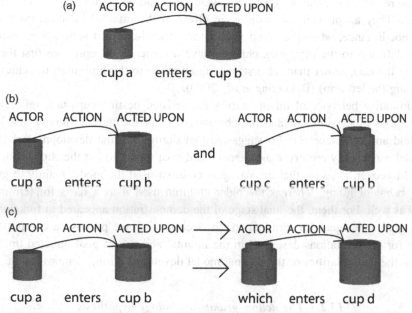

Figure 13.2 Formal correspondence between developmental manipulation strategies and sentence types. In each frame, action relations are shown on the top, the imitative behavior is shown in the center, while a descriptive sentence on the bottom illustrates the corresponding grammatical relation. (a) Strategy 1: the pairing method corresponds to a grammatical relation of type *subject–verb–object*. (b) Strategy 2: the pot method corresponds to a relation of type *subject–verb–object* and *subject–verb–object*. (c) Strategy 3: the subassembly method, corresponds to the composition with ellipsis *subject–verb–object → which[subject]–verb–object*. (Adapted after Greenfield *et al.*, 1972.)

stage 2 consist in forming two pairs or one pair and a structure of three embedded cups. Children at 16–24 months old seem to follow only the contiguity criterion (i.e., never reaching behind a nearer cup to use a more distant cup), while the 28- to 36-month-olds seems to follow the size criterion.

The consistency of each individual's strategic behavior is defined in terms of the dominant or most frequent strategy. A child's dominant strategy accounted on average for 80% of his structures. On the other hand, each infant uses at least once an intermediate manipulative method and consistency of the strategic behavior slightly decreases with the age.

13.2.2 Development of goal-directed imitation

The seriated cups experiment is of relevance to the theory of goal-directed imitation (Bekkering *et al.*, 2000). In order to imitate complex behaviors, one must recognize goals, understand how individual actions are embedded in a hierarchy of subgoals, and recognize recursive structures. Normally developing children do not copy exactly the act of adults. They adapt their imitation as an effect of the inferred intended goal of the action. For instance, when presented with a simple goal-directed action (e.g., reaching with the left arm to the right ear), children have a tendency to reproduce first the goal (touching the ear), rather than (at first) paying attention to the limb used to achieve the goal (using the left arm) (Bekkering *et al.*, 2000).

The imitative behavior of infants during the seriated nesting cups task reflects both the capacities of the imitator and the characteristics of the internal model developed. Greenfield and colleagues (1972) suggested that during the first developmental stages, the model was mainly preserved as a generalized goal state: "to put the cups inside each other." Moreover, it seems that the youngest children used the model mainly to get the nesting behavior going, whereas the older children used it as a basis for terminating activity as well. For them, the final stage of the demonstration appeared to function as a precise goal and signal for termination. In Section 13.5 we propose a computational account for the limitations described in the infants' abilities of goal-directed imitation, based on the particularities of the internal model developed during demonstration.

13.2.3 The action–grammar analogy hypothesis

A related objective of Greenfield *et al.* (1972) was to investigate the question of a formal homology between strategies for cup construction and certain grammatical constructions. Figure 13.2 illustrates the analogy they proposed between the three action strategies and specific grammatical constructions. When a cup "acts upon" another cup to form a new structure, there is a relation of *actor–action–acted upon*. Such a relation is realized in sentence structures like *subject–verb–object*. The pairing strategy is dominant at 11- and 12-month-old infants, and corresponds to the use at this age of simple sentences, formed from two or three words. The second and third strategies, on the other hand, allow the formation of multiple *actor–action–acted upon* sequences, and, as such, would corres-pond to the usage of more complex sentences. The difference is that in the second stage the child performs a *conjunction* of the sequences/words, while in the last stage the embedding of the cups is accomplished, paralleling the capacity to use *relative clauses* in language. Sources of evidence on the relative ordering of these types of grammatical constructions are provided by experimental studies showing that conjunction of sentences was frequent and preceded relative clauses in the speech of children aged 18 months to 3 years (Slama-Cazacu, 1962; Smith, 1970).

Greenfield (1991) put forward the hypothesis of a neural structural homology, rather than just an analogy, between action strategies and grammatical constructions. She adduced evidence from neurology, neuropsychology, and animal studies to support the view that object combination and speech production are built upon an initially common neurological foundation, which then divides into separate specialized areas as develop-ment progresses. Her hypothesis is that early in a child's development Broca's region may serve the dual function of coordinating object assembly and organizing the production of structured utterances. Computational support to this hypothesis was brought by Reilly (2002) (see also Section 13.6.2).

13.3 General overview of the computational model

The task of the present section is to introduce the main concepts of the computational framework. In our view, learning to seriate nesting cups and to generate grammatical constructions have some common needs: (a) the capacity to represent categorical information in a subsymbolic manner; (b) the operation of a mechanism for grounding internal representations on sensorimotor processes; (c) the ability to learn from and to represent time-ordered sequences; (d) the capacity to process and satisfy multiple constraints in a parallel manner; (e) the operation of a computational mechanism that supports cross-domain bootstrapping. In this section we address the requirements (a)–(d), while the challenging issue of inter-domain transfer of information is tackled in Section 13.6 in the description of the language modeling framework.

13.3.1 A decompositional approach to knowledge representation

The subsymbolic paradigm states that the brain represents cognitive symbols through their decomposition into smaller constituents of meaning, usually called *semantic features* or *microfeatures* (Sutcliffe, 1992). Furthermore, concepts in the brain are grounded on semantic neural networks, involved in the perception or execution of the corresponding symbols (see the grounding paradigm: Harnad, 1990). That is because, when the meaning of a concrete content word is being acquired, the learner is exposed to stimuli of various modalities related to the word's meaning, or the learner may perform actions the word refers to.

Recent neuroscientific evidence corroborates the decompositional and grounding computational approaches on language representation. Neuroimaging studies of language (Pulvermüller, 1999, 2002; Hauk *et al.*, 2004) support the idea that words are represented in the brain by distributed cell assemblies whose cortical topographies reflect aspects of word meaning (including action-based representations). There is evidence that: (1) assemblies representing phonological word forms are strongly lateralized and distributed over perisylvian cortices; (2) assemblies representing concrete content words include additional neurons in both hemispheres; (3) assemblies representing words referring to visual stimuli include neurons in visual cortices; (4) assemblies representing words referring to actions include neurons in motor cortices (see Pulvermüller (1999) for a review on the neurobiological data).

The great promise of the distributed, neurobiologically inspired approach to knowledge representation is the integration of learning and representation within the same structures. The pitfall of this approach is the increasing complexity of the biologically plausible computational models (Marian *et al.*, 2002). For a detailed introduction to biologically realistic neural architectures we refer the reader to Maas and Bishop (1999). In this work, we get inspiration from neurobiological studies and define a computational primitive, which is both simple and scalable, suited for the operation on cognitive and linguistic structures.

13.3.2 The cell assembly concept

The computational building-block of our system is a neural primitive referred to as a *cell assembly*. The concept is envisaged along the lines proposed by Hebb (1949), Pulvermüller (2002), and Frezza-Buet and Alexandre (2002). Hebb (1949) suggested that a cell assembly is a reverberatory chain of neurons, with many re-entrant loops through which activity waves can travel repeatedly. More recently, Pulvermüller (2002) uses the concept of a *neural set*, to refer to functional webs characterized by a great variety of activity states. Because of its strong internal connections, the neuronal set is assumed to act as a functional unit, which can be primed, ignite, and reverberate.

   Our approach is inspired by Pulvermüller's definition of neuronal set. A cell assembly is a neuronal set (i.e., a selection of neurons that are strongly connected to each other, act as a functional unit, can be primed and ignite) with additional special properties that are relevant to category information processing. Each cell assembly receives input from external sensorial units and can have an activating effect on other cell assemblies directly connected to it (Fig. 13.3). A cell assembly can: (a) learn a subsymbolic feature representation of an external concept or event (feedforward flow in Fig. 13.3), and (b) become a node in a sequence detector (precedence flow in Fig. 13.3). A third learning mechanism (relation flow in Fig. 13.3) extracts and stores information concerning systematic relations among sensorial or semantic features that constitute specific categories. Anti-Hebbian learning (Levy and Desmond, 1985) is used to extract the invariants between the items that are in the focus of attention.

13.3.3 Learning framework

The general approach to learning is similar to the biologically inspired model of the cortex described by Frezza-Buet and Alexandre (2002), including elementary functions in the perceptive, motor, or associative domain. The basic computational unit in their work is a maxicolumn, consisting of a set of cortical-like columns. The maxicolumn is grounded on sensorial and motor maps, has associative functions on higher-order maps, and allows three types of learning: (a) learning of the external event that the column represents, through feedforward connections; (b) associative learning within a map, through lateral connections; and (c) sequence learning. Causal sequence learning is based on a rule that finds the conditional stimulus that is most frequently associated with the unconditional stimulus, and which predicts it. The mechanism is employed to find the neural component in a map whose activity predicts the satisfaction of another neural component in the map, referred to as a goal (Frezza-Buet and Alexandre, 2002).

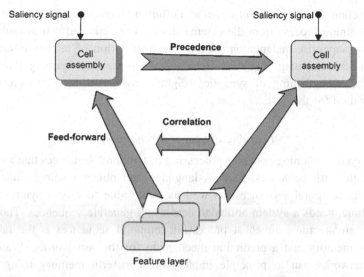

Figure 13.3 Representation of feedforward and lateral information flows of a cell assembly. Each kind of link represents a type of learning that a cell assembly has to manage. The saliency signal is received from the attention module (not shown), and it is inextricably bound up with the creation and satisfaction of a cell assembly.

Category learning

The connectivity patterns represented in Fig. 13.3 define the kinds of information flows, and thereby, the kinds of learning that a cell assembly can be involved in. The first type of learning (i.e., feedforward connections in Fig. 13.3) tunes the cell assembly to respond to a specific distribution of its input information. The learning algorithm is inspired by Adaptive Resonance Theory (ART) (Grossberg, 1976; Carpenter and Grossberg, 1987). ART networks were designed to overcome the stability–plasticity dilemma, that is, how can a system be stable against noisy or irrelevant data and yet remain plastic enough to learn novel inputs without affecting already learned knowledge. A central feature of all ART systems is a pattern-matching process that compares an external input with the internal memory of an active code. ART matching leads either to a *resonant* state, which persists long enough to permit learning, or to a parallel memory search. If the search ends at an established code, the memory representation may either remain the same or incorporate new information from matched portions of the current input. If the search ends at a new code, the memory representation learns the current input. The criterion of an acceptable match is defined by a dimensionless parameter called *vigilance*. Vigilance weights how close an input must be to the prototype for resonance to occur. Low vigilance leads to broad generalization and more abstract prototypes than high vigilance.

The cell assembly unit represents a category from ART with additional properties that are relevant for temporal pattern processing (i.e., graded activation and memory decay) and for extraction of correlation information. Different kinds of information are extracted during the learning process from the external input. The activated cell assemblies learn through an unsupervised adaptation process the distribution of feature information in the external sensorial map. Classical Hebbian, non-supervised learning (Hebb, 1949) based on the strengthening of synaptic weights between co-activated units, can be successfully used for this task.

Learning of temporal sequences

The second type of learning concerns processing of temporal sequences that are inextricably bound up with behaviors, such as language and object manipulation. Learning structure from temporal sequences, as well as being able to output symbols that are ordered in time, needs a system ability to detect and generate sequences. The common need of the architectures aimed at processing temporal sequences is the presence of a short-term memory and a prediction mechanism for the serial order (Wang, 2002). Recurrent networks can in principle implement short-term memory using feedback connections (Billard and Hayes, 1999). Temporal sequences are learned as a set of associations between consecutive components (see Chappelier *et al.* (2001) for a recent review of the fundamental types of temporal connectionist models).

The initial implementation of the cell assembly concept was inspired by previous work of ours on sequential learning with time-delay networks (Billard, 2002). A cell assembly was represented there by a node to which self-recurrent connections have been added and a time decay rate was used to simulate a short-term memory of the assembly. The cell assembly can in this case transit several states (Fig. 13.4a): (a) an *ignition* state corresponding to the maximal activation of the cell assembly, (b) an *activation* state corresponding to the decaying memory of its ignition, and (c) an *inactive* state, when its activation is below an arbitrary set threshold.

Once information from the sensorial external map reaches the cell assembly, it is then further memorized for a period of time, during which it can be associated with any incoming event in any other sensory system. This leads to a system capable of associating events delayed in time with a maximal time delay equal to the length of the working memory. Time-delay networks (Billard and Hayes, 1999) or simple recurrent networks (Elman, 1993) can be employed with success for learning the sequential order of events.

Storing temporal dependencies using graded activation states

To deal with temporal dependencies beyond consecutive components different solutions were explored. Recent models propose different ways to enhance the memory capacity, by carefully designing the basic computational unit (Hochreiter and Schmidhuber, 1997) or by exploiting learning in hierarchical structures (Tan and Soon, 1996).

In this model, learning of temporal dependencies is supported by the graded activation of the cell assemblies, and is facilitated by the layered architecture of the model. Each cell assembly can transit an increased number of activation states. This is achieved by

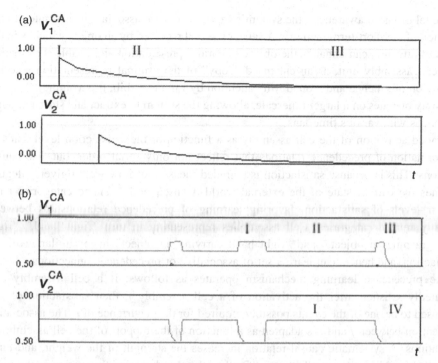

Figure 13.4 (a) Cell assembly activation curves characterized by three operation regimes: *ignition* (I), *memory decay* (II), and *inactivation* (III). (b) Activation curves with four functional regimes: *maximum activation* (I), corresponding to maximal satisfaction of the cell assembly's feature constraints; *sustained activation* (II), when the cell assembly feature constraints partially satisfy the state of the external world; *memory decay* (III), when the cell assembly is no longer satisfied and its activation decays as a function of the decay rate; and *inactivity* (IV). The second regime (sustained activation) was introduced to deal with time dependencies larger than the time constant of the decay rate.

implementing at the cell assembly level a computational mechanism that keeps the CA active for larger periods of time. The mechanism relies on the concept of a cell assembly's *degree of satisfaction* and it is explained in more detail in Section 13.4.5. Briefly, the activation of a cell assembly becomes a function of how well its subsymbolic feature representation of the learned event satisfies the current state of the external world. Consequently, a cell assembly can be in one of the following states: (a) *maximally satisfied*, leading to a temporary state of maximal activity (regime I in Fig. 13.4b); (b) *partially satisfied*, allowing a prolonged activation regime (regime II in Fig. 13.4b); (c) *unsatisfied*, when its activation decays as a function of the memory rate (regime III in Fig. 13.4b); or (d) *inactive*.

A layered neural architecture favors temporal integration at different timescale levels. In our view, the creation of a cell assembly is inextricably bound up with the receipt of a saliency signal (Fig. 13.3; see also Section 13.4.4). The sensorial external maps operate on a short timescale, in the sense that they respond spontaneously to external inputs. At this level of visual awareness, the system's capacity to store associations is limited to the duration of the short-term memory. A saliency signal received by an object which is in the focus of attention can enhance the object's neural representation and enable the creation of a cell assembly unit, as an enhanced "copy" of the original representation (see the discussion of salience and "top–down" attention by Itti and Arbib, this volume). The cell assembly operates on a larger timescale, allowing the system to extract and store temporal sequences with various time lags.

Graded activation of the cell assembly as a function of the satisfaction level leads to the formation of precedence relations that reflect not only serial order, but also causal relations. This is because satisfaction is a graded measure of how well a given category matches the current state of the external world. Consequently, related categories have similar levels of satisfaction, favoring learning of precedence relationships between causally related categories. Cell assemblies representing in turn "right hand," "right hand grasping an object," and "right hand carrying an object" have similar levels of satisfaction, and hence, can learn a set of associative or precedence connections.

The precedence learning mechanism operates as follows: if a cell assembly i is repeatedly satisfied after the activation of a cell assembly j, then satisfaction of j is supposed to be one of the events possibly required for the occurrence of i. The associative connection between j and i is adapted as a function of the rapport of the cell assemblies' activations. A systematic causal relation increases the strength of the weight, and turns the cell assembly j into a predictor of i. On the other hand, a large fluctuation in the order in which i and j are satisfied will decrease the weight, to an insignificant value. This type of mechanism can account for unary precedence relations of the type $A \rightarrow C$ or $B \rightarrow C$, where A and B are learned predictors of C.

When would a cell assembly coding for concept C respond to a binary sequence $A \rightarrow B$ and not to $B \rightarrow A$? We envisage the operation of a sequence detector mechanism similar to that described in Pulvermüller (2002). The mechanism is inspired from the functioning of the movement detectors in the visual cortex (Hubel, 1995) and is based on the following idea: sensitivity to the sequence $A \rightarrow B$ may involve low-pass filtering of the signal A, thereby delaying and stretching it over time. Integration at the site of a third neuron k of the delayed and stretched signal from i and the actual signal from j yields values that are large when activation of i precedes that of j, but small values when the activation of i and j occurs simultaneously or in reverse order. This yields strong weights between j and k but not between i and k (after Pulvermüller, 2002). A similar mechanism is implemented for the retrieval of precedence relationships during the simulation of the seriated nesting cups task (see more details in Section 13.4.6). For our goals, sequence detectors operating on no more than two or three nodes are sufficient.

13.3.4 A multiple constraints satisfaction framework

Sequential learning is sufficient neither for the acquisition of action grammars nor of language grammars. A basic property of language is the translation of a hierarchical

Table 13.1 *Steps of the retrieval process*

1. Focus attention on the most salient objects and words from the environment.
2. Cell assemblies corresponding to the distribution of features that are in the focus of attention become satisfied.
3. Hidden cell assemblies are primed through the lateral connections and precedence constraints are computed.
4. Primed, unsatisfied cell assemblies compete for setting the goals of the system.
5. Action is initiated to satisfy the most important goal and to minimize the internal global dissonance.
6. The objects of action are chosen as a function of the internal and external constraints existent in the system.
7. Termination of action occurs when all (or the most important) goals of the system are satisfied.

conceptual structure into a temporally ordered structure of actions, through a constraint satisfaction process, that integrates several types of constraints in a parallel manner (i.e., phonological, syntactic, semantic, contextual) (Seidenberg and MacDonald, 1999). In a neural network, constraints are encoded by the same machinery as knowledge of language itself, and this is clearly an advantage over approaches in which symbolic knowledge is represented separately from the system that makes use of it.

The computational framework developed in this work is based on the assumption that the reproduction of a sequence of motor or linguistic actions represents the product of the interplay between sequence detectors operating on short temporal dependencies, and a general constraints satisfaction framework. This framework controls the way different types of constraints (including sequence detection) are integrated. In particular, we applied this framework to the reproduction of the sequence of actions demonstrated in the seriated nesting cups task and we are currently exploring its application to the generation of well-formed, descriptive sentences. As stressed in the introduction, imitation is addressed as a means for disassembling and reassembling the structure of the observed behavior. We state that the reassembling task can best be described as *a multiple constraints satisfaction process*.

During the retrieval of a series of actions (either motor or linguistic) the system acts in a constrained manner, in accordance with the knowledge stored in all sets of weights. Most generally, the retrieval process consists of seven steps, which are summarized in Table 13.1.

Initiation of any type of action must be preceded by the activation of an internal goal. A cell assembly represents a *goal* of the system, if the cell assembly is not satisfied and if its precedence constraints are met over a certain threshold. Further on, the behavior of the agent is driven by a process of *dissonance minimization*. That is, the system computes the total dissonance (i.e., difference) between its goals and the current state of the environment, and acts towards the minimization of this dissonance state. The working definition of dissonance is inspired by the work of Schultz and Lepper (1992) on modeling the cognitive dissonance phenomena described by Festinger (1957).

13.4 A model of the seriated nesting cups task

13.4.1 The simulation environment

The current implementation of the model was created within the Xanim dynamic simulator (Schaal, 2001) designed to model a pair of 30 degrees of freedom (head 3, arms 7×2, trunk 3, legs 3×2, eyes 4) humanoid robots (Fig. 13.5). The external force applied to each joint is gravity. Balance is handled by supporting the hips; ground contact is not modeled. There is no collision avoidance module. The dynamics model is derived from the Newton–Euler formulation of rigid body dynamics. The simulated robot is controlled from Cartesian coordinates through inverse dynamics and inverse kinematics servos. A motor servo is used to read the current state of the robot/simulation (i.e., position, color, orientation and rotation angles, and motion speed) and to send commands to the robot/simulation. The environment is controlled, in other words, only a predefined set of objects and end-effectors are visually recognized and manipulated.

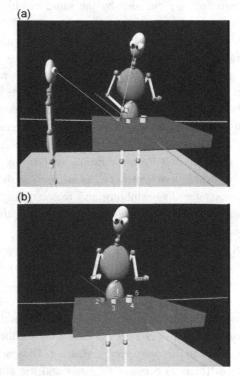

Figure 13.5 The Xanim dynamic simulation of a child–caregiver pair of humanoid robots. (a) The Caregiver demonstrates the seriated cup task. The shared focus of attention of Child and Caregiver is highlighted by the crossing of their gaze directions. (b) The Robot Child imitates the seriated cups task.

13.4.2 Developmental constraints

In modeling the seriate nesting cups behavior, we investigated the effects of varying a number of parameters of the computational model, as a way of accounting for systematic differences in child behavior. In particular, we considered the effects of shared attention, memory capacity, categorization, and development of the object concept on the child's or robot's ability to compose manipulation and linguistic steps. Each developmental stage is modeled through the combination of the effects of several developmental constraints, as shown in Fig. 13.6. The integration of each developmental constraint at a given simulated

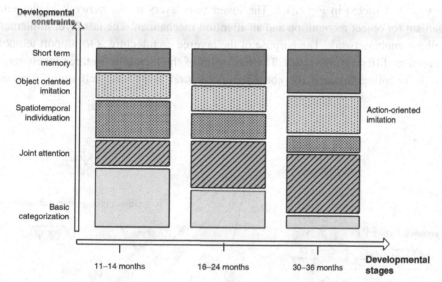

Figure 13.6 Hypothetical model of the developmental stages of the seriate cups ability. On the Y-axis is shown the contribution of each developmental constraint to the learning process for a given age. The height of the boxes indicates the weight of the corresponding process, in relation to the learning of the serriated cups task. The earliest developmental stage is accounted for through a basic categorization process, accompanied by spatiotemporal individuation and a limited short-term memory. Mervis (1987) showed that 2-year-olds form basic-level categories and suggested that infants under 2 years of age attend to the function and shape of objects to categorize. There is also evidence that infants up to 1 year rely almost exclusively on spatiotemporal information (i.e., location) to build distinct representations of the objects (Carey and Xu, 2001). Object-directed imitation and an increase of the attention and memory resources for object-related information characterize the second developmental stage. By the end of the first year of life, infants show long-term memory for serial order, and can retain in visual short-term memory about three to four items (Rose *et al.*, 2001). A small number of attended objects may be indexed in time, the indexed individuals tracked through time and space, and the spatial relations among indexed individuals represented (Carey and Xu, 2001). When objects are involved as goals, infants learn first about them, and movements are imitated if there is no object goal, or if this is ambiguous (Bekkering *et al.*, 2000). The third developmental stage is described in terms of learning the details of the hand movements, due to increased vigilance and memory resources.

age is based on the review of psychological developmental literature (see also Section 13.7 for a discussion of the experimental sources of evidence). The hypothesized model is validated through the replication of the behavioral manifestations of human infants with a pair of humanoid robots, as described in Section 13.5. However, it represents only a possible explanation for the development of nesting cups ability, and it cannot be generalized to other developmental issues.

13.4.3 Architecture and functioning principles

The model consists of a hierarchy of connectionist architectures, whose basic computational unit is depicted in Fig. 13.7. The lower two layers of the network implement a mechanism for object recognition and an attention mechanism. The last layer implements the cell assembly network. The purpose of the layered architecture is to support temporal integration at different timescales. The reduction of the hierarchy to three levels results in a tight coupling between the conceptual/structural layer (i.e., cell assemblies) and

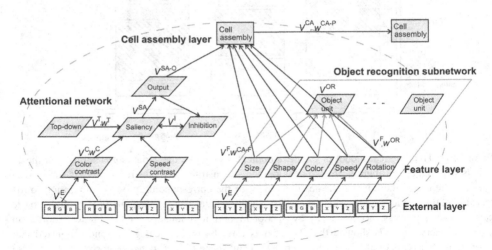

Figure 13.7 Basic computational unit of the hierarchical network consisting of three interconnected components: an attention network SA, an object recognition subnetwork OR, and a cell assembly layer CA. The saliency signal of the attention network is computed by integrating feature contrast with top–down cues. Each feature contrast unit receives input from a pair of three external units, corresponding to the object and to the context surrounding the object. An external unit encodes the color components (R, G, B) or the three-dimensional projections on X, Y, Z of the object feature. The activity of a salient winning unit is modulated by inhibition of return. An object recognition unit is fed from five feature units. An object recognition subnetwork is formed from OR units corresponding to all co-located objects. The cell assembly unit receives a saliency output signal and has a number of feature constraints grounded in the visual layer. Lateral links in the cell assembly layer learn precedence relations between cell assemblies' activations.

the perception layer. The output sent to the motor system is represented by the coordinates of the target object, which are transformed by Xanim servos (see Section 13.4.1) into commands to the simulated robot links. A one-to-one mapping is imposed in this case, between what the agent sees and what it executes.

Let us describe how the system processes visual input. At the first parsing of the visual scene, the object recognition subnetworks (OR) from all locations are activated and remain active as long as the objects remain visible. No segmentation of the visual scene is required, because the simulated agent can recognize a predefined number of objects: the end-effectors, the objects located on the table, and the table. For each object, the color, size, shape, rotation angle, and the speed of motion are read out from the robot's sensors. These values are fed in the network through the activation of the external layer's units.

Saliency (SA) is computed in a distributed manner at all locations of the visual scene and the unit with the highest activation wins the focus of attention. The winning unit enhances the object representation and allows the creation of a "copy" of the information from that location in space. This copy is referred to as a cell assembly (CA) and it is created whenever a significant variation (i.e., event) in one sensor has been detected and there is no other cell assembly which can accommodate the novel information perceived by the system.

At each developmental stage, the assumptions described in Fig. 13.6 constrain the means by which the infant robot extracts and remembers information. The following assumptions are integrated: (1) static objects situated at distinct locations are mapped into distinct categories; (2) objects or events that have a common neural structure (i.e., are co-located) can become the subject of generalization. The logic of the second assumption is simple: if two objects are co-located (i.e., two embedded cups) their visual neural representations are partially shared, and generalization over their subsymbolic feature representation is favored. Similarly, when different objects are held by the hand, the shared neural representation of the hand allows generalization over the feature representations of the held objects.

According to the individuation-by-location hypothesis, a small number of attended objects may be indexed in time, the indexed individuals tracked through time and space, and the spatial relations among indexed individuals, represented (Carey and Xu, 2001). These indexes depend upon spatiotemporal information in order to remain assigned to individuals. In our implementation, each object initially located on a distinct location is mapped into a distinct cell assembly. The spatiotemporal individuation hypothesis is responsible mainly for the modeling of youngest infants' behavior (11- to 18-month-olds) (Fig. 13.6).

As explained above, a central function of the cell assembly module is to map new events into existent memory states or to create a new cell assembly representation, if the distance between the external input and the memory codes is higher than the vigilance parameter. In our model, the vigilance parameter is set in such a way that changes in the orientation and motion speed of an object are accepted as variations within the cell assembly corresponding to that object. Thus, when the object is moving, the corresponding cell assembly updates the location coordinates of the object, until this becomes occluded. Only when a cup is grasped by the hand, or when two cups are brought together, a new cell assembly is created (see the detailed description of the learning process in Section 13.4.6).

A cup becomes occluded if it is embedded in a larger cup. In this case, a decaying memory of the cup's existence at that location is preserved in the system. There is experimental evidence that the objects' indexes can be placed in the short-term memory, and that infants can create and store more than one memory model of sets of objects, and can compare them numerically in memory (Carey and Xu, 2001). With the increase in the developmental age of the robot infant, several cup representations can be stored in memory at the same location in space, and can be compared with respect to their feature properties (i.e., size, shape).

13.4.4 The attention module

Development of goal-directed imitation and object manipulation skill is supported by selective and joint attention mechanisms. The function of the visual attention module is to direct gaze towards objects of interest in the environment. A two-component framework for attention deployment has been implemented, inspired by recent research in modeling of visual attention (Itti and Koch, 2001; see also Itti and Arbib, this volume). *Bottom–up attention* is computed in a preattentive manner across the entire visual image. It has been suggested that the contrast of the features with respect to the contextual surround, rather than the absolute values of the features, drives bottom–up attention (Nothdurft, 2000). In this model, saliency is computed based on the linear integration of contrast of two features: color and motion. *Top–down attention* is deliberate and more powerful in directing attention. The robot has a pre-wired capacity to follow the gaze of another agent and to recognize the skin color of the hand. Skin color preference is used as an indicator of where are located the hands of the demonstrator. The weights of bottom–up and top–down constraints are set to satisfy a set of attention constraints, described in Table 13.2.

A two-dimensional saliency map is used to control the deployment of attention on the basis of bottom–up saliency and top–down cues. The focus of attention is deployed to the most salient location in the scene, which is detected using a winner-take-all strategy. Once the most salient location is focused, the system uses a mechanism of *inhibition of return* to inhibit the attended location and to allow the network to shift to the next most salient object (Itti and Koch, 2001).

Fig. 13.8 illustrates the functioning of the attention mechanism. Figure 13.8a illustrates the case when the imitator focuses its attention on the demonstrator's hands. Fig. 13.8b shows the time evolution of the saliency map output vs. inhibition, corresponding to the locations of the hands. After shutting down the salient unit, the inhibitory unit preserves a memory of its activation, which decays in time and allows the unit to win again further in future. Different locations can be attended to due to the inhibition mechanism.

Table 13.2 *Visual attention constraints*

Skin color preference For any static scene, the bottom–up saliency of the hand should be higher than that of any object.

Preference for moving stimuli For any moving object, its bottom–up saliency should be higher than that of any static object, including the hands.

Motion versus skin color preference Saliency of a moving object should be higher than that of a hand moving at a slower speed, but smaller than the saliency of a hand moving at comparable speed.

Gaze following versus moving objects The global saliency of any static object located in the focus of attention should be higher than the bottom–up saliency of any moving object located outside the focus.

Gaze following versus skin color The global saliency of any static object located in the focus of attention should be higher than the bottom–up saliency of any static hand located outside the focus.

Gaze following versus moving hand The bottom–up saliency of a moving end-effector should be higher than the global saliency of any static object placed in the focus of attention but smaller than the saliency of an object moving in the focus of attention.

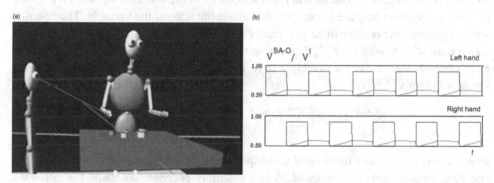

Figure 13.8 Operation of the attention module is illustrated for the case when the learner's focus of attention is driven by bottom–up attention constraints only (i.e., in the absence of the gaze signal). (a) The deployment of the focus of attention on one of the demonstrator's hands. (b) The time course of saliency output vs. the level of inhibition activation, corresponding to the shift of focus between the two hands. Inhibition regularly shunts down the saliency unit, allowing the shift of attention.

13.4.5 The cell assembly module

The cell assembly module consists of the object recognition (OR) network and the cell assembly CA layer (Fig. 13.7). The OR module implements a visual memory function on a short timescale for the representation of external objects. Each object is represented by its decomposition in five features: color, shape, size, rotation angle, and motion speed. Each feature unit receives input from three external units corresponding to the values read from the robot sensors and projected on the components X, Y, Z. For the detection of an event, the signal variation of a feature unit's activation is integrated over time and compared with a positive, arbitrary set threshold.

The learning algorithm consists of four main steps: (a) deployment of the focus of attention at the most salient location; (b) detection of new events occurring in the focused area; (c) search for the internal representation that matches the new event; and (d) adaptation of the weights as a function of the learning rules. If the best matching cell assembly does not satisfy the vigilance threshold, that is, if novelty cannot be accommodated to the knowledge retrieved from the memory, a new cell assembly is created for the corresponding category.

During learning, the activity $V_i^{CA}(t)$ of the cell assembly is a function f of the degree of satisfaction of the feature constraints $S_i^{CA-F}(t)$ between the cell assembly (CA) and the feature (F) layer and of the memory of its previous activation:

$$V_i^{CA}(t) = f\left(S_i^{CA-F}(t), \tau_i \cdot V_i^{CA}(t-1)\right) \tag{13.1}$$

where τ_i is the time decay rate of unit i. For simplicity of presentation we will give for the following equations only the parameters that affect the state of the variable. The function which computes the result will be generically noted with f or s. Satisfaction of the feature constraints, w_{ji}^{CA-F}, with $j \in F_i^{CA}$, depends on the current state of the external environment E and on whether the cell assembly i is in the focus of attention as measured by the saliency output $V_i^{SA-O}(t)$:

$$S_i^{CA-F}(t) = s\left(\sum_{j \in F_i^{CA}} w_{ji}^{CA-F} \cdot V_i^F(t), E, \theta_s, V_i^{SA-O}(t)\right) \tag{13.2}$$

where E is computed as a function of the output of the object recognition network $V_j^{OR}(t)$ and θ_s is an arbitrarily set threshold. A cell assembly becomes unsatisfied if either the difference between its feature constraints $\sum_{j \in F_i^{CA}} w_{ji}^{CA-F} \cdot V_j^F(t)$ and E is higher than θ_s or if $V_i^{SA-O}(t) = 0$. Thus, the satisfaction degree is a positive, symmetric measure of the distance between the cell assembly's feature constraints and the current state of the environment E.

During retrieval, the activation of a cell assembly depends on the satisfaction of feature and precedence constraints. The activity of CA_i during retrieval is given by

$$V_i^{CA}(t) = f\left(S_i^{CA-F}(t), S_i^{CA-P}(t), \tau_i \cdot V_i^{CA}(t-1)\right) \tag{13.3}$$

where $S_i^{CA-P}(t)$ represents the level of satisfaction of precedence (P) constraints. Satisfaction of the precedence constraints is defined as a function of the saliency output and of the summed activation of all predecessor units in P_i^{CA}

$$S_i^{CA-P}(t) = s\left(\sum_{j \in P_i^{CA}} w_{ji}^{CA-P} \cdot V_j^{CA}(t), \Theta_i^{CA-P}, V_i^{SA-O}(t) \right) \qquad (13.4)$$

where w_{ji}^{CA-P} are the weights of precedence links, and Θ_i^{CA-P} is the precedence threshold for the CA_i. Precedence is met by ensuring that the threshold Θ_i^{CA-P} is reached only by summing up the inputs in the order they have been learned (i.e., a strong connection requires a high input activation value in order to trigger an output). There is a supplementary condition which requires that only satisfied cell assemblies can activate their successors. This is necessary, in order to force the system to act externally (as opposed to an internal simulation of action) towards minimizing the distance between its goals and the current state of the external world.

13.4.6 Learning the seriated nesting cups task

Demonstration of the nesting cups task consists in the seriation of four cups by the demonstrator agent, using the subassembly strategy. Cups are placed as follows: cup1 (the smallest) in front of the demonstrator and cup2 to cup5 are arranged anticlockwise starting from the first cup. The demonstration order is cup1→cup2, cup1+cup2→cup3, cup1+cup2+cup3→cup4. For each cup the proximal hand is used to grasp and carry it.

During the first developmental stages, learning is driven by a basic-categorization process. The width of the categories is given by the value of the vigilance threshold ρ. A low value causes the formation of general categories, by forcing new states to be mapped into previously built cell assemblies. Each object initially located on a distinct location is mapped into a distinct cell assembly (see Sections 13.4.2. and 13.4.3). The vigilance threshold is set up based on the correlation matrix computed between the feature representations of all the objects in the system in all the possible states. The state of an object or end-effector changes as a function of the values of the following properties: orientation (i.e., rotation angle), motion (i.e., speed value), relation to the hand (i.e., whether it is held in the demonstrator's hand or not), and relation to another object (i.e., whether the objects are situated at the same location). The vigilance parameter is set in such a way that cell assemblies are resistant to variation on speed and rotation features. When the hand and an object or two objects are brought together, a new cell assembly is created. Co-located objects and end-effectors are subject to generalization and formation of more general categories.

The cell assemblies formed with a vigilance threshold ρ = 0.25 are shown in Fig. 13.9. Whether during each demonstration action, the system extracts specific goals of the type: "hand grasps cup1" and "hand places cup1 into cup2", until the end of the demonstration, these are gradually refined to most general categories: "hand manipulates cup" and "place cup into cup." At the end of the demonstration, "manipulates" stands for "grasp" and "carry," "hand" can be any of the hands, and "cup" stands for any of the *five* action cups.

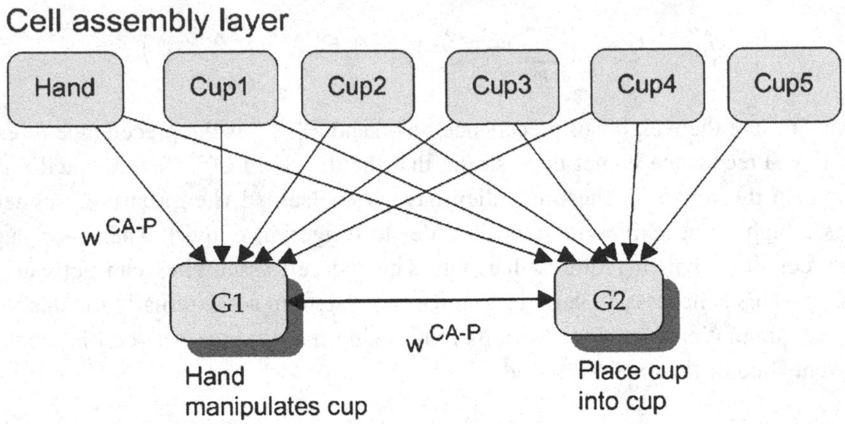

Cell assembly layer

Figure 13.9 Structure of cell assemblies developed by the system during the simulation of the earliest developmental stages of the nesting cups task. Six basic categories are formed, corresponding to the five cups and hand. Distinct cell assemblies are created for all objects and end-effectors that are initially perceived at distinct locations. Each of these cell assemblies is resistant to variations of the speed and rotation features. Two more cell assemblies are formed, corresponding to "hand manipulates cup" and "place cup into cup." We refer to these cell assemblies as *hidden*, as opposed to *visible* cell assemblies, because at the initiation of imitation they do not correspond to any state in the external world. Each hidden cell assembly learns a set of predecessors that can trigger its activation. If the precedence constraints are met over a certain threshold, the hidden cell assembly becomes a goal.

Feature weights learning

At the creation of a cell assembly CA_i, all the weights w_{ji}^{CA-F} received from the feature units $j \in F_i^{CA}$ are initialized to the values of the weights of the object recognition network. The object recognition network is locked to the object's position while this is tracked in time and space. If two objects are brought together, their OR subnetworks are updated correspondingly. A satisfied cell assembly (i.e., a cell assembly whose features satisfy the current state of the external world above a certain threshold) learns the distribution of the sensorial features that constitute the objects existent at the cell assembly location. The satisfaction condition ensures that the cell assemblies remain distinct and do not end up by storing the same representation. During the demonstration of the task, the weights of the satisfied cell assemblies are subject to a Hebbian adaptation rule, and converge to the last values $V_j^F(t)$ perceived on feature units j. A better solution would be to adapt weights such as to converge to a mean value of the values received from the OR network during demonstration.

Correlation weights learning

The imitative behavior of children indicates that infants as young as 12 months of age possess a size concept that is manifested in the capability to form simple nested structures.

At this age, children operate as though size is a binary concept, with one cup treated as the "biggest" while all the others belong to the category "little". Only infants of 28 to 36 months old seem to follow the size criteria for all structures formed.

We propose here a solution for how one can simulate the gradual development of the size concept, grounded on the interaction of the artificial system with the external world. The solution is based on the system's ability to learn invariant relationships between the features that constitute specific objects. Correlation weights w_{ji}^{CA-R} between the features of two objects j and i are adapted using an anti-Hebbian learning rule, which weakens the strength of the connection when only pre- or postsynaptic neurons are activated. Invariant features or unsystematic variations of the feature values cause weights to decay. Only systematic variations, such as the relation between the sizes of the acting and recipient cup, lead to an increase of the weights.

When the hand and an object or two objects are brought together for the first time, a new cell assembly is created for each of these external states. The resulted internal representation should persist long enough to permit learning of the relation existent between the features of the co-located objects. In our model, the external states corresponding to "place cup into cup" and "hand manipulates cup" receive a top–down attention signal, which preserves focus of attention on the corresponding objects and allows the internal comparison of their representations.

Figure 13.10 shows the time evolution of the correlation weights corresponding to the goals extracted during the demonstration of the nested cups task.

Precedence weights learning

With the creation of new cell assemblies, early cell assemblies tend to satisfy in a smaller manner the current state of the environment, and have a lower level of activity. A satisfied cell assembly CA_i learns a set of precedence links w_{ji}^{CA-R} from other activated cell assemblies CA_j. Precedence in the system is encoded in the relative order between the construction and satisfaction of the cell assemblies. Temporal learning between co-activated cell assemblies is favored by their graded activation as a function of the satisfaction level. The precedence links learned during the first developmental stage are shown in Fig. 13.9.

13.5 Reproduction of the first developmental stages of nesting cups behavior

Modeling of the developmental path of the ability to seriate cups necessitates the reproduction of two behavioral manifestations. First, the model has to account for the systematic differences existing between the infants' strategies. We consider as responsible for this, the learning process whose parameters are developmentally constrained and which leads to the construction of specific internal models. Second, the model must replicate the variety of imitative behaviors characteristic to human infants. This is achieved through a process of probabilistic satisfaction of multiple types of constraints.

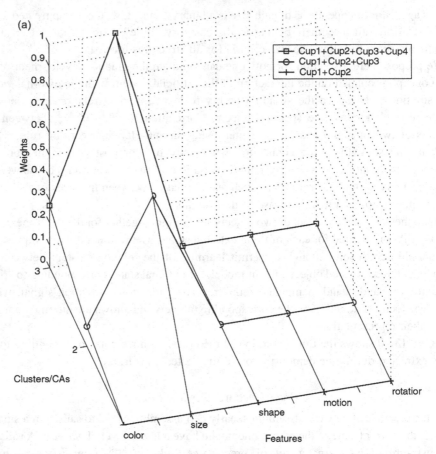

(a)

Figure 13.10 Evolution of the correlation weights corresponding to different clusters of sensorial features, during the demonstration of the nested cups task. (a) Shown on the z-axis are correlation weights for successive states when two or more cups are embedded (i.e., from state 1 to 3 on the y-axis). The size weight increases from 0 in the state "cup1 into cup2" to a maximal value in state "cup1 + cup2 + cup3 into cup4." This increase of the weight reflects the systematic relationship existent between the sizes of the acting and the recipient cup. (b) Correlation weights on the z-axis corresponding to the states when the hand grasps a cup (i.e., from 1 to 3 on y-axis). At the end of the demonstration, the weight values corresponding to size, shape and rotation features are maximal, indicating the existence of an invariant relationship between the hand and the acting cup for each of these features. These constraints will be integrated at retrieval during the third developmental stage.

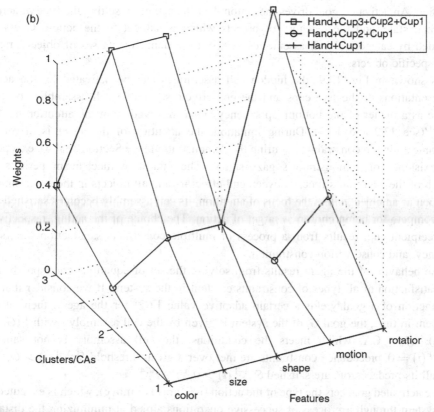

Figure 13.10 (cont.)

During retrieval, three types of constraints are operating. The *conservation* C constraint reflects the tendency of the system to minimize the amount of effort during the execution of the task. Its operation is reflected in the choice of proximal cups and in the system's preference to rest in a state with minimal energy (i.e., where all goals are satisfied). The operation of the *saliency* (SA) constraint is reflected in the choice of objects with a high color or motion contrast. Finally, *size* (S) constraints are satisfied when the imitator embeds a smaller cup within a larger cup. The size constraints are learned in the set of weights w_{ji}^{CA-R}, while saliency constraints are applied as a function of the attention module's output $V^{SA-O}(t)$. The conservation constraints C are applied: (a) by computing the physical distance between the hand and a given object; and (b) by ending the imitative behavior as soon as all the system goals are satisfied.

During imitation, the actions of the agent are driven by its internal model, more specifically, by those goals which are not yet satisfied. The internal model has the role of activating and deactivating the goal states: "hand manipulates cup" (G_1) and "place cup into cup" (G_2). Activation of a goal occurs when the precedence constraints are met and deactivation of the goal takes places when its feature constraints are maximally satisfied. An activated goal drives the action A of the agent, by setting the type of action, the final state, and the category of objects which is subject to the action. Goals are encoded by categorical cell assemblies, hence, they stand for classes of objects, rather than specific objects.

As shown in Fig. 13.9, the hidden cell assemblies can be activated by the active representations of the five cups and the end-effectors. Visible cell assemblies become active as a matter of the bottom–up saliency signal received from the attention module V_i^{SA-O}(see 13.2 and 13.3). During imitation, the attention of the agent is driven by the same attention constraints as during the demonstration (see Section 13.4.4), excepting the existence of demonstrator's gaze signal. The inhibition mechanism permits the switch of the attention's focus between end-effectors and all objects in the environment. As soon as an object enters the focus of attention, its cell assembly becomes satisfied and can compete for being chosen as target of action. The choice of the acting, respectively the recipient cup, results from a process of multiple constraints satisfaction (i.e., size, saliency, and conservation constraints).

The behavior of the agent results from solving the set of equations corresponding to the satisfaction of all types of constraints existent in the system. If we consider that the satisfaction of a goal yields a certain adaptive value $\Gamma(G)$ for the agent, then, at any moment in time, the goal G_i of the system is given by the cell assembly i with $\Gamma(G_i) > \Gamma(G_j)$, $\forall j \in$ CA, which meets the conditions: the cell assembly is not satisfied $S_i^{CA-F}(t) = 0$, precedence constraints are met over a given threshold $S_i^{CA-P}(t) > \Theta_i^{CA-P}$, and all its predecessors are satisfied $S_j^{CA-F}(t) > 0, \forall j \in P_i^{CA}$.

The activated goal sets the type of the action (i.e., grasp or move), which is executed by the system through a process of successive operations aimed at minimizing the distance between the current state of the world and the desired state corresponding to the goal. Elsewhere (Goga and Billard, 2004) we discussed how the system could learn the sequence of actions required to grasp the object (i.e., rotate the end-effector, move it towards the object, lift the object). Here, we follow a similar approach, based on the computation of the difference between the current and the desired value for each feature, and sending an appropriate command to the robot's actuators for the minimization of this distance.

The acting cup i and recipient cup j are set as a function of the maximal consonance computed over all constraints in the system, $\max_{i,j} \chi_{ij}(\sum_{k=1}^{3} p_k \cdot C_k)$ where p_k are the probabilities of size, saliency and conservation constraints C_k and χ_{ij} is the consonance computed between objects i and j (see Section 13.5.2). The agent has a pre-wired drive towards the maximization of the internal consonance computed. The system also possesses a regulating mechanism in order to decide when to terminate behavior. The infant robot stops imitating when the goal G_i with the highest value $\Gamma(G_i)$ is maximally satisfied.

In the following, we present several behavioral scenarios, corresponding to the probabilistic combination of the criteria for choosing the acting and recipient cups: size, saliency, and conservation of energy. Behavior is consistent for any one setting of the constraints.

13.5.1 Saliency constrained behavior

In the first behavioral scenario, constraints are applied as follows. The acting cup is set to the first cup that wins the focus of attention, which is the most salient object in the environment (i.e., has the highest color contrast). For all satisfied cell assemblies, size (S) and conservation (C) constraints are computed, and the recipient cup is set to the first cup that satisfies an arbitrary combination of these. Figure 13.11 shows different imitative behaviors, corresponding to different settings of the probabilities of size and conservation constraints. In Fig. 13.11a the most salient cup becomes the acting cup and minimization of the path is applied, leading to the formation of one pair. In Fig. 13.11b, after the acting cup is chosen, size constraints are computed for the cups being in the focus of attention (see Section 13.5.2 on how size consonance is computed), and the cup is nested in a larger cup. In both scenarios, after the acting cup is chosen, the hidden cell assembly corresponding to "hand manipulates cup" becomes the goal of the system, and triggers the set of

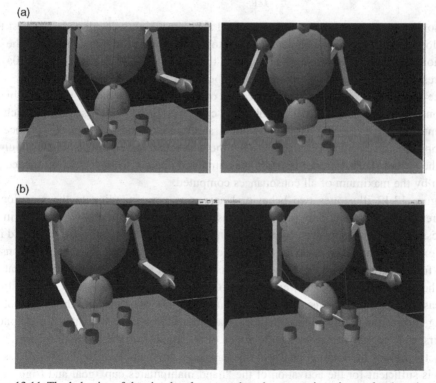

Figure 13.11 The behavior of the simulated agent, when the system is acting under the primacy of saliency constraints. (a) The most salient object (cup3) wins the focus of attention and becomes the acting cup. The recipient cup (cup2) is chosen in order to minimize the path. (b) The most salient object (cup3) becomes the acting cup and the recipient cup (cup4) is chosen after size constraints are applied for the satisfied cell assemblies.

actions required for its achievement. The robot's actuators receive the coordinates of the acting and target cup, and the end-effector is moved to the desired location through the inverse kinematics and dynamic motor servos.

13.5.2 Size constrained behavior

When the probability of the size constraint is higher than those of conservation and saliency constraints, the system chooses the acting and the recipient cups as a function of the global consonance χ computed based on the internal correlation constraints. The term *consonance* is used to reflect the fact that the behavior of the agent is constrained by its internal model on how different objects can be combined together. The consonance between two objects i and j is obtained through the summation over all features $k \in F$ of the products between the correlation weight $w_{ji}^{\text{CA-R}}$ and the difference between the activation values on the corresponding feature:

$$\chi_{ji} = \sum_{k \in F} w_{ji}^{\text{CA-R}} \cdot (V_j^k - V_i^k) \tag{13.5}$$

Consonance is computed as a function of all relation weights developed. The first two developmental stages are characterized by top–down attention biases towards the extraction of relational information concerning the size feature of objects (Fig. 13.10a). In this respect, we describe the corresponding behavior as being size constrained.

The system acts towards the maximization of the global consonance. For each object i its consonance is computed with all objects j corresponding to cell assemblies which are currently satisfied. For the pairing and pot strategy model, the maximal consonance for any object is given by the product between the size relation weight (i.e., highest weight) and the largest size difference between the compared objects. The global consonance χ is given by the maximum of all consonances computed.

Figure 13.12 illustrates two behavioral scenarios corresponding to the seriation of different cups. Note that only satisfied cell assemblies can be internally compared in the process of global consonance computation. In Fig. 13.12a, the smallest cup is placed into the next largest cup. The behavior corresponds to the computation of global size consonance for all demonstrated cups (the largest cup was not used during the demonstration) and the acting and target cups were chosen to maximize this value. Fig. 13.12b illustrates the case when internal consonance is computed for only two cups and the choice of the acting and recipient cups reflects the local maximum found. The two cups correspond to the first objects that attract the agent's focus of attention. Initiation of action depends on the activation of the internal goals. In this case, the satisfaction of the cups' cell assemblies is sufficient for the activation of the "hand manipulates cup" goal and triggers the initiation of action.

Various nesting behaviors (i.e., behaviors that satisfy size constraints, but do not maximize global consonance) can be simulated as a function of three parameters: the *ignition value*, the *precedence threshold* and the *number of internal comparisons*

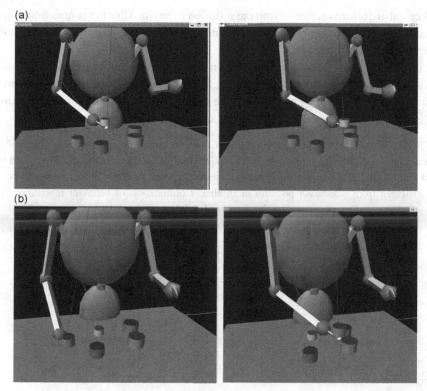

Figure 13.12 Two behavioral scenarios where action is primarily driven by the satisfaction of size constraints. (a) Internal consonance is computed for all satisfied cups (see text for explanations) and the acting and recipient cups are chosen to maximize its value. This will correspond to the movement of cup1 (the smallest cup) into cup4 (the next largest cup). (b) Internal consonance is computed for the first two cups, which enter the focus of attention. The choice of the acting and recipient cups reflects the local maximum found (see text for explanations on the role of the internal parameters on the reaction time of the robot). Accordingly, the agent moves cup2 into cup4.

performed. The precedence threshold Θ^{CA-P} (13.4) is an internal, emergent parameter of the model, while the ignition value and the number of internal comparisons are developmentally constrained parameters. The ignition value represents the percent of precedence constraints that should be met for a goal to become activated. These parameters affect the time elapsed until the activation of the first goal of the system, and represent the *reaction time (RT)* of the system.

13.5.3 One pair, two pairs, transfer of the cup

The model described above can replicate the behavioral manifestations characteristic to human infants using the pairing cups strategy. The simulated agent can form a nested structure or a tower, as a matter of the probability of size constraints satisfaction. The model can also decide whether to stop the seriation behavior after the completion of a pair or to continue it. The infant robot can form one, two, or several pairs, as a function of the termination condition imposed.

Let us analyze the state of the system when the hand carrying a cup reaches the target location. At this moment, both goals, G_1 corresponding to "hand manipulates cup" and G_2 corresponding to "place cup into cup," can be satisfied. If we consider that satisfaction of a goal yields an adaptive value $\Gamma(G)$ for the agent, two situations can occur:

- If $\Gamma(G_1) > \Gamma(G_2)$, that is, if grasping the cup is more important than forming a nested cups structure, the system will act to maximize the satisfaction of G_1. In this case, the cup is held in the hand and G_2 does not become satisfied. It corresponds to an imitative behavior where one cup is transferred from one target position to another. Continuation of behavior is assured by the non-satisfaction of the G_2 goal. See the flow on the right of Fig. 13.13.

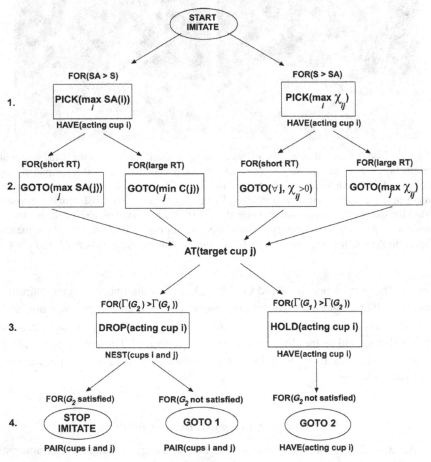

Figure 13.13 Planning graph showing the states of the system during the reconstruction of the demonstrated actions and the parameters that determine the transition from one state to another.

- If $\Gamma(G_2) > \Gamma(G_1)$, the system will act towards the maximization of the satisfaction of G_2. Accordingly, the cup will be dropped at the target location. The system stops if the pair formed satisfies the goal G_2, or it continues if the pair does not satisfy the features of G_2 (i.e., if the formed pair has never been perceived before) (see the flows on the left of Fig. 13.13).

13.5.4 The pot strategy

Hypothetically, the transition from the pairing to the pot strategy can be modeled in two ways: (1) by increasing the number of goals that are learned during the demonstration of the task (i.e., the formation of a sequence of subgoals corresponding to the embedding of several cups into the pot cup); or (2) by preserving the main structure of the internal model, and increasing the information stored within each goal representation. The second alternative comes more naturally within our developmental model, which acts towards the reduction of the amount of information stored in the memory.

We consider that during the second stage of development, cognitive resources (i.e., attention and memory) are employed to extract more information concerning the goal "place cup into cup." This leads to an increased capacity to store and retrieve the information concerning the number of cups that are embedded during demonstration. In the cell assembly structure, this acquisition is reflected in the specialization of the state goal "place cup into cup," by storing a symbolic representation of all the cups that can be embedded. The behavior of the agent evolved in this second developmental stage towards the maximization of the size consonance, computed for all the objects stored in the representation of the state goal "place cup into cup." With each new pair formed, the global consonance increases, and makes less probable with behaviors such as the transfer of one cup through several positions.

Various behaviors can be modeled through the manipulation of the factors mentioned above (i.e., precedence threshold, number of internal comparisons) and as a matter of how global consonance is computed: a tower vs. a nest; a pot containing all cups vs. two or three cups nested. Figure 13.14 illustrates an imitative behavior corresponding to the nesting of three cups using the pot strategy.

Caption for Figure 13.13 (*cont.*)
At top of each state, the precondition is shown (i.e., FOR) and at bottom the post condition (i.e., HAVE). Actions at level 1 correspond to the choice of the acting cup as a function of the saliency (SA) and size constraints (S) applied. Actions at level 2 describe the criteria considered in the choice of the target cup. The system can react faster (short reaction time RT) or slower (large RT) and conservation (C) constraints can also be integrated. If the adaptive value of goal G_2 "place cup into cup" is higher then that of goal G_1, the system forms one pair (level 3), after which it can continue with the pairing behavior or stop the imitation (level 4). If the adaptive value of goal G_1, "hand manipulates cup" is higher then that of goal G_2, the system holds the acting cup, and the resulting behavior corresponds to the transfer of one cup from a target position to another.

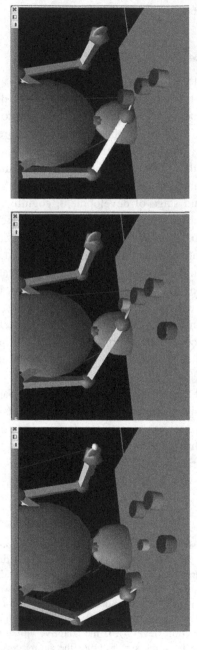

Figure 13.14 Imitative behavior of the simulated agent corresponding to the formation of a pot at the position of cup4. The first cup embedded is cup2, followed by the smallest cup from position 1, followed by the cup from position 3. Size constraints are satisfied with respect to the acting cup and the recipient pot cup. A tower may result (not shown), because consonance is computed between visible cups, and the size of the already nested cups is not taken into account.

13.6 From nesting cups model to an action–language model

Our task now is to provide a computational framework that extends the action system towards the integration of a language system, and to discuss the possibility that previously developed structures for goal-directed imitation and object manipulation can provide a substrate for language bootstrapping. The working hypothesis is that the capacity for language and the ability to combine objects and to imitate novel behavior follow a common developmental path, and may share a common neural architecture, where they influence each other.

13.7 Lexical grounding on subsymbolic knowledge

An action–language computational model has at its foundation a mechanism for grounding the conceptual information on subsymbolic structures. Our work builds on previous modeling attempts on the world-to-word mapping.

Regier (1995) described a system capable of learning the meaning of spatial relations such as "on" and "under" from pictorial examples, each labeled with a spatial relation between two of the objects in the scene. Learning spatial knowledge was based on a tripartite trajectory representation of type *source–path–destination*, aimed at grasping the event logic in both motion and language. Bailey (1997) developed the VerbLearn model, where actions are represented using executing schemas, while words are encoded using structures of features, given by the parameters of the motor system: acceleration, posture/shape of the hand, elbow joints, target object. The interface between language and action levels is played by a linking feature structure, which binds the words bidirectionally with the actions. Siskind (1995, 2001) proposed a conceptual framework for grounding the semantics of events for verb learning, using visual primitives which encode notions of support, contact, and attachment. Event logic has been recently applied by Dominey (2003) for learning grammatical constructions in a miniature language from narrated video events, and by Billard *et al.* (2003) to the learning and reproduction of a manipulation task by a humanoid robot.

Our previous work on learning a synthetic protolanguage in an autonomous robot (Billard, 2002) used a time-delay associative network, consisting of a Willshaw network (Willshaw *et al.*, 1969), to which self-recurrent connections have been added, to provide a short-term memory of activation of units. Sensor and actuator information is memorized for a fixed duration to allow association to be made between time-delayed presentations of two inputs. For the work described here, we used a similar architecture consisting of two time-delay networks, one feeding in sensorimotor information and the other linguistic input, connected through a set of bidirectional, associative links. The model takes as input pictures of the environment (i.e., an agent demonstrates how an object can be moved on a table) and short sentences about them. The aim is to ground, using the set of associative, bidirectional links, the meaning of a small lexicon in the structure of sensorimotor features available to the simulated agent.

There are a number of challenges that a system faces in learning word meanings. Among these are: (a) multi-word utterances, where the agent has to figure out which words are responsible for which parts of the meaning, and (b) bootstrapping, i.e., when first words are learned, the system cannot use the meaning of words in an utterance to narrow down the range of possible interpretations. To solve the former problem, we followed the approach described in Siskind (1996) by applying cross-situational learning to the symbols that make up the meaning of a word. *Cross-situational learning* refers to the idea that children narrow the meaning of words by retaining only those elements of meaning that are consistently plausible across all the situations in which the word is used.

Hebbian and anti-Hebbian learning rules adapt the associative links between the sensorial and linguistic time-delay layers, in such a way that each word develops a set of associative links with the feature units. Words can compete for labeling an associated feature. The winner is the unit with the highest connection strength that passes an arbitrary confidence threshold. Autonomous bootstrapping, consisting in extracting tiny bits of linguistic knowledge and using them for further analysis of the inputs, is applied to simplify and urge the learning process.

As a result of the specifics of Hebbian learning, frequently used words (e.g., the verb "move") develop stronger weights, and the systematic association between a word and a set of features is reflected in the development of a rich and strong set of feature weights. Anti-Hebbian learning is meant to decrease the weights for variant, unsystematic associations between a word and world features (e.g., for the word "look").

13.7.1 Attention-directing interactions during the seriated cups task

The preliminary experiments described above have been conducted with an artificially generated lexicon. The next step in the development of the action–language computational framework will be to train the model with input from natural language. We are currently running a set of experimental studies for the systematic observation of the interaction between a human caregiver and the child during the seriated nesting cups task (see Fig. 13.15). In this respect, the original experiment of the seriated nesting cups task is modified in several ways. The demonstrator is replaced by the child caregiver, who interacts with the infant in three different conditions. During the first condition the infant and the caregiver are playing with several toys and nested cups. The second condition is similar to the experimental setting of the original seriated nesting cups task (Fig. 13.15a). The caregiver demonstrates the seriation task and also interacts with the child during the imitation phase, by directing her attention and providing feedback on the success of the task. In the third condition, the caregiver is allowed to freely interact with the infant during the demonstration and imitation periods, using linguistic imperatives and gestures, in order to help the infant build the seriated structure using the most advanced strategy (Fig. 13.16b). Infants between 12 and 36 months of age are observed and all sessions are recorded on video and audio.

(a)

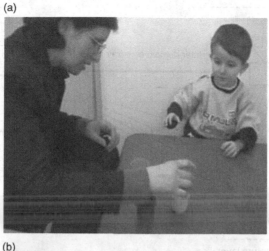

(b)

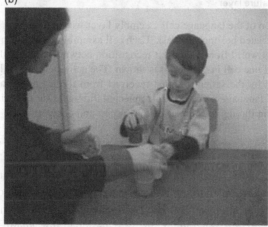

Figure 13.15 Human child–caregiver experimental setting. (a) The caregiver demonstrates the seriated nesting cups task. (b) The child imitates the caregiver, while the latter/guides the child's attention using linguistic cues and gestures.

The video and audio recordings will be transcribed using CHAT system (Codes for Human Analysis of Transcripts) (MacWhinney, 2000). CHAT was developed and used as the standard transcription system for the CHILDES project (Child Language Data Exchange System) aimed at researching infants' directed speech. The transcripts of the records will be analyzed and segmented, and relevant linguistic sequences from the human child–caregiver interaction will be selected and used as training input for an action–language computational model. We envisage the existence of two stages in the development of the linguistic function in robot infants. During the first stage, the model will be trained with simple descriptive sentences in order to learn the meaning of a basic set of words. This will enable the imitator to communicate with the demonstrator during the seriated cups task.

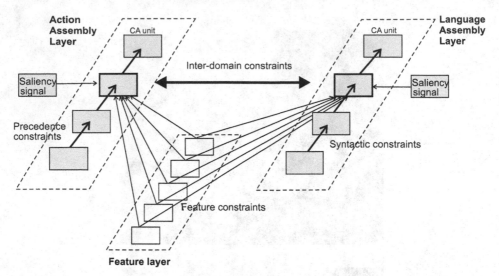

Figure 13.16 Integration of the language cell assembly layer within the action model developed for the simulation of the seriated nesting cups task. Each cell assembly layer is grounded on the internal mapping of the external world from the object recognition network. For simplicity of presentation, only the connectivity of one cell in each layer is shown. The activity of the action and linguistic cell assembly is modulated by the saliency signal received from the attention module (not shown in figure). The action and language networks are connected through a set of bidirectional, associative links, subject to learning through asymmetric cooperation and competition.

When the agent becomes capable of naming the objects and the events occurring in the environment, the input provided to the imitator during the demonstration task can become more complex. Relevant attention-directing sequences from the human child–caregiver interaction will be fed into the model, by the demonstrator agent. The demonstrator robot will accompany the seriation actions with linguistic cues and short descriptions of what it is doing, in order to direct the imitator's attention and to explain the goal of the demonstration. In the current model of seriated nesting cups task, the joint attention mechanism is meant to spotlight objects and events being attended to, through the integration of a number of bottom–up and top–down saliency cues (e.g., color contrast, motion contrast, gaze follow, and pointing). In the action–language model we intend to enhance the power of the joint attention mechanism, and to investigate the role of linguistic and gestural attention cues in scaffolding the robot's early language and goal-directed imitation development.

13.7.2 A computational view on the action–language reuse hypothesis

Having established a basic action and imitation system, we want to go beyond the "mirror hypothesis" and to investigate how the acquisition of language capability may interact with previously constructed structures for goal-directed imitation and object manipulation. Further development of the action model is towards the integration of the language network within the cell assembly architecture developed for the seriated nesting cups task. The model will be trained with sequences of gestural and symbolic attention-directing interactions between the human child and caregiver, as described above.

The architecture of the envisaged action–language model is shown in Fig. 13.16. The language conceptual layer develops in a similar manner with the construction of the action cell assembly layer. The adaptation of the language network's sets of weights is governed by similar learning rules. The basic categorization process may be responsible for the mapping of words into linguistic cell assemblies. The vigilance parameter can be used to tune the generality of the linguistic categories formed. Each linguistic cell assembly is grounded in a subsymbolic sensorial representation from the object re cognition network. In this way, the semantic features of a word are grounded on the feature constraints of a linguistic cell assembly. Short serial-order dependencies between words may be learned in the precedence constraints at the cell assembly level. The cell assembly layers are connected through bidirectional, associative weights, which play a major role in applying the neural patterns of activity established in the action layer to the developing processes taking place in the language layer. To give an account of how this process of interdomain bootstrapping may take place, we introduce the reader to the theory of Cortical Software Re-Use (CSRU) (Reilly, 2001; Reilly and Marian, 2002).

The CSRU theory states that the sensorimotor areas of the brain provide the computational building-blocks for higher-level functions. A computational account of how motor programs developed for object manipulation might be reused for language syntax was provided by Reilly (1997) in responding to the work of Greenfield *et al.* (1972). Reilly (1997) selected the seriated nesting cups task and an equivalent speech task and implemented them within a connectionist framework. The hierarchical structures of the motric and linguistic representations were first encoded using a recursive auto-associative memory (Pollack, 1990), then the goal representations were fed into a simple recurrent network, which had to generate the appropriate sequence of actions as output. The reusability hypothesis was consistent with the finding of a training advantage when a recurrent network was pretrained on the object assembly task prior to learning a language task.

The neural building-block proposed by CSRU theory is the *collaborative cell assembly* (CCA). Re-use is operationalized through the operation of different types of computational collaborations between cell assemblies. *Structural isomorphic reuse* occurs from exploiting approximately the same cortical region (see Greenfield's (1991) hypothesis on the action–language homology on the Broca's area). *Symmetric collaboration* occurs between equally developed cell assemblies, and employs an "indexing" operation of a neural population by another neural population (Reilly and Marian, 2002). An example can be a system that learns to foveate a visual target and then "reuse" the saccade map to achieve ballistic reaching (Scassellati, 1999). Finally, *asymmetric collaboration* occurs where a less well-developed cell assembly exploits the functionality of a more developed one. We believe that the latter form of collaboration can play an important role in explaining the neural patterns of competition and cooperation, responsible for the integration of proprioceptive, sensorimotor and linguistic information.

Our approach to the modeling of neural grammars is inspired by the concept of *word-sensitive sequence detectors* proposed by Pulvermüller (2003). These can be assumed to operate on *pairs* of elementary units, for the detection of frequently occurring sequences of two words. Sequence detectors could operate on webs representing words from given lexical categories, and could support generalization of syntactic rules to novel word strings. For instance, separate sequence detectors could respond to pairs of the types: pronoun–verb, verb–verb suffix, and preposition–noun. By operating on short rather than long temporal dependencies, the word-sequence detectors are scalable and composable. They allow temporal processing of complex structures through the assemblage of simple sequence detectors in a hierarchical, recursive, and compositional fashion.

Operationalization of the re-use hypothesis comes naturally in our model. First, the word's semantic is grounded in the sensorial representation of the associated object/event. This enables us to investigate the means by which word understanding and production can be affected by the operation of previously constructed sensorimotor structures. Second, the robot must process, store, and retrieve syntactical information from the transcribed sequences of the human child–caregiver interactions. These processes will take place during the simulation of a social interaction protocol, when the infant robot must also execute different sequences of movements, according to its internal model developed.

The operation of the action schema for goal-directed imitation introduces a number of constraints on learning by the robot infant how to generate correct and meaningful sentences. We intend to investigate the dynamic computational means for cooperation and competition between the neural patterns established in the action network and those developing in the language network. The goal is to study not only how the linguistic function can make use of the sequence detectors developed for the seriation ability, but also how goal-directed action can use top–down linguistic cues to discover "what" to imitate.

13.8 Discussion

In this chapter we have presented ongoing work on modeling the action–language analogy hypothesis. We have used the seriated nesting cups task (Greenfield *et al.*, 1972) as a developmental framework for investigating the interaction between goal-directed imitation, object manipulation, and language acquisition. A conceptual framework drawing on neurobiological and psychological experimental data was proposed to account for the development of the goal structure that lies at the foundation of object manipulation and language usage in infants and robots. In modeling the seriated nesting cups behavior, we investigated the effect of varying a number of parameters of the computational model, as a way of accounting for systematic differences in child behavior.

A central tenet of our model is that development of goal-directed imitation and object manipulation skill is supported by joint attention mechanisms. These mechanisms play an important role in coordinating the behavior and sharing a similar perceptual context necessary for the learner to infer the demonstrator's intentions and goals (Tomasello, 1998). During the last decade, several researchers have attempted to build mechanisms of joint attention in robots (Scassellati, 1999; Kozima and Yano, 2001; Nagai *et al.*, 2003). Our approach is unique in that it focuses on the integration of multiple attention constraints in order to obtain more complex behavior for the agents.

Our model also contributes to the theory of goal-directed imitation, and the experimental data showing that infants learn *what* to imitate before learning *how* to imitate (Bekkering *et al.*, 2000). Modeling of the pairing and pot strategies in seriating cups was based on the assumption that the robot infant learns first about objects, then about the relations between them, and finally about the events and movements. The internal model developed during the first stages reflects the primacy of object related information. By learning *how* to imitate, we mean learning how the goal of the imitation task (i.e., the pot of cups) is achieved using a certain strategy (i.e., subassembly strategy) by imitating the way the hand movements are sequenced relative to the acting objects.

Modeling of the seriated cups task was challenging in that the resulting model had to reproduce the consistency of infant strategic behavior, as well as the variety of human behavior. The systematic differences between infants' strategies were accounted for through a learning process whose parameters are developmentally constrained (i.e., basic categorization, spatiotemporal object representation, joint attention mechanism). In the model presented here, the basic categorization process plays a crucial role in modeling the pairing and pot strategies.

With the increasing resources available to the learning process, the system is able to extract, store, and remember more information from the sequence of demonstrated actions. As pointed out in the methodological approach outlined in Section 13.1, the system gradually develops to accommodate more complex behaviors, by integrating in a cumulative manner prior structures that provide competencies to be reused by later developed skills.

The developmental model of the seriated nesting cups task is to our knowledge the first computational proposal of this type. A central tenet of this approach is that modeling of the seriated cups task can best be addressed in a multiple constraints satisfaction framework. The variety of imitative behaviors was replicated through a process of probabilistic satisfaction of three types of constraints (conservation, size, and saliency). The choice of the constraints was inspired by the experimental observations of Greenfield and colleagues. They describe the existence of three criteria in the infants' choice of the objects for action: proximity, size, and contiguity. In our model, conservation C stands for the satisfaction of proximity and contiguity. Saliency was introduced to account partially for the variability of the behavior observed in the group of human infants.

At this point, the model lacks the simulation of the third and most advanced sub-assembly strategy for nesting cups. We envisage modeling of this stage along the lines described above, based on several developmental assumptions: (a) the imitator's focus of attention shifts from the objects to the movements executed by the hand; (b) the increase of the vigilance parameter leads to the specialization of the goal "hand manipulates cup"; and (c) during retrieval other constraints than those given by the size weights may be taken into account in computing the internal consonance of the model (see Fig. 13.10b).

As a drawback, the model does not have a consistent way to handle noise in the environment.

13.8.1 Relevance to the Mirror Neuron System hypothesis

Computational modeling can significantly contribute to the understanding of the wider impact that the mirror neuron system has on brain computation. Evidence of the mirror neuron system and its role in driving action, language, and imitation is still new (Rizzolatti and Arbib, 1998; Iacoboni *et al.*, 1999). With respect to this, computational modeling significantly contributes to the understanding of the wider impact that the mirror neuron system has on brain computation (Oztop and Arbib, 2002; Oztop *et al.*, this volume). The work presented here can bring computational evidence for the hypothesis that the abilities to imitate and to manipulate objects lie at the foundation of language development in humans.

Greenfield *et al.* (1972) suggested that the internal model of seriated cups was used by the youngest children mainly to get the nesting behavior going, whereas it was used by the older children as a basis for terminating activity as well. For them, the final stage of the demonstration appeared to function as a precise goal and signal for termination. Similarly, during the first developmental stage, the robot infant extracts a generalized goal state of the form "put the cups inside each other," while during the second stage, the goal representation is restructured to "put several/all cups into the largest cup." Only during the third developmental stage is the agent capable of extracting detailed information on the order in which the hand manipulates the cups, and to infer the demonstrator's goal "put all cups in the largest cup by placing each cup into the next largest cup."

Acknowledgments

This research was carried out during the visit of Ioana (Marian) Goga to the Autonomous System Laboratory, L'Ecole Polytechnique Fédérale de Lausanne, and was supported by the Swiss National Science Foundation through grant 620-066127 of the SNF Professorships Program. The authors are very grateful to Stefan Schaal for providing access to the Xanim simulation environment for the experiments presented here.

References

Arbib, M., 2003. The evolving mirror system: a neural basis for language readiness. In M.H. Christiansen and S. Kirby (eds.) *Language Evolution: The States of the Art*. Oxford, UK: Oxford University Press, pp. 182–200.

Bailey, D.R., 1997. When push comes to shove: a computational model of the role of motor control in the acquisition of action verbs. Ph.D. dissertation, University of California, Berkeley, CA.

Bekkering, H., Wohlschlager, A., and Gattis, M., 2000. Imitation is goal-directed. *Q.J. Exp. Psychol.* **53A**: 153–164.

Billard, A., 2002. Imitation: a means to enhance learning of a synthetic proto-language in an autonomous robot. In K. Dautenhahn and C.L. Nehaniv (eds.) *Imitation in Animals and Artifacts*. Cambridge, MA: MIT Press, pp. 281–311.

Billard, A., and Dautenhahn, K., 2000. Experiments in social robotics: grounding and use of communication in autonomous agents, *Adapt. Behav.* **7**. 411–434.

Billard, A., and Hayes, G., 1999. DRAMA, a connectionist architecture for control and learning in autonomous robots. *Adapt. Behav. J.* **7**: 35–64.

Billard, A., and Mataric, M., 2001. Learning human arm movements by imitation: evaluation of a biologically-inspired connectionist architecture. *Robot. Auton. Syst.* **941**: 1–16.

Billard, A., Epars, Y., Schaal, S., and Cheng, G., 2003. Discovering imitation strategies through categorization of multi-dimensional data. *Proceedings Int. Conference on Intelligent Robots and Systems*, Las Vegas, NV, pp. 2398–2403.

Breazeal (Ferrell), C., and Scassellati, B., 2002. Challenges in building robots that imitate people. In K. Dautenhahn and C.L. Nehaniv (eds.) *Imitation in Animals and Artifacts*. Cambridge, MA: MIT Press, pp. 363–390.

Brooks, R.A., Breazeal, C., Irie, R., *et al.*, 1998. Alternative essences of intelligence. *Proceedings 15th National Conference on Artificial Intelligence*, Madison, WI, pp. 961–968.

Byrne, R.W., and Russon, A.B., 1998. Learning by imitation: a hierarchical approach. *Behav. Brain Sci.* **21**: 667–721.

Calinon, S. and Billard, A., in press. Learning of gestures by imitation in a humanoid robot. In K. Dautenhahn and C.L. Nehaniv (eds.) *Imitation and Social Learning in Robots, Humans and Animals: Behavioural, Social and Communicative Dimensions*. Cambridge, UK: Cambridge University Press.

Carey, S., and Xu, F., 2001. Infant's knowledge of objects: beyond object files and object tracking? *Cognition* **80**: 179–213.

Carpenter, G. A., and Grossberg, S., 1987. ART 2: Self-organization of stable category recognition codes for analog input patterns. *Applied Optics* **26**: 4919–4930.

Chappelier, J.C., Gori, M., and Grumbach, A., 2001. Time in connectionist models. In R. Sun and C.L. Giles (eds.) *Sequence Learning: Paradigms, Algorithms, and Applications*. New York: Springer-Verlag, pp. 105–134.

Clark, A., 1997. *Being There: Putting Brain, Body and World Together Again*. Cambridge, MA: MIT Press.

Demiris, J., and Hayes, G.M., 2002. Imitation as a dual-route process featuring predictive and learning components: a biologically plausible computational model. In K. Dautenhahn and C.L. Nehaniv (eds.) *Imitation in Animals and Artifacts*. Cambridge, MA: MIT Press, pp. 321–361.

Dominey, P.F., 2003. Learning grammatical constructions from narrated video events for human–robot interaction. *Proceedings IEEE Humanoid Robotics Conference*, Karlsruhe, Germany

Elman, J.L., 1993. Learning and development in neural networks: the importance of starting small. *Cognition* **48**: 71–99.

Festinger, L.A., 1957. *A Theory of Cognitive Dissonance*. Stanford, CA: Stanford University Press.

Frezza-Buet, H., and Alexandre, F., 2002. From a biological to a computational model for the autonomous behavior of an animal. *Inform. Sci.* **144**: 1–43.

Glenberg, A.M., and Kaschak, M.P., 2002. Grounding language in action. *Psychonom. Bull. Rev.* **9**: 558–565.

Goga, I., and Billard, A., 2004. *A Computational Framework for the Study of Parallel Development of Manipulatory and Linguistic Skills*, Technical Report. Lausanne, Switzerland: Autonomous Systems Laboratory, Swiss Institute of Technology.

Goodson, B.D., and Greenfield, P.M., 1975. The search for structural principles in children's manipulative play. *Child Devel.* **46**: 734–746.

Greenfield, P., 1991. Language, tool and brain: the ontogeny and phylogeny of hierarchically organized sequential behavior. *Behav. Brain Sci.* **14**: 531–550.

Greenfield, P.M., and Schneider, L., 1977. Building a tree structure: the development of hierarchical complexity and interrupted strategies in children's construction activity. *Devel. Psychol.* **13**: 299–313.

Greenfield, P., Nelson, K., and Saltzman, E., 1972. The development of rulebound strategies for manipulating seriated cups: a parallel between action and grammar. *Cogn. Psychol.* **3**: 291–310.

Grossberg, S., 1976. Adaptive pattern classification and universal recoding. II. Feedback, expectation, olfaction, and illusions. *Biol. Cybernet.* **23**: 187–202.

Iacoboni, M., Woods, R.P., Brass, M., *et al.*, 1999. Cortical mechanisms of human imitation. *Science* **286**: 2526–2528.

Itti, L., and Koch, C., 2001. Computational modeling of visual attention. *Nature Rev. Neurosci.* **2**: 194–203.

Iverson, J.M., and Thelen, E., 1999. Hand, mouth, and brain: the dynamic emergence of speech and gesture. *J. Consci. Stud.* **6**: 19–40.

Harnad, S., 1990. The symbol grounding problem. *Physica D: Nonlin. Phenom.* **42**: 335–346.

Hauk, O., Johnsrude, I., and Pulvermüller, F., 2004. Somatotopic representation of action words in human motor and premotor cortex. *Neuron* **41**: 301–307.

Hebb, D.O., 1949. *The Organization of Behavior: A Neuropsychological Theory*. New York: John Wiley.

Hochreiter, S., and Schmidhuber, J., 1997. Long short-term memory. *Neur. Comput.* **9**: 1735–1780.

Hubel, D., 1995. *Eye, Brain, and Vision*, 2nd edn. New York: Scientific American Library.

Kozima, H., and Yano, H., 2001. A robot that learns to communicate with human caregivers. *Proceedings 1st Int. Workshop on Epigenetic Robotics: Modeling Cognitive Development in Robotic Systems*, Lund, Sweden, p. 85.

Levy, W.B., and Desmond, N.L., 1985. The rules of elemental synaptic plasticity. In W. Levy and J. Anderson (eds.) *Synaptic Modifications, Neuron Selectivity and Nervous System Organization*. Hillsdale, NJ: Lawrence Erlbaum, pp. 105–121.

Lieberman, P., 2000. *Human Language and Our Reptilian Brain: The Subcortical Bases of Speech, Syntax, and Thought*. Cambridge, MA: Harvard University Press.

Maas, W., and Bishop, C.M. (eds.) 1999. *Pulsed Neural Networks*. Cambridge, MA: MIT Press.

MacWhinney, B., 2000. *The CHILDES Project*, 3rd edn., vol. 1, *Tools for Analyzing Talk: Transcription Format and Programs*. Mahwah, NJ: Lawrence Erlbaum.

Marian (Goga), I., Reilly, R.G., and Mackey, D., 2002. Efficient event-driven simulation of spiking neural networks. *Proceedings 3rd WSEAS International Conference on Neural Networks and Applications*, Interlaken, Switzerland.

Mervis, C.B., 1987. Child-basic objects categories and early lexical development. In U. Neisser (ed.) *Concepts and Conceptual Development: Ecological and Intellectual Factors in Categorization*. Cambridge, UK: Cambridge University Press, pp. 201–233.

Metta, G., Sandini, G., and Konczak, J., 1999. A developmental approach to visually guided reaching in artificial systems. *Neur. Networks* **12**: 1413–1427.

Nagai, Y, Hosoda, K., and Asada, M., 2003. How does an infant acquire the ability of joint attention? A constructive approach. *Proceedings 3rd Int Workshop on Epigenetic Robotics: Modeling Cognitive Development in Robotic Systems*, pp. 91–98.

Nothdurft, H.C., 2000. Salience from feature contrast: additivity across dimensions. *Vision Res.* **40**: 3181–3200.

Oztop, E., and Arbib, M., 2002. Schema design and implementation of the grasp-related mirror neuron system. *Biol. Cybernet.* **87**: 116–140.

Piaget, J., 1970. *Genetic Epistemology*. New York: Columbia University Press.

Pollack, J.B., 1990. Recursive distributed representations. *Artif. Intell.* **46**: 77–105.

Pulvermüller, F., 1999. Words in the brain's language. *Behav. Brain Sci.* **22**: 253–336.

2002. A brain perspective on language mechanisms: from discrete neuronal ensembles to serial order. *Prog. Neurobiol.* **67**: 85–111.

2003. *The Neuroscience of Language: On Brain Circuits of Words and Serial Order*. Cambridge, UK: Cambridge University Press.

Regier, T., 1995. A model of the human capacity for categorizing spatial relations. *Cogn. Ling.* **6**: 63–88.

Reilly, R.G., 2001. Collaborative cell assemblies: building blocks of cortical computation. In *Emergent Neural Computational Architectures Based on Neuroscience: Towards Neuroscience-Inspired Computing*. New York: Springer-Verlag, pp. 161–173.

2002. The relationship between object manipulation and language development in Broca's area: a connectionist simulation of Greenfield's hypothesis. *Behav. Brain Sci.* **25**: 145–153.

Reilly, R.G., and Marian (Goga), I., 2002. Cortical software re-use: a computational principle for cognitive development in robots. *Proceedings 2nd Int. Conference on Development and Learning*.

Rizzolatti, G., and Arbib, M.A., 1998. Language within our grasp. *Trends Neurosci.* **21**: 188–194.

Rose, S.A., Feldman, J.F., and Jankowski, J.J., 2001. Visual short-term memory in the first year of life: capacity and recency effects. *Devel. Psychol.* **37**: 539–549.

Sauser, E., and Billard, A., 2005. Three-dimensional frames of references transformations using recurrent populations of neurons. *Neurocomputing* **64**: 5–24.

Scassellati, B., 1999. Imitation and mechanisms of shared attention: a developmental structure for building social skills. In C. Nehaniv (ed.) *Computation for Metaphors, Analogy, and Agents*. New York: Springer-Verlag, pp. 176–195.

Schaal, S., 1999. Is imitation learning the route to humanoid robots? *Trends Cogn. Sci.* **3**: 223–231.

2001. *The SL Simulation and Real-Time Control Software Package*, Computer Science Technical Report. Los Angleles, CA: University of Southern California.

Schultz, T., and Lepper, M., 1992. A constraint satisfaction model of cognitive dissonance phenomena. *Proceedings 14th Annual Conference of the Cognitive Science Society*, Bloomington, IN, pp. 462–467.

Seidenberg, M.S., and MacDonald, M.C., 1999. A probabilistic constraints approach to language acquisition and processing. *Cogn. Sci.* **23**: 569–588.

Siskind, J.M., 1995. Grounding language in perception. *Artif. Intell. Rev.* **8**: 371–391.

1996. A computational study of cross-situational techniques for learning word-to-meaning mappings. *Cognition* **61**: 1–38.

2001. Grounding the lexical semantics of verbs in visual perception using force dynamics and event logic. *Artif. Intell. Rev.* **15**: 31–90.

Slama-Cazacu, T. (1962). Particularități ale însușirii structurii gramaticale de către copil, între doi și trei ani. In *Culegere de Studii de Psihologie, Vol. IV*, Editura Academiei Republicii Populare Române.

Smith, C., 1970. An experimental approach to children's linguistic competence. In J.R. Hayes (ed.) *Cognition and the Development of Language*. New York: John Wiley, pp. 109–135.

Steels, L., 2003. Evolving grounded communication for robots. *Trends Cogn. Sci.* **7**: 308–312.

Sutcliffe, R.F.E., 1992. Representing meaning using microfeatures. In R. Reilly and N.E. Sharkey (eds.) *Connectionist Approaches to Natural Language Processing* Englewood Cliffs, NJ: Lawrence Erlbaum, pp. 49–73.

Tan, A., and Soon, H., 1996. Concept hierarchy memory model: a neural architecture for conceptual knowledge representation, learning, and commonsense reasoning. *Int. J. Neur. Syst.* **7**: 305–319.

Tomasello, M., 1988. The role of joint attentional processes in early language development. *Lang. Sci.* **1**: 69–88.

Wang, D., 2002. Temporal pattern processing. In M. Arbib (ed.) *The Handbook of Brain Theory and Neural Networks*, 2nd edn. Cambridge, MA: MIT Press, pp. 1163–1167.

Weng, J., McClelland, J., Pentland, A., *et al.*, 2001. Autonomous mental development by robots and animals. *Science* **291**: 599–600.

Willshaw, D., Buneman, O., and Longuet-Higgins, H., 1969. Non-holographic associative memory. *Nature* **222**: 960–962.

Ziemke, T., 1999. Rethinking grounding. In A. Riegler, M. Peschl and A. von Stein (eds.) *Understanding Representation in the Cognitive Sciences*. New York: Plenum Press, pp. 177–190.

Zlatev, J., and Balkenius, C., 2001. Introduction: Why epigenetic robotics? *Proceedings 1st Workshop on Epigenetic Robotics*, Lund, Sweden, pp. 1–4.

14

Assisted imitation: affordances, effectivities, and the mirror system in early language development

Patricia Zukow-Goldring

14.1 Introduction

Rizzolatti and Arbib (1998) argue in their exposition of the Mirror System Hypothesis that brain mechanisms underlying human language abilities evolved from our non-human primate ancestors' ability to link self-generated actions and similar actions of others (see Arbib, Chapter 1, this volume). On this view, communicative gestures emerged eventually from a shared understanding that actions one makes oneself are indeed like those made by conspecifics. Thus, what the self knows can be enriched by an understanding of the actions and aims of others, and vice versa. From this perspective, the origins of language reside in behaviors not originally related to communication. That is, this common understanding of action sequences may provide a "missing link" to language.

In answering the question "What are the sources from outside the self that inform what the child knows?", the basic idea is that negotiating a shared understanding of action grounds what individuals know in common, including foregrounding the body's part in detecting that the actions of the self are "like the other." Given this footing, what then might the evolutionary path to language and the ontogeny of language in the child have in common? This perspective roots the source of the emergence of language in both as arising from perceiving and acting, leading to gesture, and eventually to speech.

I report here on an ongoing research program designed to investigate how perceiving and acting inform achieving a consensus or common understanding of ongoing events hypothesized to underlie communicating with language. This effort entails an analysis of the influences of the environment and, in particular, of the ways in which caregivers attune infants to that environment ("what the head is inside of": Mace, 1977). Building on what infants might "know" from birth, my work delineates the interplay of perceptual processes with action that might allow them to come to know "what everyone else already knows" (Schutz, 1962), including word meaning. Further, these caregiver practices may illuminate how automata might detect and learn new actions by observing and interacting with other intelligent agents.

Action to Language via the Mirror Neuron System, ed. Michael A. Arbib. Published by Cambridge University Press.

14.2 Affordances, effectivities, and assisted imitation:
caregivers educate attention

The ability to imitate has profound implications for learning and communication as well as playing a crucial role in attempts to build upon the Mirror System Hypothesis (Iacoboni *et al.*, 1999; Arbib, 2002). The focus here is to understand how imitation, especially assisted imitation, contributes to communicative development.

In the present study, I focus on how caregivers may assist infants as they gradually learn to adeptly engage in new activities that they have observed others doing. This interactive process may provide as well the basis for a shared understanding of events. The key notions for this study are "affordances" and "effectivities." J. J. Gibson (1979) proposed the notion of *affordances*, referring to the ability of creatures to perceive opportunities for action in their environment. The classic example is that as we walk across a room we see more, and that the more that we see tells us which surfaces will support our walking, what objects block our way, and so on. Some characterize the relation as emergent properties of the animal–environment system in which affordances are specified relative to an agent (Stoffregen, 2003), others add the notion of *effectivities* – the repertoire of what the body can do – as a dual complement of affordances (Shaw and Turvey, 1981; Turvey *et al.*, 1981; Turvey, 1992). I suggest that effectivities expand as an individual gains skill participating in new activities, thus concomitantly differentiating further what the environment affords for action.

The question, however, is how does an infant become more adept? What to do with objects, beyond the most rudimentary self-directed actions such as sucking and grasping, presents a challenge. Can infants, novice members of their culture, detect affordances and consummate an activity without assistance, if the action is not present at birth? (See Oztop, Arbib, and Bradley, this volume, for the development of grasping that is present at birth.) I have argued that objects cannot "tell us" what they afford (Zukow-Goldring, 1997). Nor can caregivers of young infants *tell them* in so many words, since verbal instructions directed to infants before they know "what and that" words mean, prove most ineffective (Zukow-Goldring, 1996, 2001). Novices do learn affordances as they engage in daily life in a particular time, place, and culture (Zukow, 1990; Costall, 1995). However, even careful non-verbal guidance – pointing out aspects of elements, configuring them in time/space, and/or modeling actions – may not be sufficient, because the body's work is left implicit. Shaw (2001) made that work explicit. He has argued that effectivities transform potential experiences into actual ones; that is, *affordances dispose*, while *effectivities deliver* (actualize). As Kadar and Shaw (2000, p.161) suggest, the fit between affordance and effectivity "is not between equals, but between promise and performance." This position suggests that affordances will go unrealized and, perhaps, undetected unless the creature has the requisite bodily ability within its repertoire. Elaborating further, Witherington (2005) argues that the range of what affordances offer an individual is not available to a creature prior to interaction with a particular object; rather the novice discovers them in and through agent–environment coupling. Therefore, in contrast to other views of imitation, I suggest that unless the novice perceives or knows the "body's part," affordances will not be picked up.

I propose that the child at first lacks the ability to detect by observation alone how the body relates to the physical layout and to the furniture of the world, except for the most rudimentary actions. From an ecological realist view, we might say that directing attention facilitates detecting a field of practice in which perceptual learning can take place. E. J. Gibson (1969) investigated the development of affordances present at birth in terms of perceptual differentiation. Studies of such affordances are important for understanding pre-attunements underlying basic modes of functioning as well as the subsequent role of experience. However, E. J. Gibson and Rader (1979) noted that most human affordances are learned, stressing observation of others over instruction by them (E. J. Gibson, 2002). Elaborating, E. J. Gibson and Pick (2000) discussed how the social context "acts back," responding to the infants' spontaneous exploratory activity, but did not discuss whether or how caregiver and infant interact in the learning of affordances. Filling in the gap, I stress that the caregiver *educates attention*. She guides the infant to discover for herself the means to realize the aim of an action (its goal) through the inextricable link between detecting the perceptual structure specifying affordances and the gradual differentiation of her own effectivities.

What do infants have to learn about the world? Infants must learn the most basic things, e.g., taking a bath, eating with utensils, walking. During mundane activities, infants must detect and participate in assembling the structure and organization of everyday events before they can communicate with others about these events (Zukow, 1989). In part, the child can discover affordances and effectivities as she explores the world on her own, developing a useful repertoire of skills by trial and error (Oztop *et al.*, 2004). However, I stress the role of the caregiver in directing the child's attention in ways that greatly expand this basic repertoire of affordances and effectivities, potentially reducing the search space and thus speeding learning. In contrast, many studies and theories assume that children know and/or learn autonomously how their bodies move in space and in relation to animate and inanimate things (Piaget, 1962; Thelen and Smith, 1994) and thus do not explore what experiences might underlie eventual adept performance. Most research investigating the development and implications of imitation ("learning to do something from seeing it done": Thorndike, 1898), focuses on what the child knows, rather than how the child comes to know.

Accounting for these achievements usually takes the form of proposing some combination of maturing modules, socio-pragmatic knowledge, or cognitive precursors hypothesized to be necessary for the activity (Uzgiris, 1991; Tomasello *et al.*, 1993; Meltzoff and Moore, 1995). This literature documents the age at which the average child can observe someone else's action and repeat it accurately either promptly or after a delay. The actions investigated, such as tongue protrusion or grasping, transferring, and stacking objects, are part of the child's repertoire. What is novel may be the object acted on and/or the sequence of acts, not the action itself. This literature may underestimate sources of the infants' accomplishments located in the caregiving environment. Greenfield (1972, this volume) observed that children imitate those actions that are entering their repertoire. Why might these particular actions be ripe for imitation and not others? Are the children's imitations usually autonomous accomplishments or do they have a robust history of assistance from others? Piaget (1962, p.18) asserted that only the infant's independent achievements contribute to cognitive development. In the same vein, he referred to the "pedagogical mania" of those who tend children as interfering at best. Yet, as many researchers have documented, learning a new skill by observation alone can be a very slow, trial-and-error process whether in human (Vygotsky, 1978), robot (McGovern and Barto, 2001), or non-human primate (K. R. Gibson, 1993). Fortuitously, caregivers do invite infants to imitate, and I suggest that this is for the better. *Assisted imitation*, informed by an integrative view of action and perception, delineates how educating attention may contribute to achieving a consensus or common understanding of events hypothesized to be a prerequisite for communicating with language.

I focus on learning a new activity (not in the infant's repertoire) which may not be easily attained by observation alone. However, Dautenhahn and Nehaniv (2002a) asserted that the possibility of confirming whether or not an activity/behavior is in an intelligent agents' repertoire is very problematic, because no one has a complete history of someone's actions from birth. They claim that "new" is always a matter of degree. Nevertheless, an infant's behavior often can and does tell those who know them best how to interpret what the infant does or does not know. The infant's behavior and the caregiver's response to that behavior display how the caregiver evaluates the infant's current level of skill. In particular, if an infant initially misunderstands a caregiver's messages inviting her to do something, the caregiver has evidence that the infant does not know how to respond, i.e., that the behavior is not in the infant's repertoire (Zukow-Goldring, 1996, 2001).

The key idea here is *assisted imitation*: I argue that caregiver practices guide infants to perceive possibilities for action (Zukow-Goldring, 1997).[1] Out of the unceasing perceptual flow, which is quite unlike the highly edited cuts of most movies, *caregivers* continuously educate attention to aspects of ongoing events. Further, our normal experience is highly multisensory, not restricted to the limited perceptual input of, say, a videoclip. Indeed, Stoffregen and Bardy (2001) have argued that multisensory perception is not merely the primary type of perception; it's the only type of perception. Caregivers and children detect "the something that something is happening to" as well as "the something that is happening" through vision, smell, hearing, taste, movement, and touch (Michaels and Carello, 1981; Zukow-Goldring, 1997). Especially relevant to this idea is the young infant's known ability to detect amodal regularities or invariants in the continuous stream of perceptual information (Spelke, 1979; Bahrick and Pickens, 1994). Caregivers guide

---

[1] In the same vein, Ingold (2000, p.190) has documented how the perceptually more adept make what they know available to the less skilled, guiding the attention of others along the same paths.

infants to notice key elements of what persists or remains invariant over time and what changes.

In what follows, I illustrate the findings from a number of studies of infant development with some qualitative examples (Zukow-Goldring, 1996, 1997, 2001).

14.3 The naturalistic investigations

The infant's initially unsuccessful attempts at imitation with a toy or food item often display some familiarity with the cultural use of objects, as she attempts to engage in the relevant sequence of activities. Grasping an object is usually the fulcrum around which novel action grows. Even though such objects and their uses are not entirely novel, what is required to imitate apparently is. That is, the ability to notice the relevant affordances and coordinate them with particular effectivities that are necessary to accomplish these tasks is not available to the infant without assistance. Infants' fragmentary, flawed attempts to imitate actions observed in the past elicit very careful and elaborate tutoring on the part of the caregivers to direct attention to relevant affordances and effectivities. Going further, we need to understand how picking up the perceptual information that the caregiver has highlighted allows the infant to get a grip on what to do. Getting a feel for what to do can provide the basis for detecting the affordances that will guide infants in their attempts to imitate that action.

In these studies the following hypotheses were tested:

(1) Providing a child with more perceptual structure specifying either effectivities or affordances will assist caregiver and child to achieve the consensus or common understanding needed for communication, including, where appropriate, explicit guidance of the child's movements
(2) Additional or more specific verbal information will not enhance understanding when no basis for that understanding has been embodied.

The examples from the studies reported illuminate how a human child learns about the world. Of course, I do not deny the utility of verbal instruction for older children or their own autonomous learning. Rather, the purpose is to illustrate how the fundamental link between perception and action provides the information upon which communication can build. It is a separate study to understand the later "bootstrapping" that occurs when words can take a far greater role in advancing what the child knows.

14.3.1 Sample and data collection

Five Euro-American middle-class families and six Latino working-class families with an infant of 6 months were followed monthly through the one-word period. Monthly 20-minute videos of naturalistic interaction at home, field notes, diaries, and check-lists of lexical development were collected, as well as interviews following each videotaping session to ascertain the caregiver's interpretation of ongoing events and of the infant's utterances.

14.3.2 Procedure: attention-directing interactions

Situations were selected in which caregivers directed infants to notice the content of messages, such as specific elements, relations, or events over the myriad other possibilities available. This collection of *attention-directing interactions* included all instances of perceptual imperatives expressed by caregivers, such as *look!/¡mira!, listen!/¡oye!*, and so on, as well as the accompanying gestures, and the gestures alone as well as the infants' subsequent actions. Zukow-Goldring noticed this set of perceptual imperatives when doing fieldwork in Mexico (1981–2). They occurred massively. Caregivers constantly said, *¡mira!/look!*. Surely there is no *mira* neuron; however, the use of perceptual imperatives may help draw the child's attention to the gestures that co-occur with them. These gestures may, in turn, provide specific support for imitation by directing attention to the perceptual information (in touch, smell, taste, vision, movement, and hearing) that may lay the groundwork for knowing that the self is like others. Thus, this ability opens up the possibility of learning the actions that others display.

A key issue was to determine the relative importance for the young child of verbal messages versus gestures that educate attention and action. The distinction here is between words that provide explicit instructions, such as *Peel the orange!*, and the pairing of perceptual imperatives like *¡mira!* and gestures that direct the child to attend to opportunities for action during ongoing events. The conclusion was that explicit verbal instructions are ineffective in the early stages. (For further details, see Zukow-Goldring, 1996, 1997, 2001.)

Targets of attention

Caregiver messages combine gestures with *targets of attention* throughout the prelinguistic and one-word periods. In messages, caregivers express what persists and changes as events coalesce and disperse. The targets vary in semantic complexity, which we categorized in four levels: non-dynamic animate beings or inanimate objects; simple action or change of state of the former; more complex relations within an event (Greenfield and Smith, 1976); and relations between events. These targets of attention bring to life structural and transformational invariants in the environment across space/time.

Attention-directing gestures

Five gestures that direct attention often accompany caregivers' verbal messages (Fig. 14.1). These gestures direct attention to perceptual information/semantic content through action. The gestures encompass varying degrees of other- to self-regulation of attention to the effectivities of the body and the affordances of the environment. When a caregiver *embodies* his infant, he puts her through the motions of some activity, so the two move as one (e.g., the caregiver takes the child's hand, using his own hand on top of the child's to press down a lever as he says, *¡por abajo!/down!*). The infant may already "know" the movement, but this activity calls for a new way to match or fit the effectivity of the hand/arm movement to the affordance of an object or the action

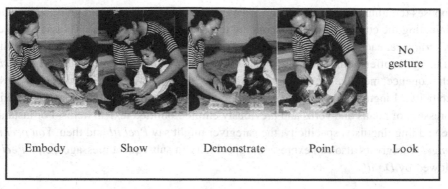

| Embody | Show | Demonstrate | Point | Look |

Figure 14.1 Attention-directing gestures. The child was about 2 years old, an age at which some children approach the end of the one-word period.

may be completely novel. Intriguingly, an inanimate object (like the spring of the vibrating toy described below) can also *embody* in this sense by putting the child "through the motions" in some way. During *shows*, a caregiver regulates the infant's line of sight. Sometimes the path of a gesture generated by a translational motion of an object through interactional space successively occludes other information in a scene or can magnify detail by bringing the object closer. For instance, the caregiver looms a toy dog toward the infant, saying *gwow-wow/wow-wow*. Other times, the caregiver performs some action using a familiar bodily effectivity to introduce a new possibility for action with an unfamiliar object or affordance. For instance, she pushes a button on a new toy, saying *¡empújalo!/push it!* The caregiver does not request that or leave interactional space for the infant to join in, but moves on to some subsequent activity. In contrast to *shows*, during *demonstrations* an infant is invited to act/imitate through action and/or speech *(¿Y tu?/You do it!)*. The infant who watches closely must detect or pick up in the perceptual flow a familiar coupling of effectivity and affordance to be duplicated. For example, the caregiver may synchronize rhythmically retracting fingers of an upright palm with saying *adiocito/bye-bye* when catching gaze and smiling. Alternately, an infant may be asked to pretend to avoid the sharp spines of a prickly pear while the caregiver mimes approaching and pulling away from the fruit's surface, saying, *¡espinoso!/prickly!* Regarding *points*, the infant must detect where a gesture's trajectory through space converges with some target of attention (the caregiver pointing to and saying *p'acá/over here*). For *looks*, no gestures accompany the caregiver's speech. Instead, only the caregiver's words and gaze direct the infant to correlate attention with that of the caregiver.

Assessing consensus or shared understanding

We determined whether or not caregivers treated their infants' responses to each message in a sequence as adequate or not. Action or speech indicating approval or embellishing the ongoing activity followed acceptable attempts (understanding), whereas inadequate responses (continued misunderstanding) were followed by repeated and revised messages, terminating the current activity, or noticing the child's lack of interest. To assess *perceptual structure*, each message in a sequence after the initial one was scored as providing or failing to offer additional perceptual information. To evaluate *linguistic specificity*, each sequence, message by message, was assessed for increases or decreases in linguistic specificity. Linguistic messages in ensuing turns can contain more or less explicit expression of nouns and verbs and previously ellipted/omitted lexical items. For instance, when adding linguistic specificity, the caregiver might say *Peel it!* and then, *You peel the orange!* Caregivers also can express less specificity in subsequent messages, e.g., *Peel it!* followed by *Do it!*

14.3.3 Qualitative examples: assisted imitation

Caregivers regularly take care to arrange the physical layout, so that the configuration in space of caregiver, infant, and object(s) makes them suitably aligned so that action is within reach. In addition, optimal proximity makes perceptually prominent what the object or some aspect of it affords for action. Sometimes the caregiver *demonstrates* the action(s) and then gives the infant a chance to act. Quite often the infant's attempt is inadequate. Frequently the caregiver embodies the infant, so that the child can perceive the relation of his or her body in terms of posture, motor actions, and rhythm to the caregiver as they move as one to accomplish the action or action sequence. Of the three examples, the middle one focuses less on tutoring effectivities and affordances and more on learning a sequence of actions to consummate an activity (vibrating toy).

Pop beads (13 months): caregiver tutoring of effectivities and affordances when concatenating beads

Pop beads, easily graspable by infants and toddlers, have affordances that allow concatenation. Play with this toy consists of (i) orienting toward each other the parts of each bead that afford concatenation (the dual complements of protrusion and opening), (ii) moving the appropriately oriented beads toward each other on a converging path, and (iii) applying enough force when the parts meet to embed the protrusion of one in the opening of the other.

The infant, Angela, begins by pressing a block lacking the appropriate affordances and a pop bead together. She displays an understanding that completing the task requires two small graspable objects and the application of some force to bring them together (Fig. 14.2, PB1). Her behavior provides no evidence that she knows that a set of objects with specific parts must come together, nor that they must sustain an orientation as they meet along a converging path. Her mother, Cecilia, provides perceptual information to Angela, gradually foregrounding the affordances of the objects and the effectivities of the body required to put the beads together. At first, Cecilia provides a bit of both, *point-touching* the opening on one bead (but not the protrusion in the other) as she directs

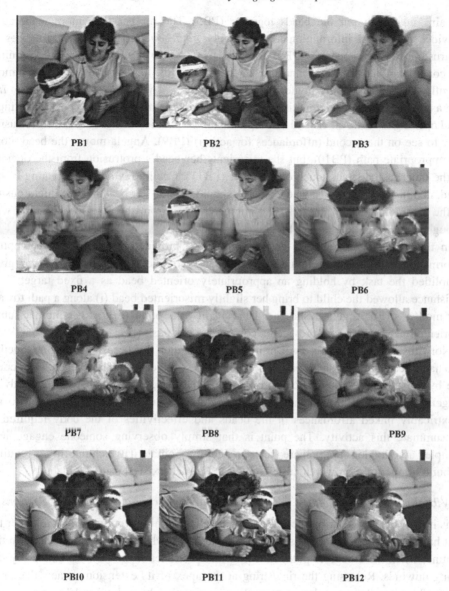

Figure 14.2 Pop beads.

attention to an affordance (PB2) and then reorients another bead (PB3), enacting a movement that aligns the beads on the same converging path. Lacking at this point, however, is information displaying the path itself or the force required to push the converging objects together. After an unsuccessful attempt by Angela, Cecilia *shows* her what the body must do to move the beads along a path with the required orientation as she pushes the protrusion into the opening with appropriate force (PB4). The infant remains unable to put the beads together (PB5) The mother then more elaborately provides perceptual information as she slowly *point-touches* both the affordances of protrusion (PB6) and opening (PB7), followed by *demonstrating* the effectivities required for connecting and then disconnecting of the beads (PB8). As the infant watches more intently, Cecilia eventually invites Angela to imitate: *¿A ver, tu?/Let's see, you (do it)?* She assists her daughter's imitation by *partially demonstrating* what to do by orienting a bead opening toward Angela and holding it in a fixed position as she makes the protrusion easy to see on the second (affordances for action) (PB9). Angela moves the bead along the appropriate path (PB10), but she misorients her bead's protrusion from the opening of the other as the beads touch one another (PB11). Cecilia realigns the opening of her bead, making prominent just where Angela should push in her bead. As the infant pushes in the protruding end of her bead, the mother pushes from the opposite direction with enough force to link the beads (PB12).

In this case, the caregiver's gestures gradually provided increments in perceptual information that guided the infant to concatenate two objects. Eventually, the caregiver simplified the task by holding an appropriately oriented bead as a fixed target. This assistance allowed the child to bring her slightly misoriented bead (i) along a path toward her mother's (ii). Angela pushed her bead against the other (iii), as her mother subtly reoriented her bead (i) and provided a complimentary push (iii).

Notwithstanding Angela's noteworthy improvement on this occasion, bringing together two hands, each grasping a properly oriented object, was not within her "reach." Nor could she by herself apply enough force to connect her beads. It is possible that embodying Angela, putting her through the motions, might have drawn maximal attention to the inextricably linked affordances of the beads and effectivities of the body required to consummate this activity. The point is that simply observing someone engage in a complex activity does not prepare the young child to imitate it. Much tutoring is required to build a repertoire on which "true" imitation (if such exists!) can take place.

Vibrating toy (14.5 months): caregiver and "toy" tutoring of a sequence of actions

This infant, Elsa, and her mother, Kathy, engage in a familiar routine with a reindeer toy that has a hidden affordance, a spring inside the toy to which a string is attached. In this routine, when the caregiver pulls on a ring that protrudes from the back of the toy, the string unwinds. Releasing the ring/string at the apex of its extension retracts the string so quickly that the toy vibrates strongly accompanied by a loud pulsing noise. Elsa expresses delight when she feels the vibrating toy placed on her stomach. Elsa, however, cannot make the toy vibrate by herself. This "game" entails a sequence of actions: (i) someone grasps the string by the ring, (ii) the string unwinds as it is pulled, and (iii) retracts within the toy as the tension on the string lessens. Finally, (iv) someone places the vibrating toy on the infant's stomach.

Family members had played this "game" with Elsa quite frequently during the prior 11 months. However, she had never attempted to imitate the others. In this example, more emphasis is placed on the sequence of actions than on the briefly noted affordances

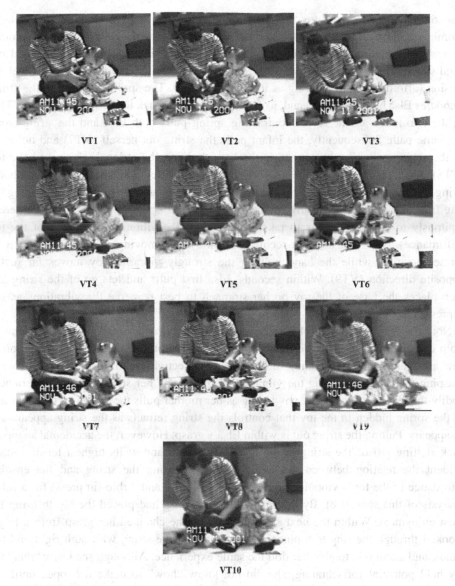

VT1 VT2 VT3

VT4 VT5 VT6

VT7 VT8 VT9

VT10

Figure 14.3 Vibrating toy.

and effectivities that mother and toy make perceptually available. For conciseness' sake, this sequence has been abbreviated by omitting repetitions and variations leading to the child's final adept enactment of this activity.

Elsa, sitting in front of her mother, turns to give her mother, Kathy, the toy that she wants her to animate (Fig. 14.3, VT1). Kathy pulls the string out by the ring (VT2), releases the string, and places the vibrating toy on Elsa's stomach (VT3). When Elsa wants her mother to continue, Kathy says *You do it!*, as she *partially demonstrates* by orienting the back of the toy toward Elsa, making the ring for pulling (affordance) prominent and within the infant's reach. The infant grasps the ring (VT4). The mother *embodies* by putting her hand/fingers on top of Elsa's ring-grasping hand to hold her hand steady as she pulls the toy away from them, presumably so the infant can feel the tension (affordance) of the string as it unwinds (VT5). The spring attached to the string *embodies* Elsa by pulling her hand, as it holds the string, back toward the toy (VT6). That is, the two move as one as the contracting spring pulls both Elsa and the string along the same path. Subsequently, the infant pulls the string out herself (VT7) and holds on as the spring *embodies* by retracting the hand holding the string back toward the toy (VT8). At this point, the toy vibrates weakly, if at all. A few seconds later, Elsa pulls the string so quickly and fully out, that the tension on the string *embodies* her by snapping the ring from her fingers. Vigorously pulling the string (effectivity) allows her quite seren-dipitously to experience how to take advantage of the vibratory properties of the toy (affordance). Note that her arm recoils from the force moving quickly away from its former position, while the hand holding the strongly vibrating toy moves far in the opposite direction (VT9). Within seconds, Elsa first pulls and lets go of the string and then places the base of the toy on her stomach to best perceive the vibrations as she expresses evident joy (VT10).

Note the free building-up of a sequence, as the child understands each new element. Both caregiver and toy educate Elsa's attention, cultivating perceptual differentiation of new affordances and the refining of her actions (effectivities) to fill in the gap between grasping the ring and feeling the vibration of the toy on her stomach. Elsa experiences bodily the tension of the string unwinding as her mother pulls the toy away from her and as the spring hidden in the toy that controls the string retracts as the string appears and disappears. Pulling the string out is within Elsa's grasp. However, the accidental snapping back (letting go) of the string at the apex of its path and at its highest tension made evident the relation between the effectivity of releasing the string and the ensuing affordance of the toy's vibrations (see Zukow-Goldring and Arbib (in press) for a fuller analysis of this sequence). By the second attempt, Elsa had placed the toy to bring the most enjoyment. Within the next several minutes, she changed her grasp from a finger crooked through the ring to a pincer grip, could pull the string with both right and left hands, and attempted to give her doll the same experience. Although she knew "that" the toy held potential for vibrating, she did not know "how" to make it happen until she received very careful tutoring. This educating of attention and action contrasts sharply with the effort it takes to tutor the attention and action of monkeys as well as autistic children (Nadel *et al.*, 1999; L. Fogassi, personal communication).

Orange peeling (16 months): caregiver tutoring
"when actions speak louder than words"

Peeling an orange with the hand entails penetrating the peel (both zest and pith), grasping the pulled away edge in a pincer grip, pulling the peel away from the flesh, and separating that portion of the peel from the fruit. Tearing off a piece of peel may involve yanking

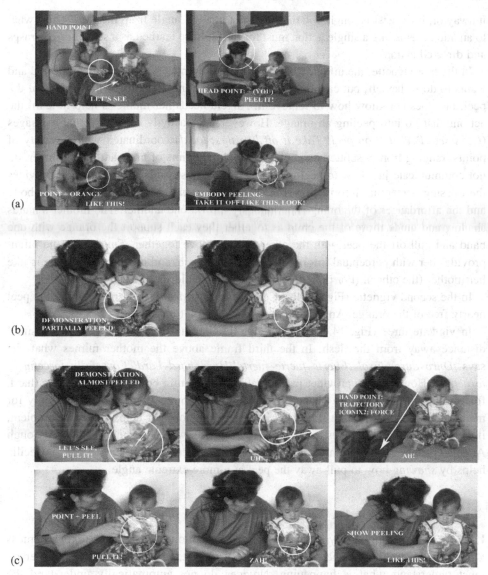

Figure 14.4 Orange peeling. (a) Hand, head, and finger points and increasingly specific verbal messages do not convey just how child and orange engage in "you peel it." Cecilia embodies her child to provide the missing perceptual information. (b) Cecilia partially demonstrates by lifting the peel almost entirely free of the flesh. She invites Angela to join in, saying *A ver, estírale/Let's see, pull it*. Now, Angela knows how. (c) Cecilia makes the task more difficult, peeling only part way through the pith. Angela cannot tear through the zest. Cecilia mimes "strong and hard" *as well as* change in direction. Angela imitates, but still needs help.

it away or, if the zest is tough, rotating the wrist at a 90° angle from the fruit. Thus, what to an adult seems like a single action must for the child be learned as a sequence of grasps and directed actions.

In the first vignette, the infant, who is quite fond of oranges, has seen them peeled and wants to do so herself, but cannot. She scratches quite ineffectually at the surface of the peel, but does not know how to remove it. Angela needs her mother's help to learn the actions that go into peeling an orange. However, increasingly explicit verbal messages (*Let's see. Peel it! You peel! Take it off like this, look!*) coordinated with a variety of points, ranging from a subtle head point to proximal points of hand and index finger, do not communicate just how to peel an orange. In contrast, *embodying* the child provides the missing perceptual information (Fig. 14.4a) regarding how effectivities of her body and the affordances of the orange continuously inform one another. The mother's hands shadow and guide those of the child as together they each support the orange with one hand and pull off the peel with the other as they move together. *Embodying* the infant provides her with perceptual information in vision, touch, and movement that she is like her mother (the other). (For more detail, see Zukow-Goldring, 2001.)

In the second vignette (Fig. 14.4b), Cecilia *partially demonstrates* by pulling the peel nearly free of the orange. Angela easily removes that bit of peeling.

In vignette three (Fig. 14.4c), Cecilia ups the ante as she lifts the peel just a small distance away from the flesh. In the third frame above the mother mimes what she says *¡Duro–duro–duro! ¡Fuerte–fuerte–fuerte!/Hard–hard–hard! Forcefully–forcefully–forcefully!* However, she provides perceptual information for both manner of action (hard, forcefully) and trajectory (a change in direction of about 90°) in gesture, but only for manner in speech. Note that trajectory is the information that helps solve the problem, illustrating when and that gestures "speak" louder than words (Kendon, 2004). Although Angela changes the direction of her pulling about 45° (compare frames 2 and 5), Cecilia helps by *showing* how to pull away the peel at a more extreme angle.

14.3.4 The naturalistic investigations in perspective

I have argued that ambiguity is the rule, not the exception. Words cannot explain unless a person already knows what words mean. To understand what words mean a person must understand what is happening. Novices do not automatically understand the organization and structure of daily events. Experts show them. These qualitative examples emphasize the finding that caregivers work hard to cultivate their infants' understanding of ongoing events by providing perceptual information in the form of attention-directing interactions to disambiguate their initially misunderstood messages. When there is lack of consensus, providing more perceptual structure tends to resolve misunderstanding, whereas adding specificity to verbal messages does not reduce ambiguity. Further, as infants respond to caregiver guidance, the infants' misdoings that may ensue pinpoint possible misperceiving as well as lack of bodily skill that, in turn, may inform the caregiver's subsequent seeing and doing. Perceptual restructuring of messages following communicative breakdowns led to achieving a common understanding in both working-class Spanish-speaking and middle-class English-speaking families in the United States.

From being a body to becoming a cultural being "like the other"

Some attention-directing interactions may have another equally important function: caregivers foreground the correspondence between the effectivities of the infant-actor's body and that of their own, providing an opportunity to learn the self is "like the other." In contrast, Meltzoff and Moore (1995, 1999) and many others propose that infants already know that others are "like me," giving the infant the basis for imitating.[2] Instead, I suggest the reverse. As the infant knows relatively little, another "like me" would be of little assistance. Knowing the self is "like the other," on the other hand, would provide access to more skillful behavior. When caregivers *embody* their infants, the infants have a chance to see and feel that their own movements are "like the other." That is, as caregiver and child move as one, the infant can detect amodal invariants (proprioceptively, kinesthetically, visually, tactilely, and so on) in the synchronous onset/offset, rhythm, and tempo of the action that may specify the correspondence between caregiver and infant movements. Embodying experiences may provide the basis for the infant's later ability to benefit from *shows* and *demonstrations* as the infant must already know that the self can move "like the other."

Byrne (2003) has drawn attention to the fact that scholars use the term *imitation* in two ways:

- *the transfer of skill problem* (how can the child acquire *novel, complex behavior* by observing?)
- *the correspondence problem* (how can the child match *observed actions* with *self-executed actions?*).

Thus, if a child knows that she is herself "like the other" (e.g., the caregiver), perhaps she can learn to do what the other does to achieve similar benefits or avoid risks. But can a novice (who does not know she is "like the other") spontaneously or after a delay imitate just any novel and/or complex "developmentally appropriate" behavior not yet in her repertoire without assistance? From a correspondence view of imitation (knowing the other is "like me"), the expectation would be that the novice can do so (Meltzoff and Moore, 1995, 1999). However, examples of *assisted imitation* provide evidence that the answer is "probably not." Instead, I suggest that investigating the *transfer of skill problem* can provide answers to the *correspondence problem* as well. That is, understanding the caregiver methods that promote the *transfer of skill* can illuminate how the infant comes to know the self is "like the other." Thus, as the child learns new bodily skills, she may literally get in touch with both sides of the *correspondence problem* perceiving the match between the other's actions and those of the self. (See Oztop and Arbib (2002) and Oztop *et al.* (this volume) for modeling of the development of grasp-related mirror neurons,

---

[2] For a cogent critique of prevailing cognitive approaches, see Moore's (2005) alternative cognitive proposal explaining how infants achieve an understanding of "self-other equivalence."

suggesting how the mirror system grounds imitation as a core component of communicative and linguistic development within an action/perception framework.) I stress again the role of the caregiver in directing attention to *effectivities* as well as *affordances* of the self and others – the two sides of the mirror system.

Challenges for the future

A challenge to these findings is the following question: do caregivers from all walks of life across the world's cultures engage in these methods on a daily basis? Evidence from studies in Indonesia, England, and France (Bril, 2004), the Solomon Islands (Watson-Gegeo and Gegeo, 1986, 1999), the Maya of Mexico (Maynard, 2002), rural and technological sites in Mexico (Zukow, 1989), and the Marquesas (Martini and Kirkpatrick, 1992) suggest that a broad range of both sibling and adult caregivers engage in very similar imitative sequences. (For a brief treatment of similar caregiving received by the deaf–blind Helen Keller, see Zukow-Goldring, 2001.)

An additional question entails whether or not "pedagogical mania" makes a difference: do infants who engage in assisted imitation most frequently develop communicative skills more rapidly than those who do not? Future research can reveal the answer to this question and enrich our understanding of these phenomena. However, sufficient may be a more likely answer than "most." In contrast, evidence from a clinical case study suggests that too little of these caregiving practices apparently is detrimental. In the case of a retarded caregiver who did not carefully coordinate speech and actions in terms of content and timing, her cognitively normal children displayed arrested language abilities (N. Rader, personal communication).

In sum, caregivers establish an understanding of what is happening. They gather and direct attention to perceptual structure that makes prominent the relations among animate beings, objects, and their actions. These dynamic relations specify the organization and structure of the most mundane daily activities. Caregivers introduce their infants to new effectivities or bodily capabilities and affordances for action and interaction on a daily basis. They assist them to link sequences of actions that comprise more and more complex activities. Caregivers also set aside language training when communication breaks down and, instead, focus on providing the perceptual information that will lead to a consensus. As caregivers educate attention, infants gradually learn to perceive, act, and know in culturally relevant ways.

14.4 Child development and the development of imitation in robots

Before turning to the discussion, the relation of this naturalistic research to developmental robotics is explored. Zukow-Goldring and Arbib (in press) have looked at the implications of these findings, structured around a number of questions raised by Breazeal and Scassellati (2002).

How do robots know what to imitate? While much work in robotics involves explicit programming of robot behavior, Zukow-Goldring and Arbib's concern with "robot development and learning" focuses attention on studies in which researchers prime robots to monitor preselected, simplified aspects of the visual scene, such as certain colors, areas of two-dimensional space, human movement, and so on to ensure the identification of "what to do" (Billard, 2002). Goga and Billard (this volume) offer an explicit model of caregiver–child interaction, using artificial neural networks to control the behavior of two robots, but here the "child" simply attends to the actions of the "caregiver," without specific attentional guidance on the part of the caregiver. However, much work in developmental psychology has established that sharing the target of attention during interaction plays a crucial role in cognitive and language development (Bates, 1976; Bruner, 1983; Zukow, 1990; Moore and Dunham, 1995). Rather than focusing on what the child already knows how to do, some of these scholars investigate the conditions under which infants learn new behaviors during social interaction. Several developmental approaches explicitly examine interactive processes that can lead a novice to learn a new task, including Vygotsky's (1978) groundbreaking work on the zone of proximal development, followed by investigations of scaffolding (Wood *et al.*, 1976), other-regulation (Wertsch *et al.*, 1980), and fine-tuning (Snow *et al.*, 1987). This research investigated what children need to know about objects of action and how caregivers guide them with verbal and non-verbal messages. However, the body's work is left implicit. In contrast, during assisted imitation, unless the individual perceives or knows the "body's part," affordances will not be picked up.

How do robots map the perception (of other's actions) onto their own action repertoire to replicate it? Dautenhahn and Nehaniv (2002; Nehaniv and Dautenhahn, 2002) suggest that solving the correspondence problem (matching *observed actions* to the *self's actions*) answers this question. For instance, Atkeson and Schaal (1997) have endowed automata with a priori knowledge of the goal of some action, the means to match movements observed to movements in the robot's repertoire, and provided reinforcement schedules that resulted in rapid matching of behaviors (see Schaal, this volume). However, children are not so endowed from birth, but must learn the aim of an action as well as the means, e.g., movements that result in achieving the target activity.

Bearing on these issues, Breazeal and Scassellati (2002) contrast descriptive studies of what biological agents can do and when with generative investigations of artificial systems. In such studies, actions are built up piece by piece, a very complex problem for those who design automata. The authors, however, do not make explicit in their critique that many studies of biological agents take the body's work for granted. In the main, researchers assume that the creature has in its repertoire whatever movements are necessary to imitate the activity observed. However, cultural knowledge of object use is not present at birth; such knowing must be learned. As studies of assisted imitation demonstrate, infant novices must learn to use their bodies in new ways as behaviors emerge during interaction with the environment.

Dautenhahn and Nehaniv (2002a) and Schaal (1999) have called for addressing imitation from new perspectives, emphasizing the import of perception–action coupling to the creation of autonomous humanoid robots. Exchanging skills entails at least two embodied agents in a particular environment who display information to each other via verbal and/ or non-verbal communications. In one use of forward models,[3] the agent can interpret the visual scene by anticipating the consequences of her/his actions and so guide ensuing action toward that end. Oztop *et al.* (this volume) discuss the relevance of forward and inverse models to the operation of the mirror system, and Skipper, Nusbaum, and Small (this volume) carry this theme into the study of brain mechanisms for audiovisual speech production and recognition. From a computational perspective, the success of training results from the child's ability to develop a forward model (Jordan and Rumelhart, 1992) that can link the (unobservable) internal processes to their observable consequences. The solution must relate to developing a means to discover the goals that characterize the success of diverse actions. From this computational perspective, one might say that the caregiver draws the "infant's" attention to the affordances in the environment to provide goals for the infant's/robot's developing effectivities. From an ecological realist view, we would say that directing attention provides practice, so that perceptual learning or differentiation can take place. Thus, whether in infant or robot, the actual specification of the processes guiding the learner are beyond the reach of even the most explicit instruction.

Zukow-Goldring and Arbib (in press) stressed the importance of *assisted imitation*, and the utility of the caregiver *embodying* the infant. With a robot, the challenge is greater since one can no longer guarantee so direct a map of action–perception coupling between "caregiver" and robot as there is between the adult human and child. Thus the robot situation poses extra challenges in finding the novel effectivities that match the affordances to which the caregiver draws attention. However, with robot or child, the skill of the caregiver resides in assisting the novice to detect the inseparable coupling of its/her own effectivities with the affordances of the environment.

Clearly the robots to be studied in future developmental robotics are very different from the highly programmed robots of today's assembly lines. However, note an interesting divergence between developmental robots used as models of child development and those being prepared for application in the workplace. In the former, the exploration of affordances and effectivities provides the basic substructure for language. In the latter, basic symbol structures can readily be programmed on the underlying hardware so that directing a robot's attention to a specific part of its body can be done symbolically in a way closed to the infant.

What would be gained if an instructor notices the robot is performing incorrectly? When the behavior of the robot displays an inadequate attempt, the instructor should redirect or educate attention by repeating or modifying an action. The catch is that the human's degrees of freedom might not map at all well onto those of the robot (for example, see Bicchi (2000) for the variety of forms taken by the dexterous robot's hands). In some ways, though, the difference between child and robot is not so great. Even though

---

[3] In forward models, information from both the body and the environment inform emerging action.

the adult human shares general bodily architecture with the child, the adult cannot share or access the child's neural processes nor those of the robot. (As is evident in these passages a challenge, only partially met, is to unpackage aspects of ecological realist and computational approaches that do not at first converge.)

From a computational perspective, the success of training results from the child's ability to develop a forward model that can link the (unobservable) internal processes to their observable consequences. The solution must relate to developing a means to discover the goals that characterize the success of diverse actions. Given that a robot can receive symbolic messages, telling the robot how to link what is observed to the knowledge with which the robot has been endowed might facilitate finding a better solution. Such messages would make explicit what caregivers must do by ostension during assisted imitation of human infants. A whole other issue, of course, for practical robots is what happens when the "caregiver" is another robot. When does an "adult" robot need to train a "young" robot rather than simply downloading the program and parameters that encode the fruits of its own experience?

14.5 Discussion

These "naturalistic investigations" delineated some of the ways in which imitation, especially assisted imitation, might contribute to communicative development. Engaging in action sequences with the caregiver cultivated a precursor to language: negotiating a common understanding of ongoing events. Implications from this research may clarify documented differences in the communicative and didactic behaviors of non-human and human primates, and inform studies of early lexical development, as well as the transition from single-element communications to two-word utterances.

14.5.1 Differentiating the difference between non-human primates and humans

The empirical literature documents that monkeys imitate each other very infrequently, if at all (Whiten and Ham, 1992; Bard and Russell, 1999; Visalberghi and Fragaszy, 2002). Chimpanzees imitate some actions of others in the wild (Quiatt and Itani, 1994; Greenfield *et al.* 2000), imitate the sequential and hierarchical structure of action in experimental, laboratory settings (Whiten, 1992), but learn a broader range of complex actions with objects when raised by humans (Savage-Rumbaugh, 1986; Tomasello *et al.*, 1993; Savage-Rumbaugh and Lewin, 1994; Tomasello and Call, 1997). However, the pace and extent of their imitation is very limited with respect to that of humans (see Stanford (this volume) for a critique of this literature; Greenfield (this volume) for a more expansive view of the commonalities between monkeys, apes, and humans). Indeed, the vast majority of human children do imitate, albeit to varying degrees at different ages and for behaviors that differ in modality and complexity of content (Eckerman, 1993; Nadel and Butterworth, 1999).

Why does teaching occur so massively among humans, whereas very sparse evidence documents that chimpanzees or any other non-human primates do so? What basic abilities allow humans to tutor their prelinguistic infants that remain out of reach or, perhaps, out of sight for non-human primates? Human caregivers invite their offspring to imitate new behaviors, and intervene and prompt them when their attempts go awry. To teach effectively, a caregiver must know how to engage in a proffered activity and see whether the novice can use similar methods to achieve such an aim. Further, as infants respond to caregiver guidance, their embodied misunderstandings of caregiver messages often inform caregivers' subsequent feedback. Detecting what infants cannot do and the affordances they miss perceiving is inextricably linked to effectively educating infants' attention, so that their subsequent attempts fall closer to the mark.

Some have noted that humans learn most affordances in a social/cultural context (Heft, 1989), but have had little to say about that process, except to stress that practice enhances perceptual differentiation (E. Gibson, 2002). Gibson and Rader (1979) reasoned that infants and young children pick up information not relevant to the task at hand as they simply may not know what to notice. In the same vein, Sharrock and Coulter (1998) noted that animals and prelinguistic infants distinguish one thing from another, but they do not know their significance within cultural activities. This present study of assisted imitation documents that caregivers actively, not indirectly, fill in this gap day in and day out. During the most humdrum activities they educate attention to what just such a body can do in concert with the objects on hand to achieve a specific aim while situated in a particular environment.

Some scholars have suggested that Theory of Mind abilities serve as a basis for teaching/tutoring (Meltzoff and Moore, 1995, 1999; Bråten, 2002). as well as language evolution and development (see Gordon (this volume) for discussion of the relation between the evolution of Theory of Mind and of language). In this view, the expert can *mind read*, inferring what the novice knows and does not know. Reboul (2004), however, argues that precursors of Theory of Mind, rather than full-blown abilities, may be sufficient to engage in complex social–cultural activities. That is, detecting the visual appearance of body and face, direction of gaze, and the target or aim of another's action may contribute to *behavior reading*, understanding another's actions (Sterelny, 2002). Numerous recent investigations of non-human primate perception document that chimpanzees engage in behavior reading. That is, they can engage in joint attention and observational learning (Greenfield *et al.*, 2000) and understand that if another faces or orients toward a particular configuration of objects and/or creatures, the other may act in particular ways (Boesch, 1993; Povinelli *et al.*, 2000; Sterelny, 2002). According to Hare *et al.* (2001), chimpanzees can prospectively perceive differences in the aims of others' actions. For example, if the subordinate chimpanzee notices that the more dominant one cannot see something desirable, the less dominant often will take the object or food quickly before the other notices. Finally, bonobos can orient others with a declarative hand point to an individual who is out of their line of sight (author's field notes: Kanzi, July 1987). In all these examples, I suggest that chimpanzees *perceive what the other perceives;* they watch *what* others do and can "see" *what* another is likely to do next. However, do such achievements, in turn, provide enough information or display the abilities necessary for teaching new actions with objects to a novice? Apparently not, as non-human primate infants rarely receive "active" teaching (for two recorded examples, see Boesch, 1993), and to date there is no evidence for "hands on" guidance (embodying of another) while manipulating objects. Non-human primates who observe and sometimes imitate new activities, albeit not at the rate of humans nor in quite the same way, take many years to learn skills, such as tool use (Boesch, 1993; K. R. Gibson, 1993). In contrast, human infants and enculturated chimpanzees who benefit from the guidance of caregivers learn similar skills quite rapidly (Savage-Rumbaugh, 1994).

To teach or tutor, experts must *perceive how others perceive.* More competent individuals do not simply perceive *what* others do or do not perceive. Teachers observe the public and emergent process of just *how* infants' knowing/learning what the body is and does educates perceiving affordances.[4] As action unfolds, the infants' misdoings, such as misaligning hand and object, misorienting of one object to another, deviations in path of action, and so on, pinpoint possible misperceiving of affordances as well as lack of bodily skill. That is, infants' embodied misunderstandings of caregiver messages display both *what* and *how* they "miss" perceiving as more adept members do. Thus, human caregivers can and do detect the (culturally) unskilled behavior of novices which frequently informs their subsequent feedback. At this point, caregivers often embody infants, so they can help them get and keep in touch with *how* the body and the objects of action work together. In addition, as the interaction unfolds dynamically, the infant–novice can learn how to "look" at the ways something can be seen/perceived in order to achieve some aim. For example, an infant first sees and feels in arm and hand the resistance of the protrusion on the pop bead before just the right amount of force at a particular angle pushes it through the opening on another bead. Without that perceptual information, the infant does not have the embodied understanding of what body and beads do together in and over time to get them to concatenate. Caregiver assistance may be the source that speeds up learning and increases skill by educating *what and how* the infant perceives as body and environment meet.

To get things done, members continuously constitute a joint understanding of what is happening.[5] These interactions also may be central to communicative development. As a by-product of the constant monitoring and engagement in interaction during mundane daily activities, these practices may pave the way to early word learning. In particular, engaging in these activities may provide the means to grasp important prerequisites that underlie communicating with language. These basics include knowing that words have an instrumental effect on the receiver of a message (Braunwald, 1978, 1979), words refer (Bates *et al.*, 1979; Schlesinger, 1982; Zukow-Goldring, 1997;

---

[4] People can perceive with accuracy where another is looking (J. J. Gibson and Pick, 1963) as well as detect affordances for others (Stoffregen *et al.*, 1999).
[5] See Goodwin (2000) for a recent edited collection of studies from linguistic anthropology that explores how people make the unfolding structure and organization of events visible and meaningful to one another.

Zukow-Goldring and Rader, 2001), and coparticipants share or negotiate a common understanding of ongoing events (Moerman, 1988; Zukow, 1990; Macbeth, 1994; Zukow-Goldring, 1997).

14.5.2 *Perceiving reference: a key to early lexical development*

How do infants unlock the puzzle of reference, the mundane fact that words refer to objects, actions, attributes, and more? That is, how do infants learn the relation between the stream of speech and the unceasing flow of events? While this sounds like a simple question, its answer remains elusive.

The preponderance of research investigating lexical development has tested hypotheses proposing that innate predispositions (Markman, 1989), cognitive prerequisites (Clark, 1993; Golinkoff *et al.*, 1994), and/or social abilities (Tomasello, 1988, 2001) underlie lexical development, usually from 15 months onward. Some 6 months earlier, however, infants begin to display comprehension and production of their first words. A large body of work documents which types of words and their tokens emerge when and in what order (Fenson *et al.*, 1994). By and large, this literature assesses what the infant already knows, not how the infant comes to know. Caregivers talk about what they are doing as they do it. Given these circumstances, how is consensus achieved as the child becomes an adept member of the community? My approach integrates discovering the interactional methods or practices that inform perceptual differentiation, the assembling of action sequences, and the detecting of word meaning, despite the fact that many studies of language acquisition assume that gestures entail ambiguity of reference (Schlesinger, 1982; Markman, 1989). These authors rely on Quine's classic essay (1960) in which he discussed the ambiguity of reference entailed in, say, speaking about and pointing to a rabbit. But caregivers tend to focus attention with precision. They do not simply say an unfamiliar word (such as Quine's *gavagai*) while pointing. Instead, caregivers may rub a rabbit's fur while saying "fur"; trace the topography of its ears while saying "ear," stroke the entire rabbit or rotate the whole animal when saying "rabbit," etc. (Zukow 1990; Reed, 1995; Zukow-Goldring, 1996). Successful teaching entails marking the correspondence between what is said and what is happening.

Zukow-Goldring and colleagues (Zukow, 1990; Zukow-Goldring, 1997; Zukow-Goldring and Rader, 2001) derived hypotheses from naturalistic, longitudinal studies of early word learning, in Mexico and the United States that have been confirmed by Gogate and colleagues (Gogate *et al.*, 2000, 2001). Zukow-Goldring has shown how caregivers bracket ongoing actions with gestures that direct the child's attention to perceptual information embodied in action sequences as well as the perceivable correspondence (in offset/onset, tempo, and rhythm) between word and referent. Attention to movement and synchrony in gesture and speech facilitates infants' detection of the correspondence between two fundamentally different kinds of things: words and aspects of ongoing events. Caregivers nest their messages in higher-order amodal regularities with a temporal basis that make word and referent "stand in" relation to each other and "stand out" from other possibilities. For example, a caregiver may "say-and-do" synchronously by saying *sti::cky*, in a rough, raspy voice while simultaneously displaying the irregular surface below her fingers as she runs them over uneven broom bristles, assisting the infant to "see-and-hear" the correspondence between the two. Understanding infants' perceptual abilities and caregiver practices underlying the emergence of early word comprehension is key, because language development without knowing what and that words mean cannot proceed.

Enlarging on the import of these practices, Falk (2004) has explored the possibility that the multi-modal characteristics of contemporary mother–infant gesture and vocalizations in both chimpanzees and humans may have derived from the behavior of early hominins. He hypothesized that early communicative behaviors arose in the form of a "primal" song. That is, when foraging for food, early hominins may have kept in touch with their infants in a non-tactile way by using a melodic, comforting vocalizations as a sort of vocal rocking or soothing after putting them down. Elaborating from these early communicative behaviors, he speculated that these messages might have the power to explain the evolution of protolanguage that would eventually lead to conventionalized meanings and language itself.

Returning to present-day mothers of human infants, some indirect evidence suggests that assisted imitation may also cultivate production of early words and the comprehension of caregiver messages. The longitudinal data from both English- and Spanish-speaking families confirm that caregiver messages communicate perceptual structure and/or semantic functions that are a step or two ahead of those expressed in infants' speech (Zukow-Goldring, 1997). These findings document a tie between caregiver input and subsequent production of words. Further, caregivers of less advanced infants (not necessarily younger infants) other-regulate their infants' attention by *embodying* and *show*ing most frequently, shifting to *demonstrations, points*, and eventually *looks* as the infants gradually self-regulate their own attention. When caregivers provide more perceptual structure after initial misunderstandings of caregiver messages, infants do comprehend what is meant significantly more often (Zukow-Goldring, 1996, 2001). A challenge for the future is to determine if there is a relation between this set of caregiver gestures and infant gestural production. (For other aspects of learning deictic and symbolic gestures, see Capirci *et al.*, 1996; Iverson *et al.*, 1994.) However, *embodying* would likely appear last in the infant's repertoire rather than first, as the embodier must detect *how* the other might misperceive and then redirect attention to a more informative action–perception coupling. In contrast, pointing to *what* you want is not as complex a task.

14.5.3 Communicative action and gesture precede and inform two-word utterances

Greenfield and Smith (1976) emphasized that language development arises from and builds on the infants' non-verbal understanding of events, their "world of experience," a world embedded in social interaction (see also Greenfield, this volume). That world is "*in* culture"; it is the only world the infant could ever know (Costall, 1989, p.19; Zukow-Goldring, 1997). This section delineates some of the links between early action sequences and two-word utterances, the continuity between the content and structure of these messages as the infant develops, and the role of dialogue during social interaction.

Some infant actions, although unintentionally communicative, do have a perlocutionary force or effect on the caregiver, when noticed. If a caregiver notices an infant pointing for the self to, say, a toy dog, the caregiver will often orient to the toy and say, *dog*, although the infant has not engaged in pointing for the caregiver's benefit. On the other hand, an infant who reached and whined for a piece of *bolillo* (roll) remained empty-handed when he did not catch his aunt's gaze. In contrast, 6 weeks later, under the same interactional and situational circumstances, he waited to catch and alternate gaze to the desired food. This time he very quickly received some roll (author's field notes, 1981–2). Thus, beginning in the months just prior to the emergence of speech, infants request action and/or objects by reaching and whining while alternating gaze or by placing objects in the caregiver's hands for manipulation: *proto-imperatives*. They also direct the attention of others by pointing to elements in the environment while alternating gaze with caregivers: *proto-declaratives* (Bates *et al.*, 1979). Most infants begin to produce such communicative actions between 8 and 10 months of age (Bates, 1976; Volterra *et al.*, 2005). Other single-element messages comprised of a symbolic gesture or a one-word utterance emerge somewhat later, nearer to the end of the first year of life or early in the second year (Acredolo and Goodwyn, 1985; Butcher and Goldin-Meadow, 2000; Volterra *et al.*, 2005).[6] Single-element action or gestural communications predominate until around 16 months of age; thereafter words gradually become more numerous (Iverson *et al.*, 1994).

During the single-element period, infant communications express semantic functions, such as agents, actions/states, objects, recipients of action, locations, instruments, and more (Greenfield and Smith, 1976; Zukow-Goldring, 1997). In the interim between single-element messages and two-word utterances, infants begin to combine communicative actions, symbolic gestures, and single words in a variety of ways from about 17–18 months of age (Greenfield and Smith, 1976; Greenfield *et al.*, 1985; Butcher and Goldin-Meadow, 2000; Volterra *et al.*, 2005). The messages may convey *redundant* (semantic) information by referring to the same referent (waving bye-bye while saying *ciao*), *complementary* information by enacting a deictic[7] gesture directing attention to a referent (pointing to a cup, while saying *cup*), and *supplementary* or *syntagmatic*[8] information by expressing different semantic functions that add information to each other (pointing to a record player, while saying *on*) (Greenfield and Smith, 1976; Greenfield

---

[6] Note that definitions of gesture vary from one researcher to another. For instance, Volterra and her colleagues (2005) include communicative actions in which the infant is touching something or someone, whereas Butcher and Goldin-Meadow (2000) exclude such actions.

[7] *Deictic* refers to a word or gesture that depends on the context of its production for its meaning, e.g., *I* refers to the speaker; the trajectory of a point intersects with a particular target of attention.

[8] Butcher and Goldin-Meadow (2000) use the term *complementary* instead of *supplementary* or *syntagmatic* to designate such combinations, e.g., pointing to a cup and saying *mine*.

et al., 1985; Capirci *et al.*, 1996). In addition, the infant may produce successive single-word utterances first (*Bib. Off.*), then two-element messages combining gesture and word (reaching and whining for a banana, while saying *nana*), and eventually multi-word utterances (*eat nana*), suggesting that these transitional forms may underpin syntactic expression (Greenfield and Smith, 1976; Greenfield *et al.*, 1985).

Both Greenfield and Volterra and their colleagues explicitly elaborate some potential links between two-element *supplementary* messages and the emergence of syntax. Further, both research groups have explored the ways that dialogue with the caregiver nurtures these communicative developments. Multi-word caregiver messages and question–answer sequences about a particular activity may set the stage for subsequent messages on the "same" topic built from contributions of both the caregiver and the infant. These jointly expressed messages conveying different aspects of the ongoing interaction may precede and promote two element messages that the infant later will produce autonomously.

Butcher and Goldin-Meadow's research (2000) documents that the onset of combinations in which gesture and speech expressed different information correlated significantly with the onset of two-word utterances. The ability to concatenate two elements conveying different aspects of a situation within a single communicative act appears necessary but not sufficient to guarantee two-word speech as no infant in their study produced two-word speech without first expressing such gesture–word combinations. During the latter part of the second year, nearly all infants had made the transition to multi-word utterances.

This body of research highlights the continuity between prelinguistic and linguistic communication, underscoring the crucial role of action sequences embedded in social interaction and dialogue to the emergence of gestural communication and ultimately for the emergence of language.

14.5.4 Summary

Caregivers both direct attention to aspects of the ongoing events and tutor actions to "achieve consensus." In a sense, they bring the child to share prospectively with the caregiver how someone may discover which effectivities can "deliver" the affordances offered by objects in the environment. These interactional opportunities give infants crucial practice in (and a refining of) what to notice and do, and when to do it.

In the context of understanding how the methods of the caregiver correspond to the expanding capabilities of the child, note that often developmental researchers and scholars study affective, motor, perceptual, and cognitive development separately. Caregivers do not. During the prelinguistic and one-word periods, caregivers prepare infants to imitate by assisting them "to see what to do" before they can "do what they see" others doing. Day in and day out, they cultivate imitation within mundane daily activities with gestures. They animate and direct their infants' attention to their own and others' bodily movements as well as making prominent what the environment offers for action.

In the process of learning a new skill, especially when embodied or put through the motions by the caregiver, infants directly experience that the self is "like the other." Thus, embedded in grasping the *transfer of skill* problem are opportunities to see and feel solutions to the *correspondence* problem, detecting the match between self and other. Humans who eventually learn/understand that the self is "like the other" cultivate abilities in their young that contribute to imitating, tutoring, and communicating. The mirror system offers a means to clarify in what manner human and non-human primates as well as intelligent automata understand what they see other intelligent agents doing, what abilities and perceptual information underlie learning to do what they see others do, and much more.

Acknowledgments

The Spencer Foundation and the Michelle F. Elkind Foundation provided funds for this research. I wish to express my gratitude to colleagues who generously commented on prior versions of this manuscript: Harry Heft, Thomas Stoffregen, Dankert Vedeler, Cathy Dent-Read, Alan Costall, Blandine Bril, Michael Arbib (our indefatigable editor), and three anonymous reviewers. I extend my thanks to the caregivers and infants who participated in these investigations as well as Sarah Anderson and Heewoon Chung for their assistance.

References

Acredolo, L. P., and Goodwyn, S. W., 1985. Symbolic gesture in language development: a case study. *Hum. Devel.* **28**: 40–49.

Arbib, M. A., 2002. The Mirror System, imitation, and evolution of language. In C. L. Nehaniv and K. Dautenhahn (eds.) *Imitation in Animals and Artifacts*. Cambridge MA: MIT Press, pp. 229–280.

Atkeson, C. G., and Schaal, S., 1997. Robot learning from demonstration. *Int. Conference on Machine Learning*, San Francisco, CA, pp. 12–20.

Bahrick, L. E., and Pickens, J. N., 1994. Amodal relations: the basis for intermodal perception and learning in infancy. In D. J. Lewkowicz and R. Lickliter (eds.) *The Development of Intersensory Perception: Comparative Perspectives*. Hillsdale, NJ: Lawrence Erlbaum, pp. 205–233.

Bard, K. A., and Russell, C. L., 1999. Evolutionary foundations of imitation: social, cognitive and developmental aspects of imitative processes in non-human primates. In J. Nadel and G. Butterworth (eds.) *Imitation in Infancy*. Cambridge, UK: Cambridge University Press, pp. 89–123.

Bates, E., 1976. *Language and Context: The Acquisition of Pragmatics*. New York: Academic Press.

Bates, E., Begnini., L., Camaioni, L., Bretherton, I., and Volterra, V., 1979. *The Emergence of Symbols: Cognition and Communication in Infancy*. New York: Academic Press.

Bicchi, A., 2000. Hands for dexterons manipulation and robust grasping: a difficult road towards simplicity. *IEEE Trans. Robot. Automat.* **16**: 652–662.

Billard, A., 2002. Imitation: a means to enhance learning of a synthetic proto-language in an autonomous robot. In C. L. Nehaniv and K. Dautenhahn (eds.) *Imitation in Animals and Artifacts.* Cambridge, MA: MIT Press, pp. 281–310.

Boesch, C., 1993. Aspects of transmission of tool-use in wild chimpanzees. In K. R. Gibson and T. Ingold (eds.) *Tools, Language, Cognition in Human Evolution.* Cambridge, UK: Cambridge University Press, pp. 171–183.

Bråten, S., 2002. Altercentric perception by infants and adults in dialogue. In M. I. Stamenov and V. Gallese (eds.) *Mirror Neurons and the Evolution of Brain and Language.* Amsterdam, Netherlands: Benjamin, pp. 273–294.

Braunwald, S., 1978. Context, word and meaning: toward a communicational analysis of lexical acquisition. In A. Lock (ed.) *Action, Gesture, and Symbol: The Emergence of Language.* London: Academic Press, pp. 285–327.

  1979. On being understood: the listener's contribution to the toddler's ability to communicate. In P. French (ed.) *The Development of Meaning.* Hiroshima, Japan: Bunko Hyonron Press, pp. 71–113.

Breazeal, C., and Scassellati, B., 2002. Robots that imitate humans. *Trends Cogn. Sci.* **6**: 481–487.

Bril, B., 2004. Learning to use scissors: adult's scaffolding and task properties. *Proceedings Int. Congress of Psychology,* Beijing, China, pp. 545–546.

Bruner, J., 1983. *Children's Talk: Learning to Use Language.* New York: Norton.

Butcher, C., and Goldin-Meadow, S., 2000. Gesture and the transition from one- to two-word speech: when hand and mouth come together. In D. McNeill (ed.) *Language and Gesture.* Cambridge, UK: Cambridge University Press, pp. 235–257.

Byrne, R. W., 2003. Imitation as behaviour parsing. *Phil. Trans. Roy. Soc. London B* **358**: 529–536.

Capirci, O., Iverson, J. M., Pizzuto, E., and Volterra, V., 1996. Gestures and words during the transition to two-word speech. *J. Child Lang.* **23**: 645–673.

Clark, E. V., 1993. *The Lexicon in Acquisition.* Cambridge, UK: Cambridge University Press.

Costall, A., 1989. A closer look at "direct perception." In A. Gellatly, D. Rodgers, and J. A. Sloboda (eds.) *Cognition and Social Worlds.* Oxford, UK: Clarendon Press, pp. 10–21.

  1995. Socializing affordances. *Theory Psychol.* **5**: 1–27.

Dautenhahn, K., and Nehaniv, C. L., 2002. The agent-based perspective on imitation. In C. L. Nehaniv and K. Dautenhahn (eds.) *Imitation in Animals and Artifacts.* Cambridge, MA: MIT Press, pp. 1–40.

Eckerman, C. O., 1993. Toddlers' achievement of coordinated action with conspecifics: a dynamic systems perspective. In L. B. Smith and E. Thelen (eds.) *A Dynamic Systems Approach to Development.* Cambridge, MA: MIT Press, pp. 333–357.

Falk, D., 2004. Prelinguistic evolution in early hominins: whence motherese? *Behav. Brain Sci.* **27**: 491–503.

Fenson, L., Dale, P. S., Reznick, J. S., *et al.*, 1994. Variability in early communicative development. *Monogr. Soc. Res. Child Devel.* **59**.

Gibson, E. J., 1969. *Principles of Perceptual Learning and Development.* New York: Appleton-Century-Crofts.

  2002. *Perceiving the Affordances: A Portrait of Two Psychologists.* Mahwah, NJ: Lawrence Erlbaum.

Gibson, E. J., and Pick, A. D., 2000. *An Ecological Approach to Perceptual Learning and Development.* New York: Oxford University Press.

Gibson, E. J., and Rader, N., 1979. Attention: the perceiver as performer. In G. A. Hale and M. Lewis (eds.) *Attention and Cognitive Development.* New York: Plenum Press, pp. 1–21.

Gibson, J. J., 1979. *The Ecological Approach to Visual Perception.* Boston, MA: Houghton Mifflin.

Gibson, J. J., and Pick, A. D., 1963. Perception of another person's looking behavior. *Am. J. Psychol.* **76**: 386–394.

Gibson, K. R., 1993. Introduction: generative interplay between technical capacities, social relations, imitation and cognition. In K. R. Gibson and T. Ingold (eds.) *Tools, Language, Cognition in Human Evolution.* Cambridge, UK: Cambridge University Press, pp. 131–137.

Gogate, L. J., Bahrick, L. E., and Watson, J. D., 2000. A study of multimodal motherese: the role of temporal synchrony between verbal labels and gestures. *Child Devel.* **71**: 878–894.

Gogate, L. J., Walker-Andrews, A. S., and Bahrick, L. E., 2001. The intersensory origins of word comprehension: an ecological–dynamic systems view. *Devel. Sci.* **4**: 1–18.

Golinkoff, R., Mervis, C., and Hirsh-Pasek, K., 1994. Early object labels: the case for a developmental lexical principles framework. *J. Child Lang.* **21**: 125–155.

Goodwin, C. (ed.), 2000. Vision and inscription in practice. *Mind, Culture, and Activity* **7**.

Greenfield, P. M., 1972. Cross-cultural studies of mother–infant interaction: toward a structural–functional approach. *Hum. Devel.* **15**: 131–138.

Greenfield, P. M., and Smith, J., 1976. *The Structure of Communication in Early Language Development.* New York: Academic Press.

Greenfield, P. M., Reilly, J., Leaper, C., and Baker, N., 1985. The structural and functional status of single-word utterances. In M. D. Barrett (ed.) *Children's Single-Word Speech.* Chichester, UK: John Wiley, pp. 233–267.

Greenfield, P. M., Maynard, A. E., Boehm, C., and Schmidtling, E. Y., 2000. Cultural apprenticeship and cultural change: tool learning and imitation in chimpanzees and humans. In S. T. Parker, J. Langer, and M. L. McKinney (eds.) *Biology, Brains and Behaviour.* Santa Fe, NM: SAR Press, pp. 237–277.

Hare, B., Call, J., and Tomasello, M., 2001. Do chimpanzees know what conspecifics know? *Anim. Behav.* **61**: 139–151.

Heft, H., 1989. Affordances and the body: an intentional analysis of Gibson's ecological approach to visual perception. *J. Theory Soc. Behav.* **19**: 1–30.

Iacoboni, M., Woods, R. P., Brass, M., *et al.*, 1999. Cortical mechanisms of human imitation. *Science* **286**: 2526–2528.

Ingold, T., 2000. *The Perception of the Environment: Essays in Livelihood, Dwelling and Skill.* London: Routledge.

Iverson, J. M., Capirci, O., and Caselli, M. C., 1994. From communication to language in two modalities. *Cogn. Devel.* **9**: 23–43.

Jordan M. I., and D. E. Rumelhart, 1992. Forward models: supervised learning with a distal teacher. *Cogn. Sci.* **16**: 307–354.

Kadar, E., and Shaw, R. E., 2000. Toward an ecological field theory of perceptual control of locomotion. *Ecol. Psychol.* **12**: 141–180.

Kendon, A., 2004. *Gesture: Visible Action as Utterance.* Cambridge, UK: Cambridge University Press.

Macbeth, D., 1994. Classroom encounters with the unspeakable: "Do you see, Danelle?" *Discourse Proc.* **17**: 311–335.

Mace, W., 1977. Ask not what's in your head, but what your head's inside of. In R. E. Shaw and J. Bransford (eds.) *Perceiving, Acting, and Knowing*. Hillsdale, NJ: Lawrence Erlbaum, pp. 43–65.

Markman, E. M., 1989. *Categorization and Naming in Children: Problems of Induction*. Cambridge, MA: MIT Press.

Martini, M., and Kirkpatrick, J., 1992. Parenting in Polynesia: a view from the Marquesas. In J. L. Roopnarine and D. B. Carter (eds.) *Parent–Child Socialization in Diverse Cultures*. Norwood, NJ: Ablex, pp. 199–223.

Maynard, A. E., 2002. Cultural teaching: the development of teaching skills in Zinacantec Maya sibling interactions. *Child Devel.* **73**: 969–982.

McGovern, A., and Barto, A. G., 2001. Automatic discovery of subgoals in reinforcement learning using diverse density. *Proceedings 18th Int. Conference on Machine Learning*, San Francisco, CA, pp. 361–368.

Meltzoff, A. N., and Moore, M. K., 1995. Infants' understanding of people and things: from body imitation to folk psychology. In J. Bermúdez, A. J. Marcel, and N. Eilan (eds.) *The Body and the Self*. Cambridge, MA: MIT Press, pp. 43–69.

1999. Persons and representation: why infant imitation is important for theories of human development. In J. Nadel and G. Butterworth (eds.) *Imitation in Infancy*. Cambridge, UK: Cambridge University Press, pp. 9–35.

Michaels, C. F., and Carello, C., 1981. *Direct Perception*. Englewood Cliffs, NJ: Prentice-Hall.

Moerman, M., 1988. *Talking Culture: Ethnography and Conversation Analysis*. Philadelphia, PA: University of Pennsylvania Press.

Moore, C., 2005. *Discussion: Imitation, Identification, and Self–Other Awareness*. Atlanta, GA: Society for Research on Child Development.

Moore, C., and Dunham, P. (eds), 1995. *Joint Attention: Its Origins and Role in Development*. Hillsdale, NJ: Lawrence Erlbaum.

Nadel, J., and Butterworth, G. (eds.), 1999. *Imitation in Infancy*. Cambridge, UK: Cambridge University Press.

Nadel, J., Guérini, C., Pezé, A., and Rivet, C., 1999. The evolving nature of imitation as a format for communication. In J. Nadel and G. Butterworth (eds.) *Imitation in Infancy*. Cambridge, UK: Cambridge University Press, pp. 209–234.

Nehaniv, C. L., and Dautenhahn, K., 2002. The correspondence problem. In K. Dautenhahn and C. L. Nehaniv (eds.) *Imitation in Animals and Artifacts*. Cambridge, MA: MIT Press, pp. 41–61.

Oztop, E., and Arbib, M. A., 2002. Schema design and implementation of the grasp-related mirror neuron system. *Biol. Cybernet.* **87**: 116–140.

Oztop, E., Bradley, N. S., and Arbib, M. A., 2004. Infant grasp learning: a computational model. *Exp. Brain Res.* **158**: 480–503.

Piaget, J., 1962. *Play, Dreams, and Imitation in Childhood*. New York: Norton.

Povinelli, D. J., Bering, J. M., and Giambrone, S., 2000. Toward a science of other minds: escaping the argument by analogy. *Cogn. Sci.* **24**: 509–541.

Quiatt, D., and Itani, J., 1994. *Hominid Culture in Primate Perspective*. Niwot, CO: University Press of Colorado.

Quine, W. V. O., 1960. *Word and Object*. New York: John Wiley.

Reboul, A., 2004. Evolution of language from theory of mind or coevolution of language and theory of mind? Available at http://www.interdisciplines

Reed, E. S., 1995. The ecological approach to language development. *Lang. Communi.* **15**: 1–29.

Rizzolatti, G., and Arbib, M. A., 1998. Language within our grasp. *Trends Neurosci.* **21**: 188–194.

Savage-Rumbaugh, E. S., 1986. *Ape Language: From Conditioned Response to Symbol.* New York: Columbia University Press.

Savage-Rumbaugh, E. S., and Lewin, R., 1994. *Kanzi: The Ape at the Brink of the Human Mind.* New York: John Wiley.

Schaall, S., 1999. Is imitation learning the route to humanoid robots? *Trends Cogn. Sci.* **3**: 233–242.

Schlesinger, I. M., 1982. *Steps to Language: Toward a Theory of Native Language Acquisition.* Hillsdale, NJ: Lawrence Erlbaum.

Schutz, A., 1962. *Collected Papers*, vol. 1, *The Problem of Social Reality.* Leiden, Netherlands: Martinus Nijhoff.

Sharrock, W., and Coulter, J., 1998. On what we can see. *Theory Psychol.* **8**: 147–164.

Shaw, R., 2001. Processes, acts, and experiences: three stances on the problem of intentionality. *Ecol. Psychol.* **13**: 275–314.

Shaw, R., and Turvey, M., 1981. Coalitions as models of ecosystems: a realist perspective on perceptual organization. In M. Kubovy and J. R. Pomerantz (eds.) *Perceptual Organization.* Hillsdale, NJ: Lawrence Erlbaum, pp. 343–415.

Snow, C. E., Perlmann, R., and Nathan, D., 1987. Why routines are different: toward a multiple-factor model of the relation between input and language acquisition. In K. E. Nelson and A. van Kleeck (eds.) *Children's Language*, vol. 6. Hillsdale, NJ: Lawrence Erlbaum, pp. 65–97.

Spelke, E. S., 1979. Perceiving bimodally specified events in infancy. *Devel. Psychol.* **15**: 626–636.

Sterelny, K., 2002. Primate worlds. In C. Heyes and L. Huber (eds.) *The Evolution of Cognition.* Cambridge, MA: MIT Press, pp. 143–162.

Stoffregen, T. A., 2003. Affordances as properties of the animal–environment system. *Ecol. Psychol.* **15**: 115–134.

Stoffregen, T., and Bardy, B., 2001. On specification and the senses. *Behav. Brain Sci.* **24**: 195–261.

Stoffregen, T. A., Gorday, K. M., Sheng, Y., and Flynn, S. B., 1999. Perceiving affordances for another person's actions. *J. Exp. Psychol. Hum. Percept. Perform.* **25**: 120–136.

Thelen, E., and Smith, L., 1994. *A Dynamic Systems Approach to the Development of Cognition and Action.* Cambridge, MA: MIT Press.

Thorndike, E. L., 1898. Animal intelligence: an experimental study of the associative processes in animals. *Psychol. Rev. Monogr.* **2**.

Tomasello, M., 1988. The role of joint attentional processes in early language development. *Lang. Sci.* **10**: 69–88.

   2001. Perceiving intentions and learning words in the second year of life. In M. Bowerman and S. C. Levinson (eds.) *Language Acquisition and Conceptual Development.* Cambridge, UK: Cambridge University Press, pp. 132–158.

Tomasello, M., and Call, J., 1997. *Primate Cognition.* New York: Oxford University Press.

Tomasello, M., Kruger, A. C., and Ratner, H. H., 1993. Cultural learning. *Behav. Brain Sci.* **16**: 495–552.

Tomasello, M., Savage-Rumbaugh, S., and Kruger, A. C., 1993. Imitative learning of actions on objects by children, chimpanzees, and enculturated chimpanzees. *Child Devel.* **64**: 1688–1705.

Turvey, M., 1992. Affordances and prospective control: an outline of the ontology. *Ecol. Psychol.* **4**: 173–187.

Turvey, M., Shaw, R., Reed, E., and Mace, W., 1981. Ecological laws of perceiving and acting. *Cognition* **9**: 237–304.

Uzgiris, I., 1991. The social context of imitation. In M. Lewis and S. Feinman (eds.) *Social Influences and Socialization in Infancy*. New York: Plenum Press, pp. 215–251.

Visalberghi, E., and Fragaszy, D., 2002. "Do monkeys ape?" – ten years later. In K. Dautenhahn and C. L. Nehaniv (eds.) *Imitation in Animals and Artifacts*. Cambridge, MA: MIT Press, pp. 471–499.

Volterra, V., Caselli, M. C., Capirci, O., and Pizzuto, E., 2005. Gesture and the emergence and development of language. In M. Tomasello and D. Slobin (eds.) *Beyond Nature–Nurture: Essays in Honor of Elizabeth Bates*. Mahwah, NJ: Lawrence Erlbaum, pp. 3–40.

Vygotsky, L. S., 1978. *Mind in Society*. (M. Cole, V. John-Steiner, S. Scribner, and E. Souberman, eds.) Cambridge, MA: Harvard University Press.

Watson-Gegeo, K. A., and Gegeo, D. W., 1986. Calling-out and repeating routines in Kwara'ae children's language socialization. In B. Schieffelin, and E. Ochs (eds.) *Language Socialization across Cultures*. New York: Cambridge University Press, pp. 17–50.

1999. (Re)modeling culture in Kwara'ae: the role of discourse in children's cognitive development. *Discourse Stud.* **1**: 241–260.

Wertsch, J. V., McNamee, G. D., McLane, J. B., and Budwig, N., 1980. The adult–child dyad as a problem-solving system. *Child Devel.* **51**: 1215–1221.

Whiten, A., 2002. Imitation of sequential and hierarchical structure in action: experimental studies with children and chimpanzees. In K. Dautenhahn and C. L. Nehaniv (eds.) *Imitation in Animals and Artifacts*. Cambridge, MA: MIT Press, pp. 192–209.

Whiten, A., and Ham, R., 1992. On the nature and evolution of imitation in the animal kingdom: reappraisal of a century of research. In P. J. B. Slater, J. S. Rosenblatt, C. Beer, and M. Milinski (eds.) *Advances in the Study of Behavior*, vol. 21. San Diego, CA: Academic Press, pp. 239–283.

Witherington, D., 2005. A comparison of two conceptualizations for the concept of affordance. *Proceedings Int. Conference on Perception and Action*, Asilomar, CA.

Wood, D., Bruner, J. S., and Ross, G., 1976. The role of tutoring in problem solving. *J. Child Psychol. Psychiatr.* **17**: 89–100.

Zukow, P. G., 1989. Siblings as effective socializing agents: evidence from Central Mexico. In P. G. Zukow (ed.) *Sibling Interactions across Cultures: Theoretical and Methodological Issues*. New York: Springer-Verlag, pp. 79–105.

1990. Socio-perceptual bases for the emergence of language: an alternative to innatist approaches. *Devel. Psychobiol.* **23**: 705–726.

1996. Sensitive caregivers foster the comprehension of speech: when gestures speak louder than words. *Early Devel. Parent.* **5**: 195–211.

1997. A social ecological realist approach to the emergence of the lexicon: educating attention to amodal invariants in gesture and speech. In C. Dent-Read and P. Zukow-Goldring (eds.) *Evolving Explanations of Development: Ecological*

Approaches to Organism-Environment Systems. Washington, DC: American
Psychological Association, pp. 199–250.

2001. Perceiving referring actions: Latino and Euro-American infants and caregivers
comprehending speech. In K. E. Nelson, A. Aksu-Koc, and C. Johnson (eds.)
Children's Language, vol. 11. Hillsdale NJ: Lawrence Erlbaum, pp. 139–163.

Zukow-Goldring, P., and Arbib, M. A., in press. Affordances, effectivities, and assisted
imitation: caregivers and the direction of attention. *Neurocomp*.

Zukow-Goldring, P., and Rader, N., 2001. Perceiving referring actions: a commentary on
Gogate, Walker-Andrews, and Bahrick. *Devel. Sci.* **4**: 28–30.

15

Implications of mirror neurons for the ontogeny and phylogeny of cultural processes: the examples of tools and language

Patricia Greenfield

15.1 Introduction

In this chapter I explore two qualities of the mirror neuron system that are critical for the evolution of tool use and language, central characteristics of human culture. The two characteristics of the mirror system are: (1) the ability of the system to respond both to one's own act and to the same act performed by another and (2) the system's selective response to intentional or goal-directed action (Fogassi et al., 2005). The ability to respond neurally both to one's own act and to the same act performed by another constitutes the neural foundation of imitation on the behavioral level (Iacoboni et al., 1999) and of repetition on the linguistic and cognitive levels (Ochs (Keenan), 1977). The selective response of the mirror neuron system to goal-directed action constitutes the neural facilitation of goal-directed action on the behavioral level and of intentionality on the cognitive level (Greenfield, 1980). My purpose is then to demonstrate the importance of these neurally grounded behavioral competencies for the evolution and ontogenetic development of two key aspects of human culture, tool use and language. In so doing, my larger goal is to contribute to understanding the neural underpinnings for the ontogeny and phylogeny of human culture.

In order to provide data on phylogeny, I draw upon my own research and that of others to compare chimpanzees (Pan troglodytes), bonobos (Pan paniscus), and humans (Homo sapiens). The Pan line and the hominid line diverged in evolutionary history approximately 5 million years ago (Stauffer et al., 2001). The two species of Pan later separated from each other about 2 million years ago (Zihlman, 1996). By cladistic logic, if we find the same characteristic in all three species, it is very likely to constitute an ancestral trait that was present before the phylogenetic divergence.

Cladistics refers to a taxonomic analysis that emphasizes the evolutionary relationships between different species. A clade – the basic unit of cladistic analysis – is defined as the group of species that all descended from a common ancestor unique to that clade. Cladistic analysis separates ancestral traits, which are inherited from the ancestors of

Action to Language via the Mirror Neuron System, ed. Michael A. Arbib. Published by Cambridge University Press. © Cambridge University Press 2006.

the clade, from derived traits, which are possessed by only some of the clade members (Boyd and Silk, 2000). Derived traits arose through natural selection after the divergence of the clade from the common ancestor. Ancestral traits, in contrast, have a genetic foundation in the common ancestor. This genetic foundation drives the development of neural and all other biological systems.

This focus on what was present *before* the divergence of *Pan* and *Homo* is especially relevant to the role of mirror neurons in the evolution of human culture because mirror systems are known to exist in both macaque monkeys (Gallese *et al.*, 1996; Rizzolatti *et al.*, 1996) and humans (Iacoboni *et al.*, 1999) (see Arbib, Chapter 1, this volume for a review). Since Old World monkeys (of which the macaque is one) diverged about 23 million years ago from the hominoid line that became *Homo* and *Pan* (Stauffer *et al.*, 2001) and because the relevant behaviors mirror neurons subserve (reaching and grasping) are possessed by chimpanzees and bonobos, as well as by macaque monkeys and humans, it is likely (and I assume this for purposes of this chapter) that chimpanzees and bonobos also possess mirror neuron systems. There is also a bootstrap element to my evidence: in so far as I find that the behaviors subserved by a mirror system in humans also exist in *Pan*, this provides behavioral evidence suggesting the presence of a mirror system in chimpanzees and bonobos. For ethical and pragmatic reasons, it has not been possible up to now to investigate the presence or absence of a mirror system in *Pan*. One cannot implant electrodes in apes for ethical reasons; one cannot put an ape in a functional magnetic resonance imaging (fMRI) machine for pragmatic reasons – an ape will not lie still in the scanner! However, we may expect better adapted forms of brain mapping to greatly increase knowledge of ape brain function in the coming decades.

Gorillas and orangutans diverged earlier than chimpanzees from the hominoid line; the gorilla more recently at about 6 million years ago, orangutan about 11 million years ago (Stauffer *et al.*, 2001). However, the omission of detailed data on these species in the present chapter is in no way an assertion that they lack tools or the ability to learn a humanly devised communication system. For example, it is known that gorillas spontaneously make tools (Fontaine *et al.*, 1995); and it is equally known that gorillas and orangutans have developed a human protolanguage under the tutelage of human sign-language teachers (Patterson, 1978; Miles, 1990). Hence the capabilities for protolanguage and even tools may have been present in the common ancestor of the whole great ape and hominoid line, a minimum of about 11 million years ago.

Indeed, the most recent evidence from monkeys indicates that, through extended experience in watching human tool behavior, macaque monkeys can develop mirror neurons that respond selectively to observing human beings use tools to act on objects (Ferrari *et al.*, 2005). This implies that a basic cognitive capacity to associate hand and tool is present in the common ancestor of Old World monkeys, apes, and hominoids, a minimum of 23 million years ago (Stauffer *et al.*, 2001). Hence one would expect tool understanding, if not behavior, throughout the great ape line.

In sum, I will use similarities among chimpanzees, bonobos, and humans as clues to what foundations of human language may have been present in our common ancestor

5 million years ago. Such foundations would then have served as the basis from which distinctive traits of human language could have evolved in the following millions of years. While species differences are as important as similarities in determining the evolution of human culture (compare the chapters by Arbib and Stanford, this volume), my focus here is on the similarities. These similarities provide clues as to the capabilities in our common ancestor present 5 million years ago. This focus on the 5-million-year-old foundation of human evolution contrasts with the focus of the Arbib and Stanford chapters on the evolutionary elaboration of human tools and language that occurred in the last 5 million years.

15.1.1 Connection between the evolution of culture and the brain

Although phenotypic variation is the basis for natural selection, there has to be an underlying biological structure in order to have something genetic to select for. Mirror (and other) neurons specialized for different types of cultural learning provide biological structures that could have been selected for in the course of human evolution. Often culture, seen to be a part of the environment, is considered the opposite of biology. However, the notion of neural capacities that make cultural learning possible avoids the either/or dichotomy between biology and environment, biology and culture. Mirror and canonical neurons (the distinction is explained below) provide some key capacities for cultural learning that make an important link between the brain and culture. They provide something biological that could have been selected for as culture evolved.

Mirror neurons

Mirror neurons were originally discovered in monkeys (Gallese *et al.*, 1996; Rizzolatti *et al.*, 1996). Mirror systems were later found in humans (Fadiga *et al.*, 1995; Rizzolatti *et al.*, 1996; Iacoboni *et al.*, 1999). (The term "neuron" is used for the monkey research because the researchers utilize single-cell recording methods. The term "system" is used for the human research, because fMRI and other brain imaging methods used with humans cannot resolve single neurons.) Mirror neurons discharge when a goal-directed action is enacted or observed. In contrast, they do *not* discharge when the same movements are enacted or observed outside the context of the goal. Their activation pattern differs in the context of different goals. An important subset will discharge before the final goal is observed, indicating the perception of intentionality. Finally, mirror neurons do *not* discharge in the presence of a goal-object alone.

Canonical neurons

Like mirror neurons, canonical neurons were originally discovered in monkeys (Gallese *et al.*, 1996). Later, they were found in humans (Garbarini and Adenzato, 2004). They discharge not only when a goal-directed action is enacted (like mirror neurons), but also when a goal-object is observed and may be acted upon. They therefore represent a connection between a goal-object and its associated motor action. However (unlike mirror neurons) they are not active when the subject observes the actions of another.

15.1.2 Connection between ontogeny and phylogeny?

Ontogeny does *not* recapitulate phylogeny. However, there are important theoretical and evolutionary connections between ontogeny and phylogeny. First, earlier stages of development are more universal within a species than are later stages of development. Second, earlier stages of development are more similar among phylogenetically related species than are later stages of development. In other words, as phylogenetic divergence progresses, the evolutionarily later developments are more likely to occur later rather than earlier in an ontogenetic sequence. In that way, phylogenetic changes interfere less with subsequent ontogeny. This is because evolutionary change in earlier stages may compromise development in later stages that depend on them. Note that this formulation is contrary to the evolutionary myth that adult chimpanzees resemble human children. The notion is simply that human and chimpanzee babies will be more alike than human and chimpanzee adults. This is a relative statement concerning ontogenetic trends and in no way precludes differences between human and chimpanzee babies from having evolved.

Indeed, one must always remember that *Pan* has undergone evolutionary change in the last 5 million years, just as *Homo* has. We cannot assume either species of *Pan* is closer in form and behavior to the common ancestor than is *Homo sapiens*, although evidence has been presented to support the idea that the bonobo may be closest to the common ancestor of the three species – that is that the bonobo has evolved in the last 5 million years less than humans or chimpanzees (Zihlman, 1996).

Third, as we saw above, similarity of a characteristic among groups of phylogenetically related species indicates that a characteristic was likely to be part of the common ancestor of those species. It therefore follows that early stages of development (ontogeny) provide clues about phylogenetic foundations at the evolutionary point of species divergence. The clues are even stronger if the early stages are shared among a family (or clade) of closely related species. Correlatively, differences in the later stages of development among members of a clade will provide important clues as to species differences in adult capabilities. Again, my theoretical and empirical focus in this chapter is on the nature of the evolutionary foundation that existed 5 million years ago before the divergence of the three species. We are most likely to find clues in the early development in all three species.

These are the reasons why ontogeny can help us understand the phylogeny of cultural processes. In what follows, I examine the development across species of behavioral and cognitive capabilities that correspond to capabilities shown in prior research to be subserved by mirror neurons (and to a much lesser extent, canonical neurons). I then make the case that these capabilities are crucial to the ontogeny and phylogeny of cultural processes.

15.2 Imitation, observation, and cultural learning: ontogeny and phylogeny

In this section, I start with the assumption that observation and imitation are two central mechanisms for cultural learning. I then try to show that both are present from the beginning of life in both *Homo sapiens* and *Pan* (both species, bonobos and chimpanzees). From there, I illustrate the use of these mechanisms in the cultural learning of tool use across species. However, data concerning newborn imitation in humans have a strong traditional opponent in Piaget.

15.2.1 Piagetian theory, mirror neurons, and newborn imitation in humans

Piaget's (1962) theory of imitation is basically a visual one. Piaget theorized that action imitation requires seeing both one's own action and the action of the other in order to make the cross-modal correspondence between visual stimulus and motor movement. Yet Meltzoff and Moore (1977) reported that newborns imitate tongue movements, even though the baby cannot see its own tongue (Fig. 15.1). This is cross-modal imitation (the visual model of the tongue being stuck out is responded to motorically in the action response of sticking out one's own tongue).[1] In Piaget's conceptualization, all imitation is cross-modal, linking the sensory to the motor, and therefore had to await what he thought of as a later developmental stage when cross-modal cognitive correspondence would become possible for a baby. It was once thought that cross-modal imitation was a fairly high-level cognitive skill; one needed actively to make cognitive (and presumably neural) connections between visual stimulus and motor response. However, mirror neurons provide a theoretical and neural construct that can explain this seeming contradiction between the young age of a newborn baby and the sophisticated imitation response. Note, for future reference, that I consider the baby's imitation to have an intentional structure, even though it is automatic. Indeed, Miller *et al.* (1960) have made a strong theoretical case for the goal-directed nature of reflexes.

My point here is that imitation is basic, not derived (Favareau, 2002). It is basic because it stems from a neural identity between observing and responding. Learning is not required to imitate (although imitation facilitates learning).

As action sequences become more cortically controlled with increasing age, I theorize that the same basic mechanism can be used to activate imitation of increasingly complex action sequences (Greenfield *et al.*, 1972; Goodson and Greenfield, 1975; Childs and Greenfield, 1980).[2] The reader should also be forewarned that, while neonatal

---

[1] Some authors reserve the term "cross-modal" for the integration of different *sensory* modalities, but given the importance of corollary discharge and proprioceptive feedback in motor control, the situation here may be seen as falling under this apparently more restrictive definition.

[2] Oztop *et al.* (this volume) take a somewhat different view. While they would not, I assume, deny that the Meltzoff–Moore observations implicate a basic class of mirror neurons in neonatal imitation, they do argue that the mirror neurons for grasping (e.g., that distinguish precision pinches from power grasps) are themselves the result of a developmental process that stretches over the first year (for the human timetable) of the infant's life. They thus distinguish neonatal imitation from what they view

Figure 15.1 A newborn imitates Andrew Meltzoff's tongue protrusion movement. (Photograph courtesy of Andrew Meltzoff.)

imitation undercuts Piaget's notion that imitation cannot take place without seeing one's own response, other parts of Piagetian theory provide important insight into both ontogenetic and phylogenetic aspects of imitation. Thus, I will later have occasion to draw heavily on another aspect of Piaget's theory of imitation, the notion that one imitates (or transforms) a model in line with one's cognitive understanding of the model's actions.

The discovery of mirror neurons suggests that, ontogenetically, imitation does not begin as the relatively high-level cognitive process that Piaget (1962) posited. The explanation of newborn imitation by means of mirror neurons is that observation of adult tongue movement by the newborn triggers the baby's mirror neurons that control his/her own tongue movement. The imitation therefore occurs when observation of the tongue movement excites a series of mirror neurons, which discharge as motor neurons. The discovery of mirror neurons makes the ontogenetic basis of imitation more reflexive and less cognitive than Piaget thought. Clearly at least some mirror neurons are there from birth. But most important, the basic connection between observation and action does not have to be learned through an associative process. The substrate for a cross-modal connection between visual stimulus and action is already present, internal to each mirror neuron of this initial set.

as "real" imitation which requires more cognitive attention to the structure of the imitated action. I accept this distinction, but see the former as the developmental foundation for the latter.

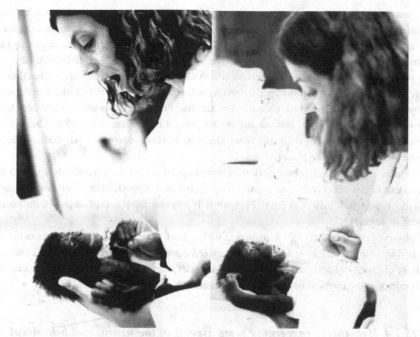

Figure 15.2 In the video frame on the right, a newborn chimpanzee imitates Kim Bard's mouth opening movement shown in an earlier frame on the left. (Photograph courtesy of Kim Bard.)

15.2.2 Newborn imitation in chimpanzees

Given that early stages of ontogeny are much more likely to be shared with sibling species than later stages, we might expect that newborn apes would also be able to do facial imitation. And indeed this is the case: chimpanzee newborns are also capable of cross-modal facial imitation (Bard and Russell, 1999). In Fig. 15.2, we see a newborn chimpanzee imitating Kim Bard's mouth-opening gesture.

15.2.3 The role of imitation in the ontogeny of tool use: intergenerational transmission

It is not much later in development that observation and imitation begin to be used for tool learning. By 1 year of age, these processes can be used for the transmission of *human* tool culture. To illustrate this point I summarize a videoclip in which NF (age 11 months, 14 days) and his grandmother are out with the stroller and both are involved with cups of water. The clip, which one can think of as video ethnography, is part of a longitudinal corpus designed for cross species comparative purposes. I will also provide theoretical interpretation of the behavior in terms of the hypothesized operation of mirror and canonical neurons.

(1) NF is in the stroller; his grandmother is next to him with two cups in her hand, an adult cup and an infant's "sippy cup" (a sippy cup is a baby cup with a no-spill top and a built-in straw). NF preferred an adult cup (which he has observed being used by others), over his sippy cup (which he has not observed being used by others). (Here the cups are considered to be goal-objects that can potentially activate canonical neurons, which in turn activate the relevant goal-action.) This preference for the adult cup is hypothesized to reflect a predisposition to imitate, reflecting the operation of canonical neurons. Canonical rather than mirror neurons are invoked here because this is a self-initiated action without any model present to imitate. (See Arbib (Chapter 1, this volume) for an analysis of the many other elements beyond mirror neurons that are necessary for such an action to take place.)

(2) As soon as his grandmother hands NF the empty cup, he responds immediately with a drinking action that he has observed occurring with cups in the past (hypothesized operation of canonical neurons rather than mirror neurons because it is elicited by the goal-object, with no action model to imitate).

(3) NF observes grandmother's drinking action with a similar cup and immediately puts his cup to his lips in a similar drinking action (hypothesized operation of mirror neurons in concert with activity of canonical neurons). (Here there is a model to imitate as well as a goal-object; hence, both mirror and canonical neurons are hypothetically called into play.)

15.2.4 Hypotheses concerning some aspects of the neural and behavioral development of imitation

The immediate translation from observation to action seen in this clip seems to result from lack of cortical inhibition. It has been observed that patients with prefrontal cortical lesions may have problems inhibiting imitative responses (Brass *et al.*, 2005). In (normal) development, prefrontal cortical circuits do not connect with more posterior parts of the brain, such as motor areas, until about age 2 (Greenfield, 1991). Hence, it is logical to hypothesize that a 1-year-old may resemble the patients with cortical lesions in terms of the inability to inhibit imitative responses. Because of the lack of inhibition in the first year or two of human life, the links between observation and manual motor response inherent in the canonical and mirror neurons are more overtly reflected in behavior at this very young age. In the above-described scene with the cups, NF's behavior at 11 months of age suggests the utility of the canonical and motor neuron systems for acquiring skill with cultural tools (e.g., a cup).

15.3 The role of imitation in subsequent development of object-oriented manual activity

With increasing age and development, more complex motor activities can be observed and imitated. For example, we used imitation procedures to elicit a developmental sequence of grammars of action in construction activities from children ranging from 11 months to 7 years of age (Greenfield *et al.*, 1972; Goodson and Greenfield, 1975). By grammar of action, I mean a consistent strategy that is homologous to some element of

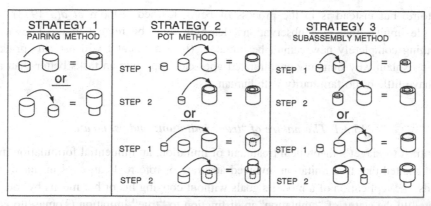

Figure 15.3 Strategy 3 was modeled for children from 11 to 36 months of age. The youngest children "imitated" the model with Strategy 1, the next oldest with Strategy 2, and the oldest with Strategy 3. From Greenfield *et al.* (1972).

linguistic grammar. One of these grammar of action tasks is presented in Fig. 15.3 (Greenfield *et al.*, 1972). In the case of Strategy 1 for example, we can see it as analogous to a linguistic combination of agent (active cup) and object (passive cup) in a simple sentence. Evidence for homology – the involvement of the same neural structures for both the manual and linguistic grammar – will be presented later in the chapter. In addition, Goga and Billard (this volume) present a model of the Greenfield *et al.* (1972) linkage of seriation to language.

For present purposes, note that children of all ages from 11 to 36 months of age were presented with the same model – Strategy 3 below – by an adult experimenter. However, younger children systematically transformed the model in their imitations. The youngest children "imitated" the model with Strategy 1, the pairing method. The next oldest children tended to "imitate" the model with Strategy 2, the pot method. Only some of the oldest children accurately replicated the model, responding with Strategy 3, the subassembly method.

These tasks show that a transition from simple reflexive imitation in the newborn to the imitation of complex action sequences has taken place. However the sequence of stages leading up to the most complex and complete mode of replicating the model also provides evidence for the developmental model of imitation posited by Piaget (1962). That is, at each stage, the child transforms the model by interpreting it through the lens of his or her stage of understanding, i.e., cognitive development. This process also occurs in language acquisition (Slobin and Welsh, 1973).

At the same time, the nesting cup study also provides an example of how observation and imitation might provide scaffolding to bring a child to the next stage of development. This principle has been empirically demonstrated in human language acquisition. In that domain, imitation of a particular linguistic structure is most frequent when the structure in question is neither completely present or completely absent from a child's behavioral repertoire, but instead is in the process of being learned (Bloom *et al.*, 1974). This principle implies that new observational learning must be related to old knowledge; something completely new cannot be imitated. (In a moment I will use this principle to help explain why humanly enculturated apes can copy human tasks better than apes who have little or no familiarity with human tasks.)

15.3.1 The nature of "true" imitation: end vs. means

In contrast to Piaget's theoretical treatment of imitation, an influential formulation in the 1980s posited that true imitation involved accurate or rote replication of means as well as ends; the replication of a model's goals without copying his or her means by rote was demoted to the status of "emulation" in distinction to "true" imitation (Tomasello *et al.*, 1987; Tomasello, 1989). Contrary to this formulation, the development of human imitation is not a question of rote imitation, as we have seen. For each task, it consists of transformations of a model that follow the sequence of understandings and interpretations of a particular task.

Nor is human imitation a question of imitating means rather than ends. In the ontogeny of human imitation, babies often understand and therefore imitate the goal before the means (Bruner, 1974). However, at a later point in development, they will be able successfully to replicate the means as well as the end. In other words, a stage of "emulation" is an intrinsic component of human imitation. As another example, Gergely *et al.* (2002) show that if an adult demonstrates a new way to execute a task to a group of infants aged 14 months, the children will use this action to achieve the same goal only if they consider it to be the most rational alternative. In other words, "emulation" is an important strategy in human imitation. The results of Gergely and colleagues also indicate that imitation of goal-directed action by preverbal infants is a selective, interpretative process, very much in line with the Piagetian framework. In sum, for developmental and pragmatic reasons, one cannot differentiate human and ape imitation according to whether the means is accurately imitated or not; both replication of a goal and transformation of means are important components of human imitation and its role in learning and development.

While not the only learning processes, observation (visual attention on a model) and imitation (attempt to replicate a model's actions) are keys to cultural transmission for humans, and, as such, they first appear early in development. In line with the principle that early stages of ontogeny are most likely to be observed in sibling species, we would expect them in chimpanzees as well. In the next section, we turn to this issue in a species-comparative perspective.

15.3.2 Observation and imitation are keys to cultural transmission for chimpanzees as they are for human beings

Tools are cultural traditions for groups of chimpanzees, as they vary from group to group in ways that cannot be reduced to ecological availability or usefulness (Whiten *et al.*,

1999; see Stanford, this volume). Videotapes made by Christopher Boehm showed that infant and juvenile chimpanzees at Gombe Reserve, Jane Goodall's field station, system-atically observed experienced adults use probes (stems or vines) to fish for termites in mounds of dirt. Their gaze often followed the probe from dirt to mouth (Greenfield *et al.*, 2000). The video footage also indicated that imitation can follow observation: an infant or juvenile chimpanzee often grabs the mother's abandoned fishing tool when she gets up to leave; the young chimpanzee will then use the tool to fish for termites, often with no success. The learning process takes years, beginning with playful experi-mentation with sticks, then moving to observation of models, and finally independent practice (Greenfield *et al.*, 2000). Our future research will elaborate the process of expert–novice apprenticeship in enculturated chimpanzees and bonobos.

I apply the Piagetian perspective on imitation (Piaget, 1962; Greenfield *et al.*, 2000) to non-human primates. As we have seen, this perspective emphasizes the importance of cognitive understanding of the observed model that is to be imitated. Hence, when animals are too young to understand a means–end relationship or lack motor skill to successfully imitate an action, this theoretical perspective implies that an attempted imitation will only partially replicate the model. This developmental principle can explain why it takes chimpanzees so long to learn to crack nuts; they can only imitate what they are developmentally ready to learn both cognitively and physically.

Why do imitative abilities in chimpanzees not lead to rapid diffusion of innovations within a group? Given that 14-month-old babies, in the sensorimotor period of develop-ment, rationally evaluate the functional appropriateness of a model's actions for their own situation before imitating it (Gergeley *et al.*, 2002) and given that chimpanzees manifest the same basic stages of human sensorimotor development (Parker and McKinney, 1999), we would expect chimpanzees to be equally selective in their imita-tions. This selectivity might explain the slow movement of a tool or other cultural innovation within a chimpanzee group.

15.3.3 Observation, imitation, and object combination in monkeys: a comparative and phylogenetic analysis

Experimental study of object combination in four species indicates that the tendency to imitate manipulative strategies for object combination, a cornerstone of tool use, exists in *Cebus* monkeys, a New World monkey, humans, and both species of *Pan* (Johnson *et al.*, 1999). In that study, all four species were shown the model of Strategy 3 (Fig. 15.3) to combine seriated nesting cups. All four species were able to replicate the model when given their own cups, although the monkeys required more training than the other species to do so. (But see Visalberghi and Fragaszy (2002) and Arbib (Chapter 1, this volume) for another interpretation of the monkey data.) In so far as complex sequences were modeled and imitated across all of these species in this experiment, one can see that reflexive imitation at birth grows, in a wide variety of primate species, with increasing age and experience, into skill in intentionally imitating a sequence of acts directed both to a sequence of subgoals and integrated into an overall goal of the sequence itself. Note that for the monkeys and apes in particular, the sequence shown in Fig. 15.3 was a novel one, a type of imitation that is considered particularly important in the human repertoire.

Although all four species learned to use the most complex strategy for nesting the cups, the strategy that was demonstrated at the outset, there were species differences. Monkeys and, to a lesser extent, apes did not construct structures. Instead they would utilize the same strategy to combine the cups in a sequence of moves, but they would then take apart the structure that they had just constructed. Perhaps this is why humans build big complex buildings and other primate species are limited to much simpler technologies and constructions. This ability to make object combination yield complex "permanent" structures is a major achievement of human culture that must have evolved in the last 5000 years, in the period since *Homo* and *Pan* diverged.

Another important difference between chimpanzee and human cultural learning seems to be the cumulative quality of the latter. The cumulative quality of human culture may have to do with increasing memory capacity that is a function of increased brain size that has evolved in *Homo* but not *Pan* in the last 5 million years. It may also have to do with the uniquely human use of symbol systems to transmit or teach cultural skills to the next generation (Greenfield *et al.*, 2000).

Following cladistic logic, the presence of observation and imitation in the transmission of cultural tool traditions in both *Homo* and *Pan* presents the possibility that these processes of cultural learning may go back in evolution to the common ancestor of humans and chimpanzees and perhaps even to our common ancestor with monkeys.

15.3.4 Can apes ape? Mirror neurons resolve a paradox

Apes clearly do ape in the wild: young chimpanzees at Gombe imitate more experienced chimpanzees in learning to termite (Greenfield *et al.*, 2000). However, in the laboratory, only humanly enculturated apes show an ability to imitate the means to a goal in a human tool task, according to Tomasello *et al.* (1993). If we accept for a moment Tomasello and colleagues' emphasis on the accurate replication of means in the analysis of imitation, what conclusion can we draw? Tomasello *et al.* conclude from their data that apes cannot imitate without human enculturation and that imitation is therefore phylogenetically new with humans, rather than part of our primate heritage.

Mirror neurons, however, challenge this interpretation and resolve the paradox between field and laboratory in the following way. Because mirror neurons are part of specific action systems (Buccino *et al.*, 2001), the implication is that an animal or human being would be able to imitate only what he or she was able to do motorically. Imitation proceeds from some understanding of what is being done – whether the understanding comes from the existence of a similar motor response (as in neonatal imitation of mouth movements) or from a cognitive understanding (as in the nesting cup task).[3] Thus enculturated apes have an advantage over wild apes when tested on their ability to imitate a human tool task, since enculturated apes are more likely to have had experience related to this sort of task. Hence, when Tomasello *et al.* (1993) used a human tool task with non-human primates, apes not enculturated by humans were unfamiliar with this sort of task and could not imitate its solution. Humanly enculturated apes, in contrast, were familiar with the genre and could successfully imitate the details of its solution. The hypothesis is that their success was due to familiarity, which led to understanding. Because the non-enculturated apes lacked this familiarity and understanding, they could not imitate the solution to the tool task.

In other words, it is not that apes lack the ability to imitate. Like humans, they can imitate what they can understand. Further evidence on this point comes from careful experimental work (Whiten, 1998). When a human model showed a humanly enculturated chimpanzee in Whiten's study how to open artificial fruit, the chimpanzee at first copied the model's every action, including acts irrelevant to the goal. As the chimpanzees practiced the task and understood its means–end relations better, the irrelevant acts dropped out of the sequence. In other words, chimpanzee imitation, like human imitation, is driven by understanding, not by a motive for rote imitation of a sequence of acts. On the neural level, I believe that future research will show these cross-species behavioral similarities to be driven by similarities on the level of neural functioning, specifically similarities in basic properties of the mirror systems possessed by each species (although evolution may have wrought changes to expand the mirror system to support faster and more flexible imitation in humans: M. A. Arbib, personal communication).

In conclusion, mirror neurons do not provide a *general ability* to imitate in either apes or humans. Instead, mirror neurons provide a set of *specific abilities* to imitate particular actions that are encoded in the motor component of various somatopically organized mirror systems (Buccino *et al.*, 2001) and are therefore understandable on the motor level.

For all of these reasons, I conclude that observation and imitation skills are held in common between humans and apes. These skills are therefore likely to be part of the phylogenetic heritage from our common ancestor and a prerequisite for the evolution of culture. According to my theoretical analysis, these imitation skills are subserved by mirror neurons in both apes and humans.

15.3.5 Mirror neurons, monkey culture, and human culture: what has evolved in the last 5 million years?

Mirror neurons could also contribute to monkey culture (Perry *et al.*, 2003), as they do to human and ape culture. The question then arises as to how mirror neurons can contribute

---

[3] Oztop, Bradley, and Arbib (this volume) make the point that since we develop new skills, there should be "quasi-mirror neurons" that learn to recognize an action as part of acquiring it. It remains to be seen whether such a mechanism is in fact required or whether the notion that one can use imitation only to learn a skill that is already partly in the repertoire suffices as an explanation for the role of imitation in developing new skills.

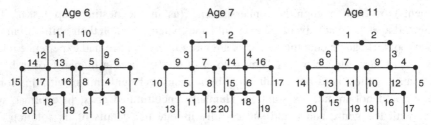

Figure 15.4 Age-typical strategies used to construct a three-level tree structure made of construction straws and presented as a model to copy. The 6-year-old uses a chain strategy beginning at the lowest level in which each successive straw is placed near the one before. The 7-year-old organizes the replication by branches beginning at the highest level. The 11-year-old organizes the replication by levels beginning at the highest level. Each age-typical strategy is considered to be a grammar of action. (From Greenfield and Schneider, 1977.)

to our understanding of the distinctive cumulative quality of human culture, even though monkeys have these neurons, but do not create a cumulative culture. In general, my answer is that mirror neurons are necessary but not sufficient for complex and cumulative cultural learning.

Transmission mechanisms

Only human beings use symbolic means to instruct their young in tool use (Greenfield *et al.*, 2000). This would be an explanation based on the evolution of more powerful transmission systems (Greenfield *et al.*, 2000). The use of symbols to instruct in tool use is a strong candidate for a skill that has evolved in the last 5 million years.

The complexity of human neural programs

In addition, mirror neuron systems are always attached to particular neural programs – e.g., for manual action, mouth action, foot action, emotion – and occur in different, corresponding parts of the brain (Buccino *et al.*, 2001; Carr *et al.*, 2003). In other words they are somatotopic and participate in different circuits.

The neural programs for human action are more complex than those of monkeys. This complexity – for example, the ability to create more hierarchically complex constructions with objects – also would have facilitated the evolution of human culture. Figure 15.4 portrays a complex structure composed of construction straws that was used as a model (already built) and shown to children of different ages with a request to replicate it. No ape or monkey would be able to create a structure of such complexity.

Each age group used a different (and age-typical) strategy to build the model successfully. We conceptualized each strategy as a grammar of action because each involves a different systematic ordering (parallel to a linguistic surface structure) for creating the underlying, hierarchically organized tree-structure (parallel to a linguistic deep structure).

This level of complexity makes human building of permanent structures possible, something that is shared with no other species. Hence, to posit that neural programs subserving more complex action sequences evolved since our split with *Pan* is far from trivial.

15.3.6 A scientific contradiction

Ironically, in the light of the controversy surrounding ape imitation in the domain of tool use, imitative abilities of apes when it comes to language have never been questioned (Terrace *et al.*, 1979). Indeed, Terrace and colleagues used the *presence* of linguistic imitation in a sign-using chimpanzee as evidence that ape learning differs from human learning. In direct contradiction, Tomasello and colleagues use the *absence* of imitation learning in unenculturated apes as evidence that ape learning differs from human learning. The scientific community has shown itself equally ready to accept each assertion, without noticing the contradiction. Why has this happened? My hypothesis is that human beings and science as a human activity are always more willing to accept a conclusion that draws a firm line between human beings and other species. And the use of imitation to draw such a line is what Tomasello and Terrace have in common, despite the fact that the two lines of reasoning are logically incompatible with each other. My enterprise is the opposite: to establish an understanding of the continuities in human evolution.

15.3.7 The role of intentions and goal-directedness in tool use and construction activity

While this section has focused on imitation as an important mechanism in stimulating and guiding tool use and construction activity, it is necessary, before leaving this domain and proceeding to language, to briefly call attention to the other key feature of mirror neuron systems, their goal-directedness and attunement to intentionality. Clearly action models in this domain not only stimulate imitation; they also provide a goal for tool use and construction behavior. Just as clearly, this cultural domain of tool use and construction activity is one in which goals and intentionality are central to domain-relevant behavior. To give a few examples: one intends to turn a screw with a screwdriver. The goal of building is to create a structure such as a house or bridge. Clearly, there is an important match between the goal-directed structure of this cultural domain and the sensitivity of mirror neuron systems to goal directed action, rather than to movement per se.

Up to now, I have tried to establish mirror neurons as the neural substrate for imitation, tool learning, and object combination in both the ontogeny and phylogeny of human culture. In the next section, I show how the structure of the mirror neuron mechanism can be applied to understanding mechanisms of cultural learning in the ontogeny and phylogeny of language.

15.4 Mirror neurons and language: ontogeny and phylogeny

This section is based on an important neuroanatomical link between the mirror neurons discovered for manual action and the neural substrate for language – Broca's area in the left prefrontal cortex of the brain. I begin with a discussion of this link. Next, I discuss implications of the key feature of mirror neurons, their mirroring property – that is, their potential to fire either upon the execution or the observation of an action – for language learning. Finally, I will introduce a second feature of mirror neurons, their sensitivity to goal-directed action rather than to movement per se and draw out its implications for linguistic ontogeny and phylogeny.

15.4.1 Mirror neurons in Broca's area: implications for the ontogeny and phylogeny of language

Broca's area is a key area of the human brain for language. One perspective on Broca's is that it is a programming area for the oral–facial motor area that produces speech or the manual motor area that produces sign (Greenfield, 1991; see also Emmorey, this volume). Rizzolatti and his colleagues (1996) found a mirror system for manual action in the Broca homologue of their monkey subjects' brains. Iacoboni and colleagues (1999) later found mirror neurons for simple manual imitation in the Broca's area of their human participants. There is also evidence that Broca's area (Brodmann area 44) programs (that is, directs) the mirror system responsible for manual action performed on an object (Nishitani and Hari, 2000). (Brodmann areas are a numerical system for identifying different locations on the surface of the human brain.) The location of mirror neurons for manual action in Broca's area has a number of important theoretical implications.

First, this location in Broca's area implies an intimate relation between language and manual action. This theoretical implication has received empirical support in a study of motor evoked potentials; the cortical representation of the hand muscle was excited by purely linguistic tasks, but not by auditory or visual–spatial tasks (Flöel *et al.*, 2003). Indeed, evidence for the intimate neural connections between language and manual action provides evidence for my theory that Broca's area helps program the construction of both language and manual action (Greenfield, 1991). Even in monkeys, mirror and other closely related neurons in the Broca's homologue area discharge at the sound and not merely the sight of an action (Kohler *et al.*, 2002). Understanding the meaning of sound could be considered an evolutionary foundation for language.

Humans with different kinds of cortical damage provide another kind of evidence of a neural link between action and speech. Grossman (1980) gave Broca's and Wernicke's aphasics two versions of the tree-structure model, shown in Fig. 15.5a. Broca's aphasics who are unable to construct hierarchically organized grammatical tree-structures in speech (that is, their speech often consists of a string of isolated words rather than syntactically organized sentences), also had difficulty in constructing a

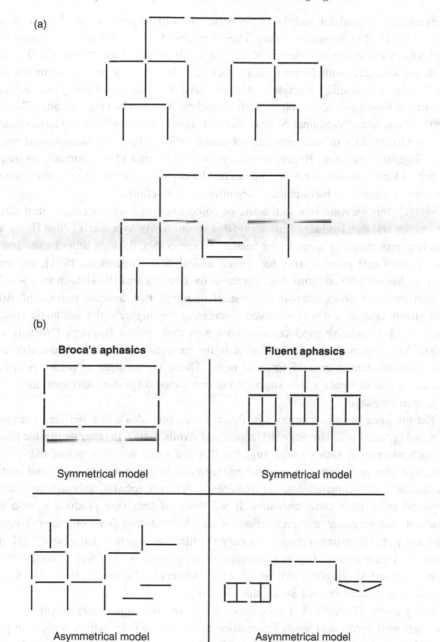

Figure 15.5 (a) Two tree-structures provided as models (constructed with tongue depressors) to Broca's and fluent aphasics (Grossman, 1980). (b) Typical "copies" of the models by Broca's and fluent aphasics. The Broca's aphasics get more detail (number of sticks) but less hierarchical (tree) structure, whereas the fluent aphasics get more hierarchical tree-structure, but less detail. (From Grossman, 1980.)

hierarchically organized multilevel tree-structure with objects, even given a model to copy (Fig. 15.5b) (Grossman, 1980). Their replicas of the model show no conception of the hierarchical structure – there is no sign of levels or branches. In contrast, Wernicke's or fluent aphasics, with Broca's area intact and the capacity to form complex (albeit meaningless) syntactic structures, do not have this same difficulty in building a hierarchically-organized structure with branches and levels (Fig. 15.5b) (Grossman, 1980). More recently, Molnar-Skocas and colleagues (unpublished data) have found that one of Greenfield's manual grammar-of-action tasks utilizes the same neural space as does linguistic grammar. Hence, Broca's area is a key part of two partially overlapping circuits. One combines words into hierarchically organized sentence structures; the other combines objects into hierarchically-organized constructions.[4]

Second, this location is also the site of some important overlapping neural circuitry for spoken and sign language (Emmorey, this volume). My hypothesis is that Broca's area provides programming input to the manual motor cortex for sign articulation, just as it does to orofacial motor cortex for speech articulation (Greenfield, 1991). Emmorey's review (this volume) confirms this hypothesis for Broca's area (Brodmann area 44). There is also evidence from cortical stimulation mapping that another portion of Broca's area (Brodmann area 45) is involved in creating the higher-order linguistic structures necessary for narrative production in both sign and spoken language (Horwitz *et al.*, 2003). My hypothesis would be that it is the grammatical aspect of narrative that is activating Brodmann area 45 in these tasks. Third, the location of mirror neurons for manual action in Broca's area suggests the cross-modal (gesture and speech) evolution of human language.

The presence of mirror neurons in Broca's area provides a link between comprehension and production in language (Rizzolatti and Arbib, 1998). In other words, the presence of such neurons in Broca's area suggests that the same neural structure that produces language also participates in comprehending it. It is therefore not surprising that both production and comprehension are impaired in Broca's aphasia, although production is impaired more than comprehension. If we think of language production as a motor function and language comprehension as an observational function, then it becomes very relevant that mirror neurons not only facilitate acting, but also observing. Furthermore, the motor theory of speech perception sees production as a way of comprehending (Liberman and Mattingly, 1985; see also the chapters by Goldstein, Byrd, and Saltzman and Skipper, Nusbaum, and Small, this volume).

In my model (Greenfield, 1991), Broca's area receives input from a syntactic area in the prefrontal cortex and sends information to the motor strip, which deals with phonological formation. Hence, impairment in Broca's area would be expected to affect both the syntactic and phonological levels. If a neuron can stimulate the same action it responds to observationally, comprehension of that action is implied – because one, in

---

[4] The partial overlap means that a minority of agrammatic Broca's aphasics can construct hierarchically organized tree structures. See Greenfield (1991, p.537) for details.

Figure 15.6 Lexigram board. Some of the boards will provide a translation in spoken English when a symbol is touched. In terms of English referents, the numerals on the top row are obvious. In terms of futher examples, the symbols on the next row, from left to right, are translated in English as: sweet potato, tickle, orange, Mary, trailer, peanut, car, raisin, hamburger, Sherman, egg, dig, Kanzi, Austin, fridge, and stand pile. (For the English translations of virtually all the other lexigrams on the board, go to www.iowagreatapes.org/bonobo/ language/pdf/lexo2.pdf.) (Photograph courtesy of Sue Savage-Rumbaugh.)

some sense, "understands" something one can already do. In other words, perhaps there are mirror neurons, yet to be discovered, that respond both upon making and hearing a sound. This possibility has started to be explored with very positive results (Aziz-Zadeh *et al.*, 2004). Finally, the presence of mirror neurons in Broca's area explains the importance of repetition (which involves observation and imitation) as a strategy in the ontogeny and phylogeny of conversational communication. I turn next to repetition.

15.4.2 Role of repetition in the ontogeny and phylogeny of language

Here I draw on an established body of research in child language (particularly Ochs Keenan, 1977), as well as my collaborative research on conversational repetition in bonobos with Sue Savage-Rumbaugh (Greenfield and Savage-Rumbaugh, 1993).

Details and background of the ape studies

The major ape data that I will present on repetition come from two bonobos, Kanzi and Mulika, exposed to a humanly devised language (perhaps better termed, protolanguage) system consisting of arbitrary (non-iconic) visual symbols (Fig. 15.6). They were exposed to the symbols in the course of naturalistic communication, rather than through formal training. Lexigrams are arbitrary visual symbols presented on a keyboard. During the lifetimes of the apes discussed here, many of the keyboards also presented the spoken

English gloss through a speech synthesizer when a particular key was pressed. Most of the lexigram symbols stand for nouns (e.g., banana, dog) or verbs (e.g., chase, bite). The same communication system was used with two chimpanzees, Sherman and Austin. However, their exposure was more in the mode of formal training.

Kanzi was raised at the Language Research Center of Georgia State University by his mother; Mulika did not have a mother present at the center. Kanzi learned to use lexigrams by watching his mother being exposed to the visual symbols. Mulika was exposed to the lexigrams by communicating directly with human caregivers and by watching Kanzi use lexigrams to communicate with the caregivers. Kanzi produced his first lexigram at age 2.6, Mulika at 1.0. At the time of the study, Kanzi was 5 years old, Mulika almost 2.

A 17-month-long investigation of Kanzi's vocabulary acquisition ended when he was 3 years 10 months old. Mulika's vocabulary was studied at a younger age, from before her very first lexigram at 12 months of age; Mulika's 11-month-long investigation of vocabulary acquisition ended when she was 22 months old. Very strict criteria for spontaneous and meaningful usage in everyday life were used to assess lexigram production (Savage-Rumbaugh *et al.*, 1986). At 46 months of age, Kanzi had 44 lexigrams in his spontaneously used, productive vocabulary that met these criteria. At 22 months of age, Mulika had six lexigrams in her spontaneously used productive vocabulary that met these criteria (Savage-Rumbaugh *et al.*, 1986). (Mulika was given the opportunity to use lexigrams at an earlier age than Kanzi, and may have been on track to acquire a larger vocabulary, as Kanzi's productive vocabulary was only seven lexigrams when his spontaneous lexigram use was assessed at 30 months of age. Unfortunately, Mulika's untimely and tragic death made it impossible to assess her later course of development.)

Pragmatic uses of repetition in the ontogeny and phylogeny of language

Ochs Keenan (1977) made the important point that children just starting to speak use linguistic imitation not only to copy but, more frequently, to fulfill many different pragmatic functions in a conversation. These functions include greetings (caregiver "hi," child "hi"), confirmation, and choosing from among alternatives (see Table 15.1). Because the term "imitation" had implications of rote imitation whose only purpose is to copy, Ochs Keenan used the term "repetition." Although not realized at the time, her research was revolutionary because it illuminated imitation/repetition as a phenomenon that was intrinsic to the conversational competence of young human beings. Later, Deborah Tannen (1989) showed how these pragmatic uses of repetition remain part and parcel of the conversational competence of adult human beings.

Ironically (especially in the light of later claims that apes cannot imitate), imitation had been used by the scientific community to minimize and even obliterate the linguistic accomplishments of apes (Terrace *et al.*, 1979). In this historical context, it seemed important to demonstrate that apes use imitation/repetition exactly as human children do. In order to establish this point, I collaborated with Sue Savage-Rumbaugh on a study of conversational repetition in young bonobos (Greenfield and Savage-Rumbaugh, 1991,

Table 15.1 *Use of repetition in discourse by children and bonobos*

Conversational use[a] of repetition to:	
Confirm/agree	Choose alternative
Human	
Twins, Toby and David, age 2.9, with their nanny Jill	Katie, age 1.2, with caregiver at infant daycare center (Caregiver pretends to pour tea for both of them and they pretend to drink it.)
Jill: *And we're going to have hot dogs.*	
Toby: *Hot dogs!* (excitedly)	
Jill: *And soup.*	Caregiver: *Are you full or do you want some more?*
David: *Mmm soup!*	Katie: *More.*
(Ochs Keenan, 1977)	(K. Leddick, unpublished data)
Bonobo	
Mulika, age 2, with human Kelly	Kanzi, age 5, with human caregiver/researcher, Rose
Kelly: *GO A-FRAME* (informing Mulika of destination)	
	Rose: *You can either PLAY or watch TV*
Mulika: *GO* (vocalizing excitedly)	Kanzi: *TV* (Kanzi watches after Rose turns it on)
(Greenfield and Savage-Rumbaugh, 1993)	

Note:
[a] For the bonobas, capitals indicate lexigrams (visual symbols) only; for the humans, capitals indicate simultaneous lexigrams and speech. Lower-case indicates speech only.
Source: Greenfield and Savage-Rumbaugh (1993).

1993). I now present cross-species data (Table 15.1) comparing our bonobo data on conversational repetition with the data of Ochs Keenan (1977).

In the case of both "confirm/agree" and "choose alternative," the repetition is used selectively to construct a pragmatic function. This table illustrates how both children and bonobos use repetition for two of the same pragmatic or conversational functions: to confirm/agree or to choose from among two alternatives. Note that the other examples in the published papers (Greenfield and Savage-Rumbaugh, 1991, 1993) show that many other pragmatic functions are expressed through repetition by both species, as well as by two chimpanzees, Sherman and Austin, who were also exposed to the lexigram system, but in a more training-like context.

For young children, repetition represents the initial ontogeny of conversational functions. When children first start to talk they have no other way to express conversational functions such as agreement (Greenfield and Smith, 1976). In general children repeat the part that will express the function that they are trying to communicate. For example, when Matthew was 14 months old, his mother said "Do you want to get up?" His selective imitation "up" was a way to express what he wanted to do and his first way of expressing an affirmative (Greenfield and Smith, 1976). Another way of looking at this is that the child repeats the most informative part of the prior utterance, that which is changing or to which there are alternatives.

Because early stages of development are more often similar within a clade or family of species than are later stages, the early appearance of repetition in child language made it likely and reasonable that repetition would be important in ape communication, and this is indeed what we found. In terms of recreating evolutionary history, which took place in the forest and savanna, not the laboratory, it is important to note that imitation/repetition is also crucial in the vocal communication of chimpanzees (third member of the clade) in the wild (for example, exchange of grunts, Goodall, 1986, p.131). Calls in the wild are culture-dependent, in that they are conventionalized in particular groups of animals (Whiten *et al.*, 1999). Because of the lack of data, it is difficult to compare the symbolic qualities of calls and lexigrams.

The evolutionary implication of the parallels between repetition in child and chimpanzee is the following. Because repetition has been found to carry communicative significance in the three species that make up the clade of *Homo sapiens*: chimpanzees, bonobos, and human beings, it is likely that the use of repetition to communicate vocally was present 5 million years ago, before the evolutionary divergence of *Homo* and *Pan* (bonobos and chimpanzees). If this is the case, it could have been one of the foundations for the subsequent evolution of more complex linguistic conversation. This is particularly potent in the light of Jane Goodall's observations of the use of vocal repetition by chimpanzees in the wild communicating at a distance (Goodall, 1986).

In the context of these similarities, Greenfield and Savage-Rumbaugh (1993) also noted two interesting differences between *Homo* and *Pan* in the conversational use of repetition: first, human children sometimes used repetition to stimulate more talk in their conversational partner; the chimpanzees, in contrast, used repetition exclusively to forward the non-verbal action. Second, the 1- and 2-symbol repetitions used by the chimpanzees to fulfill a variety of pragmatic functions were less than half the maximum length found in either the visual symbol combination addressed to them by their adult human caregivers or the oral repetitions of 2-year-old children. As suggested earlier, this species difference probably reflects the evolution of increased brain size and consequent increased memory capacity that has occurred since the phylogenetic divergence *of Homo* and *Pan* 5 million years ago (Greenfield and Savage-Rumbaugh, 1993).

Repetition functions to coordinate intended actions for children and bonobos

Before proceeding to the next section and presenting theory and data concerning intentionality, goal-directed action, mirror neurons, and language, I would like to point to the interpersonal coordination of goals and intentions as a pervasive function for both children and apes of the repetition examples in Table 15.1. For example, in the child example of confirmation/agreement, the conversational use of repetition is used by Toby, David, and their nanny Jill to coordinate intentions concerning the menu of their next meal. In the comparable ape example, Mulika and her human caregiver use repetition to coordinate intentions about what to do next. Similarly, the "choose alternative" examples for both species are all about coordinating intentions concerning what will happen next. Keep these examples in mind, as I present theoretical concepts and more detailed analysis in the domain of language, goal-directed action, intentionality, and mirror neurons.

15.5 Mirror neurons, goal-directed action, and intentionality: implications for the ontogeny and phylogeny of language

Further implications of the mirror neuron system for the ontogeny and phylogeny of language stem from the fact that both mirror neurons and language, on their different levels, privilege the encoding of goal-directed action. On the neural level, mirror neurons respond to the execution or observation of goal-directed action, not to particular physical movements or to objects in isolation. Goal-directed action, in turn, implies intentionality, which is central to language, a point to which I now turn. The point of this section and the examples that follow is twofold: (1) to show that, just as mirror neurons encode intentional action, so do children spontaneously use language to encode and communicate intentional action at the very dawn of language development; and (2) to show that, just as mirror neurons encode intentional action, so do our closest phylogenetic relatives, chimpanzees and bonobos, when given a humanly devised symbol system and the opportunity for interspecies communication, spontaneously use this system to encode and communicate intentional action. The theoretical hypothesis is that it is the presence of a mirror system in brain areas used to program both language and physical action that makes the encoding of intentional action so basic to the semantics and pragmatics of human language.

15.5.1 Intentionality and language

Let me begin with the words of Jerome Bruner (personal communication, 1979):

The more deeply I have gone into the psychology of language, the more impressed I have become with the absence in psychology of certain forms of psychological analysis that are needed in the study of language acquisition and language use generally. *One such is the role of intention and the perception of intention in others* (emphasis added). Language use is premised in a massive way upon presuppositions about intentions. . . Yet psychology, or at least positivistic "causal" psychology, ignores the role of intention. . . Such matters are most often treated as epiphenomena.

Whereas Bruner called attention to the importance of intentionality in language, Searle (1980) fleshed out this insight with an analysis of the behavioral and cognitive features of intentions. He identified two key features of intention: directedness and presentation or representation of conditions of satisfaction (i.e., presentation or representation of the goal). He further subdivided intentions into two levels. The first level he called intention-in-action. Intention-in-action involves *presentation* rather than *representation* of conditions of satisfaction. In intention-in-action, conditions of satisfaction are *implicitly* present *during* the intentional action. Prior intent, in contrast, involves a representation of the conditions of satisfaction. In prior intent, a mental model of conditions of satisfaction are *explicitly* present *before* the action begins. As the reader will see, language is a major way of externalizing goal representation.

15.5.2 Intentionality and the ontogeny of language

I begin with the case that the expression of intention-in-action is part of the very beginnings of child language (Greenfield, 1980). Here is an example from an observation of a toddler at the one-word stage. My example occurred during a children's gym class that his mother was teaching and I was observing. This was an unplanned ethnographic example that supplemented systematic formal study (Greenfield and Smith, 1976; Greenfield, 1980).

The child goes toward his mother, whining "shoes, shoes" (he has only socks on). He comes back toward me and gets his blue sandals. I try to help him while standing up, but cannot do it. So I sit down with one shoe, put him on my lap, and put his shoe on. Then I put him down, not saying anything. He walks straight to his other shoe, picks it up, and comes back to me. I put him on my lap and put his other shoe on. He then runs toward his mother still talking, saying "shoe, shoe" in an excited voice. He lifts his foot to show her. When she attends, he points to me. She understands, saying something like "The lady put your shoes on." Both are very excited.

This communication sequence involves social interfacing and coordination of observing and executing goal-directed actions. The sequence includes intention-in-action and the expression of prior intention on the part of the child, my active cooperation to fulfill his intention, and linguistic recognition of fulfilled intention on the part of the mother. Let me show how all of this played out through an analysis of the sequence.

First, the child used his language, a single-word utterance, to communicate *intention* through *explicitly representing* his *goal* ("shoes"); this is the expression of prior intention. I then observed and responded to his *intention* by my own *complementary goal-directed action* (my action of putting on his shoes). This was a goal-directed or intentional act on my part, which of course allowed the boy to fulfill his own intention and reach his goal. The child then used his language to communicate *goal achievement* ("shoe, shoe" in an excited voice). The mechanism of self-repetition, here used to express excitement, could be considered a kind of self-mirroring mechanism. Here he has shown excitement about fulfilling his *intention* and reaching his *goal*. His mother then observes and represents my goal-directed action in a full, adult sentence ("The lady put your shoes on"). This sentence also acknowledges that his intention has been fulfilled, his goal reached.

As the preceding sequence exemplifies, early language is specialized for the representation of intentional action, and early conversation is specialized for the interpersonal coordination of intentional action, as we saw in the earlier examples of conversational repetition. Just as mirror neurons are specialized for goal-directed action, it seems that language is too. The preceding conversational sequence is backed up by the entire corpus of early language studies in many languages (e.g., Bowerman, 1973; Brown, 1973). At the stage of both one-word and two-word "sentences," children encode intentional action. Common semantic functions at the one-word stage are the expression of action on object (for example *ba(ll)*, having just thrown a ball) or action of agent (for example, *up*, trying to get up on a chair) (Greenfield and Smith, 1976). At the two-word stage, again virtually all the semantic relations in child language encode intentional action, such as a relationship between agent and action (for example, *daddy* (agent) *bye-bye* (action) after his father leaves for work) or between object and action (*caca* (baby talk word for "record", the object) *ong* (baby-talk word for "on", the action), while carrying a record to the record player (Greenfield and Smith, 1976). In some instances, like *daddy bye-bye*, the child is describing the intentional action of another, which he has observed. In other cases, such as *caca ong*, the child is expressing his own intention. From the point of view of mirror neurons, the enactment and the interpretation of observed intentionality are equally important.

15.5.3 Intentionality and the phylogeny of language

Participants and background for the ape data

The data in this section come from Kanzi, his half-sister Panbanisha, and Panbanisha's constant companion, the female chimpanzee Panpanzee. All were exposed to English and lexigram communication in a naturalistic rather than a training modality.

The use of symbolic combinations to express intended action and goals by humanly enculturated apes

All three apes developed an open protogrammatical system in which they combined two or three lexigrams together (e.g., touching *playyard* lexigram followed by *Austin* lexigram (when he wanted to visit the chimp Austin in his playyard)) or combined a lexigram with a gesture (e.g., touching *balloon* lexigram followed by gesturing to Liz, who he wants to give him the balloon) to form original utterances that were not rote imitations of humans. These symbolic combinations expressed the same major semantic relations as young children's (Greenfield and Savage-Rumbaugh, 1991; Greenfield and Lyn, in press), relations such as action–object (e.g., touching *keepaway* lexigram followed by *balloon* lexigram, wanting to tease caregiver with a balloon) or agent–action (e.g., touching *carry* lexigram followed by gesture to caregiver, who agrees to carry Kanzi) (Greenfield and Savage-Rumbaugh, 1991). Each of these semantic relations also encodes intended action or an action goal.

While a corpus of combinatorial communications in the wild has not been assembled, it is clear that chimpanzees in the wild can combine gestures to express comparable complex semantic meanings and intended actions (Plooji, 1978). This evolutionary timetable is different from that proposed by Arbib, who places protolanguage after the divergence of *Pan* and *Homo*.

The encoding of intentional action characterizes child language as it dawns in the first year of life. Given that we find the greatest similarity among sibling species at the earliest points in development, it is not surprising that we would find the encoding of intentional action in ape language as well, as the examples indicate. Indeed, specialization for the encoding of intentional action is highly dominant in the spontaneous two-lexigram and lexigram-plus-gesture combinations of both bonobo and chimpanzee (Greenfield and Lyn, in press). Semantic relations that encode goal-directed action (such as agent–action and action–object), both frequent and universal in child language, have similar relative frequency in the spontaneous two-element combinations of two bonobos and a chimpanzee (Greenfield and Lyn, in press). Another example is found in Fig. 15.7. In this example, Kanzi expresses his desired goal by combining a lexigram with a gesture: he first touches the *chase* lexigram, then points to his caregiver, Rose, indicating that he would like Rose to play chase with him. Interestingly, Kanzi's open protogrammatical system also included protosyntax – for example, a creative sequencing "rule" (that is, not modeled by his human caregivers) that gesture generally follows lexigram (Greenfield and Savage-Rumbaugh, 1991).

Intentional action, language, and mirror neurons

Mirror systems clearly privilege intentional or goal-directed action over mere movement, and this holds for monkeys as well as humans. It might be something as simple as lesser brain size that is responsible for the Broca's area analogue in monkeys not entering into a complex symbolic communications system. Be that as it may, the encoding (understanding) of goal-directed action and the enacting of goal-directed action, not to mention the interindividual coordination of goal-directed action are at least as important in the wild as in captivity. This is probably the adaptational reason for the evolution of neural circuitry that both produces goal-directed action and understands it in others. Indeed, observations of chimpanzee communication in the wild have revealed many examples of the use of the vocal system to coordinate action among conspecific members of the group (Goodall, 1986). Observations of chimpanzee families with young children in the wild reveal gestural communication to coordinate action between mother and child (Plooji, 1978). Falk has recently emphasized this type of vocal and gestural coordination between mother and infant apes (Falk, 2004).

Why monkeys have mirror neurons, but do not use symbols to encode intended action

Just as with sequences of actions using objects, the reason for species differences between monkeys, apes, and humans is the following. Mirror properties occur in many kinds of neuron. Presumably they can also occur in the more complex circuits that control action sequences. Hence they are specific to particular kinds of behaviors and behavior sequences. Therefore, if monkeys are lacking certain kinds of neurons or neural circuitry, they will also lack the mirror neurons in those areas. Hence, one would not expect the behaviors to be similar from species to species even though all had mirror neurons in

Figure 15.7 (a) Kanzi touches the lexigram *chase*. (b) Immediately he points to Rose, communicating that he wants her to chase with him. (Photographs courtesy of Sue Savage-Rumbaugh.)

their nervous system. What one would expect to be similar would be the tendency to observe and imitate whatever goal-directed actions were feasible for that species and the tendency to enact and understand those same goal-directed actions.

The use of lexigrams to represent prior intent

This phenomenon has been seen in the enculturated apes, as it is in children. For example, Panpanzee, touched the *dog* lexigram (**agent**) and then touched the *play* lexigram (**action**). She then led her caregiver over to the dog house, where she and the dogs played together (Greenfield and Lyn, in press). In this example, Panpanzee is clearly representing a goal that she intends to attain. Sometimes an ape will combine gesture with lexigram to communicate prior intent. For example, Panpanzee pointed to a tree (**goal**) and then touched the *play* lexigram (**action**). Her caregiver said *yes* and Panpanzee climbed the tree to play. Again, her utterance communicated a goal that she intended to attain through her action. The expression of prior intention is equally central to bonobo communication. For example, Panbanisha touched the *open* lexigram (**action**) and then touched the *dog* lexigram (**goal**). She was asking her caretaker to open the door so they could visit the dogs. Clearly, she has expressed an intended action that she will subsequently try to enact.

In sum, intentional action is as much a focus of the language of *Pan* as it is of *Homo sapiens*. Thus, it is a feature that reaches across a whole family of species. The expression of intentional action is not unique to our own studies of two bonobos and a chimpanzee. Any trait that is found in all branches of an evolutionary tree is a good candidate to be an ancestral trait, that is, a trait found in the common ancestor of all the species in the clade. Although humans have intervened to teach various symbol systems to these animals, in all cases the representation of intended action was a spontaneous use of that symbol system. We tentatively conclude that language has at very least been overlaid on a foundation of intentional action, which it is used to represent (Greenfield and Lyn, in press). Indeed, symbolic communication of and about intended action is particularly valuable for the interindividual coordination of goal-directed action, as in the examples above and in the section on repetition.

15.5.4 Hypothesis concerning the relation between mirror neurons and intentionality in the ontogeny and phylogeny of language

Mirror neurons encode the execution and observation of goal-directed action – such as we saw in a non-verbal action when the boy acted as though his goal was to get his shoes put on *(intention-in-action)* and in the linguistic representation of *prior intent* by child, chimpanzee, and bonobo. An (intentional) agent carrying out action on objects is implicit in children's single-word or telegraphic utterances and more explicit in the adult's longer sentences (as in "The lady put your shoes on"). This latter describes the intentional action of another, based on observation. Indeed, the neural and cognitive linking of observed goal-directed action with enacted goal-directed action is the essence of the mirror system. My hypothesis is that mirror neurons subserve both the expression and interpretation (comprehension) of intentional action in language. Once we take action to the level of symbolic representation we can represent many actions that we cannot carry out ourselves. However, what is important for present purposes is that the comprehension and expression of intentional action is a bedrock of language development and evolution, a foundation from which other functions can develop in both ontogeny and phylogeny. Ontogenetically, language builds gradually on the intentional structure of action, i.e., the structure that is coded by the mirror neurons. That is, mirror neurons may highlight intentionality, but the particular neurons and neuronal circuits that have mirror properties are responsible for diverse kinds of intentional action, including interspecies diversity in the particular action systems that have evolved.

Early language is used to encode intentional action, but action structures are initially more complex than the language structures available to the child at that time. Hence, in both child language and ape language, more and more of the intentional action can be encoded linguistically as language acquisition advances in both child and ape. Here is an example of such a developmental progression in child language:

Step 1: Representation of intentional action in a **single word utterance** (Greenfield and Smith, 1976)
Matthew (15 : 17)[5]: *crecor* (record) **(prior intent)**
went over to the record player and started pushing the buttons **(intention-in-action)**
Interpretation as an intention by his mother: "I want the record on"
Step 2: Representation of intentional action in a **two-word combination** (Greenfield and Smith, 1976)
Matthew (17 : 18): *caca* (record) *ong* (on) **(prior intent)**
while carrying a record **(instrumental action)** to the record player **(goal location)**
Interpretation as an intention by his mother: "I want the record on"

Similarly, in the ontogeny of ape language, symbolic structure builds gradually over time on the intentional structure of action, i.e., the structure that is coded by the mirror neurons. Thus, Kanzi could use a single symbolic element to communicate intended action such as "chase" at an earlier point in his development than he formed two-element combinations to get across the same message concerning his goal in the situation (for example, *chase* (lexigram) *you* (pointing to Rose) (Fig. 15.7). Similar ontogenetic sequences from single-element to two-element encodings of goal-directed action have been observed in the bonobo Panbanisha, and the chimpanzee Panpanzee.

Phylogenies are simply sequences of ontogenies that are modified over evolutionary time (Parker *et al.*, 2000). Therefore, the common ontogenetic sequence by which linguistic structures come to encode increasingly explicit structures of intentional action across species has potential evolutionary significance. Again, while phylogeny does not recapitulate the ontogeny of one species, common ontogenies across species in the same clade or family imply that the sequence was present, at least potentially or in some form, in the common phylogenetic ancestor.

15.6 Conclusions

Developmental research, cross-species comparison, and mirror neuron studies are converging to provide clues to the neural foundation of cultural learning and transmission in ontogeny and phylogeny. In this chapter, I focus on two central features of human culture, language and tools (Greenfield, 1991). Mirror neurons and canonical neurons provide

---

[5] The two numbers represent the age in months and days.

neural mechanisms for the centrality of observation and imitation in the ontogeny and phylogeny of the intergenerational transmission of cultural tool systems. The location of mirror neurons in Broca's area provides a neural mechanism for the role of repetition in the ontogeny and phylogeny of conversational discourse, an important component of language evolution. The sensitivity of mirror neurons to goal-directed action also provides a neural mechanism for the role of intentional action in the ontogeny and phylogeny of symbol combinations and their interpretation. Behavioral similarities between apes and humans in these areas suggest that imitation of tool use, repetition as a conversational device, and communication about intentional action constituted a foundation for the evolution of human culture present in the ancestor shared by *Pan* and *Homo* 5 million years ago. Future research at the neurobehavioral level needs to establish empirically these theoretical links between the mirror system and these basic characteristics of the ontogeny and phylogeny of language and tools. A particular need is for the neural investigation of the mirror system in young children and apes of different ages.

References

Aziz-Zadeh, L., Iacoboni, M., Zaidel, E., Wilson, S., and Mazziotta, J., 2004. Left hemisphere motor facilitation in response to manual action sounds. *Eur. J. Neurosci.* **19**: 2609–2612.

Bard, K. A., and Russell, C. L., 1999. Evolutionary foundations of imitation: social cognitive and developmental aspects of imitative processes in non-human primates. In J. Nadel and G. Butterworth (eds.) *Imitation in Infancy*. New York: Cambridge University Press, pp. 89–123.

Beagles-Roos, J., and Greenfield, P. M., 1979. Development of structure and strategy in two dimensional pictures. *Devel. Psychol.* **15**: 483–494.

Bloom, L., Hood, L., and Lightbown, P., 1974. Imitation in language: if, when, and why. *Cogn. Psychol.* **6**: 380–420.

Bowerman, M., 1973. *Early Syntactic Development: A Cross-Linguistic Study with Special Reference to Finnish*. Cambridge, UK: Cambridge University Press.

Boyd, R., and Silk, J. B., 2000. *How Humans Evolved*. New York: W. W. Norton.

Brass, M., Derrfuss, J., and von Cramon, D. Y., 2005. The *inhibition* of imitative and overlearned responses: a functional double dissociation. *Neuropsychologia* **43**: 89–98.

Brown, R., 1973. *A First Language*. Cambridge, MA: Harvard University Press.

Bruner, J. S., 1974. The organization of early skilled action. In M. P. M. Richards (ed.) *The Integration of a Child into a Social World*. Cambridge, UK: Cambridge University Press, pp. 167–184.

Buccino, G., Binkofski, G. R., Fink, G. R., *et al.*, 2001. Action observation activates premotor and parietal areas in a somatotopic manner: an fMRI study. *Eur. J. Neurosci.* **13**: 400–404.

Carr, L., Iacoboni, M., Dubeau, M., Mazziotta, C., and Lenza, L., 2003. Neural mechanisms of empathy in humans: A relay from neural systems for imitation to limbic areas. *Proc. Natl Acad. Sci. USA* **100**: 5497–5502.

Childs, C. P., and Greenfield, P. M., 1980. Informal modes of learning and teaching: the case of Zinacanteco weaving. In N. Warren (ed.) *Studies in Cross-Cultural Psychology*, vol. 2. London: Academic Press, pp. 269–316.

Falk, D., 2004. Prelinguistic evolution in early hominins: whence motherese? *Behav. Brain Sci.* **27**: 491–503.

Favareau, D., 2002. Beyond self and other: the neurosemiotic emergence of intersubjectivity. *Sign Systems Stud.* **30**: 57–100.

Ferrari, F., Rozzi, S., and Fogassi, L., 2005. Mirror neurons responding to observation of actions made with tools in monkey ventral premotor cortex. *J. Cogn. Neurosci.* **17**: 212–226.

Flöel, A., Ellger, T., Breitenstein, C., and Knecht, S., 2003. Language perception activates the hand motor cortex: implications for motor theories of speech perception. *Eur. J. Neurosci.* **18**: 704–708.

Fogassi, L., Ferrari, P. F., Gesierich, B., *et al.*, 2005. Parietal lobe: from action organization to intention understanding. *Science* **308**: 662–667.

Fontaine, B., Moisson, P. Y., and Wickings, E. J., 1995. Observations of spontaneous tool making and tool use in a captive group of western lowland gorillas (*Gorilla gorilla gorilla*). *Folia Primatol.* **65**: 219–223.

Gallese, V., Fadiga, L., Fogassi, L., and Rizzolatti, G., 1996. Action recognition in the premotor cortex. *Brain* **119**: 593–609.

Garbarini, F., and Adenzato, M., 2004. At the root of embodied cognition: cognitive science meets neurophysiology. *Brain Cogn.* **56**: 100–106.

Gergely, G., Bekkering, H., and Király, I., 2002. Rational imitation in preverbal infants. *Nature* **415**: 755–756.

Goodall, J., 1986. *The Chimpanzees of Gombe: Patterns of Behavior.* Cambridge, MA: Harvard University Press.

Goodson, B. D., and Greenfield, P. M., 1975. The search for structural principles in children's manipulative play. *Child Devel.* **46**: 734–746.

Greenfield, P. M., 1980. Towards an operational and logical analysis of intentionality: the use of discourse in early child language. In D. Olson (ed.) *The Social Foundations of Language and Thought: Essays in Honor of J.S. Bruner.* New York: Norton, pp. 254–279.

1991. Language, tools, and brain: the ontogeny and phylogeny of hierarchically organized sequential behavior. *Behav. Brain Sci.* **14**: 531–551.

Greenfield, P. M. and Lyn, H., in press. Symbol combination in *Pan*: language, action, culture. In D. Washburn (ed.) *Primate Perspectives on Behavior and Cognition.* Washington, DC: American Psychological Association.

Greenfield, P. M., and Savage-Rumbaugh, E. S., 1991. Imitation, grammatical development, and the invention of protogrammar. In N. Krasnegor, D. Rumbaugh, M. Studdert-Kennedy, and R. Schiefelbusch (eds.) *Biological and Behavioral Determinants of Language Development.* Hillsdale, NJ: Lawrence Erlbaum, pp. 235–258.

1993. Comparing communicative competence in child and chimp: the pragmatics of repetition. *J. Child Lang.* **20**: 1–26.

Greenfield, P. M. and Schneider, L., 1977. Building a tree structure: the development of hierarchical complexity and interrupted strategies in children's construction activity. *Devel. Psychol.* **3**: 299–313.

Greenfield, P. M., and Smith, J. H., 1976. *The Structure of Communication in Early Language Development.* New York: Academic Press.

Greenfield, P. M., Nelson, and Saltzman, 1972. The development of rulebound strategies for manipulating seriated cups: a parallel between action and grammar. *Cogn. Psychol.* **3**: 291–310.

Greenfield, P. M., Maynard, A. E., Boehm, C., and Yut Schmidtling, E., 2000. Cultural apprenticeship and cultural change: tool learning and imitation in chimpanzees and humans. In S. T. Parker, J. Langer, and M. L. McKinney (eds.) *Biology, Brains, and Behavior: The Evolution of Human Development*. Santa Fe, NM: SAR Press, pp. 237–277.

Grossman, M., 1980. A central processor for hierarchically structured material: evidence from Broca's aphasia. *Neuropsychologia* **18**: 299–308.

Horwitz, B., Amunts, K., Bhattacharyya, R., *et al.*, 2003. Activation of Broca's area during the production of spoken and signed language: a combined cytoarchitectonic mapping and PET analysis. *Neuropsychologia* **41**: 1868–1876.

Iacoboni, M., Woods, R. P., Brass, M., *et al.*, 1999. Cortical mechanisms of human imitation. *Science* **286**: 2526–2528.

Johnson Pynn, J., Fragaszy, D. M., *et al.*, 1999. Strategies used to combine seriated cups by chimpanzees (*Pan troglodytes*), bonobos (*Pan paniscus*), and capuchins (*Cebus apella*). *J. Comp. Psychol.* **113**: 137–148.

Keenan (Ochs), E., 1977. Making it last: uses of repetition in child language. In S. Ervin-Tripp and C. Mitchell-Kernan (eds.) *Child Discourse*. New York: Academic Press, pp. 125–138.

Kohler, E., Keysers, C., Umiltà, M. A., *et al.*, 2002. Hearing sounds, understanding actions: action representation in mirror neurons. *Science* **297**: 846–848.

Liberman, A. M., and Mattingly, I. G., 1985. The motor theory of speech perception revised. *Cognition* **21**: 1–36.

Meltzoff, A. N., and Moore, M. K., 1977. Imitation of facial and manual gestures by human neonates. *Science* **198**: 75–78.

Miles, H. L., 1990. The cognitive foundations for reference in a signing orangutan. In S. T. Parker and K. R. Gibson (eds.) *"Language" and Intelligence in Monkeys and Apes: Comparative Developmental Perspectives*. New York: Cambridge University Press, pp. 511–539.

Miller, G. A., Galanter, E., and Pribram, K. H., 1960. *Plans and the Structure of Behavior*. New York: Holt.

Molnar-Szakacs, I., Kaplan, J. T., Greenfield, P. M., and Iacoboni, M. (submitted for publication) Observing complex action sequences: The role of the fronto-parietal mirror neuron system.

Nishitani, N., and Hari, R., 2000. Temporal dynamics of cortical representation for action. *Proc. Natl Acad. Sci. USA* **97**: 913–918.

Parker, S. T., Langer, J., and McKinney, M. L. (eds.), 2000. *Biology, Brains, and Behavior: The Evolution of Human Development*. Santa Fe, NM: SAR Press.

Parker, S. T., and McKinney, M. L., 1999. *Origins of Intelligence: The Evolution of Cognitive Development in Monkeys, Apes, and Humans*. Baltimore, MD: Johns Hopkins University Press.

Patterson, F. P., 1978. The gesture of a gorilla: langauge acquisition in another pongid. *Brain Lang.* **5**: 72–97.

Perry, S., Baker, M., Fedigan, L., *et al.*, 2003. Social conventions in wild white-faced capuchin monkeys: evidence for traditions in a neotropical primate. *Curr. Anthropol.* **44**: 241–268.

Piaget, J., 1962. *Play, Dreams, and Imitation in Childhood*. New York: Norton.

Plooji, F. X., 1978. Some basic traits of language in wild chimpanzees. In A. Lock (ed.) *Action, Gesture, and Symbol: The Emergence of Language*. New York: Academic Press, pp. 111–131.

Rizzolatti, G., and Arbib, M. A., 1998. Language within our grasp. *Trends Neurosci.* **21**: 188–194.

Rizzolatti, G., Fadiga, L., Matelli, M., *et al.*, 1996. Localization of grasp representations in humans by positron emission tomography. I. Observation versus execution. *Exp. Brain Res.* **111**: 246–252.

Savage-Rumbaugh, E. S., McDonald, K., Sevcik, R. A., Hopkins, W. D., and Rupert, E., 1986. Spontaneous symbol acquisition and communicative use by pygmy chimpanzees (*Pan paniscus*). *J. Exp. Psychol. General* **115**: 211–235.

Searle, J. R., 1980. The intentionality of intention and action. *Cogn. Sci.* **4**: 47–70.

Slobin, D. I., and Welsh, C. A., 1973. Elicited imitation as a research tool in developmental psycholinguistics. In C. A. Ferguson and D. I. Slobin (eds.) *Studies of Child Language Development*. New York: Holt, Rinehart and Winston, pp. 485–497.

Stauffer, R. L., Walter, A., Ryder, O. A., Lyons-Weiler, M., and Blair-Hedges, S., 2001. Human and ape molecular clocks and constraints on paleontological hypotheses. *J. Hered.* **92**: 469–474.

Tannen, D., 1989. *Talking Voices: Repetition, Dialogue, and Imagery in Conversation*. Cambridge, UK: Cambridge University Press.

Terrace, H. S., Pettito, L. A., Sanders, R. A., and Bever, T. G., 1979. Can an ape create a sentence? *Science* **206**: 891–900.

Tomasello, M., 1989. Chimpanzee culture? *Newslett. Soc. Res. Child Devel.* **Winter**: 1–3.

Tomasello, M., Davis-Dasilva, M., Camak, L., and Bard, K., 1987. Observational learning of tool use of young chimpanzees. *J. Hum. Evol.* **2**: 175–183.

Tomasello, M., Savage-Rumbaugh, E. S., and Kruger, A. C., 1993. Imitative learning of actions on objects by children, chimpanzees, and enculturated chimpanzees. *Child Devel.* **64**: 1688–1705.

Visalberghi, E., and Fragaszy, D., 2002. "Do monkeys ape?" Ten years after. In C. Nehaniv and K. Dautenhahn (eds.) *Imitation in Animals and Artifacts*. Cambridge, MA: MIT Press, pp. 471–499.

Whiten, A., 1998. Imitation of the sequential structure of actions in chimpanzees (*Pan troglodytes*). *J. Comp. Psychol.* **112**: 270–281.

Whiten, A., Goodall, J., McGrew, W. C., *et al.*, 1999. Cultures in chimpanzees. *Nature* **399**: 683–685.

Zihlman, A., 1996. Reconstructions reconsidered: chimpanzee models and human evolution. In W. C. McGrew and L. F. Marchant (eds.) *Great Ape Societies*. New York: Cambridge University Press, pp. 293–304.

Index